Elliptically Symmetric Distributions in Signal Processing and Machine Learning

Jean-Pierre Delmas · Mohammed Nabil El Korso ·
Stefano Fortunati · Frédéric Pascal
Editors

Elliptically Symmetric Distributions in Signal Processing and Machine Learning

Springer

Editors
Jean-Pierre Delmas
laboratoire Samovar
Institut Polytechnique de Paris, Telecom SudParis
Palaiseau, France

Mohammed Nabil El Korso
Laboratoire des signaux et systèmes
Université Paris-Saclay, CNRS, CentraleSupélec
Gif-sur-Yvette, France

Stefano Fortunati
Laboratoire des signaux et systèmes
Université Paris-Saclay, CNRS, CentraleSupélec & IPSA
Gif-sur-Yvette, France

Frédéric Pascal
Laboratoire des signaux et systèmes
Université Paris-Saclay, CNRS, CentraleSupélec
Gif-sur-Yvette, France

ISBN 978-3-031-52115-7 ISBN 978-3-031-52116-4 (eBook)
https://doi.org/10.1007/978-3-031-52116-4

This Springer imprint is published by the registered company Springer Nature Switzerland AG
The registered company address is: Gewerbestrasse 11, 6330 Cham, Switzerland

Paper in this product is recyclable.

To our families

Preface

Aim of the Book

One of the main goals of Statistical Signal Processing and, more recently, Statistical Learning approaches is to infer/learn information from a set of acquired data. Nowadays, datasets could be more and more complex, e.g., high dimensional, heterogeneous, with missing data. Remarkably, statistical approaches for estimating a set of parameters or testing predefined hypotheses using a collection of noisy measurements have triggered key progress in many fields, such as communication, remote sensing, image processing, radar and sonar processing, seismology, and finance. A critical prerequisite for the inference procedure is the choice of a statistical model for the measured data, i.e., a data distribution. One of the most common assumptions in engineering practice is to assume a Gaussian distribution. While the Gaussianity assumption may be well justified in several scenarios where the Central Limit Theorem can be invoked, everyday practice shows how frequently the acquired data stream can deviate from such a Gaussian modeling. Specifically, extensive analyses have highlighted the impulsive, heavy-tailed nature of the data. One can also mention the potential presence of outliers or missing data in the received dataset. Since it is well known that the performance of Gaussian-based inference procedures can decrease dramatically when the Gaussian assumption is violated, the need for a more general statistical data model takes center stage. The elliptical family of distributions provides a reliable generalization of the Gaussian distribution. The elliptical model can characterize a broad class of heavy- and light-tailed distributions, including the Gaussian one as a particular case. An elliptical distribution is fully characterized by its location vector and scatter/covariance matrix and by an additional functional parameter, called the density generator, that describes the distribution shape, such as the impulsiveness of the data distribution. One can cite, among others, the t-distribution or the generalized Gaussian distribution as part of the elliptical distribution family. Due to their flexibility and ability to model various data behaviors for different and heterogeneous applications, the elliptical distributions have started attracting a lot of attention from the Engineering community, specifically the Signal Processing one.

The aim of this book is thus to provide the interested reader with an up-to-date overview of the recent developments and practical uses of the elliptical model in classical and emerging Engineering areas, such as graph learning, robust clustering, linear shrinkage, information geometry, subspace-based algorithms, and semiparametric and misspecified estimation.

Outline of This Book

This book consists of ten chapters which are organized into an introductory Chap. Background on Real and Complex Elliptically Symmetric Distributions and three parts as follows:

Chapter 1 - Background on Real and Complex Elliptically Symmetric Distributions, by J-P. Delmas.

This chapter presents a short overview of real elliptically symmetric (RES) distributions, complemented by circular and noncircular complex elliptically symmetric distributions as complex representations of RES distributions. These distributions are both an extension of the multivariate Gaussian distribution and a multivariate extension of univariate symmetric distributions. They are equivalently defined through their characteristic functions and their stochastic representations, which naturally follow from the spherically symmetric distributions after affine transformations. Particular attention is paid to the absolutely continuous case and the subclass of compound Gaussian distributions. Results related to moments, affine transformations, marginal and conditional distributions, and summation stability are also presented. Some well-known instances of RES distributions are provided with their main properties. Finally, the estimation of the symmetry center and scatter matrix is briefly discussed through the sample mean, sample covariance matrix estimate, maximum likelihood estimators, M-estimators, and Tyler's M-estimators. Particular attention is given to the asymptotic (Gaussian) distribution of the M-estimators of the scatter matrix, and some hints about the Slepian–Bangs formula are provided.

Theoretical Developments

Chapter 2 - The Fisher–Rao Geometry of CES Distributions, by F. Bouchard, A. Breloy, A. Collas, A. Renaux, and G. Ginolhac;

Chapter 3 - Linear Shrinkage of Sample Covariance Matrix or Matrices Under Elliptical Distributions: A Review, by E. Ollila;

Chapter 4 - Robust Estimation with Missing Values for Elliptical Distributions, by A. Hippert-Ferrer and M. N. El Korso.

The second part of this book, which contains Chaps. The Fisher–Rao Geometry of CES Distributions–Robust Estimation with Missing Values for Elliptical Distributions, focuses on the theoretical aspects of the elliptical distributions. Although

Chap. Background on Real and Complex Elliptically Symmetric Distributions furnishes an extensive review of elliptical distributions, this part aims at discussing recent advances in the elliptical distributions theory. First, Chap. The Fisher–Rao Geometry of CES Distributions derives the Fisher–Rao geometry of the complex elliptically symmetric distributions. The Fisher metric associated with a Riemannian manifold allows for leveraging many tools from differential geometry, thus offering alternative and somehow more powerful approaches for parameter learning. Then, Chap. Linear Shrinkage of Sample Covariance Matrix or Matrices Under Elliptical Distributions: A Review addresses the fundamental problem of estimator regularization that can be encountered, *e.g.*, in high-dimensional settings. This chapter presents how to shrink the sample covariance matrix under elliptical distributions, providing optimal strategies for learning this shrinkage parameter. Finally, Chap. Robust Estimation with Missing Values for Elliptical Distributions deals with the problem of missing data under an elliptical distribution framework. This chapter reviews various robust estimation methods based on the expectation–maximization algorithm, designed for handling various patterns of missing values. Specifically, robust covariance matrix estimation under RES distribution (with a possible low-rank structure) and robust probabilistic component analysis (possibly under a mixed-effects model) will be studied.

Performance Analysis

Chapter 5 - Semiparametric Estimation in Elliptical Distributions, by S. Fortunati;

Chapter 6 - Estimation and Detection Under Misspecification and Complex Elliptically Symmetric Distributions, by C. D. Richmond and A. S. Bondre;

Chapter 7 - Performance Analysis of Subspace-Based Algorithms in CES Data Models, by J-P. Delmas and H. Abeida.

The third part of this book, which contains Chaps. Semiparametric Estimation in Elliptical Distributions–Performance Analysis of Subspace-Based Algorithms in CES Data Models is devoted to statistical performance analysis of some estimation and detection problems in the context of elliptically symmetric distributions. Chapter Semiparametric Estimation in Elliptical Distributions puts the problem of the joint estimation of the scatter matrix and of the location vector of an elliptically distributed set of observations in the general framework of the semiparametric theory. After having provided an introduction to this fundamental subject in modern robust statistics, the chapter aims at providing a semiparametric lower bound and the related semiparametric efficient estimator of the scatter matrix in the presence of an unknown density generator. After an extensive review of classic Cramér–Rao-type bounds, Chap. Estimation and Detection Under Misspecification and Complex Elliptically Symmetric Distributions aims at discussing their extensions when the assumed probability distribution for the observed data differs from the true distribution. Thus the theory of the misspecified parameter bounds is developed in this chapter. The generalized likelihood ratio test (GLRT) is also derived and its performances are assessed along with its finite sample receiver operating characteristic (ROC) in radar detection under well-specified complex compound Gaussian distributed data. Finally, Chap. Performance Analysis of Subspace-Based Algorithms in CES Data Models focuses on complex elliptically symmetric noisy linear mixture models whose parameters of the mixing matrix characterize its column subspace. This chapter unifies the asymptotic distribution (w.r.t. the number of samples) of subspace-based algorithms adapted to different models of the data and covariance matrix estimates and proves several invariance properties that have impacts on the parameters to be estimated.

Applications to Machine Learning

Chapter 8 - Robust Bayesian Cluster Enumeration for RES Distributions, by F. K. Teklehaymanot, C. Schroth, and M. Muma;

Chapter 9 - FEMDA: A Unified Framework for Discriminant Analysis, by P. Houdouin, M. Jonckheere, and F. Pascal;

Chapter 10 - Learning Graphs from Heavy-Tailed Data (with application to finance data), by J. V. de M. Cardoso, J. Ying, and D. P. Palomar.

The fourth part of this book, which contains Chaps. Robust Bayesian Cluster Enumeration for RES Distributions–Learning Graphs from Heavy-Tailed Data (with application to finance data), presents some supervised and unsupervised machine learning methods based on elliptically symmetric distributed data that advantageously replaces the classic Gaussian distribution to take into account heavy-tailed noise, artifacts, and outliers. First, Chap. Robust Bayesian Cluster Enumeration for RES Distributions revisits some recent developments on robust statistical model-based cluster analysis even when the number of clusters is unknown. In particular, Bayesian robust cluster enumeration criteria are discussed. Such approaches formulate the problem of estimating the number of clusters as a maximization of the posterior probability of the candidate models. Robust clustering algorithms are derived and applied to real-data examples. Chapter FEMDA: A Unified Framework for Discriminant Analysis presents a new approach that overcomes the limitations of the linear and quadratic discriminant analysis under Gaussian assumption. In the considered model, the data in each cluster share the same shape matrix with arbitrary scale parameters with its own elliptically symmetric distribution. The structure of the proposed algorithm involves recursively estimating the parameters of each cluster and then classifying test observations based on the derived decision rule where the unknown scatter and location parameters are estimated by a Bayesian approach with the maximum marginal likelihood estimators. Finally, Chap. Learning Graphs from Heavy-Tailed Data (with application to finance data) investigates the problem of learning graph matrices whose structure follows that of a Laplacian matrix of an undirected weighted graph for which the data-generating process is assumed to be Student's t-distributed. The underlying learning problem is solved via an algorithm based on alternating direction method of multipliers and applied to real-world data.

Audience of This Book

Although the elliptically symmetric distributions were introduced into the statistical literature in the 1970s and a tutorial article in the IEEE Transaction on Signal Processing introduced them to the signal processing community in 2012, no book has been devoted to the elliptically symmetric distributions to the signal processing and/or machine learning community. Since this tutorial article, there has been a proliferation of research, applications, and results in this field. An in-depth book that provides a timely overview, a systematic account, and a foundational theory is essential.

This book is mainly targeted at researchers and graduate students who rely on the theory of probability and statistics to conduct their work in signal processing (e.g., communications, radar, sonar, etc.) and machine learning.

The necessary background which is supposed to take profit from reading this book is sufficient knowledge in probabilities, statistics, analysis, and matrix algebra at the postgraduate level.

Evry, France — Jean-Pierre Delmas
Gif-sur-Yvette, France — Mohammed Nabil El Korso
Ivry-sur-Seine, France — Stefano Fortunati
Gif-sur-Yvette, France — Frédéric Pascal

Contents

Chapter 1
Background on Real and Complex Elliptically Symmetric Distributions

Jean-Pierre Delmas

Abstract This chapter presents a short overview of real elliptically symmetric (RES) distributions, complemented by circular complex elliptically symmetric (C-CES) and noncircular CES (NC-CES) distributions as complex representations of RES distributions. These distributions are both an extension of the multivariate Gaussian distribution and a multivariate extension of univariate symmetric distributions. They are equivalently defined through their characteristic functions and their stochastic representations, which naturally follow from the spherically symmetric distributions after affine transformations. Particular attention is paid to the absolutely continuous case and to the subclass of compound Gaussian distributions. Results related to moments, affine transformations, marginal and conditional distributions, and summation stability are also presented. Some well-known instances of RES distributions are provided with their main properties. Finally, the estimation of the symmetry center and scatter matrix is briefly discussed through the sample mean (SM), sample covariance matrix (SCM) estimate, maximum estimate (ML), M-estimators, and Tyler's M-estimators. Particular attention will be paid to the asymptotic Gaussianity of the M-estimators of the scatter matrix. To conclude, some hints about the Slepian–Bangs formula are provided.

1.1 Introduction

Until 50 years ago, most of the procedures in multivariate analysis were developed under the Gaussian assumption, mainly for mathematical convenience. However, in many applications, Gaussianity is a poor approximation of reality. As a consequence, elliptically symmetric distributions have been widely used in various applications due to their flexibility and capability to better model various data behaviors. These distributions form a natural extension of the Gaussian one by allowing for both heavier-than-Gaussian and lighter-than-Gaussian tails while maintaining the elliptical geometry of the underlying equidensity (when it exists) contours. These real

J.-P. Delmas (✉)
laboratoire Samovar, Institut Polytechnique de Paris, Telecom SudParis, 91120 Palaiseau, France
e-mail: jean-pierre.delmas@it-sudparis.eu

J.-P. Delmas et al. (eds.), *Elliptically Symmetric Distributions in Signal Processing and Machine Learning*, https://doi.org/10.1007/978-3-031-52116-4_1

elliptically symmetric (RES) distributions were equivalently defined in the statistical literature [9, 17, 18, 33] through their characteristic functions and their stochastic representations, which naturally follow from the spherically symmetric distributions after affine transformations. A first systematic treatment of circular complex elliptically symmetric (C-CES) distributions was provided in the engineering literature [37] and further fully studied in [55], and in the tutorial paper [49]. Then, the general complex representation of the RES distributions, called noncircular CES (NC-CES) distributions, was introduced in [16].

The aim of this chapter is twofold. At first, a short overview of RES, C-CES, and NC-CES distributions (as complex representations of the RES distributions) is introduced with the aim of providing a common background for the other chapters of this book. Secondly, the main definitions and properties of these distributions are listed and shortly discussed.

This chapter is organized as follows. Section 1.2 defines the RES distributions equivalently through their characteristic functions and to their stochastic representations. Particular attention is paid to the absolutely continuous case and to the subclass of compound Gaussian distributions. Section 1.3 defines the C-CES and NC-CES distributions as complex representations of the RES distributions. Section 1.4 presents basic properties related to moments, affine transformations, marginal and conditional distributions, and summation stability. Then some well-known instances of RES distributions are provided with their main properties in Sect. 1.5. The joint estimation of the symmetry center and scatter matrix is briefly discussed in Sect. 1.6 through the SM and SCM estimators, ML, M-estimators, and Tyler's M-estimators, asymptotic Gaussian distribution of scatter M-estimators, and Slepian–Bangs formula. Finally, Sect. 1.7 briefly concludes this chapter.

The following notation are used throughout this chapter. Matrices and vectors are represented by bold upper case and bold lower case characters, respectively. Vectors are by default in column orientation, while the superscripts T, H, $*$, and # stand for transpose, conjugate transpose, conjugate, and Moore–Penrose inverse, respectively. $(\mathbf{a})_k$ and $(\mathbf{A})_{k,\ell}$ denote the k and (k, ℓ)th element of the vector $\mathbf{a}$ and the matrix $\mathbf{A}$, respectively. E(.), |.|, and Tr(.) are the expectation, determinant, and trace operators, respectively. $\mathbf{I}$ is the identity matrix with the appropriate dimension. vec($\mathbf{A}$) denotes the "vectorization" operator that turns a matrix $\mathbf{A}$ into a vector by stacking the columns of the matrix one below another and $v(\mathbf{A})$ is the vector that is obtained from vec($\mathbf{A}$) by eliminating all supradiagonal elements of $\mathbf{A}$. These vectors are used in conjunction with the Kronecker product $\mathbf{A} \otimes \mathbf{B}$ as the block matrix whose (i, j) block element is $a_{i,j}\mathbf{B}$, and with the commutation matrix $\mathbf{K}$ and the duplication matrix $\mathbf{D}$ of appropriate dimension such that $\text{vec}(\mathbf{C}^T) = \mathbf{K}\text{vec}(\mathbf{C})$ and $\text{vec}(\mathbf{A}) = \mathbf{D}v(\mathbf{A})$ where $\mathbf{A}$ is symmetric. The acronyms r.v., p.d.f., and c.d.f. for, respectively, random variable, probability density function, and cumulative distribution function are used. Finally, $\Gamma(u) \stackrel{\text{def}}{=} \int_0^\infty t^{u-1}e^{-t}dt$ is the Gamma function with $\Gamma(k) = (k-1)!$ for $k \in \mathbb{N}$, $B(k, \ell)$ denotes the Beta function with $B(k, \ell) = \frac{\Gamma(k)\Gamma(\ell)}{\Gamma(k+\ell)}$ and Gam(k, θ) is the Gamma distribution of scale θ with p.d.f. $p(x) = \frac{1}{\Gamma(k)\theta^k}x^{k-1}\exp(-x/\theta)$. $x =_d y$, $x_n \rightarrow_d D$ and $x \sim D$ mean that the r.v. x and y have the same distribution, the sequence of r.v. x_n converges in distribution to D and x follows the distribution D, respectively. The subscripts r and c are used to refer to the real and complex data cases, respectively.

1.2 Definition of the Real Elliptically Symmetric Distributions

1.2.1 Characteristic Function

After earlier works on this topic, RES distributions were formalized by [33] and further studied by [9, 17, 18]. They were first defined as affine transformations of spherically distributed r.v. Then, by the uniqueness theorem (see, e.g., [56, pp. 346–351]), they were alternatively defined by their characteristic functions.

Definition 1.1 An r.v. $\mathbf{x} \in \mathbb{R}^m$ is said to have a RES distribution if there exists a vector $\boldsymbol{\mu} \in \mathbb{R}^m$, an $m \times m$ symmetric positive semi-definite matrix $\boldsymbol{\Sigma}$ of rank $k \leq m$ and a function $\phi_r(.) : \mathbb{R}^+ \to \mathbb{R}$ called *symmetry center*, *scatter matrix*, and *characteristic generator*, respectively, such that the characteristic function of $\mathbf{x}$ is of the form

$$\Phi_x(\mathbf{t}) \stackrel{\text{def}}{=} \mathrm{E}[\exp(i\mathbf{t}^T\mathbf{x})] = \exp(i\mathbf{t}^T\boldsymbol{\mu})\phi_r(\mathbf{t}^T\boldsymbol{\Sigma}\mathbf{t}), \qquad \mathbf{t} \in \mathbb{R}^m. \tag{1.1}$$

We shall write $\mathbf{x} \sim \mathrm{RES}_m(\boldsymbol{\mu}, \boldsymbol{\Sigma}, \phi_r)$ and note that the couple $(\boldsymbol{\Sigma}, \phi_r(.))$ does not uniquely identify the distribution of $\mathbf{x}$ because $(c^2\boldsymbol{\Sigma}, \phi_r(./c^2))$ gives the same distribution. This scale ambiguity is easily avoided by restricting the function $\phi_r(.)$ in a suitable way (e.g., by fixing a moment as it is explained in Sect. 1.4.1), or by putting a constraint on the scatter matrix $\boldsymbol{\Sigma}$ (e.g., $\mathrm{Tr}(\boldsymbol{\Sigma}) = m$). Note that, for $m = 1$, these distributions coincide with the class of one-dimensional symmetric distributions w.r.t. the symmetry center.

1.2.2 Stochastic Representation

Equivalently to the definition (1.1), the RES distributed r.v. $\mathbf{x}$ can be defined from an affine function

$$\mathbf{x} \stackrel{\text{def}}{=} \boldsymbol{\mu} + \mathbf{A}\mathbf{x}_s \tag{1.2}$$

of a k-dimensional spherically distributed r.v. $\mathbf{x}_s$, where $\mathbf{A} \in \mathbb{R}^{m \times k}$ is any square root ($\boldsymbol{\Sigma} = \mathbf{A}\mathbf{A}^T$) of the scatter matrix $\boldsymbol{\Sigma}$ of rank k, and thus full column rank. Such spherical distributions are defined equivalently in the following [17, Chap. 2].

Definition 1.2 An r.v. $\mathbf{x}_s \in \mathbb{R}^k$ is spherically distributed i.f.f.

- $\mathbf{x}_s =_d \mathcal{O}\mathbf{x}_s$ for arbitrary real-valued k-dimensional orthonormal matrix $\mathcal{O}$.
- There exists a function $\phi_r(.) : \mathbb{R}^+ \mapsto \mathbb{R}$, such that the characteristic function of $\mathbf{x}_s$ is given by
$$\Phi_{x_s}(\mathbf{t}) = \phi_r(\|\mathbf{t}\|^2), \ \mathbf{t} \in \mathbb{R}^k. \tag{1.3}$$
- For every $\mathbf{h} \in \mathbb{R}^k$, $\mathbf{h}^T\mathbf{x}_s =_d \|\mathbf{h}\| x_{s_i}$ with $\mathbf{x}_s = (x_{s_1}, \ldots, x_{s_i}, \ldots, x_{s_k})^T$.
- If $\mathbf{x}_s$ is absolutely continuous w.r.t. Lebesgue measure on $\mathbb{R}^k$, there exists a function[1] $g_r(.)$: $\mathbb{R}^+ \mapsto \mathbb{R}^+$ such that the p.d.f. of $\mathbf{x}_s$ w.r.t. this measure is of the form
$$p(\mathbf{x}_s) = g_r(\|\mathbf{x}_s\|^2), \tag{1.4}$$
where
$$\delta_{r,k} \stackrel{\text{def}}{=} \int_0^\infty t^{k/2-1} g_r(t)dt = \frac{\Gamma(k/2)}{\pi^{k/2}}, \tag{1.5}$$
ensuring that $p(\mathbf{x}_s)$ integrates to one.
- There exists a non-negative r.v. $\mathcal{Q}_{r,k}$, and $\mathbf{u}_{r,k}$ that are independent where $\mathbf{u}_{r,k}$ is uniformly distributed on the unit real k-sphere ($\mathbf{u}_{r,k} \sim U(\mathbb{R}S^k)$) such that
$$\mathbf{x}_s =_d \sqrt{\mathcal{Q}_{r,k}}\mathbf{u}_{r,k}. \tag{1.6}$$

Consequently, from (1.2) and (1.3), we find the characteristic function (1.1) which defines the RES distribution since

$$\Phi_x(\mathbf{t}) = \exp(i\mathbf{t}^T\boldsymbol{\mu})\Phi_{x_s}(\mathbf{A}^T\mathbf{t}) = \exp(i\mathbf{t}^T\boldsymbol{\mu})\phi_r(\mathbf{t}^T\boldsymbol{\Sigma}\mathbf{t}).$$

From (1.2) and (1.6), we obtain the following.

Theorem 1.1 $\mathbf{x}$ *is* $\text{RES}_m(\boldsymbol{\mu}, \boldsymbol{\Sigma}, \phi_r)$ *distributed, i.f.f. it admits the following stochastic full-rank representation*

$$\mathbf{x} =_d \boldsymbol{\mu} + \sqrt{\mathcal{Q}_{r,k}}\mathbf{A}\mathbf{u}_{r,k} = \boldsymbol{\mu} + \mathcal{R}_{r,k}\mathbf{A}\mathbf{u}_{r,k}. \tag{1.7}$$

The r.v. $\mathcal{Q}_{r,k}$ and $\mathcal{R}_{r,k} \stackrel{\text{def}}{=} \sqrt{\mathcal{Q}_{r,k}}$ are the second-order modular and modular (or generating) variates of the r.v. $\mathbf{x}$, respectively. We note that there is a one-to-one mapping between the characteristic generator ϕ_r and the c.d.f. F_{R_r} of $\mathcal{R}_{r,k}$ (called generating c.d.f.). Thus, we can also write $\mathbf{x} \sim \text{RES}_m(\boldsymbol{\mu}, \boldsymbol{\Sigma}, F_{R_r})$ and, equivalently to Definition 1.1, we retrieve the scale ambiguity in the couple $(\mathbf{A}, \mathcal{R}_{r,k})$ in (1.7). Theorem 1.1 provides an obvious mechanism to generate r.v. $\mathbf{x} \sim \text{RES}_m(\boldsymbol{\mu}, \boldsymbol{\Sigma}, F_{R_r})$: it only involves generating $\mathcal{R}_{r,k}$ according to its c.d.f. F_{R_r} and $\mathbf{u}_{r,k} = \frac{\mathbf{n}_{r,k}}{\|\mathbf{n}_{r,k}\|}$ where

[1] Both functions $\phi_r(.)$ and $g_r(.)$ are generally parameterized by the dimension k and in practice by a finite-dimensional parameter (see examples in Sect. 1.5).

$\mathbf{n}_{r,k}$ is k-dimensional zero-mean Gaussian distributed r. v. with covariance $\mathbf{I}$ ($\mathbf{n}_{r,k} \sim \mathbb{R}N_k(\mathbf{0}, \mathbf{I})$). Moreover, the following important property follows from Theorem 1.1:

$$\mathcal{Q}_{r,k} =_d (\mathbf{x} - \boldsymbol{\mu})^T \boldsymbol{\Sigma}^{\#} (\mathbf{x} - \boldsymbol{\mu}). \tag{1.8}$$

1.2.3 The Absolutely Continuous Case

From (1.2) the r.v. $\mathbf{x}$ is absolutely continuous w.r.t. Lebesgue measure on $\mathbb{R}^m$, i.f.f. $\mathbf{x}_s$ is too. From (1.6), it is immediate to verify that this condition is satisfied i.f.f. $\mathcal{Q}_{r,k}$ or $\mathcal{R}_{r,k}$ is absolutely continuous w.r.t. Lebesgue measure on $\mathbb{R}^+$. In this case, the p.d.f. of $\mathbf{x}$ is defined on the k-dimensional subspace of $\mathbb{R}^m$ spanned by the range space of $\mathbf{A}$.

In the particular case where $k = m$, i.e., $\text{rank}(\boldsymbol{\Sigma}) = m$, $\mathbf{A}$ is a non-singular $m \times m$ square matrix. From the one-to-one mapping (1.2) between $\mathbf{x}$ and $\mathbf{x}_s$, and the p.d.f. (1.4), the p.d.f. of $\mathbf{x}$ on $\mathbb{R}^m$ can be expressed as

$$p(\mathbf{x}) = |\boldsymbol{\Sigma}|^{-1/2} g_r[(\mathbf{x} - \boldsymbol{\mu})^T \boldsymbol{\Sigma}^{-1} (\mathbf{x} - \boldsymbol{\mu})]. \tag{1.9}$$

We note that unlike the notation used in, e.g., [49], (1.9) does not explicitly include the usual p.d.f. normalizing constant. This is a consequence of the definition of $g_r(.)$ introduced in (1.4).

Alternatively, if in (1.9) $g_r(.)$: $\mathbb{R}^+ \mapsto \mathbb{R}^+$ is an arbitrary function, called *density generator*, such that

$$\delta_{r,m,g} \stackrel{\text{def}}{=} \int_0^\infty t^{m/2-1} g_r(t) dt < \infty, \tag{1.10}$$

$\delta_{r,m,g}$ depends now on g_r and the couple $(\boldsymbol{\Sigma}, g_r(.))$ in (1.9) does not uniquely identify the distribution of $\mathbf{x}$ because $(c^2\boldsymbol{\Sigma}, c^m g_r(.\, c^2))$ gives the same distribution.

We adopt the notation $\text{RES}_m(\boldsymbol{\mu}, \boldsymbol{\Sigma}, g)$ instead of $\text{RES}_m(\boldsymbol{\mu}, \boldsymbol{\Sigma}, \phi)$. The level sets of $p(\mathbf{x})$ are a family of hyper ellipsoids in $\mathbb{R}^m$ symmetrically centered at $\boldsymbol{\mu}$, where shape and orientation are determined by $\boldsymbol{\Sigma}$. This justifies the terminology of symmetrical elliptical distributions. Furthermore, the stochastic representation (1.7) reduces to

$$\mathbf{x} =_d \boldsymbol{\mu} + \sqrt{\mathcal{Q}_{r,m}} \boldsymbol{\Sigma}^{1/2} \mathbf{u}_{r,m} = \boldsymbol{\mu} + \mathcal{R}_{r,m} \boldsymbol{\Sigma}^{1/2} \mathbf{u}_{r,m}. \tag{1.11}$$

Here too, note that $(\boldsymbol{\Sigma}, \mathcal{Q}_{r,m})$ and $(c^2\boldsymbol{\Sigma}, c^{-2}\mathcal{Q}_{r,m})$ give the same distribution of $\mathbf{x}$. Equation (1.11) implies that (1.8) simplifies to

$$\mathcal{Q}_{r,m} =_d (\mathbf{x} - \boldsymbol{\mu})^T \boldsymbol{\Sigma}^{-1} (\mathbf{x} - \boldsymbol{\mu}). \tag{1.12}$$

From (1.9), (1.11), and (1.12), the p.d.f. of $\mathcal{Q}_{r,m}$ and $\mathcal{R}_{r,m}$ are, respectively,

$$p(q) = \delta_{r,m}^{-1} q^{m/2-1} g_r(q) \text{ and } p(r) = 2\delta_{r,m}^{-1} r^{m-1} g_r(r^2). \tag{1.13}$$

Finally, we note that the RES distributions do not necessarily possess a p.d.f. w.r.t. Lebesgue measure on $\mathbb{R}^m$ even when $\boldsymbol{\Sigma}$ is not singular. Such an example is the $U(\mathbb{R}S^m)$ distribution which belongs to $\text{RES}_m(\mathbf{0}, \mathbf{I}, \phi_r)$ distributions, where the explicit (but somewhat involved) form of ϕ_r can be found in [18, Theorem 2.51].

1.2.4 The Subclass of Compound-Gaussian Distributions

An important subclass of RES distributions are the compound-Gaussian (CG) distributions, whose circular complex representations (denoted C-CCG) have been widely used in the engineering literature, for example, for modeling radar clutter [65]. An r.v. having CG distributions with zero symmetry center is also referred to as *spherically invariant random vectors* (SIRV) in the engineering literature (see, e.g., in [11, 12, 54, 66]) and as *scale mixtures of normal distributions* in the statistics literature [4, 25]. These distributions are defined by their stochastic representation.

Definition 1.3 An r.v. $\mathbf{x} \in \mathbb{R}^m$ is said to have a real CG distribution (RCG) if it admits the following representation:

$$\mathbf{x} =_d \boldsymbol{\mu} + \sqrt{\tau_r}\mathbf{n}, \tag{1.14}$$

for some positive real r.v. τ_r with c.d.f. F_τ (not related neither to dimension m nor to rank k), called the *texture* independent of $\mathbf{n} \sim \mathbb{R}N_m(\mathbf{0}, \boldsymbol{\Sigma})$, called the *speckle*. The r.v. $\sqrt{\tau_r}$ is often called *mixing variable* with mixing distribution in the statistical literature (see e.g., [25]). We write $\mathbf{x} \sim \text{RCG}_m(\boldsymbol{\mu}, \boldsymbol{\Sigma}, F_\tau)$ to denote this case.

Note that (1.14) can be rewritten as

$$\mathbf{x} =_d \boldsymbol{\mu} + \sqrt{\tau_r}\mathbf{A}\mathbf{n}_0, \tag{1.15}$$

where $\mathbf{n}_0 \sim \mathbb{R}N_k(\mathbf{0}, \mathbf{I})$. Then by recalling that $\mathbf{n}_0 = \|\mathbf{n}_0\|\mathbf{u}_{r,k}$ with $s \overset{\text{def}}{=} \|\mathbf{n}_0\|^2 \sim \chi_k^2 = \text{Gam}(k/2, 2)$ and $\mathbf{u}_{r,k} \sim U(\mathbb{R}S^k)$, and where s and $\mathbf{u}_{r,k}$ are independent. It follows that the stochastic representation (1.14) can also be written as

$$\mathbf{x} =_d \boldsymbol{\mu} + \mathcal{R}_{r,k}\mathbf{A}\mathbf{u}_{r,k}, \tag{1.16}$$

where the modular variate $\mathcal{R}_{r,k} \overset{\text{def}}{=} \sqrt{\tau_r s}$ and $\mathbf{u}_{r,k}$ are independent. Consequently, the RCG distributions form a subclass of the RES distributions. Furthermore, $\mathbf{x} \sim \text{RES}_m(\boldsymbol{\mu}, \boldsymbol{\Sigma}, \phi)$ belongs to the set of RCG distributions i.f.f. there exists an r.v. τ_r such that the second-order modular variate $\mathcal{Q}_{r,k} = \mathcal{R}_{r,k}^2$ satisfies $\mathcal{Q}_{r,k} = \tau_r s$, i.e., $\mathcal{Q}_{r,k}$ is a scale mixture of the $\text{Gam}(k/2, 2)$ distribution. This means that the conditional distribution of $\mathcal{Q}_{r,k}$ given $\tau_r = \tau$ is the $\text{Gam}(k/2, 2\tau)$ distribution and thus the p.d.f of $\mathcal{Q}_{r,k}$ is

$$p(q) = \int_0^\infty \frac{1}{\Gamma(k/2)(2\tau)^{k/2}} q^{k/2-1} \exp(-q/(2\tau)) dF_\tau(\tau). \tag{1.17}$$

The characteristic function of $\Phi_{x_s}(\mathbf{t})$ of a RCG distributed r.v. $\mathbf{x}$ defined by (1.14) is given straightforwardly by

$$\begin{aligned}\Phi_{x_s}(\mathbf{t}) &= \exp(i\mathbf{t}^T\boldsymbol{\mu}) \int_0^\infty \mathrm{E}(\exp(i\mathbf{t}^T\sqrt{\tau}\mathbf{n})/\tau) dF_\tau(\tau)\\ &= \exp(i\mathbf{t}^T\boldsymbol{\mu}) \int_0^\infty \exp\left(-\frac{\tau}{2}\mathbf{t}^T\boldsymbol{\Sigma}\mathbf{t}\right) dF_\tau(\tau)\\ &= \exp(i\mathbf{t}^T\boldsymbol{\mu})\phi_r(\mathbf{t}^T\boldsymbol{\Sigma}\mathbf{t}),\end{aligned} \tag{1.18}$$

where the characteristic generator

$$\phi_r(u) = \int_0^\infty \exp\left(-\frac{\tau}{2}u\right) dF_\tau(\tau) \tag{1.19}$$

does not depend on the dimension m nor on the rank k, unlike the RES distributions which are not RCG whose characteristic generator can depend on it.

In the particular case where $\boldsymbol{\Sigma}$ is not singular ($k = m$), the conditional distribution of $\mathbf{x}$ given $\tau_r = \tau$ is the $\mathbb{R}N_m(\boldsymbol{\mu}, \tau\boldsymbol{\Sigma})$ distribution from (1.14). Consequently, the distribution of $\mathbf{x}$ is always continuous w.r.t. Lebesgue measure on $\mathbb{R}^m$ and its p.d.f. is given by

$$p(\mathbf{x}) = (2\pi)^{-m/2}|\boldsymbol{\Sigma}|^{-1/2} \int_0^\infty \tau^{-m/2} \exp\left(-\frac{1}{2\tau}(\mathbf{x}-\boldsymbol{\mu})^T\boldsymbol{\Sigma}^{-1}(\mathbf{x}-\boldsymbol{\mu})\right) dF_\tau(\tau). \tag{1.20}$$

Note that the p.d.f. (1.20) can always be written in the form (1.9) with the density generator

$$g_r(t) = (2\pi)^{-m/2} \int_0^\infty \tau^{-m/2} \exp\left(-\frac{t}{2\tau}\right) dF_\tau(\tau), \tag{1.21}$$

and similarly to the RES distributions, we are faced with scale ambiguity where $(c^2\boldsymbol{\Sigma}, c^{-2}\tau_r, c^m g_r(.c^2))$ gives the same R-CG distribution. Note that (1.21) reduces to the density generator (1.69) of the Gaussian distribution when τ_r is a degenerate r.v. putting all the probability at $\tau_r = 1$. Note also that the ϵ-contaminated Gaussian distribution belongs to the class of CG distributions and is obtained when τ_r is a discrete r.v. with $P(\tau_r = a^2) = \epsilon$ and $P(\tau_r = 1) = 1 - \epsilon$, where (a^2, ϵ) are parameters that control the heaviness of the tails as compared to the Gaussian distribution.

1.3 Definition of the Complex Elliptically Symmetric Distributions

1.3.1 Characteristic Function

An r.v. $\mathbf{x} \in \mathbb{C}^m$ is said to have a noncircular complex elliptically symmetric (NC-CES) distribution (also called generalized complex elliptical in [47] if the r.v. $\bar{\mathbf{x}} \stackrel{\text{def}}{=} (\text{Re}(\mathbf{x})^T, \text{Im}(\mathbf{x})^T)^T \in \mathbb{R}^{2m}$ is RES distributed. Denote the symmetry center and the scatter matrix (of rank $k \leq 2m$) of $\bar{\mathbf{x}}$ by $\bar{\boldsymbol{\mu}} \in \mathbb{R}^{2m}$ and $\bar{\boldsymbol{\Sigma}} \in \mathbb{R}^{2m \times 2m}$, respectively. Using the one-to-one mapping $\bar{\mathbf{x}} \mapsto \widetilde{\mathbf{x}} \stackrel{\text{def}}{=} (\mathbf{x}^T, \mathbf{x}^H)^T = \sqrt{2}\mathbf{M}\bar{\mathbf{x}}$ where $\mathbf{M} \stackrel{\text{def}}{=} \frac{1}{\sqrt{2}}\begin{pmatrix} \mathbf{I} & i\mathbf{I} \\ \mathbf{I} & -i\mathbf{I} \end{pmatrix}$ is unitary, we obtain $\bar{\mathbf{t}}^T \bar{\boldsymbol{\mu}} = \text{Re}(\mathbf{t}^H \boldsymbol{\mu})$, $\bar{\mathbf{t}}^T \bar{\boldsymbol{\Sigma}} \bar{\mathbf{t}} = \frac{1}{2}\widetilde{\mathbf{t}}^H(\mathbf{M}\bar{\boldsymbol{\Sigma}}\mathbf{M}^H)\widetilde{\mathbf{t}} = \frac{1}{4}\widetilde{\mathbf{t}}^H \widetilde{\boldsymbol{\Sigma}} \widetilde{\mathbf{t}}$ with $\bar{\mathbf{t}} \stackrel{\text{def}}{=} (\text{Re}(\mathbf{t})^T, \text{Im}(\mathbf{t})^T)^T$, $\bar{\boldsymbol{\mu}} \stackrel{\text{def}}{=} (\text{Re}(\boldsymbol{\mu})^T, \text{Im}(\boldsymbol{\mu})^T)^T$, $\widetilde{\mathbf{t}} \stackrel{\text{def}}{=} (\mathbf{t}^T, \mathbf{t}^H)^T$ and $\widetilde{\boldsymbol{\Sigma}} \stackrel{\text{def}}{=} 2\mathbf{M}\bar{\boldsymbol{\Sigma}}\mathbf{M}^H$ of rank k structured as

$$\widetilde{\boldsymbol{\Sigma}} = \begin{pmatrix} \boldsymbol{\Sigma} & \boldsymbol{\Omega} \\ \boldsymbol{\Omega}^* & \boldsymbol{\Sigma}^* \end{pmatrix}, \tag{1.22}$$

where $\boldsymbol{\Sigma}$ and $\boldsymbol{\Omega}$ defined from $2\mathbf{M}\bar{\boldsymbol{\Sigma}}\mathbf{M}^H$, are positive semi-definite Hermitian and complex symmetric matrices, respectively. Consequently, we obtain the following theorem by the definition (1.1).

Theorem 1.2 *The characteristic function of an NC-CES distributed r.v.* $\mathbf{x} \in \mathbb{C}^m$ *is of the form*

$$\Phi_x(\mathbf{t}) \stackrel{\text{def}}{=} \text{E}[\exp(i\bar{\mathbf{t}}^T\bar{\mathbf{x}})] = \exp(i\text{Re}(\mathbf{t}^H\boldsymbol{\mu}))\phi_c\left(\frac{1}{2}\widetilde{\mathbf{t}}^H\widetilde{\boldsymbol{\Sigma}}\widetilde{\mathbf{t}}\right) \qquad \mathbf{t} \in \mathbb{C}^m, \tag{1.23}$$

where

$$\phi_c(u) \stackrel{\text{def}}{=} \phi_r\left(\frac{1}{2}u\right) \tag{1.24}$$

is the characteristic generator, and where $\boldsymbol{\mu} \in \mathbb{C}^m$ *and* $\widetilde{\boldsymbol{\Sigma}} \in \mathbb{C}^{2m \times 2m}$ *denote respectively the symmetric center and the extended scatter matrix of the NC-CES distributed r.v.* $\mathbf{x}$.

We shall write $\mathbf{x} \sim \text{CES}_m(\boldsymbol{\mu}, \boldsymbol{\Sigma}, \boldsymbol{\Omega}, \phi_c)$. In the particular case where $\boldsymbol{\Omega} = \mathbf{0}$, the rank of $\widetilde{\boldsymbol{\Sigma}}$ is even with $\text{rank}(\widetilde{\boldsymbol{\Sigma}}) = 2\text{rank}(\boldsymbol{\Sigma}) = k$ and $\mathbf{x}$ is C-CES distributed [37, 49, 55]. The term *circular* is often dropped in the current terminology used in signal processing where the distribution of a C-CES r.v. is usually indicated as $\mathbf{x} \sim \text{CES}_m(\boldsymbol{\mu}, \boldsymbol{\Sigma}, \phi_c)$. Moreover, from (1.23), we get:

Theorem 1.3 *The characteristic function of a C-CES distributed r.v.* $\mathbf{x} \in \mathbb{C}^m$ *is of the form*

$$\Phi_x(\mathbf{t}) = \exp(i\mathrm{Re}(\mathbf{t}^H \boldsymbol{\mu}))\phi_c(\mathbf{t}^H \boldsymbol{\Sigma}\mathbf{t}), \qquad \mathbf{t} \in \mathbb{C}^m, \tag{1.25}$$

where $\boldsymbol{\mu} \in \mathbb{C}^m$ *and* $\boldsymbol{\Sigma} \in \mathbb{C}^{m \times m}$ *denote the symmetric center, and the scatter matrix of the C-CES distributed r.v.* $\mathbf{x}$*, respectively.*

1.3.2 Stochastic Representation

From the definition of the NC-CES distribution, a simple complex-valued extension of the stochastic representation (1.7) is only possible if the rank of $\widetilde{\boldsymbol{\Sigma}}$, which is equal to the rank of $\bar{\boldsymbol{\Sigma}}$, is even (it is, in particular, the case of the C-CES distribution and the case where $\widetilde{\boldsymbol{\Sigma}}$ is not singular for which $k = 2m$). Let $2k$ be the rank of $\widetilde{\boldsymbol{\Sigma}}$. In this case, there exists an $m \times k$ full column rank matrix $\mathbf{A}$ such that $\boldsymbol{\Sigma} = \mathbf{A}\mathbf{A}^H$ and $\boldsymbol{\Omega} = \mathbf{A}\boldsymbol{\Delta}_\kappa\mathbf{A}^T$ where $\boldsymbol{\Delta}_\kappa = \mathrm{Diag}(\kappa_1, \ldots, \kappa_k)$ is a real diagonal matrix with non-negative real entries $(\kappa_i)_{i=1,\ldots,k}$ [28, Corollary 4.6.12(b)]. Furthermore, it has been proved in [16] that $0 \leq \kappa_i \leq 1$. This parameterization allows us to state that the stochastic representation of this distribution, proved in [1], is a multivariate extension of the univariate generation of NC-CES r.v. presented in [46, Sect. IV.C].

Theorem 1.4 $\mathbf{x}$ *is* $\mathrm{CES}_m(\boldsymbol{\mu}, \boldsymbol{\Sigma}, \boldsymbol{\Omega}, \phi_c)$ *distributed i.f.f. it admits the following stochastic representation*

$$\mathbf{x} =_d \boldsymbol{\mu} + \sqrt{\mathcal{Q}_{c,k}}\mathbf{A}[\boldsymbol{\Delta}_1\mathbf{u}_{c,k} + \boldsymbol{\Delta}_2\mathbf{u}^*_{c,k}], \tag{1.26}$$

where $\mathcal{Q}_{c,k} \stackrel{\text{def}}{=} \frac{1}{2}\mathcal{Q}_{r,2k}$ *and* $\mathbf{u}_{c,k} \sim U(\mathbb{C}S^k)$ *are independent,* $\boldsymbol{\Delta}_1 \stackrel{\text{def}}{=} \frac{\boldsymbol{\Delta}_+ + \boldsymbol{\Delta}_-}{2}$ *and* $\boldsymbol{\Delta}_2 \stackrel{\text{def}}{=} \frac{\boldsymbol{\Delta}_+ - \boldsymbol{\Delta}_-}{2}$ *where* $\boldsymbol{\Delta}_+ \stackrel{\text{def}}{=} \sqrt{\mathbf{I} + \boldsymbol{\Delta}_\kappa}$ *and* $\boldsymbol{\Delta}_- \stackrel{\text{def}}{=} \sqrt{\mathbf{I} - \boldsymbol{\Delta}_\kappa}$.

In the particular case of C-CES distributions, $\boldsymbol{\Omega} = \mathbf{0}$, which is equivalent to $\boldsymbol{\Delta}_\kappa = \mathbf{0}$, i.e., $\boldsymbol{\Delta}_1 = \mathbf{I}$ and $\boldsymbol{\Delta}_2 = \mathbf{0}$ and consequently the stochastic representation (1.26) reduces to the well known stochastic representation reported in [49]:

$$\mathbf{x} =_d \boldsymbol{\mu} + \sqrt{\mathcal{Q}_{c,k}}\mathbf{A}\mathbf{u}_{c,k}. \tag{1.27}$$

Note that similarly to the RES distribution, the C-CES distribution can be defined from the affine function (1.2), where here $\mathbf{x}_s$ is k-dimensional spherically distributed defined by the equality $\mathbf{x}_s =_d \mathcal{U}\mathbf{x}_s$ for arbitrary complex-valued k-dimensional unitary matrix $\mathcal{U}$, and where $\mathbf{A} \in \mathbb{C}^{m \times k}$ is any square root ($\boldsymbol{\Sigma} = \mathbf{A}\mathbf{A}^H$) of the scatter matrix [37, 55]. Such r.v. $\mathbf{x}_s$ is also characterized by the stochastic representation $\mathbf{x}_s =_d \sqrt{\mathcal{Q}_{c,k}}\mathbf{u}_{c,k}$ where the non-negative r.v. $\mathcal{Q}_{c,k}$, and $\mathbf{u}_{c,k}$ are independent, with $\mathbf{u}_{c,k} \sim U(\mathbb{C}S^k)$. $\mathbf{x}_s$ is also characterized by a characteristic function of the form $\Phi_{x_s}(\mathbf{t}) = \phi_c(\|\mathbf{t}\|^2), \mathbf{t} \in \mathbb{C}^k$. Consequently, from (1.2), we find the characteristic function (1.25) which defines the C-CES distribution since $\Phi_x(\mathbf{t}) \stackrel{\text{def}}{=} \mathrm{E}[\exp(i\mathrm{Re}(\mathbf{t}^H\mathbf{x}))] = \exp(i\mathrm{Re}(\mathbf{t}^H\boldsymbol{\mu}))\Phi_{x_s}(\mathbf{A}^H\mathbf{t}) = \exp(i\mathrm{Re}(\mathbf{t}^H\boldsymbol{\mu}))\phi_c(\mathbf{t}^H\boldsymbol{\Sigma}\mathbf{t})$.

1.3.3 The Absolutely Continuous Case

The p.d.f. of $\mathbf{x}$ is defined on $\mathbb{C}^m$ i.f.f. $\bar{\mathbf{x}}$ is absolutely continuous w.r.t. Lebesgue measure on $\mathbb{R}^{2m}$. Assuming that $\text{rank}(\widetilde{\boldsymbol{\Sigma}}) = 2m$ and using the identities $(\bar{\mathbf{x}} - \bar{\boldsymbol{\mu}})^T \bar{\boldsymbol{\Sigma}}^{-1} (\bar{\mathbf{x}} - \bar{\boldsymbol{\mu}}) = (\widetilde{\mathbf{x}} - \widetilde{\boldsymbol{\mu}})^H \widetilde{\boldsymbol{\Sigma}}^{-1} (\widetilde{\mathbf{x}} - \widetilde{\boldsymbol{\mu}})$ and $|\bar{\boldsymbol{\Sigma}}| = 2^{-2m} |\widetilde{\boldsymbol{\Sigma}}|$, the p.d.f. (1.9) becomes

$$p(\mathbf{x}) = |\widetilde{\boldsymbol{\Sigma}}|^{-1/2} g_c \left[\frac{1}{2} (\widetilde{\mathbf{x}} - \widetilde{\boldsymbol{\mu}})^H \widetilde{\boldsymbol{\Sigma}}^{-1} (\widetilde{\mathbf{x}} - \widetilde{\boldsymbol{\mu}}) \right], \tag{1.28}$$

where $g_c(t)$ is defined by

$$g_c(t) \stackrel{\text{def}}{=} 2^m g_r(2t), \tag{1.29}$$

and $g_r(t)$ is the density generator associated with the distribution $\text{RES}_{2m}(\bar{\boldsymbol{\mu}}, \bar{\boldsymbol{\Sigma}}, \phi_r)$, which satisfies

$$\delta_{c,m} \stackrel{\text{def}}{=} \int_0^\infty t^{m-1} g_c(t) dt = \delta_{r,2m,g} < \infty. \tag{1.30}$$

We note that, in this case, (1.26) is written equivalently in the form $\widetilde{\mathbf{x}} =_d \widetilde{\boldsymbol{\mu}} + \widetilde{\boldsymbol{\Sigma}}^{1/2} \widetilde{\mathbf{u}}_{c,m}$ (where $\widetilde{\mathbf{u}}_{c,m} \stackrel{\text{def}}{=} (\mathbf{u}_{c,m}^T, \mathbf{u}_{c,m}^H)^T$), then it follows that

$$\mathcal{Q}_{c,m} = \frac{1}{2} \mathcal{Q}_{r,2m} \tag{1.31}$$

and

$$\mathcal{Q}_{c,m} =_d \frac{1}{2} (\widetilde{\mathbf{x}} - \widetilde{\boldsymbol{\mu}})^H \widetilde{\boldsymbol{\Sigma}}^{-1} (\widetilde{\mathbf{x}} - \widetilde{\boldsymbol{\mu}}). \tag{1.32}$$

For C-CES distributed $\mathbf{x}$, (1.28) reduces to

$$p(\mathbf{x}) = |\boldsymbol{\Sigma}|^{-1} g_c[(\mathbf{x} - \boldsymbol{\mu})^H \boldsymbol{\Sigma}^{-1} (\mathbf{x} - \boldsymbol{\mu})], \tag{1.33}$$

and (1.27) implies

$$\mathcal{Q}_{c,m} =_d (\mathbf{x} - \boldsymbol{\mu})^H \boldsymbol{\Sigma}^{-1} (\mathbf{x} - \boldsymbol{\mu}). \tag{1.34}$$

Following the same derivation as for the RES distribution, we get the following p.d.f. of $\mathcal{Q}_{c,m}$ and $\mathcal{R}_{c,m}$

$$p(q) = \delta_{c,m}^{-1} q^{m-1} g_c(q) \text{ and } p(r) = 2\delta_{c,m}^{-1} r^{2m-1} g_c(r^2). \tag{1.35}$$

1.3.4 The Subclass of Compound Gaussian Distributions

In many engineering applications, only the circular complex case is considered. An r.v. $\mathbf{x} \in \mathbb{C}^m$ is said to be circular complex compound-Gaussian (C-CCG) distributed if the r.v. $\bar{\mathbf{x}} \in \mathbb{R}^{2m}$ is RCG distributed (see Definition 1.3) where the associated extended scatter matrix (1.22) is bloc-diagonal $\widetilde{\mathbf{\Sigma}} = \begin{pmatrix} \mathbf{\Sigma} & \mathbf{0} \\ \mathbf{0} & \mathbf{\Sigma}^* \end{pmatrix}$. Consequently, from the one-to-one mapping $\bar{\mathbf{x}} \mapsto \tilde{\mathbf{x}}$, definition (1.3) gives the following stochastic representation

$$\mathbf{x} =_d \boldsymbol{\mu} + \sqrt{\tau_c}\mathbf{n}, \tag{1.36}$$

where $\tau_c = 2\tau_r$ with τ_r, independent from $\mathbf{n}$, is associated with the $2m$-dimensional RCG distribution and $\mathbf{n} \sim \mathbb{C}N_m(\mathbf{0}, \mathbf{\Sigma})$.

As a consequence, all the properties given in Sect. 1.2.4 can be deduced from this real to complex representation. In particular with $\mathbf{\Sigma} = \mathbf{A}\mathbf{A}^H$ of rank k: $\mathbf{n}_0 \sim \mathbb{C}N_k(\mathbf{0}, \mathbf{I})$, (1.16) with now $\mathcal{R}_{c,k} \stackrel{\text{def}}{=} \sqrt{\tau_c s}$ where $s \stackrel{\text{def}}{=} \|\mathbf{n}_0\|^2 \sim \frac{1}{2}\chi^2_{2k} = \text{Gam}(k, 1)$. Then, the C-CCG distributions form a subclass of the C-CES distributions and $\mathbf{x} \sim$ C-CES$_m(\boldsymbol{\mu}, \mathbf{\Sigma}, \phi)$ belongs to the set of C-CCG distributions i.f.f. the p.d.f of $\mathcal{Q}_{c,k}$ is

$$p(q) = \int_0^\infty \frac{1}{\Gamma(k)\tau^k} q^{k-1} \exp(-q/\tau) dF_\tau(\tau) \tag{1.37}$$

and (1.20) becomes

$$p(\mathbf{x}) = \pi^{-m}|\mathbf{\Sigma}|^{-1} \int_0^\infty \tau^{-m} \exp\left(-\frac{1}{\tau}(\mathbf{x}-\boldsymbol{\mu})^H \mathbf{\Sigma}^{-1}(\mathbf{x}-\boldsymbol{\mu})\right) dF_\tau(\tau). \tag{1.38}$$

1.4 Basic Properties

In this section and throughout the rest of this chapter, we mainly consider the RES distributions, knowing that the C-CES distributions are only a particular representation of them for even m, where the complementary scatter matrix $\mathbf{\Omega}$ defined in (1.22) is zero. Consequently we drop the indices r in $\delta_{r,k}$, $\delta_{r,k,g}$, g_r, ϕ_r, $\mathcal{Q}_{r,k}$, $\mathcal{R}_{r,k}$ and $\mathbf{u}_{r,k}$ associated with the v.a. $\mathbf{x}$. We will show that these distributions benefit from most of the properties of the Gaussian distribution except the additive stability, whose conditions are more restrictive (see e.g., the quick surveys in [20, 24]).

1.4.1 Moments

From the full-rank stochastic representation (1.7), it is clear that $\mathbf{x}$ admits pth-order moments i.f.f. $\text{E}(\mathcal{R}_k^p) < \infty$. Using the characteristic function (1.1), $\text{E}(\mathcal{R}_k^p) < \infty$ i.f.f. the characteristic generator $\phi(\mathbf{t})$ is p times differentiable. In this case, the pth-order moments of $\mathbf{x}$ are given by $\text{E}(x_1^{p_1} x_2^{p_2} \dots x_m^{p_m}) = \frac{1}{i^p} \frac{\partial^p \Psi_x(\mathbf{t})}{\partial t_1^{p_1} \partial t_2^{p_2} \dots \partial t_m^{p_m}}|_{\mathbf{t}=\mathbf{0}}$ with $p = \sum_{i=1}^m p_i$ and $\mathbf{x} = (x_1, x_2, \dots, x_m)^T$.

Assuming that the correspondent moments are finite, one has

$$\mathrm{E}(\mathbf{x}) = \boldsymbol{\mu} \tag{1.39}$$

$$\mathrm{Cov}(\mathbf{x}) = \frac{\mathrm{E}(\mathcal{R}_k^2)}{k}\boldsymbol{\Sigma} = \frac{\mathrm{E}(\mathcal{Q}_k)}{k}\boldsymbol{\Sigma} = -2\phi'(\mathbf{0})\boldsymbol{\Sigma}. \tag{1.40}$$

In particular for RCG distributions, (1.40) becomes

$$\mathrm{Cov}(\mathbf{x}) = \mathrm{E}(\tau)\boldsymbol{\Sigma}. \tag{1.41}$$

We see from (1.40) and (1.41) that under finite second-order moment assumption, the covariance of $\mathbf{x}$ does not necessarily coincide with the scatter matrix $\boldsymbol{\Sigma}$, but these two matrices are proportional. Note that many second-order signal processing methodologies, such as, for example, subspace-based processing (where $k = m$), require an estimate of the covariance only at up to a multiplicative scalar. In this case, the *shape matrix*, defined as a scaled version of the scatter matrix $\mathbf{V} \stackrel{\text{def}}{=} \frac{1}{s(\boldsymbol{\Sigma})}\boldsymbol{\Sigma}$ can be adopted to characterize the correlation structure. Even if the choice of the scale functional is entirely arbitrary, in signal processing literature, the most popular scale is the one on the trace of the scatter matrix, i.e., $s(\boldsymbol{\Sigma}) \stackrel{\text{def}}{=} \mathrm{Tr}(\boldsymbol{\Sigma})/m$ leading to the following shape matrix $\mathbf{V} \stackrel{\text{def}}{=} \frac{m}{\mathrm{Tr}(\boldsymbol{\Sigma})}\boldsymbol{\Sigma}$. One can also find $s(\boldsymbol{\Sigma}) \stackrel{\text{def}}{=} |\boldsymbol{\Sigma}|^{1/m}$, leading to unit determinant for the shape matrix.

Otherwise, (1.40) and (1.41) can be used to resolve the scale ambiguity of the couple $(\boldsymbol{\Sigma}, \phi(.))$ in the definition (1.1) of the RES distribution by fixing the constraint on the characteristic generator $\phi(.)$

$$\mathrm{E}(\mathcal{R}_k^2) = \mathrm{E}(\mathcal{Q}_k) = k = \mathrm{rank}(\boldsymbol{\Sigma}) \ \text{ or } \ \mathrm{E}(\tau) = 1 \ \text{ for RCG distributions,} \tag{1.42}$$

which ensures that $\mathrm{Cov}(\mathbf{x}) = \boldsymbol{\Sigma}$.

When the r.v. $\mathbf{x}$ is absolutely continuous w.r.t. Lebesgue measure on $\mathbb{R}^m$ and $k = m$, (1.10) implies that (1.42) is equivalent to the following constraint on the density generator $g(.)$;

$$\delta_{m+2,g}/\delta_{m,g} = m. \tag{1.43}$$

If $\mathrm{E}(\mathcal{Q}_m)$ is not finite, rather imposing $\mathrm{Median}(\mathcal{R}_m) = 1$, (i.e., from (1.13) the constraint $2\delta_m^{-1}\int_0^1 r^{m-1}g(r^2)dr = \frac{1}{2}$) is a more appropriate scaling constraint as it avoids any finite moment assumptions. Similarly for RCG distributions, if $\mathrm{E}(\tau)$ is not finite, the constraint $\mathrm{Median}(\tau) = 1$ (i.e., $F_\tau(1) = \frac{1}{2}$) can be used. Indeed many RES distributions do not have finite second-order moments.

To consider higher-order multivariate central moments, let us consider $\sigma_{i_1,i_2,\ldots,i_\ell} \stackrel{\text{def}}{=} \mathrm{E}[(x_{i_1} - \mu_{i_1})(x_{i_2} - \mu_{i_2})\ldots(x_{i_\ell} - \mu_{i_\ell})]$ with $\boldsymbol{\mu} = (\mu_1, \mu_2, \ldots, \mu_m)^T$ and $(i_1, i_2, \ldots, i_\ell) \in \{1, \ldots, m\}^\ell$. By symmetry, all odd-order central moments are zero, provided that the corresponding moments do exist. As for fourth-order moments, if $\mathrm{E}(\mathcal{R}_m^4) < \infty$ (with $k = m$), they all satisfy the identity [50, p. 2]

$$\sigma_{i,j,k,\ell} = (\kappa + 1)(\sigma_{i,j}\sigma_{k,\ell} + \sigma_{i,k}\sigma_{j,\ell} + \sigma_{i,\ell}\sigma_{j,k}), \quad (i, j, k, \ell) \in \{1, \ldots, m\}^4, \tag{1.44}$$

where

$$\kappa \stackrel{\text{def}}{=} \frac{1}{3}\left(\frac{\sigma_{i,i,i,i}}{\sigma^2_{i,i}} - 3\right) = \frac{m}{m+2}\frac{\mathrm{E}(\mathcal{R}^4_m)}{(\mathrm{E}(\mathcal{R}^2_m))^2} - 1 = \frac{\phi''(0)}{(\phi'(0))^2} - 1. \tag{1.45}$$

κ is the *kurtosis* parameter of the marginal r.v. x_i [3, 6]. It usually depends on the dimension m, but, remarkably, the kurtosis of the ith component does not depend on i, nor on the scatter matrix $\mathbf{\Sigma}$. Consequently taken $\mathbf{\Sigma} = \mathbf{I}$, we get for all RCG distributions: $\sigma_{i,i,i,i} = 3\mathrm{E}(\tau^2)$ and $\sigma_{i,i} = \mathrm{E}(\tau)$ and thus

$$\kappa = \frac{\mathrm{Var}(\tau)}{[\mathrm{E}(\tau)]^2}. \tag{1.46}$$

Consequently, the kurtosis parameter does not depend on the dimension m for RCG distributions. Note that $\kappa = 0$ for the Gaussian distribution and that there exist other definitions of the kurtosis parameter or coefficient in the literature as $\kappa \stackrel{\text{def}}{=} \frac{\sigma_{i,i,i,i}}{\sigma^2_{i,i}} - 3$ and $\kappa \stackrel{\text{def}}{=} \frac{\sigma_{i,i,i,i}}{\sigma^2_{i,i}}$. Note also that the kurtosis is bounded below such that $\kappa \geq -2/(m+2)$ [7].

1.4.2 *Affine Transformations and Marginal Distributions*

Let $\mathbf{b} \in \mathbb{R}^n$ and $\mathbf{B} \in \mathbb{R}^{n\times m}$, and consider the transformed r.v. $\mathbf{y} = \mathbf{Bx} + \mathbf{b}$. Its characteristic function $\Phi_y(\mathbf{t})$ is deduced from the characteristic function (1.1) of $\mathbf{x}$ by

$$\Phi_y(\mathbf{t}) = \exp(i\mathbf{t}^T\mathbf{b})\Phi_x(\mathbf{B}^T\mathbf{t}) = \exp(i\mathbf{t}^T(\mathbf{B}\boldsymbol{\mu} + \mathbf{b}))\phi(\mathbf{t}^T\mathbf{B}\mathbf{\Sigma}\mathbf{B}^T\mathbf{t}), \quad \mathbf{t} \in \mathbb{R}^n. \tag{1.47}$$

Consequently, $\mathbf{y}$ is $\mathrm{RES}_n(\mathbf{B}\boldsymbol{\mu} + \mathbf{b}, \mathbf{B}\mathbf{\Sigma}\mathbf{B}^T, \phi)$-distributed, and thus the class of $\mathrm{RES}_m(\boldsymbol{\mu}, \mathbf{\Sigma}, \phi)$ distributions is closed under affine transformations. Note that the parameters are transformed as $(\boldsymbol{\mu}, \mathbf{\Sigma}) \mapsto (\mathbf{B}\boldsymbol{\mu} + \mathbf{b}, \mathbf{B}\mathbf{\Sigma}\mathbf{B}^T)$, which is the usual transformation for the couple (expectation, covariance) (when it exists) for the affine transformation $\mathbf{x} \mapsto \mathbf{Bx} + \mathbf{b}$. Note also that, for $n \neq m$, $\mathbf{y}$ may not belong to the same family as that of the r.v. $\mathbf{x}$ because the characteristic generator ϕ may depend on the dimension m.

From the full-rank stochastic representation of $\mathbf{x}$ (1.7), we derive

$$\mathbf{y} =_d \mathbf{B}\boldsymbol{\mu} + \mathbf{b} + \mathcal{R}_k(\mathbf{BA})\mathbf{u}_k, \tag{1.48}$$

which is not necessarily a full-rank stochastic representation of $\mathbf{y}$ because $\mathrm{rank}(\mathbf{BA}) \leq \min(\mathrm{rank}(\mathbf{B}), \mathrm{rank}(\mathbf{A})) \leq \min(n, k)$. If $\ell \stackrel{\text{def}}{=} \mathrm{rank}(\mathbf{BA})$, there exist full column rank

$n \times \ell$ matrices $\mathbf{C}$ such that $\mathbf{B}\boldsymbol{\Sigma}\mathbf{B}^T = \mathbf{C}\mathbf{C}^T$. And thus, similarly to $\mathbf{x}$, where its distribution is equivalently defined by its stochastic full-rank representation (1.7) and by its characteristic function (1.1), we get from (1.47) a stochastic full-rank representation of $\mathbf{y}$

$$\mathbf{y} =_d \mathbf{B}\boldsymbol{\mu} + \mathbf{b} + \mathcal{R}_\ell \mathbf{C}\mathbf{u}_\ell, \tag{1.49}$$

where the non-negative r.v. $\mathcal{R}_\ell$, and $\mathbf{u}_\ell$ are independent, $\mathbf{u}_\ell$ is uniformly distributed on the unit real ℓ-sphere.

Of course, it directly follows that univariate and multivariate marginals of $\mathbf{x}$ also are RES distributed with ϕ remains unchanged. If $\mathbf{x} = (x_1, \ldots, x_m)^T = (\mathbf{x}_1^T, \mathbf{x}_2^T)^T$, $\boldsymbol{\mu} = (\mu_1, \ldots, \mu_m)^T = (\boldsymbol{\mu}_1^T, \boldsymbol{\mu}_2^T)^T$ and $\boldsymbol{\Sigma} = (\sigma_{i,j})_{i,j=1}^m = \begin{pmatrix} \boldsymbol{\Sigma}_{1,1} & \boldsymbol{\Sigma}_{1,2} \\ \boldsymbol{\Sigma}_{2,1} & \boldsymbol{\Sigma}_{2,2} \end{pmatrix}$, where $\mathbf{x}_1$ and $\boldsymbol{\mu}_1$ are m_1-dimensional vectors, $\boldsymbol{\Sigma}_{1,1}$ and $\boldsymbol{\Sigma}_{2,2}$ are $m_1 \times m_1$ and $m_2 \times m_2$ matrices, respectively (with $m_1 + m_2 = m$), then $\mathbf{x}_1 \sim \text{RES}_{m_1}(\boldsymbol{\mu}_1, \boldsymbol{\Sigma}_{1,1}, \phi)$, $\mathbf{x}_2 \sim \text{RES}_{m_2}(\boldsymbol{\mu}_2, \boldsymbol{\Sigma}_{2,2}, \phi)$, and $x_i \sim \text{RES}_1(\mu_i, \sigma_{i,i}, \phi)$, $i = 1, \ldots, m$. Of course, their stochastic representations (1.48) and (1.49) also, follow. For example, for $m_1 \geq k$, if $\mathbf{A} = \begin{bmatrix} \mathbf{A}_1 \\ \mathbf{A}_2 \end{bmatrix}$ where $\mathbf{A}_1$ is a $m_1 \times k$ full column rank matrix

$$\mathbf{x}_1 =_d \boldsymbol{\mu}_1 + \mathcal{R}_k \mathbf{A}_1 \mathbf{u}_k \tag{1.50}$$

is a stochastic full-rank representation of the marginal $\mathbf{x}_1$. As for the univariate marginal x_i, the stochastic representations (1.48) and (1.49) reduce to

$$x_i =_d \mu_i + \mathcal{R}_k \mathbf{a}_i^T \mathbf{u}_k =_d \mu_i + \mathcal{R}_1 \|\mathbf{a}_i\| u_1, \tag{1.51}$$

where $\mathbf{a}_i$ is the ith column of $\mathbf{A}^T$ (thus $\|\mathbf{a}_i\|^2 = \sigma_{i,i}$), $\mathcal{R}_k$ and $\mathcal{R}_1$ are the modular variates of $\mathbf{x}$ and x_i, respectively, and where u_1 reduces to the uniform discrete r.v. $\{-1, +1\}$.

Now we take a closer look at the marginal distributions when $k = m$. In this case, the stochastic full-rank representation (1.50) and (1.51) of arbitrary univariate or multivariate marginal r.v. $\mathbf{x}_1$ and x_i reduce to

$$\mathbf{x}_1 =_d \boldsymbol{\mu}_1 + \mathcal{R}_{m_1} \boldsymbol{\Sigma}_{1,1}^{1/2} \mathbf{u}_{m_1} \quad \text{and} \quad x_i =_d \mu_i + \mathcal{R}_1 \sigma_{i,i}^{1/2} u_1, \tag{1.52}$$

where the modular variates $\mathcal{R}_{m_1}$ of $\mathbf{x}_1$ (which includes the modular variates of the univariate marginal r.v. x_i for $m_1 = 1$) and $\mathcal{R}_m$ of $\mathbf{x}$ are related by the relation [18, Corollary p. 59]

$$\mathcal{R}_{m_1} =_d \mathcal{R}_m \, \beta_{\frac{m_1}{2}, \frac{m_2}{2}}. \tag{1.53}$$

In (1.53), the r.v. $\mathcal{R}_m$ and $\beta_{\frac{m_1}{2}, \frac{m_2}{2}}$ are independent and $\beta^2_{\frac{m_1}{2}, \frac{m_2}{2}} \sim \text{Beta}(\frac{m_1}{2}, \frac{m_2}{2})$.

Moreover, in the absolutely continuous case w.r.t. Lebesgue measure on $\mathbb{R}^m$, i.e., $\mathbf{x} \sim \text{RES}_m(\boldsymbol{\mu}, \boldsymbol{\Sigma}, g)$, (1.53) allows us to relate the p.d.f. $p_{m_1}(r)$ of $\mathcal{R}_{m_1}$ to the p.d.f. $p_m(r)$ of $\mathcal{R}_m$ [18, rel. 2.5.15]

$$p_{m_1}(r) = \frac{2r^{m_1-1}}{\mathrm{B}(\frac{m_1}{2}, \frac{m_2}{2})} \int_r^{+\infty} t^{-(m-2)}(t^2 - r^2)^{\frac{m_2}{2}-1} p_m(t)dt \tag{1.54}$$

and to the density generator $g(.)$ of $\mathbf{x}$ using $p_m(r)$ given by (1.13)

$$p_{m_1}(r) = \frac{2\pi^{m/2} r^{m_1-1}}{\Gamma(\frac{m_1}{2})\Gamma(\frac{m_2}{2})} \int_{r^2}^{+\infty} (t - r^2)^{\frac{m_2}{2}-1} g(t)dt. \tag{1.55}$$

This allows us to derive the density generators $g_{m_1|m}(.)$ of the multivariate and univariate marginal r.v. $\mathbf{x}_1$ and x_i thanks to (1.13) $p_{m_1}(r) = 2\delta_{m_1}^{-1} r^{m_1-1} g_{m_1|m}(r^2)$

$$g_{m_1|m}(u) = \delta_{m_2}^{-1} \int_u^{+\infty} (t - u)^{\frac{m_2}{2}-1} g(t)dt. \tag{1.56}$$

Therefore, the p.d.f. of the multivariate marginal r.v. $\mathbf{x}_1 \sim \mathrm{RES}_{m_1}(\boldsymbol{\mu}_1, \boldsymbol{\Sigma}_{1,1}, g_{m_1|m})$ and $x_i \sim \mathrm{RES}_1(\mu_i, \sigma_{i,i}, g_{1|m})$ are given respectively by

$$p_{m_1|m}(\mathbf{x}_1) = |\boldsymbol{\Sigma}_{1,1}|^{-1/2} g_{m_1|m}[(\mathbf{x}_1 - \boldsymbol{\mu}_1)^T \boldsymbol{\Sigma}_{1,1}^{-1}(\mathbf{x}_1 - \boldsymbol{\mu}_1)]. \tag{1.57}$$

$$p_{i|m}(x_i) = \frac{1}{\sqrt{\sigma_{i,i}}} g_{1|m}\left(\frac{(x_i - \mu_i)^2}{\sigma_{i,i}}\right). \tag{1.58}$$

Note that there is no guarantee that the integral in (1.56) leads to a closed-form expression when a closed-form expression of $g(u)$ is available, except for the class of RCG distributions for which $g_{m_1|m}(u) \propto g(u)$. In particular for the Gaussian distribution, we obtain $g_{m_1|m}(u) = \frac{1}{(2\pi)^{m_1/2}} \exp(-\frac{u}{2})$.

For the RCG distributions defined in Sect. 1.2.4, it is clear from the stochastic representation (1.14) that

$$\mathbf{x}_1 =_d \boldsymbol{\mu}_1 + \sqrt{\tau}\mathbf{n}_1, \tag{1.59}$$

with $\mathbf{n}_1 \sim \mathbb{R}N_{m_1}(\mathbf{0}, \boldsymbol{\Sigma}_{1,1})$. Consequently, all the marginals belong to the same subclass of RCG distributions with the same c.d.f $F_\tau(.)$ and the density generator (1.56) reduces thanks to (1.21) to

$$g_{m_1|m}(t) = (2\pi)^{-m_1/2} \int_0^{\infty} \tau^{-m_1/2} \exp\left(-\frac{t}{2\tau}\right) dF_\tau(\tau). \tag{1.60}$$

In fact, this property characterizes the RCG distributions. This point is specified by a consistency result [32, Theorem 1]. This result states that any univariate and multivariate marginal distribution of an r.v. $\mathbf{x}$ belong to the same family as that of the r.v. $\mathbf{x}$, i.f.f. the RES distribution belongs to the class of RCG distributions. This condition is also equivalent to the characteristic generator ϕ not related to the dimension m. This consistency result shows that not all elliptically symmetric distributions can be used to define random processes. Indeed from Kolmogorov's theorem, only among the elliptically symmetric distributions, the m-dimensional

RGG distributed r.v. $\mathbf{x} = (x_1, \ldots, x_m)^T$ for all m can define a unique random process $(x_n)_{n \in \mathbb{N}}$ that is characterize by the distribution of τ, symmetric center and scatter of the process [13].

1.4.3 Conditional Distributions

Let $\mathbf{x} = (\mathbf{x}_1^T, \mathbf{x}_2^T)^T \sim \text{RES}_m(\boldsymbol{\mu}, \boldsymbol{\Sigma}, \phi)$ where $\mathbf{x}_1 \in \mathbb{R}^{m_1}$. The conditional distribution of $\mathbf{x}_2$ given $\mathbf{x}_1$ is generally more difficult to describe than its marginal distribution presented in Sect. 1.4.2. For the sake of simplicity, we consider here only the full-rank case ($k = m$); see e.g., [9] and [20, Theorem 7, Corollary 8] for more general statements. In this case, it is proved [17, Theorem 2.18], that $\mathbf{x}_2$ given $\mathbf{x}_1 = \mathbf{x}_1^0$ (denoted $\mathbf{x}_2|\mathbf{x}_1 = \mathbf{x}_1^0$) is $\text{RES}_{m_2}(\boldsymbol{\mu}_{2|1}, \boldsymbol{\Sigma}_{2|1}, \phi_{2|1}) = \text{RES}_{m_2}(\boldsymbol{\mu}_{2|1}, \boldsymbol{\Sigma}_{2|1}, F_{2|1})$ distributed with

$$\boldsymbol{\mu}_{2|1} = \boldsymbol{\mu}_2 + \boldsymbol{\Sigma}_{2,1}\boldsymbol{\Sigma}_{1,1}^{-1}(\mathbf{x}_1^0 - \boldsymbol{\mu}_1), \tag{1.61}$$

$$\boldsymbol{\Sigma}_{2|1} = \boldsymbol{\Sigma}_{2,2} - \boldsymbol{\Sigma}_{2,1}\boldsymbol{\Sigma}_{1,1}^{-1}\boldsymbol{\Sigma}_{1,2}, \tag{1.62}$$

and where $\phi_{2|1}$ and $F_{2|1}$ correspond, respectively, to the characteristic generator of $\mathbf{x}_2|\mathbf{x}_1 = \mathbf{x}_1^0$ and the c.d.f. of the conditional modular variate $\mathcal{R}_{2|1}$ defined by the following stochastic full-rank representation

$$(\mathbf{x}_2|\mathbf{x}_1 = \mathbf{x}_1^0) =_d \boldsymbol{\mu}_{2|1} + \mathcal{R}_{2|1}\mathbf{A}_{2|1}\mathbf{u}_{m_2}, \tag{1.63}$$

where $\mathbf{A}_{2|1}$ is an arbitrary square root of $\boldsymbol{\Sigma}_{2|1}$ (i.e., $\boldsymbol{\Sigma}_{2|1} = \mathbf{A}_{2|1}\mathbf{A}_{2|1}^T$ and $\mathcal{R}_{2|1}$ and $\mathbf{u}_{m_2}$ are independent. $\mathcal{R}_{2|1}$ is given by

$$\mathcal{R}_{2|1} =_d [\mathcal{R}_m^2 - (\mathbf{x}_1^0 - \boldsymbol{\mu}_1)^T\boldsymbol{\Sigma}_{1,1}^{-1}(\mathbf{x}_1^0 - \boldsymbol{\mu}_1)]^{1/2}|\mathbf{x}_1 = \mathbf{x}_1^0. \tag{1.64}$$

Otherwise, if the conditional covariance $\text{Cov}(\mathbf{x}_2|\mathbf{x}_1 = \mathbf{x}_1^0)$ exists, its expressions can be derived from (1.40) and we get

$$\text{Cov}(\mathbf{x}_2|\mathbf{x}_1 = \mathbf{x}_1^0) = \frac{1}{m_2}\left(\text{E}[\mathcal{R}_m^2|\mathbf{x}_1 = \mathbf{x}_1^0] - (\mathbf{x}_1^0 - \boldsymbol{\mu}_1)^T\boldsymbol{\Sigma}_{1,1}^{-1}(\mathbf{x}_1^0 - \boldsymbol{\mu}_1)\right)\boldsymbol{\Sigma}_{2|1}. \tag{1.65}$$

We note that the expressions of the conditional symmetry center $\boldsymbol{\mu}_{2|1}$ and conditional scatter matrix $\boldsymbol{\Sigma}_{2|1}$ are those obtained for the Gaussian distribution, but the conditional characteristic generator $\phi_{2|1}$ no longer belongs to the same family of RES, except for the Gaussian distribution.

For RCG distributions, the p.d.f. of $\mathbf{x}_2|\mathbf{x}_1 = \mathbf{x}_1^0$, (denoted $p(\mathbf{x}_2/\mathbf{x}_1^0)$, is given by

$$p(\mathbf{x}_2/\mathbf{x}_1^0) = \frac{1}{p(\mathbf{x}_1^0)}\int_0^\infty p(\mathbf{x}_2/\mathbf{x}_1^0, \tau)p(\mathbf{x}_1^0/\tau)dF_\tau(\tau), \tag{1.66}$$

where $p(\mathbf{x}_1^0/\tau) = (2\pi\tau)^{-m_1/2}|\boldsymbol{\Sigma}_{1,1}|^{-1/2}\exp(-\frac{1}{2\tau}d^2_{\boldsymbol{\Sigma}_{1,1}}(\mathbf{x}_1^0, \boldsymbol{\mu}_1))$ and $p(\mathbf{x}_2/\mathbf{x}_1^0, \tau)$ $= (2\pi\tau)^{-m_2/2}|\boldsymbol{\Sigma}_{2|1}|^{-1/2}\exp(-\frac{1}{2\tau}d^2_{\boldsymbol{\Sigma}_{2|1}}(\mathbf{x}_2, \boldsymbol{\mu}_{2|1}))$ with $d^2_{\boldsymbol{\Sigma}_{1,1}}(\mathbf{x}_1^0, \boldsymbol{\mu}_1) \stackrel{\text{def}}{=} (\mathbf{x}_1^0 - \boldsymbol{\mu}_1)^T$ $\boldsymbol{\Sigma}_{1,1}^{-1}(\mathbf{x}_1^0 - \boldsymbol{\mu}_1)$ and $d^2_{\boldsymbol{\Sigma}_{2|1}}(\mathbf{x}_2, \boldsymbol{\mu}_{2|1}) \stackrel{\text{def}}{=} (\mathbf{x}_2 - \boldsymbol{\mu}_2)^T\boldsymbol{\Sigma}_{2|1}^{-1}(\mathbf{x}_2 - \boldsymbol{\mu}_2)$. This yields

$$p(\mathbf{x}_2/\mathbf{x}_1^0) = \frac{(2\pi)^{-m_2/2}|\boldsymbol{\Sigma}_{2|1}|^{-1/2}}{\int_0^\infty \tau^{-m_1/2}\exp(-\frac{1}{2\tau}d^2_{\boldsymbol{\Sigma}_{1,1}}(\mathbf{x}_1^0, \boldsymbol{\mu}_1))dF_\tau(\tau)}$$
$$\times \int_0^\infty \tau^{-m/2}\exp\left(-\frac{1}{2\tau}d^2_{\boldsymbol{\Sigma}_{2|1}}(\mathbf{x}_2, \boldsymbol{\mu}_{2|1}\right)\exp\left(-\frac{1}{2\tau}d^2_{\boldsymbol{\Sigma}_{1,1}}(\mathbf{x}_1^0, \boldsymbol{\mu}_1)\right)dF_\tau(\tau). \quad (1.67)$$

Comparing (1.67) to (1.9), we check that $\mathbf{x}_2|\mathbf{x}_1 = \mathbf{x}_1^0$ is $\text{RES}_{m_2}(\boldsymbol{\mu}_{2|1}, \boldsymbol{\Sigma}_{2|1}, F_{2|1})$ distributed. Moreover, comparing (1.67) to (1.20), we see that $\mathbf{x}_2|\mathbf{x}_1 = \mathbf{x}_1^0$ is RCG distributed, but with a differently distributed texture τ than $\mathbf{x}$ and the marginal $\mathbf{x}_1$.

1.4.4 Summation Stability

Consider now the sum $\mathbf{y}$ of n independent r.v. $\mathbf{x}_1, \ldots, \mathbf{x}_i, \ldots, \mathbf{x}_n$ with the same scatter matrix, where $\mathbf{x}_i \sim \text{RES}_m(\boldsymbol{\mu}_i, \boldsymbol{\Sigma}, \phi_i)$. The characteristic function $\Phi_y(\mathbf{t})$ of $\mathbf{y} = \sum_{i=1}^n \mathbf{x}_i$ is

$$\Phi_y(\mathbf{t}) = \prod_{i=1}^n \Phi_{x_i}(\mathbf{t}) = \exp\left(i\mathbf{t}^T(\sum_{i=1}^n \boldsymbol{\mu}_i)\right)\prod_{i=1}^n \phi_i(\mathbf{t}^T\boldsymbol{\Sigma}\mathbf{t}) = \exp(i\mathbf{t}^T\boldsymbol{\mu})\phi(\mathbf{t}^T\boldsymbol{\Sigma}\mathbf{t}), \quad (1.68)$$

where $\boldsymbol{\mu} \stackrel{\text{def}}{=} \sum_{i=1}^n \boldsymbol{\mu}_i$ and $\phi(u) \stackrel{\text{def}}{=} \prod_{i=1}^n \phi_i(u)$, which is structured as (1.1). Consequently, the sum $\mathbf{y}$ is RES-distributed too [20]. Similarly, for independent univariate r.v. $x_1, \ldots, x_i, \ldots, x_n$ but with arbitrary scatters σ_i, where $x_i \sim \text{RES}_1(\mu_i, \sigma_i, \phi_i)$, the characteristic function $\Phi_y(t)$ of $y = \sum_{i=1}^n x_i$ is given by $\Phi_y(t) = \exp(it\mu)\phi(t^2)$ where $\mu \stackrel{\text{def}}{=} \sum_{i=1}^n \mu_i$ and $\phi(u) \stackrel{\text{def}}{=} \prod_{i=1}^n \phi_i(\sigma_i u)$ and then $y \sim \text{RES}_1(\mu, 1, \phi)$. The sum y is then symmetrically distributed w.r.t. μ.

However, it is worth underlying that if the r.v. $\mathbf{x}_1, \ldots, \mathbf{x}_i, \ldots, \mathbf{x}_n$ belong to the same family of RES distribution (i.e., with $\phi_i = \phi$, $i = 1, \ldots, n$), the sum is not of the same family except for the so-called elliptical α-stable distributions [45], (i.e., of characteristic functions $\Phi_x(\mathbf{t}) = \exp(i\mathbf{t}^T\boldsymbol{\mu})\exp(-\frac{1}{2}(\mathbf{t}^T\boldsymbol{\Sigma}\mathbf{t})^{\alpha/2})$ with $\alpha \in (0, 2]$), which includes the Gaussian distribution for $\alpha = 2$. Finally, note that if the scatter matrices $\boldsymbol{\Sigma}_i$ of $\mathbf{x}_1, \ldots, \mathbf{x}_i, \ldots, \mathbf{x}_n$ are not identical, the summation stability is generally lost for independent multivariate non-Gaussian, RES distributed r.v.

The independence condition of the v.a. $\mathbf{x}_i$ can be relaxed by the following properties proved in [30, Theorem 4.2]. If the full-rank stochastic representations (1.7) $\mathbf{x}_i =_d \boldsymbol{\mu}_i + \mathcal{R}_k^i\mathbf{A}\mathbf{u}_k^i$, $i = 1, 2$ with $\boldsymbol{\Sigma} = \mathbf{A}\mathbf{A}^T$ satisfy the condition $(\mathcal{R}_k^1, \mathcal{R}_k^2)$, $\mathbf{u}_k^1$, $\mathbf{u}_k^2$ are mutually independent, whereas $\mathcal{R}_k^1, \mathcal{R}_k^2$ may be dependent on each other, the sum $\mathbf{y}$ is also $\text{RES}_m(\boldsymbol{\mu}, \boldsymbol{\Sigma}, \phi)$-distributed, where the expression of ϕ is given in [30, Theorem 4.2]. ϕ reduces to the product $\phi_1\phi_2$ when $\mathcal{R}_k^1$ and $\mathcal{R}_k^2$ are independent. A natural application of this property is in the context of a multivariate time series.

1.5 Example of Elliptically Symmetric Distributions

In this section, we present some examples of elliptically symmetric distributions and discuss their main specific properties. Throughout this section, we mainly consider the case of distributions absolutely continuous w.r.t. Lebesgue measure on $\mathbb{R}^m$ with $\text{rank}(\boldsymbol{\Sigma}) = m$. Each distribution is defined by its density generator under a functional form parameterized by m and a finite-dimensional parameter. We mainly use their real-valued definition through the p.d.f. (1.9) where the characteristic and density generators, and the texture (1.14) are simply denoted here by respectively ϕ, g and τ, knowing that C and NC complex-valued definitions (1.28) and (1.33) are simply deduced from the real-valued definition with double dimension (1.24), (1.29). For C-CES distributions, interested readers can consult [49].

1.5.1 Gaussian Distribution

The Gaussian distribution is the best-known and widely used distribution among the RES class in classical signal processing applications. Its ubiquity is mainly due to the central limit theorem (CLT) that allows for the use of the Gaussian distribution as a good and handy approximation of the statistical behavior of a set of observations in many practical scenarios.

As a particular case of RES distribution, the Gaussian distribution denoted hereafter by $\mathbb{R}N_m(\boldsymbol{\mu}, \boldsymbol{\Sigma})$, is characterized by a *characteristic* and a *density* generators that can be expressed, respectively, as

$$\phi(u) = \exp\left(-\frac{u}{2}\right) \text{ and } g(t) = \frac{1}{(2\pi)^{m/2}} \exp\left(-\frac{t}{2}\right). \tag{1.69}$$

Using (1.13), when $\boldsymbol{\Sigma}$ is non-singular, it is immediately verified that the p.d.f. of the second-order modular variate $\mathcal{Q}$ of a Gaussian r.v. is given by

$$p(q) = \frac{\delta_m^{-1}}{(2\pi)^{m/2}} q^{m/2-1} \exp\left(-\frac{q}{2}\right) = \frac{1}{2^{m/2}\Gamma(m/2)} q^{m/2-1} \exp\left(-\frac{q}{2}\right), \quad q \in \mathbb{R}^+, \tag{1.70}$$

where $\delta_m = \frac{\Gamma(m/2)}{\pi^{m/2}}$ from (1.5). It is immediate to verify that the p.d.f. in (1.70) is the one of a central χ^2-distribution with m degrees of freedom, i.e., $\mathcal{Q} \sim \chi^2_m$. Note that this result is perfectly in line with the well-known property of the Gaussian r.v. with mean value $\boldsymbol{\mu}$ and covariance matrix $\boldsymbol{\Sigma}$ whose quadratic form $(\mathbf{x}-\boldsymbol{\mu})^T\boldsymbol{\Sigma}^{-1}(\mathbf{x}-\boldsymbol{\mu})$ is χ^2_m-distributed.

Using again (1.13), we can express the p.d.f. of the modular variate $\mathcal{R} \sim \sqrt{\chi^2_m}$ as

$$p(r) = \frac{1}{2^{m/2-1}\Gamma(m/2)} r^{m-1} \exp\left(-\frac{r^2}{2}\right), \quad r \in \mathbb{R}^+. \tag{1.71}$$

Note that from the real to complex representation given in Sect. 1.3, the density generator of complex Gaussian distributions becomes from (1.29)

$$g(t) = \frac{1}{\pi^m} \exp(-t), \tag{1.72}$$

and consequently, the p.d.f. of the circular and non-circular Gaussian distribution, denoted respectively by $\mathbb{C}N_m(\boldsymbol{\mu}, \boldsymbol{\Sigma})$ and $\mathbb{C}N_m(\boldsymbol{\mu}, \boldsymbol{\Sigma}, \boldsymbol{\Omega})$ are given from (1.33) and (1.28) by, respectively,

$$\begin{aligned} p(\mathbf{x}) &= \pi^{-m}|\boldsymbol{\Sigma}|^{-1} \exp[(\mathbf{x}-\boldsymbol{\mu})^H\boldsymbol{\Sigma}^{-1}(\mathbf{x}-\boldsymbol{\mu})], \\ p(\mathbf{x}) &= \pi^{-m}|\widetilde{\boldsymbol{\Sigma}}|^{-1/2} \exp\left[\frac{1}{2}(\widetilde{\mathbf{x}}-\widetilde{\boldsymbol{\mu}})^H\widetilde{\boldsymbol{\Sigma}}^{-1}(\widetilde{\mathbf{x}}-\widetilde{\boldsymbol{\mu}})\right], \end{aligned} \tag{1.73}$$

where $\boldsymbol{\Sigma}$ and $\boldsymbol{\Omega}$ are the covariance and complementary covariance (or pseudo-covariance) of $\mathbf{x}$, respectively.

The Gaussian distribution is usually used as a reference to define heavier-tailed and lighter-tailed distributions in the RES class. Some examples will be given below.

1.5.2 Student's t-Distribution

A popular example of heavy-tailed distribution is the Student t-distribution (or simply t-distribution). The t-distribution belongs to the RES class, and it is characterized by the following density generator [38]:

$$g(u) = \frac{\Gamma(\frac{\nu+m}{2})}{(\nu\pi)^{m/2}\Gamma(\nu/2)} \left(1+\frac{u}{\nu}\right)^{-\frac{\nu+m}{2}}, \quad u \in \mathbb{R}^+ \tag{1.74}$$

with $0 < \nu < \infty$ *degrees of freedom*. The parameter ν controls the tails of the distribution that are uniformly heavier than the Gaussian ones. In particular, for small values of ν, the sampled r.v. are highly non-Gaussian, while, as $\nu \to \infty$, the t-distribution collapses into the Gaussian one. The case $\nu = 1$ is called the Cauchy distribution.

As shown in [38], the second-order modular variate of an m-dimensional t-distributed r.v. satisfies the following property:

$$\mathcal{Q}/m \sim F_{m,\nu}, \tag{1.75}$$

where $F_{m,\nu}$ denotes the F-distribution with m and ν degrees of freedom [31]. Moreover, using (1.13), the p.d.f. of the second-order modular variate $\mathcal{Q}$ and of the modular variate $\mathcal{R}$ can be directly derived.

It is worth mentioning that the t-distribution belongs to the subclass of the CG distribution described in Sect. 1.3.4. Specifically, t-distributed r.v. can be obtained by generating the texture τ such that its inverse is distributed as a Gamma random variable,

$$\tau^{-1} \sim \text{Gam}(\nu/2, 2/\nu). \tag{1.76}$$

Since the t-distribution belongs to the subclass of the CG distributions, it holds that the marginals belong to the same family. Hence, in particular, the univariate marginals x_i are t-distributed with parameter ν, symmetry center $(\boldsymbol{\mu})_i$ and scatter $(\boldsymbol{\Sigma})_{i,i}$ with density generator given by (1.74) where $m = 1$.

This distribution has first-, second-, and fourth-order moments if, respectively, $\nu > 1$, $\nu > 2$, and $\nu > 4$. In particular, we have $\text{E}(\tau) = \frac{\nu}{2(\nu-2)}$, $\text{E}(\mathcal{Q}) = \frac{m\nu}{\nu-2}$ and $\text{Cov}(\mathbf{x}) = \frac{\nu}{\nu-2}\boldsymbol{\Sigma}$ for $\nu > 2$, while the kurtosis, defined in Sect. 1.4.1 in (1.45) is $\kappa = \frac{2}{\nu-4}$ for $\nu > 4$.

1.5.3 Generalized Gaussian Distributions

The Generalized Gaussian (GG) distribution is a RES distribution with the remarkable ability to characterize both heavy-tailed and light-tailed (with respect to the Gaussian one) data behavior. The multivariate GG distribution was introduced in [23] as a multivariate generalization of the power exponential distribution. As a member of the RES class, using the parametrization [49, rel. (27)] and (1.29), the GG distribution has a density generator given by

$$g(u) = \frac{s\Gamma(m/2)}{(2\pi)^{m/2} b^{m/2s} \Gamma(m/2s)} \exp\left(-\frac{u^s}{2^s b}\right), \quad u \in \mathbb{R}^+, \tag{1.77}$$

where $s > 0$ and $b > 0$ are two parameters generally called *exponent* (or *shape*) and *scale*, respectively. The scale parameter b is used to ensure that $p(\mathbf{x})$ integrates to 1 in (1.9), while the exponent parameter s controls the non-Gaussianity. In particular, for $0 < s < 1$, the tails of the GG distribution are heavier with respect to the Gaussian one, while, for $s > 1$, they are lighter. For s tending to ∞, this distribution converges to a uniform distribution in an ellipsoid centered on $\boldsymbol{\mu}$. Clearly, for $s = 1$, we get the Gaussian distribution. Otherwise for $s = 1/2$, we get the Laplace distribution (also called double exponential distribution). Note that (1.77) is consistent with [36, 51] with different definitions of the scale.

It was proved in [26] that, for $s \in (0, 1]$, the GG distributions belong to the subset of CG distributions whose p.d.f. of the mixing variable $\sqrt{\tau}$ is given by [26, (2.2)]. On the contrary for $s > 1$, the GG distributions no longer belong to the CG family and the marginals are no longer GG distributed. In particular, in this case, the univariate marginals of a multivariate GG distribution are not power exponential distributed, i.e., with p.d.f. given by [39]

$$p(x) = \frac{1}{\sigma \Gamma(1 + \frac{1}{2s}) 2^{1+\frac{1}{2s}}} \exp\left(-\frac{1}{2}\left(\frac{x-\mu}{\sigma}\right)^{2s}\right). \tag{1.78}$$

Regarding the second-order modular variate $\mathcal{Q}$ of an m-dimensional GG-distributed r.v., it is straightforward to deduce from (1.13) and (1.77) that

$$\mathcal{Q}^s \sim \text{Gam}\left(\frac{m}{2s}, 2^s b\right). \tag{1.79}$$

We can exploit again (1.13) to get a closed-form expression of the p.d.f. of $\mathcal{Q}$ and $\mathcal{R}$.

We note that the moments exist at all orders and $\text{Cov}(\mathbf{x}) = \mathbf{\Sigma}$ if the parameters s and b are related by the relation $b = [\frac{m}{2}\Gamma(\frac{m}{2s})/\Gamma(\frac{m/2+1}{s})]^s$. Note that, in this case, the kurtosis κ defined in (1.45) depends on the dimension m, e.g., for $m = 1$, we get $\kappa = \frac{\Gamma(5/2s)\Gamma(1/2s)}{3[\Gamma(3/2s)]^2} - 1$ which gives $\kappa = 0$ for $s = 1$ (Gaussian case).

1.5.4 The K-Distribution

Another example of RES distribution belonging to the CG-subclass is the K-distribution. An m-dimensional r.v. $\mathbf{x}$ is said to be K-distributed if it has the CG-representation (1.14) characterized by a Gamma-distributed texture,

$$\tau \sim \frac{1}{2}\text{Gam}(\nu, 1/\nu), \tag{1.80}$$

where $\nu > 0$ is a *shape* parameter. So, the p.d.f. of τ is

$$p(\tau) = \frac{2\nu^\nu}{\Gamma(\nu)}(2\tau)^{\nu-1}\exp(-2\nu\tau), \ \tau \in \mathbb{R}^+, \tag{1.81}$$

with $\text{E}(\tau) = 1/2$.

By using the definition of CG-distribution, it can be shown that the density generator is given by

$$g(u) = \frac{\nu^{m/2}}{2^{\nu-1}\pi^{m/2}\Gamma(\nu)}(2\nu u)^{(2\nu-m)/4} K_{(2\nu-m)/2}\left(\sqrt{2\nu u}\right), \ u \in \mathbb{R}^+, \tag{1.82}$$

where $K_\alpha(\cdot)$ denotes the modified Bessel function of the second kind of order α. The shape $\nu > 0$ is a parameter that controls the tails of the K-distribution: when $\nu \to 0$, the tails become heavier, while, for $\nu \to \infty$, the K-distribution collapses onto the Gaussian one. Moreover, using (1.13), the p.d.f. of the second-order modular variate $\mathcal{Q}$ and of the modular variate $\mathcal{R}$ can be directly derived.

Since the K-distribution belongs to the subclass of the CG distributions, it holds that the marginals belong to the same family. Hence, in particular the univariate marginals x_i are K-distributed with parameter ν, symmetry center $(\boldsymbol{\mu})_i$ and scatter $(\boldsymbol{\Sigma})_{i,i}$ with density generator given by (1.82) where $m = 1$.

To conclude, we note that the moments of all orders of the K-distribution exist and it can be shown that the kurtosis in (1.45) is given by $\kappa = \frac{1}{\nu}$.

1.5.5 Related Distribution: The Angular Central Gaussian Distribution

To conclude this section, let us introduce a distribution that has a strong link with the RES family, even if it does not belong to this class. An r.v. $\mathbf{x}$ is said angular central Gaussian (ACG) distributed if it admits the stochastic representation:

$$\mathbf{x} =_d \frac{\mathbf{n}}{\|\mathbf{n}\|}, \quad \text{where } \mathbf{n} \sim \mathbb{R}N_m(\mathbf{0}, \boldsymbol{\Sigma}). \tag{1.83}$$

For non-singular $\boldsymbol{\Sigma}$, the p.d.f. of this distribution is given by [63]

$$p(\mathbf{x}) = \frac{2\pi^{m/2}}{\Gamma(m/2)} |\boldsymbol{\Sigma}|^{-1/2} \left(\mathbf{x}^T \boldsymbol{\Sigma}^{-1} \mathbf{x}\right)^{-m/2}, \quad \mathbf{x} \in \mathbb{R}S^m. \tag{1.84}$$

Even if this expression may look similar to the p.d.f. of a centered RES-distributed r.v., it is worth underlying a crucial difference: the density of a RES r.v., given in (1.9), is defined w.r.t. the Lebesgue measure on $\mathbb{R}^m$, while the density of an ACG r.v. in (1.84) is defined w.r.t. the Lebesgue measure on real unit sphere $\mathbb{R}S^m$. For this reason, the stochastic representation provided in Sect. 1.2.2 does not hold for ACG r.v. It can be noted from (1.83) or (1.84) that the parameter $\boldsymbol{\Sigma}$ can only be identified up to a multiplicative scalar factor.

Furthermore, it turns out that if $\mathbf{x}$ is arbitrarily centered RES distributed (not necessarily Gaussian), i.e., if $\mathbf{x} \sim \text{RES}_m(\mathbf{0}, \boldsymbol{\Sigma}, \phi)$, we get from its stochastic representation (1.7) with $\boldsymbol{\Sigma} = \mathbf{A}\mathbf{A}^T$ of rank k and where $\mathbf{u}_k =_d \frac{\mathbf{n}_0}{\|\mathbf{n}_0\|}$ and $\mathbf{n}_0 \sim \mathbb{R}N_k(\mathbf{0}, \mathbf{I})$

$$\frac{\mathbf{x}}{\|\mathbf{x}\|} =_d \frac{\mathbf{A}\mathbf{u}_k}{\|\mathbf{A}\mathbf{u}_k\|} =_d \frac{\mathbf{A}\mathbf{n}_0}{\|\mathbf{A}\mathbf{n}_0\|} =_d \frac{\mathbf{n}}{\|\mathbf{n}\|}, \quad \text{where } \mathbf{n} \sim \mathbb{R}N_m(\mathbf{0}, \boldsymbol{\Sigma}). \tag{1.85}$$

Consequently the projection of $\mathbf{x}$ onto the unit real m-sphere is also ACG-distributed. In its definition (1.83), $\mathbb{R}N_m(\mathbf{0}, \boldsymbol{\Sigma})$ can be replaced by any $\text{RES}_m(\mathbf{0}, \boldsymbol{\Sigma}, \phi)$ distribution, and the term ACG appears to be a slight misnomer.

Finally, note that this class of distribution is closed under standardized linear transformations, i.e., $\mathbf{x} \sim \text{ACG}_m(\mathbf{0}, \boldsymbol{\Sigma})$, then $\mathbf{w} \overset{\text{def}}{=} \mathbf{Bx}/\|\mathbf{Bx}\| \sim \text{ACG}_k(\mathbf{0}, \mathbf{B}\boldsymbol{\Sigma}\mathbf{B}^T)$ for any nonzero $k \times m$ matrix $\mathbf{B}$.

1.6 Parameter Estimation

We are interested in this section to the estimation of the symmetry center $\boldsymbol{\mu}$ and scatter matrix $\boldsymbol{\Sigma}$ (assumed invertible here) of RES distributions. Suppose we have an i.i.d. sample $\mathbf{x}_1, \ldots, \mathbf{x}_i, \ldots, \mathbf{x}_n$ of size $n > m$ from a $\text{RES}_m(\boldsymbol{\mu}, \boldsymbol{\Sigma}, g)$ distribution absolutely continuous w.r.t. Lebesgue measure on $\mathbb{R}^m$ under finite second-order moments. The identifiability issue is solved here by imposing constraint (1.42) which ensures that $\text{Cov}(\mathbf{x}_i) = \boldsymbol{\Sigma}$.

1.6.1 Sample Mean and Sample Covariance Matrix

A natural estimate of the parameters $\boldsymbol{\mu}$ and $\boldsymbol{\Sigma}$ are the sample mean $\widehat{\boldsymbol{\mu}} \overset{\text{def}}{=} \frac{1}{n}\sum_{i=1}^n \mathbf{x}_i$ and the sample covariance matrix (SCM) $\widehat{\boldsymbol{\Sigma}} \overset{\text{def}}{=} \frac{1}{n-1}\sum_{i=1}^n (\mathbf{x}_i - \widehat{\boldsymbol{\mu}})(\mathbf{x}_i - \widehat{\boldsymbol{\mu}})^T$, respectively. It is well known that $\widehat{\boldsymbol{\mu}}$ and $\widehat{\boldsymbol{\Sigma}}$ are unbiased and mutually uncorrelated estimators. Under the particular case of Gaussian data, the two estimators are then independent. $\widehat{\boldsymbol{\mu}}$ is RES distributed with symmetry center $\boldsymbol{\mu}$ and scatter matrix $\frac{1}{n}\boldsymbol{\Sigma}$ by imposing constraint (1.42) on this distribution which does not necessarily belong to the same RES distribution family as $\mathbf{x}_i$ (see Sect. 1.4.4). Under finite fourth-order moments, applying the CLT to $\sum_{i=1}^n \mathbf{x}_i\mathbf{x}_i^T$ and $\sum_{i=1}^n \mathbf{x}_i$, it can be shown that $\widehat{\boldsymbol{\Sigma}}$ is asymptotically Gaussian distributed, i.e.,

$$\sqrt{n}\left(\text{vec}(\widehat{\boldsymbol{\Sigma}}) - \text{vec}(\boldsymbol{\Sigma})\right) \to_d \mathbb{R}N_{m^2}\left(\mathbf{0}, \mathbf{R}_{\Sigma_{SCM}}\right) \tag{1.86}$$

with [50, p. 5]

$$\mathbf{R}_{\Sigma_{SCM}} = (1+\kappa)(\mathbf{I}+\mathbf{K})(\boldsymbol{\Sigma} \otimes \boldsymbol{\Sigma}) + \kappa \text{vec}(\boldsymbol{\Sigma})\text{vec}^T(\boldsymbol{\Sigma}), \tag{1.87}$$

where κ is the kurtosis parameter (1.45) of the $\text{RES}_m(\boldsymbol{\mu}, \boldsymbol{\Sigma}, g)$ distribution. Note that this asymptotic distribution would also be the asymptotic distribution of the estimate $\frac{1}{n}\sum_{i=1}^n (\mathbf{x}_i - \boldsymbol{\mu})(\mathbf{x}_i - \boldsymbol{\mu})^T$ of $\boldsymbol{\Sigma}$ when $\boldsymbol{\mu}$ is known. For heavier tails than Gaussian distributions, $\kappa > 0$ and can be very large without any upper-bound (see e.g., in Sect. 1.5.2 when κ approaches 4 for Student's t-distribution) and consequently, the SCM estimate can be a very bad estimator.

1.6.2 ML Estimation

To take into account the particular RES distribution of the data, the maximum-likelihood (ML) estimator is often considered as the reference estimator because it is generally (when it exists and is unique) asymptotically efficient with a speed of convergence in $\sqrt{n}$. From Slepian–Bangs formula (1.111), the Fisher information matrix (FIM) for the parameter $(\boldsymbol{\mu}, v(\boldsymbol{\Sigma}))$ is given by

$$\text{FIM}\begin{pmatrix}\boldsymbol{\mu}\\ v(\boldsymbol{\Sigma})\end{pmatrix} = n\begin{pmatrix} a_0\boldsymbol{\Sigma}^{-1} & \mathbf{0}\\ \mathbf{0} & \mathbf{D}^T[a_1(\boldsymbol{\Sigma}^{-1}\otimes\boldsymbol{\Sigma}^{-1})+a_2\text{vec}(\boldsymbol{\Sigma}^{-1})\text{vec}^T(\boldsymbol{\Sigma}^{-1})]\mathbf{D}\end{pmatrix}, \tag{1.88}$$

where

$$a_0 = \frac{\text{E}[\mathcal{Q}\varphi^2(\mathcal{Q})]}{m}, \quad a_1 = \frac{\text{E}[\mathcal{Q}^2\varphi^2(\mathcal{Q})]}{2m(m+2)}, \quad \text{and } a_2 = \frac{1}{4}\left(\frac{\text{E}[\mathcal{Q}^2\varphi^2(\mathcal{Q})]}{m(m+2)} - 1\right), \tag{1.89}$$

assuming that g is continuously differentiable with $\varphi(t) \stackrel{\text{def}}{=} -2g'(t)/g(t)$. Hence under existence, uniqueness, and usual regularity conditions, the ML estimate $(\widehat{\boldsymbol{\mu}}, \widehat{\boldsymbol{\Sigma}})$ is asymptotically Gaussian distributed:

$$\sqrt{n}\begin{pmatrix}\widehat{\boldsymbol{\mu}} - \boldsymbol{\mu}\\ \text{vec}(\widehat{\boldsymbol{\Sigma}}) - \text{vec}(\boldsymbol{\Sigma})\end{pmatrix} \to_d \mathbb{R}N_{m+m^2}\left(\begin{pmatrix}\mathbf{0}\\ \mathbf{0}\end{pmatrix}, \begin{pmatrix}\mathbf{R}_{\mu_{ML}} & \mathbf{0}\\ \mathbf{0} & \mathbf{R}_{\Sigma_{ML}}\end{pmatrix}\right), \tag{1.90}$$

where $\mathbf{R}_{\mu_{ML}}$ and $\mathbf{R}_{\Sigma_{ML}}$ can be deduced from (1.88) using the efficiency of the ML estimate:

$$\mathbf{R}_{\mu_{ML}} = \sigma_0\boldsymbol{\Sigma} \ \text{ and } \ \mathbf{R}_{\Sigma_{ML}} = \sigma_1(\mathbf{I}+\mathbf{K})(\boldsymbol{\Sigma}\otimes\boldsymbol{\Sigma}) + \sigma_2\text{vec}(\boldsymbol{\Sigma})\text{vec}^T(\boldsymbol{\Sigma}), \tag{1.91}$$

with [60]

$$\sigma_0 = \frac{m}{\text{E}[\mathcal{Q}\varphi^2(\mathcal{Q})]}, \quad \sigma_1 = \frac{m(m+2)}{\text{E}[\mathcal{Q}^2\varphi^2(\mathcal{Q})]}, \quad \text{and } \sigma_2 = -\frac{2\sigma_1(1-\sigma_1)}{2+m(1-\sigma_1)}. \tag{1.92}$$

Note that this asymptotic distribution was first given in [60] by using the general structure of the covariance of random matrices whose distributions are radial.

The ML estimator of $\boldsymbol{\mu}$ and $\boldsymbol{\Sigma}$ are the vector $\widehat{\boldsymbol{\mu}}$ and the symmetric positive definite matrix $\widehat{\boldsymbol{\Sigma}}$ that minimize the negative log-likelihood function equal from (1.9) to

$$L(\boldsymbol{\mu}, \boldsymbol{\Sigma}) = \frac{n}{2}\log|\boldsymbol{\Sigma}| - \sum_{i=1}^{n}\log\left(g[(\mathbf{x}_i - \boldsymbol{\mu})^T\boldsymbol{\Sigma}^{-1}(\mathbf{x}_i - \boldsymbol{\mu})]\right). \tag{1.93}$$

Setting the derivatives of $L(\boldsymbol{\mu}, \boldsymbol{\Sigma})$ w.r.t. $\boldsymbol{\mu}$ and $\boldsymbol{\Sigma}$ to zero yields the following estimation equations:

$$\mathbf{0} = \frac{1}{n}\sum_{i=1}^{n} \varphi[(\mathbf{x}_i - \widehat{\boldsymbol{\mu}})^T \widehat{\boldsymbol{\Sigma}}^{-1}(\mathbf{x}_i - \widehat{\boldsymbol{\mu}})](\mathbf{x}_i - \widehat{\boldsymbol{\mu}}), \tag{1.94}$$

$$\widehat{\boldsymbol{\Sigma}} = \frac{1}{n}\sum_{i=1}^{n} \varphi[(\mathbf{x}_i - \widehat{\boldsymbol{\mu}})^T \widehat{\boldsymbol{\Sigma}}^{-1}(\mathbf{x}_i - \widehat{\boldsymbol{\mu}})](\mathbf{x}_i - \widehat{\boldsymbol{\mu}})(\mathbf{x}_i - \widehat{\boldsymbol{\mu}})^T. \tag{1.95}$$

Except for the case where the function φ is constant, the set of implicit equations (1.94)–(1.95) guarantees neither the existence nor the uniqueness of the ML estimators $\widehat{\boldsymbol{\mu}}$ and $\widehat{\boldsymbol{\Sigma}}$. Some sufficient conditions to ensure their existence and uniqueness are given in [34]. Note that for the Gaussian distribution, $\varphi(t) = 1$ from (1.69) and (1.94)–(1.95) yield the sample mean $\widehat{\boldsymbol{\mu}} = \frac{1}{n}\sum_{i=1}^{n} \mathbf{x}_i$ and the biased sample covariance matrix $\widehat{\boldsymbol{\Sigma}} = \frac{1}{n}\sum_{i=1}^{n}(\mathbf{x}_i - \widehat{\boldsymbol{\mu}})(\mathbf{x}_i - \widehat{\boldsymbol{\mu}})^T$, respectively.

1.6.3 *M*-Estimators

Since the ML estimator may be drastically affected by the presence of outliers or when the data distribution deviates slightly from the RES distribution of the model, robust estimators have been proposed. Among the different families of robust estimators, in the following, we focus our attention on the class of M-estimators. As for the ML estimator, an M-estimator can be obtained from the minimization of a function on the observation with respect to the parameters of interest. A classical example consists in replacing in the negative log-likelihood (1.93), the function $-\log(g([.])$ by a loss function $\rho(.) : \mathbb{R}^+ \mapsto \mathbb{R}$ (generally not related to g) to form another objective function. If ρ is also continuously differentiable with $u(t) \overset{\text{def}}{=} 2\rho'(t)$, an M-estimator is obtained by minimizing this new objective function. By replacing $\varphi(t)$ with $u(t)$ and setting again the derivatives of this objective function to zero, we obtain similar equations to (1.94) and (1.95). Some sufficient conditions are also given in [34] to ensure the existence and uniqueness of this M-estimator.

The seminal paper by Maronna [41] defined a more general class of M-estimator by replacing $\varphi(.)$ by the weight functions $u_1([.]^{1/2})$ and $u_2(.)$ in (1.94) and (1.95), respectively, i.e.,

$$\mathbf{0} = \frac{1}{n}\sum_{i=1}^{n} u_1\left([(\mathbf{x}_i - \widehat{\boldsymbol{\mu}})^T \widehat{\boldsymbol{\Sigma}}^{-1}(\mathbf{x}_i - \widehat{\boldsymbol{\mu}})]^{1/2}\right)(\mathbf{x}_i - \widehat{\boldsymbol{\mu}}), \tag{1.96}$$

$$\widehat{\boldsymbol{\Sigma}} = \frac{1}{n}\sum_{i=1}^{n} u_2\left((\mathbf{x}_i - \widehat{\boldsymbol{\mu}})^T \widehat{\boldsymbol{\Sigma}}^{-1}(\mathbf{x}_i - \widehat{\boldsymbol{\mu}})\right)(\mathbf{x}_i - \widehat{\boldsymbol{\mu}})(\mathbf{x}_i - \widehat{\boldsymbol{\mu}})^T. \tag{1.97}$$

The functions $u_1([.]^{1/2})$ and $u_2(.) : \mathbb{R}^+ \mapsto \mathbb{R}$ need not be the same function and hence the more general M-estimators need not be related to a minimization problem. Under sufficient conditions (called Maronna conditions), it is proved in [41, Theorem

4], the existence and uniqueness of solution $(\widehat{\boldsymbol{\mu}}, \widehat{\boldsymbol{\Sigma}})$ of (1.96), (1.97) but also [41, Theorem 2] of $(\mathbf{t}, \mathbf{V})$ solution of

$$\mathbf{0} = \mathrm{E}\left[u_1\left([(\mathbf{x}_1 - \mathbf{t})^T\mathbf{V}^{-1}(\mathbf{x}_1 - \mathbf{t})]^{1/2}\right)(\mathbf{x}_1 - \mathbf{t})\right], \tag{1.98}$$

$$\mathbf{V} = \mathrm{E}\left[u_2\left((\mathbf{x}_1 - \mathbf{t})^T\mathbf{V}^{-1}(\mathbf{x}_1 - \mathbf{t})\right)(\mathbf{x}_1 - \mathbf{t})(\mathbf{x}_1 - \mathbf{t})^T\right]. \tag{1.99}$$

Sufficient conditions are also given in [41, Theorem 5] to ensure the strong consistency of the estimate $(\widehat{\boldsymbol{\mu}}, \widehat{\boldsymbol{\Sigma}})$ solution of (1.96), (1.97) to the solution $(\mathbf{t}, \mathbf{V})$ of (1.98), (1.99) with $\mathbf{t} = \boldsymbol{\mu}$ and $\mathbf{V} = \sigma^{-1}\boldsymbol{\Sigma}$, where σ is the unique solution of $\mathrm{E}[\sigma \mathcal{Q} u_2(\sigma\mathcal{Q})] = m$ [60, Appendix 3].

Using a general result on M-estimators given in [29, Sect. 4], Maronna proved in [41, Theorem 6] the asymptotic gaussianity of $(\widehat{\boldsymbol{\mu}}, \widehat{\boldsymbol{\Sigma}})$, where $\widehat{\boldsymbol{\mu}}$ and $\widehat{\boldsymbol{\Sigma}}$ are asymptotically independent and where only the covariance of the asymptotic distribution of $\widehat{\boldsymbol{\mu}}$ was specified. Then, using the affine invariance property of any M-estimators and the general structure of the covariance of radial random matrices, the covariance of the asymptotic distribution of $\widehat{\boldsymbol{\Sigma}}$ was specified in [60, Appendix 2] to get

$$\sqrt{n}\begin{pmatrix} \widehat{\boldsymbol{\mu}} - \boldsymbol{\mu} \\ \mathrm{vec}(\widehat{\boldsymbol{\Sigma}}) - \mathrm{vec}(\boldsymbol{\sigma}^{-1}\Sigma) \end{pmatrix} \to_d \mathbb{R}N_{m+m^2}\left(\begin{pmatrix}\mathbf{0}\\ \mathbf{0}\end{pmatrix}, \begin{pmatrix}\mathbf{R}_{\mu_M} & \mathbf{0}\\ \mathbf{0} & \mathbf{R}_{\Sigma_M}\end{pmatrix}\right), \tag{1.100}$$

where

$$\mathbf{R}_{\mu_M} = \frac{\alpha}{\beta^2}\mathbf{V} = \frac{\alpha\sigma^{-1}}{\beta^2}\boldsymbol{\Sigma}, \tag{1.101}$$

where $\alpha = \frac{1}{m}\mathrm{E}[\psi_1^2(\sqrt{\sigma\mathcal{Q}})]$ and $\beta = \mathrm{E}[(1 - m^{-1})u_1(\sqrt{\sigma\mathcal{Q}}) + m^{-1}\psi_1'(\sqrt{\sigma\mathcal{Q}})]$ with $\psi_1(t) \stackrel{\mathrm{def}}{=} t u_1(t)$, and

$$\mathbf{R}_{\Sigma_M} = \sigma_1(\mathbf{I} + \mathbf{K})(\mathbf{V} \otimes \mathbf{V}) + \sigma_2 \mathrm{vec}(\mathbf{V})\mathrm{vec}^T(\mathbf{V}), \tag{1.102}$$

where $\sigma_1 = \frac{(m+2)^2 a_1}{(2a_2+m)^2}$ and $\sigma_2 = a_2^{-1}\left[(a_1 - 1) - 2(a_2 - 1)a_1\frac{m+(m+4)a_2}{(2a_2+m)^2}\right]$ with $a_1 = \frac{\mathrm{E}[\psi_2^2(\sigma\mathcal{Q})]}{m(m+2)}$ and $a_2 = \frac{\mathrm{E}[\sigma\mathcal{Q}\psi_2'(\sigma\mathcal{Q})]}{m}$ where $\psi_2(t) \stackrel{\mathrm{def}}{=} t u_2(t)$ and $\psi_2'(t) \stackrel{\mathrm{def}}{=} \frac{d\psi_2(t)}{dt}$. Furthermore, using the general structure of the covariance of radial random matrices, it is proved in [60, Theorem 1] that

$$\sigma_2 \geq -\frac{2}{m}\sigma_1. \tag{1.103}$$

Finally, note that, for $u_1(t) = u_2(t^2) = \varphi(t^2)$, the M-estimate reduces to the ML estimate for which $\sigma = 1$ and (1.100) reduce to (1.90).

To the best of our knowledge, when both $\boldsymbol{\mu}$ and $\boldsymbol{\Sigma}$ are unknown parameters, it does not exist in the literature an analysis of sufficient conditions under which an algorithm would jointly converge toward the solution of the implicit equations (1.96), (1.97). However, when $\boldsymbol{\mu}$ is known, an algorithm which is essentially a fixed-point algorithm with a scale adjustment made at each iteration, with less severe conditions than that required in [41] was presented in [64]. Without loss of generality, $\boldsymbol{\mu}$ can be taken to $\mathbf{0}$ and the following algorithm is proved in [64] to converge to the unique solution of $\widehat{\boldsymbol{\Sigma}} = \frac{1}{n}\sum_{i=1}^{n} u_2[\mathbf{x}_i^T \widehat{\boldsymbol{\Sigma}}^{-1}\mathbf{x}_i]\mathbf{x}_i\mathbf{x}_i^T$:

$$\boldsymbol{\Sigma}_0 = \frac{1}{n}\sum_{i=1}^{n}\mathbf{x}_i\mathbf{x}_i^T, \quad \boldsymbol{\Sigma}_{k+1} = \frac{1}{n}\sum_{i=1}^{n} u_2[c_k\mathbf{x}_i^T\boldsymbol{\Sigma}_k^{-1}\mathbf{x}_i]\mathbf{x}_i\mathbf{x}_i^T, \tag{1.104}$$

c_k being the unique positive scalar satisfying $\frac{1}{n}\sum_{i=1}^{n}\psi_2[c_k\mathbf{x}_i^T\boldsymbol{\Sigma}_k^{-1}\mathbf{x}_i] = m$ with $\psi_2(t) \stackrel{\text{def}}{=} tu_2(t)$.

1.6.4 Tyler's M-Estimator

When $\boldsymbol{\mu}$ is known (equal to $\mathbf{0}$ without loss of generality), the M-estimator proposed by Tyler [62] has become a very popular robust scatter estimator in the signal processing literature. This M-estimator is defined by its weight function $u_2(t) = \frac{m}{t}$ associated with the loss function $\rho(t) = \frac{m}{2}\log t$, leading to the following objective function

$$L_T(\boldsymbol{\Sigma}) = \frac{n}{2}\log|\boldsymbol{\Sigma}| + \frac{m}{2}\sum_{i=1}^{n}\log(\mathbf{x}_i^T\boldsymbol{\Sigma}^{-1}\mathbf{x}_i) \tag{1.105}$$

that is minimized by $\widehat{\boldsymbol{\Sigma}}$, solution of the implicit equation

$$\widehat{\boldsymbol{\Sigma}} = \frac{m}{n}\sum_{i=1}^{n}\frac{\mathbf{x}_i\mathbf{x}_i^T}{\mathbf{x}_i^T\widehat{\boldsymbol{\Sigma}}^{-1}\mathbf{x}_i}. \tag{1.106}$$

Note that $L_T(c^2\boldsymbol{\Sigma}) = L_T(\boldsymbol{\Sigma}) + a$ where a does not depend on $\boldsymbol{\Sigma}$. Consequently, if $\widehat{\boldsymbol{\Sigma}}$ is solution of (1.106), $c^2\widehat{\boldsymbol{\Sigma}}$ is it too. The existence proof of solution of (1.106) given in Maronna [41] does not apply here, but existence and uniqueness (up to a multiplicative constant) under continuous RES distribution in $\mathbb{R}^m$ is proved in [62], by showing that it is the limiting point of the following specific fixed point algorithm: with $\boldsymbol{\Sigma}_0$ arbitrary symmetric positive definite matrix,

$$\boldsymbol{\Sigma}'_{k+1} = \frac{m}{n}\sum_{i=1}^{n}\frac{\mathbf{x}_i\mathbf{x}_i^T}{\mathbf{x}_i^T\boldsymbol{\Sigma}_k^{-1}\mathbf{x}_i}, \quad \boldsymbol{\Sigma}_{k+1} = \frac{m}{\mathrm{Tr}(\boldsymbol{\Sigma}'_{k+1})}\boldsymbol{\Sigma}'_{k+1}. \tag{1.107}$$

Under this same condition, it is also proved in [62] that the solution of (1.106) standardized so that $\mathrm{Tr}(\widehat{\boldsymbol{\Sigma}}) = m$ is strongly consistent to the symmetric positive definite matrix $\boldsymbol{\Sigma}$ which is also solution of

$$\mathbf{\Sigma} = m\mathrm{E}\left[\frac{\mathbf{x}_1\mathbf{x}_1^T}{\mathbf{x}_1^T\mathbf{\Sigma}^{-1}\mathbf{x}_1}\right] \quad \text{with } \mathrm{Tr}(\mathbf{\Sigma}) = m. \tag{1.108}$$

Furthermore, under continuous RES distribution in $\mathbb{R}^m$, it is proved in [62] that the solution $\widehat{\mathbf{\Sigma}}$ of (1.106) constrained to $\mathrm{Tr}(\widehat{\mathbf{\Sigma}}) = m$ is asymptotically Gaussian distributed, i.e.,

$$\sqrt{n}\left(\mathrm{vec}(\widehat{\mathbf{\Sigma}}) - \mathrm{vec}(\mathbf{\Sigma})\right) \rightarrow_d \mathbb{R}N_{m^2}\left(\mathbf{0}, \mathbf{R}_{\Sigma_{Ty}}\right) \tag{1.109}$$

with

$$\mathbf{R}_{\Sigma_{Ty}} = \left(1 + \frac{2}{m}\right)(\mathbf{I} + \mathbf{K})(\mathbf{\Sigma} \otimes \mathbf{\Sigma}) - \frac{2}{m}\left(1 + \frac{2}{m}\right)\mathrm{vec}(\mathbf{\Sigma})\mathrm{vec}^T(\mathbf{\Sigma}). \tag{1.110}$$

Comparing (1.102) to (1.110), we see that $\sigma_2 = -\frac{2}{m}\sigma_1$ and therefore we have equality in (1.103).

Note that when $\mathbf{x}_i$ are SIRV distributed, i.e., $\mathbf{x}_i \sim \mathrm{RCG}_m(\mathbf{0}, \mathbf{\Sigma}, F_{\tau_i})$ where here τ_i are assumed deterministic and unknown, the ML estimate of $\mathbf{\Sigma}$ coincides with Tyler's M-estimate (1.106). This estimate was introduced in [22] and existence and uniqueness were studied in [52]. Tyler's M-estimator enjoys four interesting properties:

- Objective function (1.105) is also the negative log-likelihood of i.i.d. data from an $\mathrm{ACG}_m(\mathbf{0}, \mathbf{\Sigma})$ distribution (see (1.84)). Consequently Tyler's M-estimator of $\mathbf{\Sigma}$ is the ML estimator $\widehat{\mathbf{\Sigma}}$ (satisfying the constraint $\mathrm{Tr}(\widehat{\mathbf{\Sigma}}) = m$) under the $\mathrm{ACG}_m(\mathbf{0}, \mathbf{\Sigma})$ distribution.
- Replacing $\mathbf{x}_i$ by $\frac{\mathbf{x}_i}{\|\mathbf{x}_i\|}$ in (1.106) does not affect the solution of (1.106). Since $\frac{\mathbf{x}_i}{\|\mathbf{x}_i\|}$ is $\mathrm{ACG}_m(\mathbf{0}, \mathbf{\Sigma})$ distributed for arbitrary $\mathrm{RES}_m(\mathbf{0}, \mathbf{\Sigma}, g)$ distribution, the distribution of Tyler's M estimator $\widehat{\mathbf{\Sigma}}$ does not depend on the density generator of this RES distribution. This is the reason why Tyler referred to his estimate as a distribution-free estimator [62].
- Suppose $\mathbf{x}_1, \ldots, \mathbf{x}_i, \ldots, \mathbf{x}_n$ are independent where $\mathbf{x}_i \sim \mathrm{RES}_m(\mathbf{0}, c_i^2\mathbf{\Sigma}, g_i)$ with unknown parameters c_i^2 and arbitrary unknown density generators g_i, then the ML of $\mathbf{\Sigma}$ corresponds to Tyler's M-estimate [48, Theorem 1].
- Tyler's M-estimator can be considered as the most robust estimator of the scatter matrix of a RES distribution in the sense of minimizing the maximum asymptotic covariance w.r.t. the generator density (see [62, Remark 3.1], and [61, Theorem 1]).

Also, note that Tyler's M-estimate has been extensively studied in the statistics literature (e.g., [35] has specified the existence proof of solution of (1.106) and [21] have proved that Tyler's iterative procedure (1.107) has a linear convergence rate).

We finally note that many complementary results have been carried out on RES distributions (see e.g., [51] for ML estimate of GG scatter, [58] for Tyler's estimate of structured scatter). Furthermore, most of the definitions and results presented in this section have been extended to CES distributions with many new results (see, e.g., [10, 15, 40, 43, 44, 52, 53]).

1.6.5 Slepian–Bangs Formula

We use in this section the real to complex representation to unify the Slepian–Bangs (SB) for RES, C-CES and NC-CES distributed data. The SB formula provides a convenient way to compute the FIM and thus the Cramér–Rao bound (CRB) on the real-valued parameter $\boldsymbol{\alpha}$ parameterizing and characterizing the couple $(\boldsymbol{\mu}, \boldsymbol{\Sigma})$ of elliptically symmetric distributions. We omit this dependence in $\boldsymbol{\alpha}$ to simplify the notation. To derive this formula, it is necessary that the scatter matrix $\boldsymbol{\Sigma}$ is not singular and the second-order moments of $\mathbf{x}$ are finite, which is equivalent to the first-order moments of $\mathcal{Q}$ being finite. Furthermore, to avoid the ambiguity between the scatter matrix and the density generator, we either assume that $\boldsymbol{\Sigma} = \text{Cov}(\mathbf{x})$ or that there is a scale constraint on $\boldsymbol{\Sigma}$. This formula has been derived for the real Gaussian distribution in [5, 57], then extended to the circular complex Gaussian and non-circular Gaussian case in [14, 59], respectively. This formula has been extended to C-CES distributions in [8, 27], and recently to NC-CES distributions [1]. In all these scenarios, the density generator is assumed to be perfectly known. For RES distributed $\mathbf{x}$, all the steps of the proof of the SB formula for C-CES distributions given in [8] apply, and we get the following structured matrix SB formula:

$$\begin{aligned}\text{CRB}^{-1}(\boldsymbol{\alpha}) = {} & a_0 \frac{d\boldsymbol{\mu}^T}{d\boldsymbol{\alpha}^T}\boldsymbol{\Sigma}^{-1}\frac{d\boldsymbol{\mu}}{d\boldsymbol{\alpha}^T} \\ & + \left(\frac{d\text{vec}(\boldsymbol{\Sigma})}{d\boldsymbol{\alpha}^T}\right)^T \left(a_1(\boldsymbol{\Sigma}^{-T}\otimes\boldsymbol{\Sigma}^{-1}) + a_2\text{vec}(\boldsymbol{\Sigma}^{-1})\text{vec}^T(\boldsymbol{\Sigma}^{-1})\right) \\ & \left(\frac{d\text{vec}(\boldsymbol{\Sigma})}{d\boldsymbol{\alpha}^T}\right), \end{aligned} \tag{1.111}$$

where $a_0 = \xi_{r,1,m}$, $a_1 = \frac{1}{2}\xi_{r,2,m}$, and $a_2 = \frac{1}{4}(\xi_{r,2,m} - 1)$ with

$$\xi_{r,1,m} \stackrel{\text{def}}{=} \frac{\text{E}[\mathcal{Q}\varphi_r^2(\mathcal{Q})]}{m} \quad \text{and} \quad \xi_{r,2,m} \stackrel{\text{def}}{=} \frac{\text{E}[\mathcal{Q}^2\varphi_r^2(\mathcal{Q})]}{m(m+2)}, \tag{1.112}$$

where $\mathcal{Q} \stackrel{\text{def}}{=} \mathcal{Q}_{r,m}$ and $\varphi_r(t) \stackrel{\text{def}}{=} -\frac{2}{g_{r,m}(t)}\frac{dg_{r,m}(t)}{dt}$.

Note that because $(c^2\boldsymbol{\Sigma}, c^2\mathcal{Q}, c^m g_{r,m}(.\ c^2), c^2\varphi_r(.\ c^2))$ gives the same RES distribution, the coefficient $\xi_{r,2,m}$ is free of scale ambiguity in contrast to $\xi_{r,1,m}$ which depends on the scale factor. This is consistent with Eq. (1.111).

This SB formula allows us to directly deduce those of NC-CES distributed data obtained thanks to the real to complex representation. This SB formula is similarly structured where $\boldsymbol{\mu}$, $\boldsymbol{\Sigma}$, $\frac{d\boldsymbol{\mu}^T}{d\boldsymbol{\alpha}^T}$, $\left(\frac{d\text{vec}(\boldsymbol{\Sigma})}{d\boldsymbol{\alpha}^T}\right)^T$, and $\text{vec}^T(\boldsymbol{\Sigma}^{-1})$ in (1.111) are replaced by $\widetilde{\boldsymbol{\mu}}$, $\widetilde{\boldsymbol{\Sigma}}$, $\frac{d\widetilde{\boldsymbol{\mu}}^H}{d\boldsymbol{\alpha}^T}$, $\left(\frac{d\text{vec}(\widetilde{\boldsymbol{\Sigma}})}{d\boldsymbol{\alpha}^T}\right)^H$, and $\text{vec}^H(\widetilde{\boldsymbol{\Sigma}}^{-1})$, respectively, where $a_0 = \xi_{c,1,m}$, $a_1 = \frac{\xi_{c,2,m}}{2}$ and $a_2 = \frac{1}{4}(\xi_{c,2,m} - 1)$ with

$$\xi_{c,1,m} \stackrel{\text{def}}{=} \frac{\mathrm{E}[Q\varphi_c^2(Q)]}{m} \text{ and } \xi_{c,2,m} \stackrel{\text{def}}{=} \frac{\mathrm{E}[Q^2\varphi_c^2(Q)]}{m(m+1)}, \tag{1.113}$$

where $Q \stackrel{\text{def}}{=} \mathcal{Q}_{c,m}$ and $\varphi_c(t) \stackrel{\text{def}}{=} -\frac{1}{g_{c,m}(t)}\frac{dg_{c,m}(t)}{dt}$. On the other hand, the SB formulas for C-CES distributed data can be deduced directly by replacing $\widetilde{\mathbf{\Sigma}}$ by $\begin{pmatrix} \mathbf{\Sigma} & \mathbf{0} \\ \mathbf{0} & \mathbf{\Sigma}^* \end{pmatrix}$, yielding the SB formulas proved in [8, 27], which are also similarly structured with $a_0 = 2\xi_{c,1,m}$, $a_1 = \xi_{c,2,m}$, and $a_2 = \xi_{c,2,m} - 1$, where $\frac{d\boldsymbol{\mu}^T}{d\boldsymbol{\alpha}^T}\mathbf{\Sigma}^{-1}\frac{d\boldsymbol{\mu}}{d\boldsymbol{\alpha}^T}$, $\left(\frac{d\text{vec}(\mathbf{\Sigma})}{d\boldsymbol{\alpha}^T}\right)^T$, and $\text{vec}^T(\mathbf{\Sigma}^{-1})$ are replaced in by $\text{Re}\left(\frac{d\boldsymbol{\mu}^H}{d\boldsymbol{\alpha}^T}\mathbf{\Sigma}^{-1}\frac{d\boldsymbol{\mu}}{d\boldsymbol{\alpha}^T}\right)$, $\left(\frac{d\text{vec}(\mathbf{\Sigma})}{d\boldsymbol{\alpha}^T}\right)^H$, and $\text{vec}^H(\mathbf{\Sigma}^{-1})$, respectively.

Note that (a_0, a_1, a_2) reduces to $(1, 1/2, 0)$, $(1, 1/2, 0)$ and $(2, 1, 0)$ for real, complex noncircular and complex circular Gaussian distributions, respectively. Note also that the decoupling between the parameters $\boldsymbol{\alpha}_1$ of $\boldsymbol{\mu}$ and the parameters $\boldsymbol{\alpha}_2$ of $\mathbf{\Sigma}$ when $\boldsymbol{\mu}$ and $\mathbf{\Sigma}$ have no parameters in common for Gaussian distributions, extends to any elliptically symmetric distributions with

$$\text{CRB}(\boldsymbol{\alpha}_1) = \left(a_0 \frac{d\boldsymbol{\mu}^T}{d\boldsymbol{\alpha}_1{}^T}\mathbf{\Sigma}^{-1}\frac{d\boldsymbol{\mu}}{d\boldsymbol{\alpha}_1{}^T}\right)^{-1} \tag{1.114}$$

and

$$\text{CRB}(\boldsymbol{\alpha}_2) = \left(\left(\frac{d\text{vec}(\mathbf{\Sigma})}{d\boldsymbol{\alpha}_2{}^T}\right)^T \left(a_1(\mathbf{\Sigma}^{-T} \otimes \mathbf{\Sigma}^{-1}) + a_2\text{vec}(\mathbf{\Sigma}^{-1})\text{vec}^T(\mathbf{\Sigma}^{-1})\right) \left(\frac{d\text{vec}(\mathbf{\Sigma})}{d\boldsymbol{\alpha}_2{}^T}\right)\right)^{-1}. \tag{1.115}$$

The coefficients $\xi_{c,1,m}$ and $\xi_{c,2,m}$ are deduced from $\xi_{r,1,m}$ and $\xi_{r,2,m}$ by the general relations

$$\xi_{c,1,m} = \xi_{r,1,2m} \text{ and } \xi_{c,2,m} = \xi_{r,2,2m}. \tag{1.116}$$

For example, the coefficients $\xi_{c,1,m}$ and $\xi_{c,2,m}$ have been calculated for complex Student's t and complex generalized Gaussian distributions in [8, 27]. They are given respectively by

$$\xi_{c,1,m} = \frac{\nu/2}{((\nu/2)-1)}\frac{(\nu/2)+m}{((\nu/2)+m+1)} \text{ and } \xi_{c,2,m} = \frac{(\nu/2)+m}{(\nu/2)+m+1}, \tag{1.117}$$

$$\xi_{c,1,m} = \frac{\Gamma(2+\frac{m-1}{s})\Gamma(\frac{m+1}{s})}{(\Gamma(1+\frac{m}{s}))^2} \quad \text{ and } \xi_{c,2,m} = \frac{m+s}{m+1}. \tag{1.118}$$

Finally note that the SB formula for elliptically symmetric distributions has been extended when the data model is misspecified by the parametric probabilistic model

[19], and when the density generator is considered as an infinite-dimensional nuisance parameter [42] or parameterized by a nuisance parameter [2].

1.7 Conclusion

The aim of this chapter was to provide a short overview of the main properties of elliptically symmetric distributions and it can be used as a background for all the other chapters in this book. There is no claim to completeness in the material presented here. The reader interested in deeper discussions and investigations on specific aspects may find the references list useful to this goal. As a last concluding remark, we would like to highlight our choice to focus mainly on the RES distributions. As explained in the chapter, this class can be considered as the most general one since it encompasses the C-CES and NC-CES distributions as special cases. Some effort has been then put into showing explicitly the mapping between the RES class and all its subclass. We hope that this chapter may represent a reference for the reader helping him to not get lost while going down his path through this book.

References

1. Abeida, H., Delmas, J.P.: Slepian–Bangs formula and Cramér-Rao bound for circular and non-circular complex elliptical symmetric distributions. IEEE Signal Process. Lett. **26**, 1561–1565 (2019)
2. Abeida, H., Delmas, J.P.: Slepian–Bangs formulas for parameterized density generator of elliptically symmetric distributions. Signal Process. **205** (2023)
3. Anderson, T.W.: An Introduction to Multivariate Statistical Analysis. Wiley, New York (1984)
4. Andrews, D., Mallows, C.: Scale mixtures of normal distributions. J. Royal Stat. Soc. Ser. B **36**(1), 99–102 (1974)
5. Bangs, W.J.: Array processing with generalized beamformers. Ph.D. thesis, Yale University, New Haven, CT (1971)
6. Bentler, P.M.: Some contributions to efficient statistics in structural models: specification and estimation of moments structure. Psychometrika **48**, 493–517 (1983)
7. Bentler, P.M., Berkane, M.: Greatest lower bound to the elliptical theory kurtosis parameter. Biometrika **73**, 240–241 (1986)
8. Besson, O., Abramovich, Y.I.: On the Fisher information matrix for multivariate elliptically contoured distributions. IEEE Signal Process. Lett. **20**, 1130–1133 (2013)
9. Cambanis, S., Huang, S., Simons, G.: On the theory of elliptically contoured distributions. J. Multivar. Anal. **11**, 368–385 (1981)
10. Chitour, Y., Pascal, F.: Exact maximum likelihood estimates for sirv covariance matrix: existence and algorithm analysis. IEEE Trans. Signal Processing **56**(10), 4563–4573 (2008)
11. Conte, E., Longo, M.: Characterisation of radar clutter as a spherically invariant random process. Proc. Inst. Electr. Eng. F **134**(2), 191–197 (1981)
12. Conte, E., Lops, M., Ricci, G.: Asymptotically optimum radar detection in compound-gaussian clutter. IEEE Trans. Aerosp. Electron. Syst. **31**(2), 617–624 (1995)
13. Delmas, J.P., Abeida, H.: Generalization of Whittle's formula to compound-gaussian Processes. IEEE Signal Process. Lett. **31**, 746–750 (2024)

14. Delmas, J.P., Abeida, H.: Stochastic Cramér-Rao bound for noncircular signals with application to DOA estimation. IEEE Trans. Signal Process. **52**, 3192–3199 (2004)
15. Draskovic, G., Pascal, F.: New insights into the statistical properties of M-estimators. IEEE Trans. Signal Process. **66**(16), 4253–4263 (2018)
16. Eriksson, J., Koivunen, V.: Complex random vectors and ICA models: identifiability, uniqueness, and separability. IEEE Trans. Inf. Theory **52**, 1561–1565 (2006)
17. Fang, K., Kotz, S., Ng, K.W.: Symmetric Multivariate and Related Distributions. Chapman & Hall, New-York (1990)
18. Fang, K., Zhang, Y.: Generalized Multivariate Analysis. Springer, Berlin, Germany (1990)
19. Fortunati, S., Gini, F., Greco, M.S., Zoubir, A.M., Rangaswamy, M.: Semiparametric inference and lower bounds for real elliptically symmetric distributions. IEEE Trans. Signal Process. **67**(1), 164–177 (2019)
20. Frahm, G.: Generalized elliptical distribution: theory and applications. Ph.D. thesis, University of Koln, Germany (2014)
21. Franks, W.C., Moitra, A.: Rigorous guarantees for tyler's M-estimator via quantum expansion. In: Proceedings of Machine Learning Research, vol. 125, pp. 1–32 (2020)
22. Gini, F.: Sub-optimum coherent radar detection in a mixture of k-distributed and gaussian clutter. In: Proceedings of the Institution of Electrical Engineers, Radar, Sonar Navigat, vol. 144, pp. 39–48 (1997)
23. Gomez-Sanchez-Manzano, E., Gomez-Villegas, M., Marin, J.: A multivariate generalization of the power exponential family of distributions. Comm. Statist. Theory Methods **27**, 589–600 (1998)
24. Gomez-Sanchez-Manzano, E., Gomez-Villegas, M., Marin, J.: A survey on continuous elliptical vector distributions. Rev. Mat. Comput. **16**, 345–361 (2003)
25. Gomez-Sanchez-Manzano, E., Gomez-Villegas, M., Marin, J.: Sequences of elliptical distributions and mixtures of normal distributions. J. Multivar. Anal. **97**, 295–310 (2006)
26. Gomez-Sanchez-Manzano, E., Gomez-Villegas, M., Marin, J.: Multivariate exponential power distributions as mixtures of normal distributions with bayesian applications. Commun. Stat. - Theory Methods **37**(6), 972–985 (2008)
27. Greco, M., Gini, F.: Cramér-Rao lower bounds on covariance matrix estimation for complex elliptically symmetric distributions. IEEE Trans. Signal Process. **61**, 6401–6409 (2013)
28. Horn, R.A., Johnson, C.A.: Matrix Analysis. University Press, New York, Cambridge (1985)
29. Huber, P.J.: The behavior of maximum likelihood estimates under nonstandard conditions. In: Proceedings of the Fifth Berkeley Symposium in Mathematical Statistics and Probability. University of California Press, Berkley (1967)
30. Hult, H., Lindskog, F.: Multivariate extremes, aggregation and dependence in elliptical distributions. Adv. Appl. Probab. **34**, 587–608 (2002)
31. Johnson, N.L., Kotz, S., Balakrishnan, N.: Continuous Univariate Distributions. Wiley (1994)
32. Kano, Y.: Consistency property of elliptic probability density functions. J. Multiv. Anal. **51**(1), 139–147 (1994)
33. Kelker, D.: Distribution theory of spherical distributions and a location-scale parameter generalization. Sankhya Ser. A **32**, 419–430 (1970)
34. Kent, J.T., Tyler, D.E.: Redescending M-estimates of multivariate location and scatter. Ann. Stat. **19**(4), 2102–2119 (1991)
35. King, A.D.: Moduli of representation of finite dimension algebras. Q. J. Math. **45**(4), 515–530 (1994)
36. Kotz, S.: Multivariate distributions at a cross road. In: Statistical Distributions in Scientific Work I. Reidel, Dordrecht (1968)
37. Krishnaiah, P.R., Lin, J.: Complex elliptically symmetric distributions. Comm. Statist. - Theory. Methods **15**, 3693–3718 (1986)
38. Lange, K.L., Little, R.J.A., Taylor, J.M.G.: Robust statistical modeling using the t distribution. J. Am. Stat. Assoc. **84**(408), 881–896 (1989)
39. Lindsey, J.K.: Multivariate elliptically contoured distributions for repeated measurements. Biometrics **55**, 1277–1280 (1999)

40. Mahot, M., Pascal, F., Forster, P., Ovarlez, J.P.: Asymptotic properties of robust complex covariance matrix estimates. IEEE Trans. Signal Process. **61**(13), 3348–3356 (2013)
41. Maronna, R.: Robust M-estimators of multivariate location and scatter. Ann. Stat. **4**(1), 51–67 (1976)
42. Mennad, A., Fortunati, S., El Korso, M.N., Younsi, A., Zoubir, A.M., Renaux, A.: Slepian–Bangs-type formulas and the related misspecified Cramér-Rao bounds for complex elliptically symmetric distributions. Signal Process. **142**, 320–329 (2018)
43. Meriaux, B., Ren, C., El Korso, M.N., Breloy, A., Forster, P.: Asymptotic performance of complex m-estimators for multivariate location and scatter estimation. IEEE Signal Process. Lett. **26**(2), 363–371 (2019)
44. Meriaux, B., Ren, C., El Korso, M.N., Breloy, A., Forster, P.: Efficient estimation of scatter matrix with convex structure under t-distribution. In: IEEE International Conference on Acoustics, Speech and Signal Processing (ICASSP 2018). Calgary, Alberta, Canada (2018)
45. Nolan, J.P.: Multivariate elliptically contoured stable distributions: theory and estimation. Computational Statistics, vol. 28, pp. 2067–2089. Springer (2013)
46. Ollila, E., Eriksson, J., Koivunen, V.: Complex elliptically symmetric random variables-generation, characterization, and circularity tests. IEEE Trans. Signal Process. **59**, 58–69 (2011)
47. Ollila, E., Koivunen, V.: Generalized complex elliptical distributions. In: Proceedings of the SAM Workshop, pp. 460–464 (2014)
48. Ollila, E., Tyler, D.E.: Distribution-free detection under complex elliptically symmetric clutter distribution. In: IEEE Sensor Array and Multichannel Signal Processing Workshop - SAM 2012. Hoboken, NJ, USA (2012)
49. Ollila, E., Tyler, D.E., Koivunen, V., Poor, H.V.: Complex elliptically symmetric distributions: survey, new results and applications. IEEE Trans. Signal Process. **60**(11), 5597–5625 (2012)
50. Paindaveine, D.: Elliptical symmetry. In book Wiley Statistics Reference Online (2014)
51. Pascal, F., Bombrun, L., Tourneret, J.-Y., Berthoumieu, Y.: Parameter estimation for multivariate generalized gaussian distributions. IEEE Trans. Signal Process. **61**(23) (2013)
52. Pascal, F., Chitour, Y., Ovarlez, J.P., Forster, P., Larzabal, P.: Covariance structure maximum likelihood estimates in compound gaussian noise: existence and algorithm analysis. IEEE Trans. Signal Process. **56**, 34–48 (2008)
53. Pascal, F., Forster, P., Ovarlez, J.P., Larzabal, P.: Performance analysis of covariance matrix estimates in impulsive noise. IEEE Trans. Signal Process. **56**(6), 2206–2216 (2008)
54. Rangaswamy, M., Wainer, D., Ozturk, A.: Non-gaussian random vector identification using spherically invariant random processes. IEEE Trans. Aerospace Electr. Syst. **29**(1), 111–124 (1993)
55. Richmond, C.D.: Adaptive array signal processing and performance analysis in non-Gaussian environments. Ph.D. thesis, Massachusetts Institute of Technology (1996)
56. Shorackd, G.R.: Probability for Statisticians. Springer, New-York (2000)
57. Slepian, D.: Estimation of signal parameters in the presence of noise. Trans. IRE Prof. Group Inf. Theory **3**(85), 68–89 (1954)
58. Soloveychik, I., Wiesel, A.: Tyler's covariance matrix estimator in elliptical models with convex structure. IEEE Trans. Signal Process. **62**(20) (2014)
59. Stoica, P., Moses, R.: Spectral Analysis of Signals. Prentice Hall, NJ (2005)
60. Tyler, D.E.: Radial estimates and the test for sphericity. Biometrika **69**(2), 429 (1982)
61. Tyler, D.E.: Robustness and efficiency properties of scatter matrices. Biometrika **70**(2), 411 (1983)
62. Tyler, D.E.: A distribution-free M-estimator of multivariate scatter. Ann. Stat. **15**(1), 234–251 (1987)
63. Tyler, D.E.: Statistical analysis for the angular central gaussian distribution on the sphere. Biometrika **74**(3), 579–589 (1987)
64. Tyler, D.E.: Some results on the existence, uniqueness, and computation of the M-estimates of multivariate location and scatter. SIAM J. Sci. Stat. Comput. **9**, 354 (1988)
65. Ward, K., R.J.A.Tough, Watts, S.: Sea clutter scattering the K distribution and radar performance. IET Radar, Sonar, Navigation and Avionics, London, U.K (2006)

66. Yao, K.: A representation theorem and its applications to spherically invariant random processes. IEEE Trans. Inf. Theory **19**(5), 600–608 (1973)

Part I
Theoretical Developments

Chapter 2
The Fisher–Rao Geometry of CES Distributions

Florent Bouchard, Arnaud Breloy, Antoine Collas, Alexandre Renaux, and Guillaume Ginolhac

Abstract When dealing with a parametric statistical model, a Riemannian manifold can naturally appear by endowing the parameter space with the Fisher information metric. The geometry induced on the parameters by this metric is then referred to as the Fisher–Rao information geometry. Interestingly, this yields a point of view that allows for leveraging many tools from differential geometry. After a brief introduction about these concepts, we will present some practical uses of these geometric tools in the framework of elliptical distributions. This second part of the exposition is divided into three main axes: Riemannian optimization for covariance matrix estimation, Intrinsic Cramér–Rao bounds, and classification using Riemannian distances.

F. Bouchard · A. Renaux
L2S, CentraleSupélec, CNRS, Université Paris-Saclay, Gif-sur-Yvette, France
e-mail: florent.bouchard@centralesupelec.fr

A. Renaux
e-mail: alexandre.renaux@centralesupelec.fr

A. Breloy (✉)
EPN5, CEDRIC, CNAM, Paris, France
e-mail: arnaud.breloy@cnam.fr

A. Collas
Université Paris-Saclay, Inria, CEA, Palaiseau, France
e-mail: antoine.collas@inria.fr

G. Ginolhac
LISTIC, Université Savoie Mont-Blanc, Chambery, France
e-mail: guillaume.ginolhac@univ-smb.fr

J.-P. Delmas et al. (eds.), *Elliptically Symmetric Distributions in Signal Processing and Machine Learning*, https://doi.org/10.1007/978-3-031-52116-4_2

2.1 Introduction: From CES Distributions to Information Geometry

This section starts with reminders on complex elliptically symmetric distributions (CES).[1] This part is concluded by introducing the Fisher information matrix of this model, which acts as a transition point to information geometry. Indeed, the Fisher information matrix actually represents a metric that induces an inherent geometry for statistical models, which is referred to as the Fisher–Rao information geometry. In the specific case of CES, this will yield a particular geometry for the space of covariance matrices. After evidencing this transition point, this section concludes by outlining the rest of the chapter.

2.1.1 Reminders on CES Distributions

Circular complex elliptically symmetric (C-CES) distributions [52] refer to a large family of multivariate distributions. Very comprehensive and detailed reviews on the topic can be found in the Refs. [70, 73], and of course, the chapter "Background on Real and Complex Elliptically Symmetric Distributions" of this book. A vector $\mathbf{x} \in \mathbb{C}^p$ follows a centered (zero-mean) C-CES distribution, denoted $\mathbf{x} \sim \text{C-CES}(\mathbf{0}, \Sigma, g)$, if it admits the following stochastic representation:

$$\mathbf{x} =_d \sqrt{Q}\, \Sigma^{\frac{1}{2}}\, \mathbf{u}, \tag{2.1}$$

where:

- The notation $=_d$ means that random variables on both sides have the same cumulative distribution function.
- The vector $\mathbf{u} \in \mathbb{C}^p$ follows a uniform distribution on the complex unit sphere $\mathbb{C}\mathcal{S}^p = \{\mathbf{u} \in \mathbb{C}^p \mid \|\mathbf{u}\| = 1\}$, denoted $\mathbf{u} \sim \mathcal{U}(\mathbb{C}\mathcal{S}^p)$.
- The scalar $Q \in \mathbb{R}^+$ is non-negative real random variable of probability density function f_Q, independent of $\mathbf{u}$, and called the second-order modular variate (while $\sqrt{Q}$ is called the modular variate).
- The matrix $\Sigma^{\frac{1}{2}} \in \mathbb{C}^{p\times p}$ is a factorization of the scatter matrix $\Sigma = \Sigma^{\frac{1}{2}} \Sigma^{\frac{H}{2}}$. If the covariance matrix of $\mathbf{x}$ exists, it is proportional to the scatter matrix, i.e., $\mathrm{E}\left[\mathbf{x}\mathbf{x}^H\right] \propto \Sigma$. If we then choose the normalization convention $\mathrm{E}\left[Q\right] = 1$, these two matrices are equal. Thus, we will abusively refer to the scatter matrix Σ as the covariance matrix, as it is a more familiar terminology.

[1] Note that this chapter considers the case where the data and covariance matrix can be complex-valued for the sake of generality. However, we focus solely on the circular case (referred to as C-CES in the chapter "Background on Real and Complex Elliptically Symmetric Distributions"). Hence, most of the presented results can also be obtained in the real-valued case (RES) with proper adjustments.

We focus only on the absolutely continuous case where the covariance matrix Σ is full rank (cf. Sect. 2.3 of the chapter "Background on Real and Complex Elliptically Symmetric Distributions"). In this case, the probability density function of $\mathbf{x}$ is given as

$$f_{\mathbf{x}}\left(\mathbf{x}|\Sigma\right) \propto |\Sigma|^{-1} g\left(\mathbf{x}^H \Sigma^{-1} \mathbf{x}\right), \tag{2.2}$$

where the function $g : \mathbb{R}^+ \to \mathbb{R}^+$ is called the density generator. The density generator satisfies the finite moment condition $\delta_{p,g} = \int_0^\infty t^{p-1} g(t)dt < \infty$. This function g is directly related to the probability density function of the second-order modular variate by the relation

$$f_Q\left(Q\right) = \delta_{p,g}^{-1} Q^{p-1} g\left(Q\right). \tag{2.3}$$

Given a n-sample $\{\mathbf{x}_i\}_{i=1}^n$, assumed to be independent and identically distributed (iid) from $\mathbf{x} \sim$ C-CES($\mathbf{0}$, Σ, g), its log-likelihood is given as

$$\mathcal{L}_g\left(\{\mathbf{x}_i\}_{i=1}^n | \Sigma\right) = \sum_{i=1}^n \log\left(g\left(\mathbf{x}_i^H \Sigma^{-1} \mathbf{x}_i\right)\right) - n \log|\Sigma|. \tag{2.4}$$

The C-CES model being defined, we move to the notion of information brought by the Fisher information matrix: intuitively, the more the sample set $\{\mathbf{x}_i\}_{i=1}^n$ depends on Σ, the more sampling from the likelihood (2.4) (increasing n) will reveal information about Σ. The score vector is a tool that will help in quantifying this notion of information: to define this quantity, we now consider a parameterization of the covariance matrix through a real-valued vector $\boldsymbol{\nu}$ of appropriate dimension,[2] denoted $\Sigma(\boldsymbol{\nu})$. The score vector $\mathbf{s}$ is then defined entry-wise as

$$[\mathbf{s}]_j = \frac{\partial \mathcal{L}_g\left(\{\mathbf{x}_i\}_{i=1}^n | \Sigma(\boldsymbol{\nu})\right)}{\partial \nu_j}, \tag{2.5}$$

which therefore reflects the variation of the log-likelihood of the sample set $\{\mathbf{x}_i\}_{i=1}^n$ with respect to the parameter ν_j. Under mild regularity conditions (satisfied by $\mathcal{L}_g$ in our case), this vector has zero mean, i.e., $\mathrm{E}[\mathbf{s}] = \mathbf{0}$. However, its covariance matrix is a fundamental quantity referred to as the Fisher information matrix, denoted $\mathbf{F}$, and defined as

$$[\mathbf{F}]_{j,k} = \mathrm{E}\left[\frac{\partial \mathcal{L}_g\left(\{\mathbf{x}_i\}_{i=1}^n | \Sigma(\boldsymbol{\nu})\right)}{\partial \nu_j} \frac{\partial \mathcal{L}_g\left(\{\mathbf{x}_i\}_{i=1}^n | \Sigma(\boldsymbol{\nu})\right)}{\partial \nu_k}\right]. \tag{2.6}$$

This matrix quantifies, on average, how much information about the vector $\boldsymbol{\nu}$, we can obtain from a sample set $\{\mathbf{x}_i\}_{i=1}^n$. In practice, the entries of the Fisher information

---

[2] In this chapter, Σ is not assumed to have a specific structure, so $\boldsymbol{\nu}$ is typically of dimension p^2 and stores the entries of the diagonal and upper triangle of the covariance matrix (where the coordinates are split in terms of real and imaginary part). However, the definition extends to any parameterization, e.g., from a choice of decomposition in the case of structured matrices [62].

matrix $\mathbf{F}$ for centered C-CES can be obtained thanks to Slepian–Bangs-type formula from [16], also presented in Sect. 6.5 of chapter "Background on Real and Complex Elliptically Symmetric Distributions". The latter is briefly recalled below using an alternate expression that is consistent whit the upcoming discussions:

Theorem 2.1 (Fisher Information matrix of centered C-CES) *Let* $\Sigma \stackrel{def}{=} \Sigma(\boldsymbol{\nu})$ *be a covariance matrix parameterized by the real-valued vector* $\boldsymbol{\nu}$. *Let* $\{\mathbf{x}_i\}_{i=1}^n$ *be a n-sample of iid from* $\mathbf{x} \sim$ *C-CES*$(\mathbf{0}, \Sigma, g)$. *The entries of the Fisher information matrix are*

$$[\mathbf{F}]_{j,k} = n\alpha_g \operatorname{Tr}(\Sigma^{-1}\boldsymbol{\xi}_j\Sigma^{-1}\boldsymbol{\xi}_k) + n\beta_g \operatorname{Tr}(\Sigma^{-1}\boldsymbol{\xi}_j)\operatorname{Tr}(\Sigma^{-1}\boldsymbol{\xi}_k), \tag{2.7}$$

with

$$\boldsymbol{\xi}_j = \frac{\partial\Sigma(\boldsymbol{\nu})}{\partial\nu_j}, \tag{2.8}$$

and where the coefficients α_g *and* β_g *are defined by*

$$\alpha_g = 1 - \frac{\mathbb{E}\left[Q^2\phi'(Q)\right]}{p(p+1)} \quad \text{and} \quad \beta_g = \alpha_g - 1, \tag{2.9}$$

with $\phi(t) = g'(t)/g(t)$.

In the statistical signal processing community [55], the Fisher information matrix has been extensively leveraged thanks to the Cramér–Rao inequality:

$$\mathbb{E}\left[(\hat{\boldsymbol{\nu}} - \boldsymbol{\nu})(\hat{\boldsymbol{\nu}} - \boldsymbol{\nu})^\top\right] \succcurlyeq \mathbf{F}^{-1} \quad \Rightarrow \quad \left\|\hat{\boldsymbol{\nu}} - \boldsymbol{\nu}\right\|_F^2 \geq \operatorname{Tr}(\mathbf{F}^{-1}), \tag{2.10}$$

that yields a lower bound for the mean squared error of any unbiased estimator $\hat{\boldsymbol{\nu}}$ of $\boldsymbol{\nu}$ (built from a set of observations $\{\mathbf{x}_i\}_{i=1}^n$). On the other hand, the seminal work of Rao [79, 80] also discusses using the Riemannian geometry of the parameter space when the Fisher information matrix is used as a metric tensor. The study of such spaces is now broadly referred to as the Fisher–Rao information geometry, which is introduced in the next section. Before this, we conclude this brief reminder by the example of multivariate (Student's) t distribution (also discussed with more details in Sect. 5.2 of the chapter "Background on Real and Complex Elliptically Symmetric Distributions").

Example 2.1 The t-distribution with $d \in \mathbb{N}^*$ degrees of freedom is obtained for the C-CES representation $\mathbf{x} \sim$ C-CES$(\mathbf{0}, \Sigma, g_d)$ with

$$g_d(t) = \left(1 + \frac{t}{d}\right)^{-(d+p)}, \tag{2.11}$$

and the second-order modular variate is distributed as $Q =_d \mathbb{C}\chi^2_p / \mathbb{C}\chi^2_d/d$ where $\mathbb{C}\chi^2_x$ denotes the chi-squared distribution with x degrees of freedom. Hence, Q follows a scaled $\mathcal{F}$-distribution. We have

$$\phi(t) = -\frac{d+p}{d+t}, \tag{2.12}$$

and the expectation

$$\mathbb{E}\left[Q^2\phi^2(Q)\right] = \frac{(d+p)p(p+1)}{d+p+1}, \tag{2.13}$$

that allows to obtain the coefficients

$$\alpha_g = \frac{d+p}{d+p+1} \quad \text{and} \quad \beta_g = \frac{-1}{d+p+1} \tag{2.14}$$

for the Fisher information metric as in Theorem 2.2. The t-distribution also encompasses the well-known multivariate Gaussian model $\mathbf{x} \sim \mathcal{N}(\mathbf{0}, \Sigma)$ (of density generator $g_{\mathcal{N}}(t) = \exp(-t)$) as limit case when $d \to \infty$. The corresponding Fisher information metric coefficient are then $\alpha_g = 1$ and $\beta_g = 0$, which makes Theorem 2.1 coincide with the classical Slepian–Bangs formula [10, 84].

2.1.2 *From the Fisher Information Matrix to Information Geometry*

This section aims at linking notions of Riemannian geometry to the classical expression of the Fisher information matrix from Theorem 2.1. The goal is to shortly build the intuition on why the C-CES statistical model naturally induces a certain geometry for covariance matrices, while the corresponding framework will be presented in detail in Sect. 2.2. We first need to re-interpret the expression of the Fisher information matrix from two key points:

- **Covariance matrices belong to the smooth manifold $\mathcal{H}_p^{++}$**
  The matrix $\Sigma \overset{\text{def}}{=} \Sigma(\mathbf{v})$ is a point in the space of covariance matrices, i.e., the set of $p \times p$ positive definite Hermitian matrices

$$\mathcal{H}_p^{++} = \left\{\Sigma \in \mathcal{H}_p : \ \forall \mathbf{x} \in \mathbb{C}^p \backslash \{\mathbf{0}\}, \ \mathbf{x}^H \Sigma \mathbf{x} > 0\right\}, \tag{2.15}$$

  where $\mathcal{H}_p$ denotes the set of $p \times p$ Hermitian matrices. As it is an open of the linear space $\mathcal{H}_p$, the space $\mathcal{H}_p^{++}$ is a smooth manifold. This means that it admits a differential structure, and notably, a tangent space at each point Σ, denoted $T_\Sigma \mathcal{H}_p^{++}$. For any point Σ, this tangent space $T_\Sigma \mathcal{H}_p^{++}$ turns out to be identifiable to be $\mathcal{H}_p$ (which again, comes from the fact that $\mathcal{H}_p^{++}$ is an open subspace of $\mathcal{H}_p$). An abstract representation of these spaces is presented in Fig. 2.1.
- **The Fisher information matrix represents an inner product on $T_\Sigma \mathcal{H}_p^{++}$**
  The entries of the Fisher information matrix of Theorem 2.1 can be compactly denoted as $[\mathbf{F}]_{j,k} = \langle \boldsymbol{\xi}_j, \boldsymbol{\xi}_k \rangle_\Sigma^{\text{FIM}}$, whose expression is identified directly from (2.7).

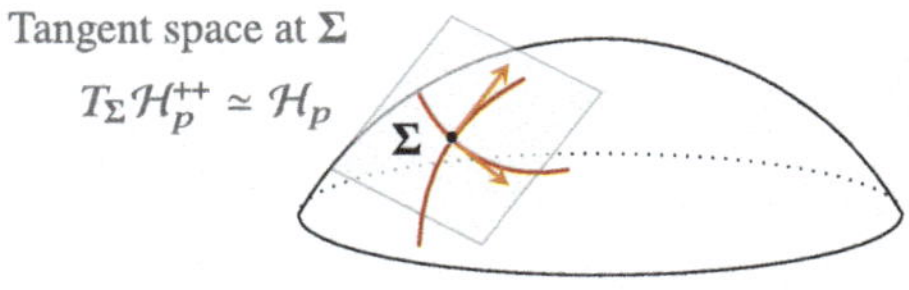

Fig. 2.1 Space of covariance matrices represented as a smooth manifold

We then remark that the matrices $\boldsymbol{\xi}_j$ and $\boldsymbol{\xi}_k$ in (2.8) are, in fact, elements of $T_\Sigma \mathcal{H}_p^{++}$. The expression in (2.7) can thus be generalized to any pair of matrices $\boldsymbol{\xi}, \boldsymbol{\eta} \in T_\Sigma \mathcal{H}_p^{++}$, which results in a bi-linear form, denoted as $\langle \cdot, \cdot \rangle_\Sigma^{\text{FIM}} : T_\Sigma \mathcal{H}_p^{++} \times T_\Sigma \mathcal{H}_p^{++} \to \mathbb{R}$. Because this bi-linear form is positive definite, it defines a metric, i.e., an inner product on the tangent space $T_\Sigma \mathcal{H}_p^{++}$. This inner product on $T_\Sigma \mathcal{H}_p^{++}$ is referred to as the Fisher information metric.[3]

These two points being stated, we can last notice that the obtained Fisher information metric $\langle \cdot, \cdot \rangle_\Sigma^{\text{FIM}}$ varies smoothly with Σ. This enables the transition from statistical models to Riemannian geometry: the branch of differential geometry studying smooth manifolds endowed with smooth local inner products (referred to as Riemannian metrics). Such framework indeed applies to parametric statistical models, as it allows us to investigate the geometry of the parameter space equipped with the Fisher information metric. The resulting Riemannian geometry is generally referred to as the Fisher–Rao information geometry. Back to our central example, we have presented enough elements to explicit that the title of this chapter "The Fisher–Rao Geometry of CES distributions" more precisely stands short for "the Riemannian geometry of Hermitian positive definite matrices (covariance matrices) induced by the Fisher information metric of centered circular complex elliptically symmetric distributions", which will be studied in the next sections.

2.1.3 Outline of the Chapter

The previous section showed why an inherent geometry of the parameter space can naturally result from a statistical model. Studying such geometry in detail requires introducing tools from the framework of Riemannian geometry, which is done in Sect. 2.2. The C-CES distributions will be used as an example throughout the exposition. Hence, we will obtain most tools related to the Fisher–Rao Geometry of C-CES distributions: the Levi-Civita connection, the geodesics (and geodesic distance) between two covariance matrices, as well as the Riemannian exponential and logarithm mappings.

---

[3] Note that the Fisher information *matrix* being obtained as $[\mathbf{F}]_{j,k} = \langle \boldsymbol{\xi}_j, \boldsymbol{\xi}_k \rangle_\Sigma^{\text{FIM}}$, it is actually a matrix representation (metric tensor) of the Fisher information *metric* $\langle \cdot, \cdot \rangle_\Sigma^{\text{FIM}}$ when the set $\{\boldsymbol{\xi}_j\}$ is chosen as a basis of coordinates for the tangent space $T_\Sigma \mathcal{H}_p^{++}$.

On a larger perspective, the second part of this chapter illustrates where tools obtained from the Fisher–Rao information geometry of C-CES can be leveraged within signal processing and machine learning tasks. In details:

- Section 2.3 addresses covariance matrix estimation problems, i.e., given a sample set $\{\mathbf{x}_i\}_{i=1}^n$, we infer Σ to perform a task (covariance analysis, filtering, metric learning, etc.). In this setup, we illustrate how the concepts of geodesic convexity and Riemannian optimization can be helpful in problems related to covariance matrix estimation.
- Still related to covariance matrix estimation problems, Sect. 2.4 presents how the statistical performance of an estimator can be evaluated with intrinsic Cramér–Rao lower bounds, which generalize the standard Cramér–Rao inequality for parameters that lie in a manifold.
- Section 2.5 discusses how the Fisher–Rao geometry of C-CES provides a measure between the distributions of samples that can be leveraged in classification methodologies. This framework is then applied to Electroencephalography (EEG) signals.

2.2 An Introduction to Riemannian Geometry Through the Fisher–Rao Geometry of CES Distributions

This section provides a short introduction to the concepts and tools of Riemannian geometry, while using the Fisher–Rao geometry of C-CES distributions as the main directive example. Some elementary notions are also assumed to be known for the sake of conciseness (e.g., basics matrix differentiation). For more detailed coverages of differential geometry, one can refer to the standard textbooks on the topic [41, 58, 59]. The notation and definitions of this section are mostly inspired from the books [1, 26], which provide very good (optimization-oriented) entry points to smooth manifolds and Riemannian geometry. The Fisher–Rao geometries of multivariate Gaussian and CES models have been studied in, e.g., [8, 15, 28, 63, 64, 83, 85].

2.2.1 $\mathcal{H}_p^{++}$ as a Riemannian Manifold

The set of $p \times p$ Hermitian positive definite matrices $\mathcal{H}_p^{++}$ is an open subspace of the space of $p \times p$ Hermitian matrices $\mathcal{H}_p$. Since $\mathcal{H}_p^{++}$ has the same dimension as its embedding space $\mathcal{H}_p$, it is a smooth manifold of dimension p^2 [26, Definition 3.10]. As every smooth manifolds, $\mathcal{H}_p^{++}$ admits a differential structure, i.e., every point $\Sigma \in \mathcal{H}_p^{++}$ possesses a tangent space $T_\Sigma \mathcal{H}_p^{++}$. The elements of $T_\Sigma \mathcal{H}_p^{++}$ are called tangent vectors, and correspond to the directional derivatives of curves in $\mathcal{H}_p^{++}$ passing through Σ (cf. Fig. 2.1 for an illustration). Since $\mathcal{H}_p^{++}$ is open in $\mathcal{H}_p$,

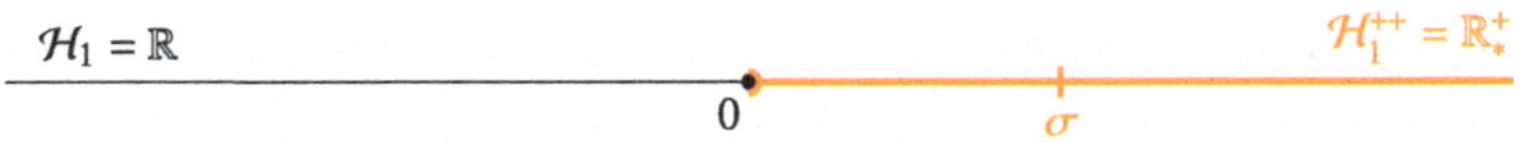

Fig. 2.2 Illustration of the manifold $\mathcal{H}_1^{++} = \mathbb{R}_*^+$, which is open in $\mathcal{H}_1 = \mathbb{R}$. We observe that the tangent space at every point $\sigma > 0$ simply corresponds to $\mathcal{H}_1 = \mathbb{R}$

the tangent space $T_\Sigma \mathcal{H}_p^{++}$ at every point Σ can be identified as $\mathcal{H}_p$ [26, Theorem 3.15]. An illustration of the 1-dimensional case $\mathcal{H}_1^{++} = \mathbb{R}_*^+$ is presented in Fig. 2.2.

Remark 2.1 The space $\mathcal{H}_p^{++}$ is often referred to as the *convex cone* of positive definite matrices. It is indeed a cone because $\Sigma \in \mathcal{H}_p^{++}$ implies that $a\Sigma \in \mathcal{H}_p^{++}$, $\forall\, a \in \mathbb{R}_+^*$. It is furthermore a convex cone because any linear combination $a\Sigma_1 + b\Sigma_2$ is also in $\mathcal{H}_p^{++}$, $\forall\, \Sigma_1, \Sigma_2 \in \mathcal{H}_p^{++}$ and $\forall\, a, b \in \mathbb{R}_+^*$. This cone visually appears in the real 2×2 case of $\mathcal{S}_2^{++}$, which is often used to represent $\mathcal{H}_p^{++}$. Still, this chapter will rely on the representation of Fig. 2.1, which is more convenient to illustrate the generic concepts and tools of Riemannian geometry.

In order to further harness the differential structure of $\mathcal{H}_p^{++}$, we endow it with a Riemannian metric. This consists in a mapping that equips every tangent space $T_\Sigma \mathcal{H}_p^{++}$ with an inner product[4] $\langle \cdot, \cdot \rangle_\Sigma$ that varies smoothly with respect to the point Σ. This allows notably for locally defining the notion of angle and length for vectors in $T_\Sigma \mathcal{H}_p^{++}$. A smooth manifold equipped with such a Riemmanian metric is then referred to as a Riemannian manifold. Notice that the definition of the metric is a choice that induces a corresponding geometry. In particular, if $\mathcal{H}_p^{++}$ is endowed with the Euclidean metric $\langle \boldsymbol{\xi}, \boldsymbol{\eta} \rangle_\Sigma^{\mathcal{E}} = \mathfrak{Re}(\mathrm{Tr}(\boldsymbol{\xi}\boldsymbol{\eta}))$, where $\mathfrak{Re}(\cdot)$ returns the real part of its argument, all the geometrical objects of the manifold $\mathcal{H}_p^{++}$ are exactly the same as those of the space $\mathcal{H}_p$. In this case, there is no distinction between $\mathcal{H}_p^{++}$ and $\mathcal{H}_p$ from a geometrical point of view, and the true structure of $\mathcal{H}_p^{++}$ cannot be exploited. This motivates the use of other metrics, that induce a more meaningful geometry on $\mathcal{H}_p^{++}$ (e.g., ensuring that the boundaries of the space are not reachable). In this scope, various options have been considered, such as the affine invariant metric [17, 65], the log-Euclidean metric [7], or the Bures–Wasserstein one [18, 43]. Overviews of the different metrics and their corresponding geometries can be found in [92, 93].

When dealing with a statistical model, the Fisher information metric is generally to be favored, as it is naturally suited to the underlying geometry of the data. Without resorting to the tedious parameterization and identification of Sect. 2.1.2, a general expression of this metric can directly be obtained following [85, Theorem 1] as

$$\langle \boldsymbol{\xi}, \boldsymbol{\eta} \rangle_\Sigma^{\mathrm{FIM}} = \mathrm{E}\left[\mathrm{D}\,\mathcal{L}_g(\mathbf{x}|\Sigma)[\boldsymbol{\xi}] \cdot \mathrm{D}\,\mathcal{L}_g(\mathbf{x}|\Sigma)[\boldsymbol{\eta}]\right] = -\mathrm{E}\left[\mathrm{D}^2\,\mathcal{L}_g(\mathbf{x}|\Sigma)[\boldsymbol{\xi}, \boldsymbol{\eta}]\right], \tag{2.16}$$

where $\mathrm{D}\,\mathcal{L}_g$ and $\mathrm{D}^2\,\mathcal{L}_g$ are the first- and second-order directional derivatives of the log-likelihood $\mathcal{L}_g$ of the distribution with respect to Σ. Recall from [44] that the first

[4] An inner product is a bilinear, symmetric, positive definite function.

and second derivatives of a function $L : \mathcal{H}_p^{++} \to \mathbb{R}$ at $\Sigma \in \mathcal{H}_p^{++}$ in directions $\boldsymbol{\xi}$ and $\boldsymbol{\eta} \in T_\Sigma \mathcal{H}_p^{++}$ are defined as

$$\begin{aligned} \mathrm{D}\, L(\Sigma)[\boldsymbol{\xi}] &= L(\Sigma + \boldsymbol{\xi}) - L(\Sigma) + o(\|\boldsymbol{\xi}\|) \\ \mathrm{D}^2\, L(\Sigma)[\boldsymbol{\xi}, \boldsymbol{\eta}] &= \mathrm{D}\, L(\Sigma + \boldsymbol{\eta})[\boldsymbol{\xi}] - \mathrm{D}\, L(\Sigma)[\boldsymbol{\xi}] + o(\|\boldsymbol{\xi}\|). \end{aligned} \tag{2.17}$$

Notice that $\mathrm{D}^2\, L(\Sigma)[\boldsymbol{\xi}, \boldsymbol{\eta}]$ is symmetrical with respect to $\boldsymbol{\xi}$ and $\boldsymbol{\eta}$. In the case of the CES distributions, the Fisher information metric was studied in [8, 15, 28, 63, 64], and its derivation is reported in the following theorem:

Theorem 2.2 (Fisher Information metric of centered CES) *Let* $\Sigma \in \mathcal{H}_p^{++}$. *Let* $\{\mathbf{x}_i\}_{i=1}^n$ *be a n-sample of iid from* $\mathbf{x} \sim C\text{-}CES(\mathbf{0}, \Sigma, g)$. *The Fisher information metric is obtained* $\forall\, \boldsymbol{\xi}, \boldsymbol{\eta} \in T_\Sigma \mathcal{H}_p^{++}$ *as*

$$\langle \boldsymbol{\xi}, \boldsymbol{\eta} \rangle_\Sigma^{\mathrm{FIM}} = n\alpha_g \operatorname{Tr}(\Sigma^{-1}\boldsymbol{\xi}\Sigma^{-1}\boldsymbol{\eta}) + n\beta_g \operatorname{Tr}(\Sigma^{-1}\boldsymbol{\xi}) \operatorname{Tr}(\Sigma^{-1}\boldsymbol{\eta}),$$

with α_g *and* β_g *defined in* (2.9).

Proof The first quantities to compute to obtain the Fisher information metric are the derivatives $\mathrm{D}\, \mathcal{L}_g(\{\mathbf{x}_i\}_{i=1}^n | \Sigma)[\boldsymbol{\xi}]$ and $\mathrm{D}^2\, \mathcal{L}_g(\{\mathbf{x}_i\}_{i=1}^n | \Sigma)[\boldsymbol{\xi}, \boldsymbol{\eta}]$ at $\Sigma \in \mathcal{H}_p^{++}$ in directions $\boldsymbol{\xi}$ and $\boldsymbol{\eta} \in T_\Sigma \mathcal{H}_p^{++}$. To do so, recall that $\mathrm{D} \log\det(\Sigma)[\boldsymbol{\xi}] = \operatorname{Tr}(\Sigma^{-1}\boldsymbol{\xi})$ and $\mathrm{D}(\Sigma^{-1})[\boldsymbol{\xi}] = -\Sigma^{-1}\boldsymbol{\xi}\Sigma^{-1}$. It follows that

$$\mathrm{D}\, \mathcal{L}_g(\{\mathbf{x}_i\}_{i=1}^n | \Sigma)[\boldsymbol{\xi}] = -n \operatorname{Tr}(\Sigma^{-1}\boldsymbol{\xi}) - \sum_{i=1}^n \phi(\mathbf{x}_i^H \Sigma^{-1} \mathbf{x}_i) \operatorname{Tr}(\Sigma^{-1}\boldsymbol{\xi}\Sigma^{-1}\mathbf{x}_i\mathbf{x}_i^H).$$

Moreover, also recall that the trace is invariant to any permutation of the product of three Hermitian matrices. Thus,

$$\begin{aligned} \mathrm{D}^2\, \mathcal{L}_g(\{\mathbf{x}_i\}_{i=1}^n | \Sigma)[\boldsymbol{\xi}, \boldsymbol{\eta}] &= \operatorname{Tr}(\Sigma^{-1}\boldsymbol{\xi}\Sigma^{-1}\boldsymbol{\eta}) \\ &\quad + 2\sum_{i=1}^n \phi(\operatorname{Tr}(\Sigma^{-1}\mathbf{x}_i\mathbf{x}_i^H)) \operatorname{Tr}(\Sigma^{-1}\boldsymbol{\xi}\Sigma^{-1}\boldsymbol{\eta}\Sigma^{-1}\mathbf{x}_i\mathbf{x}_i^H) \\ &\quad + \sum_{i=1}^n \phi'(\operatorname{Tr}(\Sigma^{-1}\mathbf{x}_i\mathbf{x}_i^H)) \operatorname{Tr}(\Sigma^{-1}\boldsymbol{\xi}\Sigma^{-1}\mathbf{x}_i\mathbf{x}_i^H) \operatorname{Tr}(\Sigma^{-1}\boldsymbol{\eta}\Sigma^{-1}\mathbf{x}_i\mathbf{x}_i^H). \end{aligned}$$

We now need to compute the expectation. To do so, we exploit the stochastic representation $\mathbf{x}_i =_d \sqrt{Q_i}\,\Sigma^{1/2}\mathbf{u}_i$. Recall that Q_i and $\mathbf{u}_i$ are independent, $\mathbf{u}_i^H\mathbf{u}_i = 1$ and $\mathrm{E}\left[\mathbf{u}_i\mathbf{u}_i^H\right] = \frac{1}{p}\mathbf{I}_p$ (since $\mathbf{u}_i \sim \mathcal{U}(\mathbb{C}S^p)$). Furthermore, from (2.3), $\mathrm{E}\left[Q_i\phi(Q_i)\right] = -p$. It follows that

$$\begin{aligned} &\mathrm{E}\left[\phi(\operatorname{Tr}(\Sigma^{-1}\mathbf{x}_i\mathbf{x}_i^H)) \operatorname{Tr}(\Sigma^{-1}\boldsymbol{\xi}\Sigma^{-1}\boldsymbol{\eta}\Sigma^{-1}\mathbf{x}_i\mathbf{x}_i^H)\right] \\ &\quad = \operatorname{Tr}(\Sigma^{-1}\boldsymbol{\xi}\Sigma^{-1}\boldsymbol{\eta}\mathrm{E}\left[\phi(Q_i)Q_i\mathbf{u}_i\mathbf{u}_i^H\right]) = \operatorname{Tr}(\Sigma^{-1}\boldsymbol{\xi}\Sigma^{-1}\boldsymbol{\eta}\mathrm{E}\left[\phi(Q_i)Q_i\right]\mathrm{E}\left[\mathbf{u}_i\mathbf{u}_i^H\right]) \\ &\quad = -\operatorname{Tr}(\Sigma^{-1}\boldsymbol{\xi}\Sigma^{-1}\boldsymbol{\eta}). \end{aligned}$$

For the second expectation, from [16], we need

$$\mathrm{E}\left[(\mathbf{u}_i^H \mathbf{A}\mathbf{u}_i)^2\right] = \frac{\mathrm{Tr}(\mathbf{A}^2) + (\mathrm{Tr}(\mathbf{A}))^2}{p(p+1)}.$$

Applying the polarization formula $\frac{1}{4}[(\mathbf{u}_i^H(\boldsymbol{A}+\boldsymbol{B})\mathbf{u}_i)^2 - (\mathbf{u}_i^H(\boldsymbol{A}-\boldsymbol{B})\mathbf{u}_i)^2]$, we get

$$\mathrm{E}\left[(\mathbf{u}_i^H \boldsymbol{A}\mathbf{u}_i)(\mathbf{u}_i^H \boldsymbol{B}\mathbf{u}_i)\right] = \frac{\mathrm{Tr}(\boldsymbol{AB}) + \mathrm{Tr}(\boldsymbol{A})\,\mathrm{Tr}(\boldsymbol{B})}{p(p+1)}.$$

Therefore,

$$\begin{aligned}
&\mathrm{E}\left[\phi'(\mathrm{Tr}(\Sigma^{-1}\mathbf{x}_i\mathbf{x}_i^H))\,\mathrm{Tr}(\Sigma^{-1}\boldsymbol{\xi}\Sigma^{-1}\mathbf{x}_i\mathbf{x}_i^H)\,\mathrm{Tr}(\Sigma^{-1}\boldsymbol{\eta}\Sigma^{-1}\mathbf{x}_i\mathbf{x}_i^H)\right] \\
&\quad = \mathrm{E}\left[Q_i^2\phi'(Q_i)(\mathbf{u}_i^H\Sigma^{-1}\boldsymbol{\xi}\mathbf{u}_i)(\mathbf{u}_i^H\Sigma^{-1}\boldsymbol{\eta}\mathbf{u}_i)\right] \\
&\quad = \mathrm{E}\left[Q_i^2\phi'(Q_i)\right]\mathrm{E}\left[(\mathbf{u}_i^H\Sigma^{-1}\boldsymbol{\xi}\mathbf{u}_i)(\mathbf{u}_i^H\Sigma^{-1}\boldsymbol{\eta}\mathbf{u}_i)\right] \\
&\quad = \frac{\mathrm{E}\left[Q_i^2\phi'(Q_i)\right]}{p(p+1)}\left(\mathrm{Tr}(\Sigma^{-1}\boldsymbol{\xi}\Sigma^{-1}\boldsymbol{\eta}) + \mathrm{Tr}(\Sigma^{-1}\boldsymbol{\xi})\,\mathrm{Tr}(\Sigma^{-1}\boldsymbol{\eta})\right).
\end{aligned}$$

From there, basic manipulations yield the result with coefficients α_g and β_g defined in (2.9). Notice that the dependency on i in Q is omitted since these parameters are assumed iid. □

The Fisher information metric of C-CES thus corresponds to a general form of the well-known affine invariant metric on $\mathcal{H}_p^{++}$ [17, 83]. Hence, if not specified otherwise the remainder of this chapter will use the more common generic denotation:

$$\langle \boldsymbol{\xi}, \boldsymbol{\eta}\rangle_\Sigma \overset{\text{def}}{=} \mathfrak{g}_\Sigma(\boldsymbol{\xi}, \boldsymbol{\eta}) = \mathfrak{Re}(\alpha\,\mathrm{Tr}(\Sigma^{-1}\boldsymbol{\xi}\Sigma^{-1}\boldsymbol{\eta}) + \beta\,\mathrm{Tr}(\Sigma^{-1}\boldsymbol{\xi})\,\mathrm{Tr}(\Sigma^{-1}\boldsymbol{\eta})) \quad (2.18)$$

and study the corresponding Riemannian geometry of $\mathcal{H}_p^{++}$ for any $\alpha \in \mathbb{R}_*^+$ and $\beta > -\alpha/p$ (necessary conditions so that $\langle\cdot,\cdot\rangle_\Sigma$ is positive definite). The Fisher–Rao information geometry of the considered C-CES model is then recovered by fixing α and β according to Theorem 2.2. We can also point out that the most studied case corresponds to $\alpha = 1$ and $\beta = 0$, which coincide with the Fisher information metric of the Gaussian distribution, as $\alpha_g = 1$ and $\beta_g = 0$ in this case [28, 85].

Remark 2.2 Taking the real part in the metric (2.18) defines a proper inner product on $T_\Sigma\mathcal{H}_p^{++}$ from the original Hermitian inner product. This way, we implicitly identify the complex space as its underlying real vector space ($\mathbb{C} \sim \mathbb{R}^2$), so that we can use the usual derivatives (defined as those used on $\mathbb{R}$). As a direct consequence, in this chapter, both $\mathcal{H}_p$ and $\mathcal{H}_p^{++}$ are of dimension p^2. Notice that, even though it is not always stated, most works that deal with complex-valued matrices (e.g., [85]) also implicitly use the real part of the Fisher information metric.

Remark 2.3 Among many other properties, the Fisher information metric from Theorem 2.2 has a notable quadratic dependence on Σ^{-1}. This makes the norm

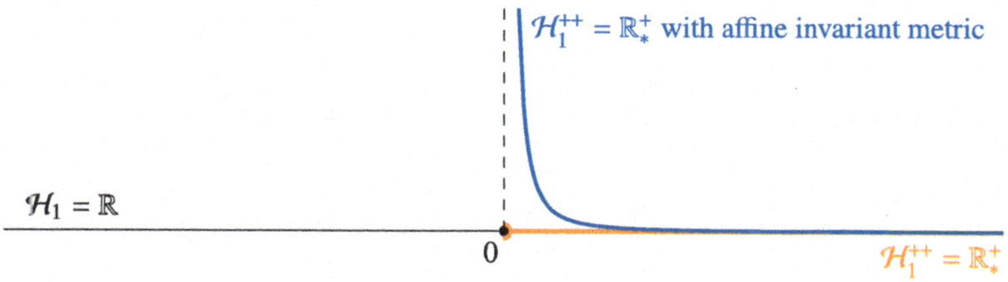

Fig. 2.3 Illustration of the effect of the affine invariant metric on $\mathcal{H}_1^{++} = \mathbb{R}_*^+$. Thanks to the metric, the excluded point 0 becomes truly unreachable

of tangent vectors $\|\boldsymbol{\xi}\|_\Sigma^2 = \langle \boldsymbol{\xi}, \boldsymbol{\xi} \rangle_\Sigma$ tend to infinity when the point Σ tends to the boundaries of the manifold (i.e., when any number of its eigenvalues tend 0). This Riemannian metric thus allows to actually perceive the boundary of $\mathcal{H}_p^{++}$ as being infinitely far, which was not the case for the Euclidean metric. An illustration of the effect of the metric is displayed for $\mathcal{H}_1^{++} = \mathbb{R}_*^+$ in Fig. 2.3.

2.2.2 *Levi-Civita Connection*

One of the most—if not the most—important tools of Riemannian geometry is the Levi-Civita connection, which generalizes the notion of directional derivatives of vector fields on manifolds. A vector field is a function that associates a unique tangent vector $\boldsymbol{\xi}_\Sigma \in T_\Sigma \mathcal{H}_p^{++}$ to every point $\Sigma \in \mathcal{H}_p^{++}$, which is illustrated in Fig. 2.4. An example of a vector field that will be involved in Sect. 2.3 is the gradient of a cost function. The set of vector fields on $\mathcal{H}_p^{++}$ is denoted $\mathfrak{X}(\mathcal{H}_p^{++})$.

To differentiate a vector field on a manifold, one needs to resort to an affine connection. This is an application from $\mathfrak{X}(\mathcal{H}_p^{++}) \times \mathfrak{X}(\mathcal{H}_p^{++})$ onto $\mathfrak{X}(\mathcal{H}_p^{++})$. The connection of $\boldsymbol{\eta}_\Sigma$ in the direction $\boldsymbol{\xi}_\Sigma$ is denoted $\nabla_{\boldsymbol{\xi}_\Sigma} \boldsymbol{\eta}_\Sigma$ and generalizes the directional derivative of $\boldsymbol{\eta}_\Sigma$ in the direction $\boldsymbol{\xi}_\Sigma$ (i.e., $\mathrm{D}\,\boldsymbol{\eta}_\Sigma[\boldsymbol{\xi}_\Sigma]$). Such generalization is needed because the tangent space changes when one moves from one point to another on a manifold. Thus, the usual directional derivative might not be properly defined, as it does not account for the structure of the manifold (constraints, Riemannian metric, etc.). This specificity is illustrated in Fig. 2.5.

Many affine connections can be defined on a manifold. However, there is a unique one that is in accordance with the chosen Riemannian metric, which is referred to as the Levi-Civita connection. This Levi-Civita connection, denoted $\nabla_{\boldsymbol{\xi}_\Sigma} \boldsymbol{\eta}_\Sigma$, is the unique solution in the tangent space $T_\Sigma \mathcal{H}_p^{++}$ to the Koszul formula:

Fig. 2.4 Illustration of a vector field on $\mathcal{H}_p^{++}$

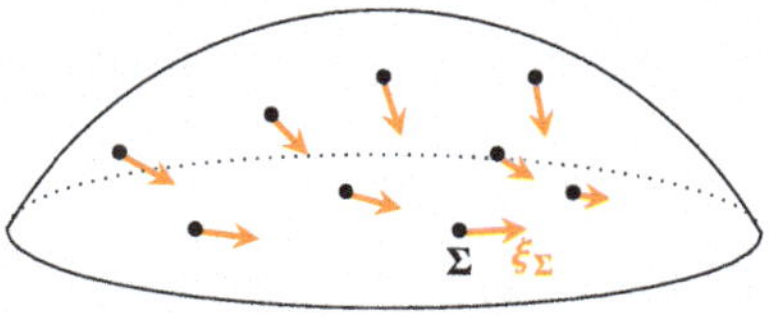

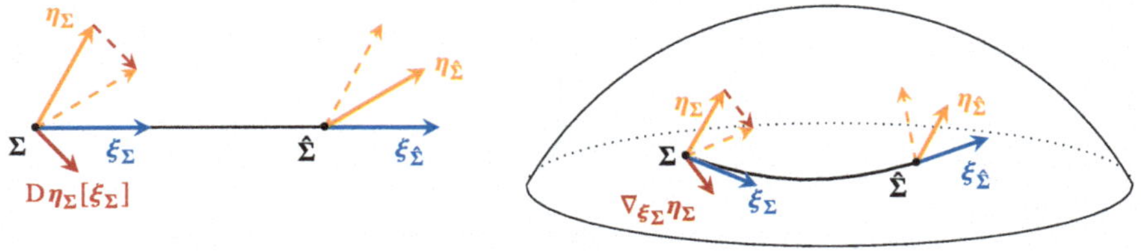

Fig. 2.5 Illustration of directional derivative $\mathrm{D}\,\boldsymbol{\eta}_\Sigma[\boldsymbol{\xi}_\Sigma]$ (left) and affine connection $\nabla_{\boldsymbol{\xi}_\Sigma}\boldsymbol{\eta}_\Sigma$ (right) of a vector field $\boldsymbol{\eta}$ in the direction $\boldsymbol{\xi}$ at Σ. As the directional derivative, an affine connection describes how the vector field $\boldsymbol{\eta}$ evolves in a given direction $\boldsymbol{\xi}$. In addition, the affine connection takes into account the structure of the manifold (curvature, and non-constant metric)

$$
\begin{aligned}
2\mathfrak{g}_\Sigma(\nabla_{\boldsymbol{\xi}_\Sigma}\boldsymbol{\eta}_\Sigma, \boldsymbol{\nu}_\Sigma) &= 2\mathfrak{g}_\Sigma(\mathrm{D}\,\boldsymbol{\eta}_\Sigma[\boldsymbol{\xi}_\Sigma], \boldsymbol{\nu}_\Sigma) \\
&\quad + \mathrm{D}\,\mathfrak{g}_\Sigma[\boldsymbol{\xi}_\Sigma](\boldsymbol{\eta}_\Sigma, \boldsymbol{\nu}_\Sigma) + \mathrm{D}\,\mathfrak{g}_\Sigma[\boldsymbol{\eta}_\Sigma](\boldsymbol{\xi}_\Sigma, \boldsymbol{\nu}_\Sigma) - \mathrm{D}\,\mathfrak{g}_\Sigma[\boldsymbol{\nu}_\Sigma](\boldsymbol{\eta}_\Sigma, \boldsymbol{\xi}_\Sigma),
\end{aligned} \tag{2.19}
$$

where we use the alternate notation of the metric, i.e., $\mathfrak{g}_\Sigma(\cdot,\cdot) = \langle\cdot,\cdot\rangle_\Sigma$. Notice that the presented formula is simpler than the general case [1]. It is because the Lie bracket is $[\boldsymbol{\xi}_\Sigma, \boldsymbol{\eta}_\Sigma] = \mathrm{D}\,\boldsymbol{\eta}_\Sigma[\boldsymbol{\eta}_\Sigma] - \mathrm{D}\,\boldsymbol{\xi}_\Sigma[\boldsymbol{\eta}_\Sigma]$ since $\mathcal{H}_p^{++}$ is an open subset of a vector space, i.e., $\mathcal{H}_p$. The Levi-Civita connection of $\mathcal{H}_p^{++}$ associated with the Riemannian metric (2.18) is provided in Theorem 2.3.

Theorem 2.3 (Levi-Civita connection) *The Levi-Civita connection on $\mathcal{H}_p^{++}$ associated with the affine invariant metric* (2.18) *is defined for $\boldsymbol{\xi}$, $\boldsymbol{\eta} \in \mathfrak{X}(\mathcal{H}_p^{++})$ and $\Sigma \in \mathcal{H}_p^{++}$, as*

$$
\nabla_{\boldsymbol{\xi}_\Sigma}\boldsymbol{\eta}_\Sigma = \mathrm{D}\,\boldsymbol{\eta}_\Sigma[\boldsymbol{\xi}_\Sigma] - \mathrm{Herm}(\boldsymbol{\eta}_\Sigma\Sigma^{-1}\boldsymbol{\xi}_\Sigma),
$$

where $\mathrm{Herm}(\cdot)$ *returns the Hermitian part of its argument.*

Proof First, recall that for $\boldsymbol{A} \in \mathcal{H}_p$ and $\boldsymbol{B} \in \mathbb{R}^{p\times p}$, $\mathrm{Tr}(\boldsymbol{A}\boldsymbol{B}) = \mathrm{Tr}(\boldsymbol{A}\,\mathrm{Herm}(\boldsymbol{B}))$. Further, recall that the trace is invariant to any permutation of the product of three Hermitian matrices. Since $\mathrm{D}(\Sigma^{-1})[\boldsymbol{\xi}] = -\Sigma^{-1}\boldsymbol{\xi}\Sigma^{-1}$, we have

$$
\mathrm{D}\,\mathfrak{Re}(\mathrm{Tr}(\Sigma^{-1}\boldsymbol{\xi}\Sigma^{-1}\boldsymbol{\eta}))[\boldsymbol{\nu}] = -2\,\mathfrak{Re}(\mathrm{Tr}(\Sigma^{-1}\boldsymbol{\xi}\Sigma^{-1}\boldsymbol{\eta}\Sigma^{-1}\boldsymbol{\nu}))
$$

and

$$
\begin{aligned}
&\mathrm{D}\,\mathfrak{Re}(\mathrm{Tr}(\Sigma^{-1}\boldsymbol{\eta}\Sigma^{-1}\boldsymbol{\nu}))[\boldsymbol{\xi}] + \mathrm{D}\,\mathfrak{Re}(\mathrm{Tr}(\Sigma^{-1}\boldsymbol{\xi}\Sigma^{-1}\boldsymbol{\nu}))[\boldsymbol{\eta}] - \mathrm{D}\,\mathfrak{Re}\,\mathrm{Tr}(\Sigma^{-1}\boldsymbol{\xi}\Sigma^{-1}\boldsymbol{\eta}))[\boldsymbol{\nu}] \\
&\quad = -2\,\mathfrak{Re}(\mathrm{Tr}(\Sigma^{-1}\boldsymbol{\xi}\Sigma^{-1}\boldsymbol{\eta}\Sigma^{-1}\boldsymbol{\nu})) = -2\,\mathfrak{Re}(\mathrm{Tr}(\Sigma^{-1}\,\mathrm{Herm}(\boldsymbol{\xi}\Sigma^{-1}\boldsymbol{\eta})\Sigma^{-1}\boldsymbol{\nu})).
\end{aligned}
$$

We also have

$$
\begin{aligned}
\mathrm{D}\left(\mathfrak{Re}(\mathrm{Tr}(\Sigma^{-1}\boldsymbol{\xi})\,\mathrm{Tr}(\Sigma^{-1}\boldsymbol{\eta}))\right)[\boldsymbol{\nu}] &= -\,\mathfrak{Re}(\mathrm{Tr}(\Sigma^{-1}\boldsymbol{\nu}\Sigma^{-1}\boldsymbol{\xi})\,\mathrm{Tr}(\Sigma^{-1}\boldsymbol{\eta})) \\
&\quad - \mathfrak{Re}(\mathrm{Tr}(\Sigma^{-1}\boldsymbol{\xi})\,\mathrm{Tr}(\Sigma^{-1}\boldsymbol{\nu}\Sigma^{-1}\boldsymbol{\eta})).
\end{aligned}
$$

It follows that

$$\begin{aligned}
&\mathrm{D}\left(\mathfrak{Re}(\mathrm{Tr}(\Sigma^{-1}\boldsymbol{\eta})\,\mathrm{Tr}(\Sigma^{-1}\boldsymbol{\nu}))\right)[\boldsymbol{\xi}] + \mathrm{D}\left(\mathfrak{Re}(\mathrm{Tr}(\Sigma^{-1}\boldsymbol{\xi})\,\mathrm{Tr}(\Sigma^{-1}\boldsymbol{\nu}))\right)[\boldsymbol{\eta}] \\
&\quad - \mathrm{D}\left(\mathfrak{Re}(\mathrm{Tr}(\Sigma^{-1}\boldsymbol{\xi})\,\mathrm{Tr}(\Sigma^{-1}\boldsymbol{\eta}))\right)[\boldsymbol{\nu}] = -2\,\mathfrak{Re}(\mathrm{Tr}(\Sigma^{-1}\boldsymbol{\nu})\,\mathrm{Tr}(\Sigma^{-1}\boldsymbol{\xi}\Sigma^{-1}\boldsymbol{\eta})) \\
&\qquad\qquad = -\,2\,\mathfrak{Re}(\mathrm{Tr}(\Sigma^{-1}\boldsymbol{\nu})\,\mathrm{Tr}(\Sigma^{-1}\,\mathrm{Herm}(\boldsymbol{\xi}\Sigma^{-1}\boldsymbol{\eta}))).
\end{aligned}$$

From there, we can deduce that

$$\begin{aligned}
&\mathrm{D}\,\mathfrak{g}_\Sigma[\boldsymbol{\xi}_\Sigma](\boldsymbol{\eta}_\Sigma, \boldsymbol{\nu}_\Sigma) + \mathrm{D}\,\mathfrak{g}_\Sigma[\boldsymbol{\eta}_\Sigma](\boldsymbol{\xi}_\Sigma, \boldsymbol{\nu}_\Sigma) - \mathrm{D}\,\mathfrak{g}_\Sigma[\boldsymbol{\nu}_\Sigma](\boldsymbol{\eta}_\Sigma, \boldsymbol{\xi}_\Sigma) \\
&\qquad = -2\mathfrak{g}_\Sigma(\mathrm{Herm}(\boldsymbol{\xi}_\Sigma\Sigma^{-1}\boldsymbol{\eta}_\Sigma), \boldsymbol{\nu}_\Sigma).
\end{aligned}$$

Injecting this into the Koszul formula yields the result. □

Remark 2.4 Notice that the Levi-Civita connection of $\mathcal{H}_p^{++}$ associated with the Riemannian metric of the metric in (2.18) does not depend on α and β. Hence, it remains the same for any underlying C-CES distribution.

2.2.3 *Geodesics, Riemannian Exponential, Logarithm, and Distance*

One of the main reasons why the Levi-Civita connection is so crucial is because it allows to define geodesics. The geodesics generalize the concept of straight lines on a manifold. These are curves $\gamma : [0, 1] \to \mathcal{H}_p^{++}$ with no acceleration, where acceleration is defined thanks to the Levi-Civita connection. They are parameterized by the choice of starting point $\gamma(0) = \Sigma \in \mathcal{H}_p^{++}$ and either initial direction $\dot{\gamma}(0) = \boldsymbol{\xi} \in \mathcal{H}_p$ or ending point $\gamma(1) = \hat{\Sigma} \in \mathcal{H}_p^{++}$. An illustration of geodesics is provided in Fig. 2.6. Formally, the geodesic $\gamma : [0, 1] \to \mathcal{H}_p^{++}$ is the solution to the differential equation

$$\nabla_{\dot{\gamma}(t)}\,\dot{\gamma}(t) = \mathbf{0}. \tag{2.20}$$

The geodesics on $\mathcal{H}_p^{++}$ according to the Levi-Civita connection of Theorem 2.3 are given in Theorem 2.4 along with the proof.

Theorem 2.4 (Geodesics) *The geodesic $\gamma : [0, 1] \to \mathcal{H}_p^{++}$ such that $\gamma(0) = \Sigma \in \mathcal{H}_p^{++}$ and $\dot{\gamma}(0) = \boldsymbol{\xi} \in \mathcal{H}_p$ is defined as*

$$\gamma(t) \;=\; \Sigma\exp(t\Sigma^{-1}\boldsymbol{\xi}) \;=\; \exp(t\boldsymbol{\xi}\Sigma^{-1})\Sigma \;=\; \Sigma^{1/2}\exp(t\Sigma^{-1/2}\boldsymbol{\xi}\Sigma^{-1/2})\Sigma^{1/2},$$

where $\exp(\cdot)$ *denotes the matrix exponential. Equivalently, one can define the geodesic* $\gamma : [0, 1] \to \mathcal{H}_p^{++}$ *such that* $\gamma(0) = \Sigma \in \mathcal{H}_p^{++}$ *and* $\gamma(1) = \hat{\Sigma} \in \mathcal{H}_p^{++}$ *by*

$$\gamma(t) = \Sigma^{1/2}\left(\Sigma^{-1/2}\hat{\Sigma}\Sigma^{-1/2}\right)^t\Sigma^{1/2},$$

where $(\cdot)^t = \exp(t\log(\cdot))$, $\log(\cdot)$ *denoting the matrix logarithm.*

Proof We only provide the proof for $\gamma(0) = \Sigma$ and $\dot{\gamma}(0) = \boldsymbol{\xi}$. The result for $\gamma(1) = \hat{\Sigma}$ is obtained by choosing

$$\boldsymbol{\xi} \;=\; \Sigma \log(\Sigma^{-1}\hat{\Sigma}) \;=\; \log(\hat{\Sigma}\Sigma^{-1})\Sigma \;=\; \Sigma^{1/2}\log(\Sigma^{-1/2}\hat{\Sigma}\Sigma^{-1/2})\Sigma^{1/2}.$$

Notice that the equality between the three versions of $\gamma(t)$ given in the theorem above and of $\boldsymbol{\xi}$ given here solely rely on the fact that $\boldsymbol{A}\exp(\boldsymbol{B})\boldsymbol{A}^{-1} = \exp(\boldsymbol{A}\boldsymbol{B}\boldsymbol{A}^{-1})$ and $\boldsymbol{A}\log(\boldsymbol{B})\boldsymbol{A}^{-1} = \log(\boldsymbol{A}\boldsymbol{B}\boldsymbol{A}^{-1})$.

The differential equation (2.20) for the Levi-Civita connection defined in Theorem 2.3 is

$$\ddot{\gamma}(t) - \dot{\gamma}(t)\gamma(t)^{-1}\dot{\gamma}(t) = \boldsymbol{0}.$$

Recall that $\frac{\mathrm{d}}{\mathrm{d}t}\exp(t\boldsymbol{A}) = \boldsymbol{A}\exp(t\boldsymbol{A})$. Thus, with $\gamma(t) = \Sigma\exp(t\Sigma^{-1}\boldsymbol{\xi})$, we have $\dot{\gamma}(t) = \boldsymbol{\xi}\exp(t\Sigma^{-1}\boldsymbol{\xi})$ and $\ddot{\gamma}(t) = \boldsymbol{\xi}\Sigma^{-1}\boldsymbol{\xi}\exp(t\Sigma^{-1}\boldsymbol{\xi})$. From there, we easily obtain $\dot{\gamma}(t)\gamma(t)^{-1} = \boldsymbol{\xi}\Sigma^{-1}$. Simple computations show that $\gamma(t)$ satisfies the differential equation above, which is enough to conclude. □

Remark 2.5 Since the Levi-Civita connection does not depend on α and β, neither does geodesics. Hence, the Fisher–Rao geometries of C-CES models share the same geodesics whatever the underlying distribution.

Geodesics allow to define the Riemannian exponential mapping. By definition, for all $\Sigma \in \mathcal{H}_p^{++}$, this is the mapping from $T_\Sigma\mathcal{H}_p^{++} \simeq \mathcal{H}_p$ onto $\mathcal{H}_p^{++}$ such that, for all $\boldsymbol{\xi} \in \mathcal{H}_p$, $\exp_\Sigma(\boldsymbol{\xi}) = \gamma(1)$, where γ is the geodesic such that $\gamma(0) = \Sigma$ and $\dot{\gamma}(t) = \boldsymbol{\xi}$. Thus, for all $\Sigma \in \mathcal{H}_p^{++}$ and $\boldsymbol{\xi} \in \mathcal{H}_p$, we have

$$\exp_\Sigma(\boldsymbol{\xi}) \;=\; \Sigma\exp(\Sigma^{-1}\boldsymbol{\xi}) \;=\; \exp(\boldsymbol{\xi}\Sigma^{-1})\Sigma \;=\; \Sigma^{1/2}\exp(\Sigma^{-1/2}\boldsymbol{\xi}\Sigma^{-1/2})\Sigma^{1/2}. \tag{2.21}$$

From there, we can define the Riemannian logarithm mapping, which is the inverse of the Riemannian exponential mapping. Given $\Sigma \in \mathcal{H}_p^{++}$, it is the mapping from $\mathcal{H}_p^{++}$ onto $T_\Sigma\mathcal{H}_p^{++} \simeq \mathcal{H}_p$ such that, for $\hat{\Sigma} \in \mathcal{H}_p^{++}$, $\log_\Sigma(\hat{\Sigma})$ is the solution to equation $\exp_\Sigma(\log_\Sigma(\hat{\Sigma})) = \hat{\Sigma}$. In our case, we have

$$\log_\Sigma(\hat{\Sigma}) \;=\; \Sigma\log(\Sigma^{-1}\hat{\Sigma}) \;=\; \log(\hat{\Sigma}\Sigma^{-1})\Sigma \;=\; \Sigma^{1/2}\log(\Sigma^{-1/2}\hat{\Sigma}\Sigma^{-1/2})\Sigma^{1/2}. \tag{2.22}$$

Illustrations of Riemannian exponential and logarithm mappings are given in Fig. 2.6.

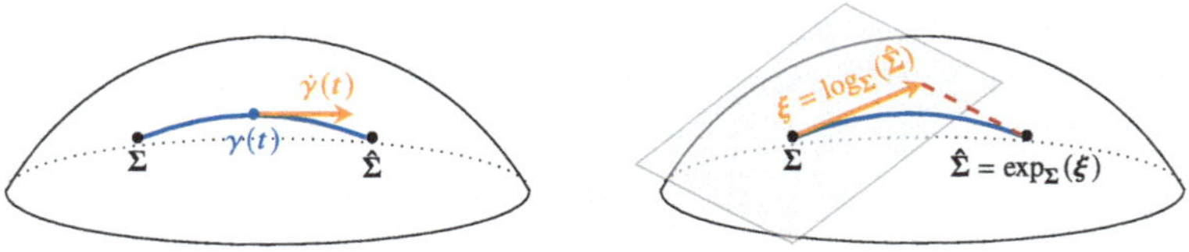

Fig. 2.6 Illustration of geodesics (left), Riemannian exponential, and logarithm mappings (right). The Riemannian distance $\delta(\Sigma, \hat{\Sigma})$ is the length of the geodesic joining Σ and $\hat{\Sigma}$

The last object from Riemannian geometry presented in this chapter is the Riemannian distance. The distance between two points corresponds to the length of the geodesic joining them. Formally, it is defined as

$$\delta(\Sigma, \hat{\Sigma}) = \int_0^1 \langle \dot{\gamma}(t), \dot{\gamma}(t) \rangle_{\gamma(t)}^{1/2} \mathrm{D}, \tag{2.23}$$

where γ is the geodesic such that $\gamma(0) = \Sigma$ and $\gamma(1) = \hat{\Sigma}$. The Riemannian distance on $\mathcal{H}_p^{++}$ associated to the metric of Theorem 2.2 is given in Theorem 2.5 along with the proof. It was derived in [28].

Theorem 2.5 (Fisher–Rao distance of C-CES distributions) *The square of the Fisher distance of C-CES distributions over* $\mathcal{H}_p^{++}$ *is defined, for all* Σ *and* $\hat{\Sigma} \in \mathcal{H}_p^{++}$, *by*

$$\delta^2(\Sigma, \hat{\Sigma}) = \alpha \| \log(\Sigma^{-1}\hat{\Sigma}) \|_F^2 + \beta (\log \det(\Sigma^{-1}\hat{\Sigma}))^2.$$

Proof From the proof of Theorem 2.4, $\dot{\gamma}(t)\gamma(t)^{-1} = \xi\Sigma^{-1}$ for all $t \in [0, 1]$. Thus, we can deduce that $\langle \dot{\gamma}(t), \dot{\gamma}(t) \rangle_{\gamma(t)} = \langle \dot{\gamma}(0), \dot{\gamma}(0) \rangle_{\gamma(0)}$ for all $t \in [0, 1]$. Therefore, $\delta^2(\Sigma, \hat{\Sigma}) = \langle \dot{\gamma}(0), \dot{\gamma}(0) \rangle_{\gamma(0)}$, with $\gamma(0) = \Sigma$ and $\dot{\gamma}(0) = \Sigma \log(\Sigma^{-1}\hat{\Sigma})$. It follows that

$$\delta^2(\Sigma, \hat{\Sigma}) = \alpha \operatorname{Tr}((\log(\Sigma^{-1}\hat{\Sigma}))^2) + \beta (\operatorname{Tr}(\log(\Sigma^{-1}\hat{\Sigma})))^2.$$

To conclude, it is enough to recall that $\operatorname{Tr}(\log(\boldsymbol{A})) = \log \det(\boldsymbol{A})$. □

Remark 2.6 We previously noticed that the Levi-Civita connection and geodesics do not depend on the coefficients α and β of the metric. However, since the Riemannian distance integrates the metric along the geodesics, it does well depend on these factors. This means that the Fisher–Rao distance (Riemannian distance according to the Fisher in formation geometry) actually depends on the underlying C-CES distribution.

2.3 Covariance Matrix Estimation with Riemannian Optimization

The estimation of the covariance matrix of a set of observations is a ubiquitous problem in signal processing and machine learning. Among many applications involving this quantity, we can mention: adaptive filtering and detection, metric learning in classification, data analysis (e.g., graph learning), and dimension reduction. This section discusses covariance matrix estimation within the class of C-CES, and illustrates how the concepts related to Fisher–Rao information geometry can be leveraged in this context. First, Sect. 2.3.1 provides some reminders on covariance matrix estimation in the C-CES framework (cf. Sect. 6 of the chapter "Background on Real and Complex

Elliptically Symmetric Distributions" for more details). Second, Sect. 2.3.2 presents an introduction to Riemannian optimization, where maximum-likelihood estimation of C-CES models is used as a driving example. Finally, Sect. 2.3.3 shortly presents how this framework can be leveraged to more general regularized covariance matrix estimation problems and points to references on the matter.

2.3.1 Reminders on Covariance Matrix Estimation within CES

Given a n-sample $\{\mathbf{x}_i\}_{i=1}^n$ assumed to be iid from $\mathbf{x} \sim$ C-CES$(\mathbf{0}, \Sigma, g)$, with unknown covariance matrix Σ, we consider inferring this matrix. The most common approach to tackle this problem consists in maximizing the log-likelihood function in (2.4). The maximum-likelihood estimator is thus obtained as a solution to the optimization problem

$$\underset{\Sigma \in \mathcal{H}_p^{++}}{\text{minimize}} \quad L(\Sigma), \tag{2.24}$$

where L denotes in short the negative log-likelihood of the sample set $\{\mathbf{x}_i\}_{i=1}^n$, i.e.:

$$L(\Sigma) = -\mathcal{L}_g\left(\{\mathbf{x}_i\}_{i=1}^n | \Sigma\right), \tag{2.25}$$

with $\mathcal{L}_g$ defined in (2.4). The solution of (2.24) yields the MLE in the form of a fixed-point equation

$$\hat{\Sigma} = \frac{1}{n}\sum_{i=1}^{n} \psi\left(\mathbf{x}_i^H \hat{\Sigma}^{-1} \mathbf{x}_i\right) \mathbf{x}_i \mathbf{x}_i^H \stackrel{\text{def}}{=} \mathcal{T}_\psi\left(\hat{\Sigma}\right), \tag{2.26}$$

where $\psi(t) = -g'(t)/g(t)$. This solution is most commonly evaluated thanks to a fixed-point algorithm

$$\Sigma_{(k+1)} = \mathcal{T}_\psi\left(\Sigma_{(k)}\right) = \frac{1}{n}\sum_{i=1}^{n} \psi(\mathbf{x}_i^H \Sigma_{(k)}^{-1} \mathbf{x}_i)\, \mathbf{x}_i \mathbf{x}_i^H. \tag{2.27}$$

The existence and uniqueness of the fixed-point solution (2.26), as well as the convergence of the fixed-point algorithm (2.27) is subject to conditions on the function ψ (resp. the density generator g) and the sample set $\{\mathbf{x}_i\}_{i=1}^n$, e.g., obtained in [73, Theorems 6 and 7]. A notable condition in the absolutely continuous case is that the sample size is required to be larger than the dimension, i.e., $n > p$.

Remark 2.7 In practice, the true density generator g may not be known or accurately specified. In the robust estimation theory, an M-estimator of the scatter matrix [60, 95] refers to an estimator built from (2.26) to (2.27) using a function $\psi(t)$ that is not necessarily linked to the density generator g (cf. Sect. 6.3 of the chapter "Background on Real and Complex Elliptically Symmetric Distributions"). In this chapter, we focus on the example of the MLE, but the tools that will be presented apply to any generic cost function L.

2.3.2 Computing MLEs with Riemannian Optimization

Riemannian optimization [1, 26] is a general framework to solve optimization problems on manifolds. This extends to Riemannian manifolds classical Euclidean optimization methods such as steepest gradient descent, conjugate gradient, Broyden–Fletcher–Goldfarb–Shanno (BFGS) algorithm, Newton method, trust region, etc. This section introduces Riemannian optimization on $\mathcal{H}_p^{++}$ as a framework to solve (2.24) that can leverage tools from the Fisher–Rao information geometry. At the end of this section, we will see that this framework actually yields the fixed point algorithm (2.27) as a special case (specifically, a Riemannian steepest gradient descent with a specific choice of metric, retraction, and step-size).

We consider an optimization problem of the from (2.24) that has no obvious closed-form solution on $\mathcal{H}_p^{++}$. In order to evaluate this solution, we resort to iterative methods, i.e., methods that yield a sequence of iterates $\{\Sigma_{(k)}\}$ in $\mathcal{H}_p^{++}$ from a starting point $\Sigma_{(0)} \in \mathcal{H}_p^{++}$. This sequence is constructed so that it eventually converges to a critical point of the objective in (2.24). When the variable is constrained to lie in the manifold $\mathcal{H}_p^{++}$, a generic first-order Riemannian optimization method operates as follows:

1. At iterate $\Sigma_{(k)} \in \mathcal{H}_p^{++}$, a descent direction in the tangent space, denoted $\boldsymbol{\xi}_{(k)} \in T_{\Sigma_{(k)}}\mathcal{H}_p^{++} \simeq \mathcal{H}_p$, is computed by leveraging the Riemannian gradient.
2. The direction descent $\boldsymbol{\xi}_{(k)}$ is used to obtain the next iterate $\Sigma_{(k+1)}$ on $\mathcal{H}_p^{++}$. This is achieved through a retraction on $\mathcal{H}_p^{++}$, which is an operator that maps tangent vectors back onto the manifold.

An illustration of such an optimization process is presented in Fig. 2.7, while the design of these two steps is discussed to solve (2.24) on $\mathcal{H}_p^{++}$ in the following.

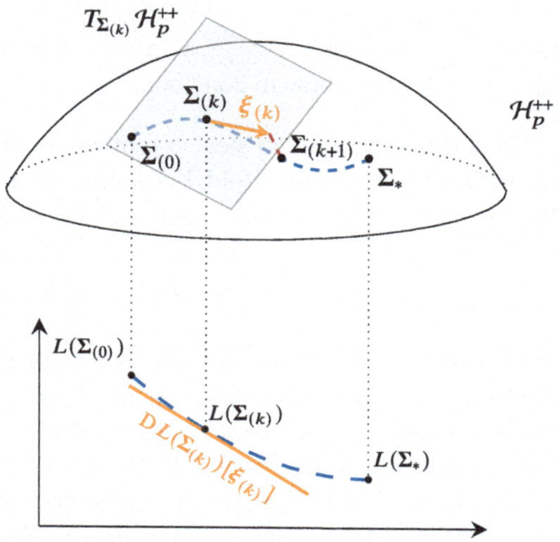

Fig. 2.7 Illustration of Riemannian optimization. Given some initialization $\Sigma_{(0)}$, the goal is to reach the minimum Σ_*. At $\Sigma_{(k)}$, the descent direction $\boldsymbol{\xi}_{(k)}$ is such that it induces a decrease in L, i.e., $\mathrm{D}\,L(\Sigma_{(k)})[\boldsymbol{\xi}_{(k)}] < 0$ (slope of the orange line)

For the first step, the steepest descent direction is given by the gradient, which is defined through the metric in the Riemannian setting. The Riemannian gradient of the negative log-likelihood L at $\Sigma \in \mathcal{H}_p^{++}$ according to the metric of Theorem 2.2 is the unique tangent vector $\operatorname{grad} L(\Sigma) \in T_\Sigma \mathcal{H}_p^{++} \simeq \mathcal{H}_p$ such that, for all $\boldsymbol{\xi} \in \mathcal{H}_p$, we have

$$\langle \operatorname{grad} L(\Sigma), \boldsymbol{\xi} \rangle_\Sigma = \mathrm{D}\, L(\Sigma)[\boldsymbol{\xi}]. \tag{2.28}$$

This Riemannian gradient is provided in Proposition 2.1.

Proposition 2.1 (Riemannian gradient of L) *The Riemannian gradient* $\operatorname{grad} L(\Sigma)$ *of the negative log-likelihood L defined in* (2.25) *at $\Sigma \in \mathcal{H}_p^{++}$ according to metric* (2.18) *is*

$$\operatorname{grad} L(\Sigma) = \left(\frac{n}{\alpha + p\beta} + \frac{\beta}{\alpha(\alpha + p\beta)} \sum_{i=1}^{n} \psi(\mathbf{x}_i^H \Sigma^{-1} \mathbf{x}_i)\, \mathbf{x}_i^H \Sigma^{-1} \mathbf{x}_i \right) \Sigma \\ - \frac{1}{\alpha} \sum_{i=1}^{n} \psi(\mathbf{x}_i^H \Sigma^{-1} \mathbf{x}_i)\, \mathbf{x}_i \mathbf{x}_i^H .$$

Proof From the beginning of the proof of Theorem 2.2, we get that the directional derivative of L at $\Sigma \in \mathcal{H}_p^{++}$ in direction $\boldsymbol{\xi} \in \mathcal{H}_p$ is

$$\begin{aligned} \mathrm{D}\, L(\Sigma)[\boldsymbol{\xi}] &= n \operatorname{Tr}(\Sigma^{-1}\boldsymbol{\xi}) + \textstyle\sum_{i=1}^{n} \frac{g'}{g}(\mathbf{x}_i^H \Sigma^{-1} \mathbf{x}_i) \operatorname{Tr}(\Sigma^{-1}\boldsymbol{\xi}\Sigma^{-1}\mathbf{x}_i\mathbf{x}_i^H) \\ &= \operatorname{Tr}\left(\Sigma^{-1}\left(n\Sigma - \textstyle\sum_{i=1}^{n} \psi(\mathbf{x}_i^H \Sigma^{-1} \mathbf{x}_i)\, \mathbf{x}_i\mathbf{x}_i^H\right)\Sigma^{-1}\boldsymbol{\xi}\right). \end{aligned}$$

Thus, for $\alpha = 1$ and $\beta = 0$, we immediately get the result by identification. To obtain the result in the general case, notice that, given $\boldsymbol{A}$ and $\boldsymbol{B}$ in $\mathcal{H}_p$, if we set $\tilde{\boldsymbol{A}} = \frac{1}{\alpha}\boldsymbol{A} - \frac{\beta}{\alpha(\alpha+p\beta)} \operatorname{Tr}(\boldsymbol{A})\boldsymbol{I}_p$, then we have $\operatorname{Tr}(\boldsymbol{AB}) = \alpha \operatorname{Tr}(\tilde{\boldsymbol{A}}\boldsymbol{B}) + \beta \operatorname{Tr}(\tilde{\boldsymbol{A}}) \operatorname{Tr}(\boldsymbol{B})$. Taking $\boldsymbol{A} = \Sigma^{-1/2}(n\Sigma - \sum_{i=1}^{n} \psi(\mathbf{x}_i^H \Sigma^{-1} \mathbf{x}_i)\mathbf{x}_i\mathbf{x}_i^H)\Sigma^{-1/2}$ and $\boldsymbol{B} = \Sigma^{-1/2}\boldsymbol{\xi}\Sigma^{-1/2}$, and basic calculations allow to conclude. □

To perform the second step, it remains to define a retraction that maps tangent vectors back onto the manifold. Formally, given $\Sigma \in \mathcal{H}_p^{++}$, a retraction is a mapping $R_\Sigma : T_\Sigma \mathcal{H}_p^{++} \simeq \mathcal{H}_p \to \mathcal{H}_p^{++}$ such that, for all $\boldsymbol{\xi} \in \mathcal{H}_p$,

$$R_\Sigma(\boldsymbol{\xi}) = \Sigma + \boldsymbol{\xi} + o(\|\boldsymbol{\xi}\|). \tag{2.29}$$

From a geometric point of view, the Riemannian exponential mapping provides the ideal retraction for a manifold equipped with a Riemannian metric (in the sense that it is the most reflective of the considered geometry). In our case, it is defined in (2.21) and illustrated in Fig. 2.6. However, this retraction involves computing the matrix exponential of some Hermitian matrix, which can be computationally costly and/or numerically unstable, as the exponential tends quickly to infinity or zero. From a practical point of view, it might thus be more advantageous to employ alternate retractions. Notice that (2.29) means that a proper retraction is (at least) a first-order approximation of the Riemannian exponential mapping. Since $\mathcal{H}_p^{++}$ is open in $\mathcal{H}_p$, a proper first-order approximation is simply obtained as

$$R_{\Sigma}^{(1)}(\boldsymbol{\xi}) = \Sigma + \boldsymbol{\xi}. \tag{2.30}$$

The main limitation of $R^{(1)}$ is that, given $\Sigma \in \mathcal{H}_p^{++}$, there are many $\boldsymbol{\xi} \in \mathcal{H}_p$ such that $R_{\Sigma}^{(1)}(\boldsymbol{\xi}) \notin \mathcal{H}_p^{++}$. This means that the iterative algorithms that employ this retraction are not guaranteed to be numerically stable. To overcome this issue, Proposition 2.2 provides a retraction that is a second-order approximation of the Riemannian exponential (2.21) (initially proposed in [51]), that does not suffer the same limitation as $R^{(1)}$.

Proposition 2.2 (Second-order retraction) *The retraction $R^{(2)}$ such that, for all $\Sigma \in \mathcal{H}_p^{++}$ and $\boldsymbol{\xi} \in \mathcal{H}_p$,*

$$R_{\Sigma}^{(2)}(\boldsymbol{\xi}) = \Sigma + \boldsymbol{\xi} + \frac{1}{2}\boldsymbol{\xi}\Sigma^{-1}\boldsymbol{\xi}$$

is a second-order approximation of the Riemannian exponential mapping (2.21). *Furthermore, for all $\Sigma \in \mathcal{H}_p^{++}$ and $\boldsymbol{\xi} \in \mathcal{H}_p$, $R_{\Sigma}^{(2)}(\boldsymbol{\xi})$ belongs to $\mathcal{H}_p^{++}$.*

Proof Recall that the matrix exponential of $\boldsymbol{A}$ is $\exp(\boldsymbol{A}) = \sum_{k=0}^{\infty} \frac{\boldsymbol{A}^k}{k!}$. Hence, the second-order approximation is $\exp(\boldsymbol{A}) = \boldsymbol{I}_p + \boldsymbol{A} + \frac{1}{2}\boldsymbol{A}^2 + o(\|\boldsymbol{A}\|^2)$. Applying this to (2.21), we obtain $\exp_{\Sigma}(\boldsymbol{\xi}) = \Sigma(\boldsymbol{I}_p + \Sigma^{-1}\boldsymbol{\xi} + \frac{1}{2}\Sigma^{-1}\boldsymbol{\xi}\Sigma^{-1}\boldsymbol{\xi}) + o(\|\boldsymbol{\xi}\|^2)$. Basic calculations yield the result. Moreover, it is obviously a proper retraction. It remains to show that we always get a matrix in $\mathcal{H}_p^{++}$. To do so, notice that

$$R_{\Sigma}^{(2)}(\boldsymbol{\xi}) = \Sigma^{1/2}(\boldsymbol{I}_p + \Sigma^{-1/2}\boldsymbol{\xi}\Sigma^{-1/2} + \frac{1}{2}(\Sigma^{-1/2}\boldsymbol{\xi}\Sigma^{-1/2})^2)\Sigma^{1/2}.$$

Let the eigenvalue decomposition $\Sigma^{-1/2}\boldsymbol{\xi}\Sigma^{-1/2} = \boldsymbol{U}\boldsymbol{\Lambda}\boldsymbol{U}^H$. Then

$$R_{\Sigma}^{(2)}(\boldsymbol{\xi}) = \Sigma^{1/2}\boldsymbol{U}(\boldsymbol{I}_p + \boldsymbol{\Lambda} + \frac{1}{2}\boldsymbol{\Lambda}^2)\boldsymbol{U}^H\Sigma^{1/2}.$$

The result follows from the fact that the second-order polynomial $\lambda \mapsto 1 + \lambda + \frac{1}{2}\lambda^2$ is strictly positive for all values of λ. □

We now have everything needed to define an iterative algorithm that solves the MLE optimization problem (2.24). Given the retraction R, we can, for instance, define the Riemannian gradient descent that yields the sequence of iterates

$$\Sigma_{(k+1)} = R_{\Sigma_{(k)}}(-\lambda_k \operatorname{grad} L(\Sigma_{(k)})), \tag{2.31}$$

where λ_k is the step size, which can be set by the user or computed through a line search; see e.g., [1, 26].

Our final point in this section is to show that the fixed point algorithm (2.27) is, in fact, a particular case of (2.31). Indeed, if we choose $\alpha = 1$ and $\beta = 0$, the Riemannian gradient of Proposition 2.1 is

$$\operatorname{grad} L(\Sigma) = n\Sigma - \sum_{i=1}^{n} \psi(\mathbf{x}_i^H \Sigma^{-1} \mathbf{x}_i)\, \mathbf{x}_i \mathbf{x}_i^H .$$

Algorithm (2.27) is then obtained from (2.31) by choosing the first-order retraction (2.30) and constant step size $\lambda_k = \frac{1}{n}$. Notice that in this particular case, the choice of the first-order retraction (2.30) is a valid choice because the particular structure of the gradient ensures that all iterates remain in $\mathcal{H}_p^{++}$. Though alternate choices of α and β in the metric, step size, and retraction could improve the convergence speed in some cases, this fixed point is generally a good all-purpose candidate to compute MLEs as in (2.26). However, having recast it from the prism of Riemannian geometry opens many perspectives, which are discussed in the next section.

2.3.3 *Beyond MLE and Fixed-Point Algorithms*

The MLEs (and M-estimators) are known for their good asymptotic performance in terms of estimation accuracy [37–39, 70, 100]. Still, they suffer from two limitations: (i) they do not exist when the sample set is lower than the dimension ($n < p$); (ii) they can be inaccurate when $n \simeq p$, as they do not leverage any bias-variance trade-off improvement. These limitations motivated the development of generalized estimation procedures by expressing new estimators as solutions to penalized optimization problems of the form:

$$\underset{\Sigma \in \mathcal{H}_p^{++}}{\text{minimize}} \quad L(\Sigma) + \lambda h(\Sigma), \tag{2.32}$$

where L is the negative log-likelihood as in (2.25), $\lambda \in \mathbb{R}^+$ is a regularization parameter, and h is a penalty function that promotes some form of regularization. Among many options considered in the literature for h, we can mention shrinkage to a target matrix [72, 74, 86], shrinkage of the eigenvalues [29, 97], promoting a sparse graphical structure [45, 101], or pooling from groups of observations [35, 71]. For appropriate choices of regularization penalty and parameters, the regularized estimators, as formulated in (2.32), can overcome the aforementioned issues of their non-regularized counterparts. In this scope, the Riemannian geometry provides useful tools to address and study (2.32), which is discussed next.

2.3.3.1 Riemannian Options for Computing Solutions of (2.32)

The optimization problems expressed in (2.32) generally do not exhibit closed-form or fixed-point solutions and, thus, require the use of iterative algorithms to be evaluated. In this setup, the Riemannian optimization framework is a good candidate in order to ensure that the variable remains in $\mathcal{H}_p^{++}$ along the iterations. Beyond the introduction of the Riemannian gradient descent presented in Sect. 2.3.2, this flexible framework extends to many other algorithms:

- Conjugate gradient, or BFGS-type algorithms, require the notion of Riemannian vector transport operator [26, Sect. 10.3], which allows to transport tangent vectors between tangent spaces at different points.
- Second-order methods, such as trust region or Newton methods require the definition of the Riemannian Hessian [26, Sect. 5.5].
- For large dimensional datasets, stochastic optimization methods can also be extended to the Riemannian setting [19, 23, 99].

A last remark is that in these algorithms, the metric is left as a choice that conditions the gradient and possibly the retraction. There are various options for $\mathcal{H}_p^{++}$ (cf. Sect. 2.2.1), with their respective pros and cons. It is still noticed that the gradient obtained from the Fisher information metric, also referred to as the natural gradient [2], is generally experienced to yield a faster convergence when dealing with a cost function related to the statistical model of the data (cf. examples in [36, 43]).

2.3.3.2 Geodesic Convexity on $\mathcal{H}_p^{++}$

The classical results on the existence and uniqueness of the MLEs [73, Theorems 6 and 7] do not directly extend to the formulation in (2.32), so one might inquire about the optimally of the solution obtained by reaching a local minimum of this problem. In this scope, the Riemannian perspective offers some answers by generalizing the property of convexity. First, we recall that the geometry induced by the Fisher information metric (2.18) yields geodesic curves $\gamma(t)$ as defined in Theorem 2.4 between any two points $\Sigma_0, \Sigma_1 \in \mathcal{H}_p^{++}$. A function f is then said to be geodesically convex (g-convex) on $\mathcal{H}_p^{++}$ if $\forall \Sigma_0, \Sigma_1 \in \mathcal{H}_p^{++}$, it satisfies the inequality

$$f(\gamma(t)) \leq (1-t) f(\Sigma_0) + t f(\Sigma_1), \ \forall t \in [0, 1]. \tag{2.33}$$

If the above inequality is strict, the function is then said to be strictly g-convex. The g-convexity enjoys properties similar to those of the convexity in the standard Euclidean case, in particular:

Theorem 2.6 (Global minimizer of g-convex functions on $\mathcal{H}_p^{++}$) *Let $f : \mathcal{H}_p^{++} \to \mathbb{R}$ be g-convex as defined in* (2.33), *then any local minimum of f over $\mathcal{H}_p^{++}$ is a global minimum. Furthermore, if f is strictly g-convex, this global minimum is unique.*

This property offers an alternate proof for the uniqueness of MLEs as in (2.24) [72], and had practical impacts for the design of regularized covariance matrix estimators as in (2.32): many examples of penalty functions (with various regularization effects) can be found in the overviews in [40, 98], and the Refs. [9, 40, 72, 96–98].

2.4 Intrinsic Cramér–Rao Bound for Covariance Matrix Estimation

The Cramér–Rao inequality is a staple tool in statistics that characterizes the optimal mean-squared error an unbiased estimator can reach given a model and setup [55]. This tool can either be used to validate estimation procedures or to design systems so that a certain level of accuracy is guaranteed to be theoretically reachable. While the Euclidean formulation of this inequality was briefly introduced in Sect. 2.1.1, the so-called *intrinsic* Cramér–Rao bounds extend it to parameters living in a manifold, and for any chosen Riemannian metric. This perspective is especially interesting as: (i) some metrics can be more meaningful to assess the estimation performance in a given application; (ii) a suitable Riemannian geometry (as opposed to the Euclidean one) can reveal hidden properties that make the bound more informative (such as curvature terms, intrinsic biases, etc.). First, Sect. 2.4.1 introduces the background on intrinsic Cramér–Rao bound from [85], where the C-CES model is used as a driving example. We also refer the reader to [25, Chap. 6] and the Ref. [14], for more details on the topic. Then, Cramér–Rao bounds are derived for various distances in the context of covariance matrix estimation within C-CES distributions [28] in Sect. 2.4.2.

2.4.1 *Introduction to Intrinsic Cramér–Rao Bounds*

This subsection will present tools that can be applied to any chosen Riemannian geometry on $\mathcal{H}_p^{++}$. The needed objects are the Riemannian metric, logarithm mapping, and square of the distance, which are denoted $\prec \cdot, \cdot \succ_{\cdot}$, $\mathfrak{log}_{\cdot}(\cdot)$ and $\mathfrak{d}^2(\cdot, \cdot)$, respectively. As in Sect. 2.3, we consider the problem of estimating the matrix Σ from a given n-sample $\{\mathbf{x}_i\}_{i=1}^n$ assumed to be iid from $\mathbf{x} \sim$ C-CES$(\mathbf{0}, \Sigma, g)$. We denote $\hat{\Sigma}$ an estimator of this parameter; e.g., the MLE presented in Sect. 2.3. We then consider the evaluation of the performance of such estimator $\hat{\Sigma}$. To do so, we exploit the chosen Riemannian metric $\prec \cdot, \cdot \succ_{\Sigma}$. Such metric can, for example, be the Fisher information one (2.18), or one of the many other options from the literature [93]. The performance criterion is the resulting square of the Riemannian distance, i.e., the error is measured through $\mathfrak{d}^2(\Sigma, \hat{\Sigma})$. The intrinsic Cramér–Rao theory from [85] then allows us to obtain a lower bound on the expectation of this error for any unbiased estimator $\hat{\Sigma}$. Eventually, this retrieves the well-known inequality "

$\mathbf{C} \succeq \mathbf{F}^{-1}$", with $\mathbf{C} \in \mathbb{R}^{p^2 \times p^2}$ being the covariance matrix of the estimation error and $\mathbf{F} \in \mathbb{R}^{p^2 \times p^2}$ being the Fisher information matrix, where $p^2 = \dim(\mathcal{H}_p^{++})$. However, these parameters have different definitions due to the specific nature of the considered objects. The point of this section is to briefly present the key ingredients to obtain such inequality and the corresponding main theorem.

First, we need to generalize the notion of estimation error vector $\boldsymbol{\varepsilon} \in \mathbb{R}^{p^2}$ to the Riemannian context. Notice that, in the Euclidean case, such vector is generally constructed by vectorizing the entry-wise subtraction of the covariance matrix Σ to its estimate $\hat{\Sigma}$, i.e., $\boldsymbol{\varepsilon}^{\mathcal{E}} = \text{vech}(\hat{\Sigma} - \Sigma)$, where $\text{vech}(\cdot)$ denotes the half-vectorization operator. As it happens, from a Riemannian geometry point of view, $\hat{\Sigma} - \Sigma$ corresponds to the Euclidean logarithm mapping at Σ. Therefore, the Riemannian logarithm $\mathfrak{log}_{\Sigma}(\hat{\Sigma})$ provides a natural way to extend the error to any geometry. It is indeed an element of the tangent space $T_{\Sigma}\mathcal{H}_p^{++}$ of Σ that "points toward" $\hat{\Sigma}$, and whose norm corresponds to the Riemannian distance. It remains to actually get an error vector $\boldsymbol{\varepsilon} \in \mathbb{R}^{p^2}$ from $\mathfrak{log}_{\Sigma}(\hat{\Sigma})$. To do so, we leverage a basis $\{\boldsymbol{\xi}_q\}_{q=1}^{p^2}$ of $T_{\Sigma}\mathcal{H}_p^{++} \simeq \mathcal{H}_p$ that is orthonormal with respect to the chosen metric $\prec \cdot, \cdot \succ_{\Sigma}$. In practice, such a basis can be obtained either analytically from mathematical calculations or numerically, thanks to the Gram–Schmidt orthonormalization process. This basis yields the decomposition

$$\mathfrak{log}_{\Sigma}(\hat{\Sigma}) = \sum_{q=1}^{p^2} \varepsilon_q \boldsymbol{\xi}_q, \tag{2.34}$$

and we denote $\boldsymbol{\varepsilon} = [\varepsilon_1, \cdots, \varepsilon_{p^2}] \in \mathbb{R}^{p^2}$ the corresponding coordinates error vector, obtained as

$$\varepsilon_q = \prec \mathfrak{log}_{\Sigma}(\hat{\Sigma}), \boldsymbol{\xi}_q \succ_{\Sigma}. \tag{2.35}$$

Moreover, the norm of this vector corresponds to the Riemannian distance between Σ and $\hat{\Sigma}$, i.e.,

$$\mathfrak{d}^2(\Sigma, \hat{\Sigma}) = \prec \mathfrak{log}_{\Sigma}(\hat{\Sigma}), \mathfrak{log}_{\Sigma}(\hat{\Sigma}) \succ_{\Sigma} = \|\boldsymbol{\varepsilon}\|_2^2, \tag{2.36}$$

which will be instrumental in the next derivations. The basis $\{\boldsymbol{\xi}_q\}_{q=1}^{p^2}$ also yields a Fisher information matrix $\mathbf{F}$, with entries

$$\mathbf{F}_{q\ell} = \langle \boldsymbol{\xi}_q, \boldsymbol{\xi}_\ell \rangle_{\Sigma}^{\text{FIM}}. \tag{2.37}$$

The matrix $\mathbf{F}$ represents the Fisher information metric of Theorem 2.2 according to this system of coordinates. Then, from [85, Corrolary 2], we obtain Theorem 2.7.

Theorem 2.7 (Intrinsic Cramér–Rao bound) *Let $\Sigma \in \mathcal{H}_p^{++}$. Let $\{\mathbf{x}_i\}_{i=1}^n$ a iid n-sample from $\mathbf{x} \sim$ C-CES$(\mathbf{0}, \Sigma, g)$. Let $\hat{\Sigma}$ an unbiased estimator of Σ with corresponding error vector $\boldsymbol{\varepsilon}$ defined in* (2.35)*. Then*

$$\mathbf{C} = E\left[\boldsymbol{\varepsilon}\boldsymbol{\varepsilon}^T\right] \succeq \mathbf{F}^{-1} + \text{curvature terms},$$

where **F** *is the Fisher information matrix in* (2.37) *and the curvature terms—which are not detailed here—depend on the Riemannian curvature tensor corresponding to the chosen geometry and on* **F***; see [24, 25, 85] for further details.*

In practice, the curvature terms can usually be neglected in Theorem 2.7. Furthermore, taking the trace of the inequality yields the desired result, i.e.,

$$\mathrm{E}\left[\mathfrak{d}^2(\Sigma, \hat{\Sigma})\right] \geq \mathrm{Tr}(\mathbf{F}^{-1}). \tag{2.38}$$

It offers a bound that can be derived for any chosen Riemannian distance $\mathfrak{d}^2(\cdot, \cdot)$ (and corresponding metric $\prec \cdot, \cdot \succ_{.}$).

Remark 2.8 The inequality in Theorem 2.7 interestingly takes into account the curvature of the manifold, which, for $\mathcal{H}_p^{++}$, only depends on the chosen metric. In the Euclidean case, such curvature term is null, and we recover the standard Cramér–Rao inequality. We also notice that the theorem in [85] also incorporates intrinsic bias terms, which was excluded here for the sake of conciseness. This intrinsic bias (expectation of the Riemannian logarithm) depends on the estimator and the chosen metric and can reveal unexpected properties. A main example is that the MLE of the covariance matrix of the Gaussian model appears unbiased in the Euclidean setting, but is, in fact, biased when using the Fisher information metric [85]. Such analysis thus opens prospects for improved estimation from the intrinsic perspective.

2.4.2 Bounds for Various Matrix Distances in C-CES Distributions

This section presents the derivation of special cases of Theorem 2.7 when considering various usual metrics. Hence, it yields intrinsic Cramér–Rao bounds for the problem of covariance matrix estimation in C-CES distributions for the corresponding Riemannian distances. Since the Fisher information metric is already obtained in Theorem 2.2, the derivation boils down to the following steps:

(a) Selecting the performance metric $\prec \cdot, \cdot \succ_{\Sigma}$ and computing $\{\boldsymbol{\xi}_q\}_{q=1}^{p^2}$, a corresponding orthonormal basis of $T_{\Sigma}\mathcal{H}_p^{++}$.
(b) Computing the elements of the Fisher information matrix with this basis, according to (2.37).
(c) Inverting the Fisher information matrix, applying Theorem 2.7, then (2.38).

These operations are conducted in the following for the Euclidean metric, the so-called natural Riemannian metric (the affine invariant metric (2.18) with $\alpha = 1$ and $\beta = 0$), and the Fisher–Rao metric of the assumed model (i.e., the metric of Theorem 2.2: (2.18) with $\alpha = \alpha_g$ and $\beta = \beta_g$, where α_g and β_g are defined in (2.9)). In order for the chosen values of α and β to be clear, in this subsection, the metric (2.18) is denoted $\langle \cdot, \cdot \rangle_{\cdot}^{(\alpha,\beta)}$ and the distance of Theorem 2.5 is denoted $\delta^2_{(\alpha,\beta)}(\cdot, \cdot)$.

2.4.2.1 Euclidean Distance

We first recall the elementary tools of the Euclidean metric for $\mathcal{H}_p^{++}$:

$$\begin{aligned}
&\text{Metric:} && \langle \boldsymbol{\xi}, \boldsymbol{\eta} \rangle_{\Sigma}^{\mathcal{E}} = \mathfrak{Re}(\mathrm{Tr}(\boldsymbol{\xi}\boldsymbol{\eta})) \\
&\text{Logarithm:} && \log_{\Sigma}^{\mathcal{E}}(\hat{\Sigma}) = \hat{\Sigma} - \Sigma \\
&\text{Distance:} && \delta_{\mathcal{E}}^2(\Sigma, \hat{\Sigma}) = \|\hat{\Sigma} - \Sigma\|_2^2.
\end{aligned} \tag{2.39}$$

A basis of the tangent space $T_\Sigma \mathcal{H}_p^{++}$ that is orthonormal with respect to the metric in (2.39) can be obtained as follows:

1. For $1 \leq i \leq p$, $\boldsymbol{\xi}_{ii}^{\mathcal{E}}$ is a $p \times p$ symmetric matrix whose ith diagonal element is one, zeros elsewhere
2. For $1 \leq i < j \leq p$, $\boldsymbol{\xi}_{ij}^{\mathcal{E}}$ is a $p \times p$ symmetric matrix whose ijth and jith elements are both $1/\sqrt{2}$, zeros elsewhere.
3. For $1 \leq i < j \leq p$, $\bar{\boldsymbol{\xi}}_{ij}^{\mathcal{E}}$ is a $p \times p$ Hermitian matrix whose ijth and jith elements are $\sqrt{-1}/\sqrt{2}$ and $-\sqrt{-1}/\sqrt{2}$, respectively, zeros elsewhere.

To shorten notation, we simply denote this basis $\{\boldsymbol{\xi}_q^{\mathcal{E}}\}_{q=1}^{p^2}$, where the p^2 elements are ordered following items (1), (2), and (3). The squared Euclidean distance between an estimator $\hat{\Sigma}$ and the true value Σ also corresponds to the summed squared errors on the coordinates in this basis. We then have the following result:

Theorem 2.8 (Cramér–Rao bound on Euclidean distance) *Let $\hat{\Sigma}$ an unbiased estimator of Σ built from iid data $\{\mathbf{x}_i\}_{i=1}^n$ drawn from $\mathbf{x} \sim$ C-CES$(\mathbf{0}, \Sigma, g)$. The Euclidean distance between $\hat{\Sigma}$ and Σ is bounded in expectation as*

$$\mathbf{C}_{\mathcal{E}} = E\left[\delta_{\mathcal{E}}^2(\hat{\Sigma}, \Sigma)\right] \geq \mathrm{Tr}(\mathbf{F}_{\mathcal{E}}^{-1}),$$

where

$$[\mathbf{F}_{\mathcal{E}}]_{q\ell} = \mathfrak{Re}(n\alpha_g \,\mathrm{Tr}(\Sigma^{-1}\boldsymbol{\xi}_q^{\mathcal{E}}\Sigma^{-1}\boldsymbol{\xi}_\ell^{\mathcal{E}})) + n\beta_g \,\mathrm{Tr}(\Sigma^{-1}\boldsymbol{\xi}_q^{\mathcal{E}})\,\mathrm{Tr}(\Sigma^{-1}\boldsymbol{\xi}_\ell^{\mathcal{E}})),$$

with α_g and β_g defined in (2.9). *Theorem* 2.2.

Proof The result is a direct application of Theorem 2.7 and (2.38) using the basis $\{\boldsymbol{\xi}_j^{\mathcal{E}}\}_{j=1}^{p^2}$. □

Remark that this corresponds to the Euclidean Cramér–Rao bounds obtained for several distributions in [16, 42, 64, 75]. Also notice that we retrieve the same result as [85, Theorem 5] for the Gaussian distribution, i.e., $\alpha_g = 1$ and $\beta_g = 0$.

2.4.2.2 Natural Riemannian Distance

The natural Riemannian distance refers to the distance induced by the affine invariant metric (2.18) with the standard choice of coefficients $\alpha = 1$ and $\beta = 0$. The elementary tools for this metric for $\mathcal{H}_p^{++}$ are

$$\begin{array}{ll} \text{Metric:} & \langle \boldsymbol{\xi}, \boldsymbol{\eta} \rangle_{\Sigma}^{(1,0)} = \mathfrak{Re}(\mathrm{Tr}(\Sigma^{-1}\boldsymbol{\xi}\Sigma^{-1}\boldsymbol{\eta})) \\ \text{Logarithm:} & \log_{\Sigma}(\hat{\Sigma}) = \Sigma \log(\Sigma^{-1}\hat{\Sigma}) \\ \text{Distance:} & \delta^2_{(1,0)}(\Sigma, \hat{\Sigma}) = \| \log(\Sigma^{-1}\hat{\Sigma}) \|_2^2. \end{array} \tag{2.40}$$

Recall that the full description of this geometry is provided in Sect. 2.2. A basis of the tangent space $T_{\Sigma}\mathcal{H}_p^{++}$ that is orthonormal with respect to the metric in (2.40) can be obtained by coloring the canonical basis of the previous section as

$$\boldsymbol{\xi}_q^{(1,0)} = \Sigma^{1/2} \boldsymbol{\xi}_q^{\mathcal{E}} \Sigma^{1/2}. \tag{2.41}$$

The whole basis is denoted $\{\boldsymbol{\xi}_q^{(1,0)}\}_{q=1}^{p^2}$. We then have the following result:

Theorem 2.9 (Cramér–Rao bound on natural Riemannian distance) *Let $\hat{\Sigma}$ an unbiased estimator of $\Sigma \in \mathcal{H}_p^{++}$ built from iid data $\{\mathbf{x}_i\}_{i=1}^n$ drawn from $\mathbf{x} \sim$ C-CES$(\mathbf{0}, \Sigma, g)$. The Riemannian distance between $\hat{\Sigma}$ and Σ is bounded in expectation as*

$$E\left[\delta^2_{(1,0)}(\hat{\Sigma}, \Sigma)\right] \geq \frac{1}{n}\left(\frac{p^2-1}{\alpha_g} + \frac{1}{\alpha_g + p\beta_g}\right), \tag{2.42}$$

with α_g and β_g defined in (2.9).

Proof Plugging the basis $\{\boldsymbol{\xi}_q^{(1,0)}\}_{q=1}^{p^2}$ of $T_{\Sigma}\mathcal{H}_p^{++}$ defined in (2.41) in (2.37) yields

$$\left[\mathbf{F}_{(1,0)}\right]_{q\ell} = \langle \boldsymbol{\xi}_q^{(1,0)}, \boldsymbol{\xi}_\ell^{(1,0)} \rangle_{\Sigma}^{\mathrm{FIM}} = \mathfrak{Re}(n\alpha_g \,\mathrm{Tr}(\boldsymbol{\xi}_q^{\mathcal{E}}\boldsymbol{\xi}_\ell^{\mathcal{E}}) + n\beta_g \,\mathrm{Tr}(\boldsymbol{\xi}_q^{\mathcal{E}})\,\mathrm{Tr}(\boldsymbol{\xi}_\ell^{\mathcal{E}})).$$

Hence, from the relations

$$\mathrm{Tr}(\boldsymbol{\xi}_q^{\mathcal{E}}\boldsymbol{\xi}_\ell^{\mathcal{E}}) = \delta_{q\ell} \qquad \text{and} \qquad \mathrm{Tr}(\boldsymbol{\xi}_q^{\mathcal{E}})\,\mathrm{Tr}(\boldsymbol{\xi}_\ell^{\mathcal{E}}) = \begin{cases} 1 \text{ if } (q,\ell) \in [\![1,p]\!]^2 \\ 0 \text{ otherwise,} \end{cases}$$

we obtain the Fisher information matrix

$$\mathbf{F}_{(1,0)} = n\alpha_g \mathbf{I}_{p^2} + n\beta_g \begin{bmatrix} \mathbf{1}_{p\times p} & \mathbf{0}_{p\times p(p-1)} \\ \mathbf{0}_{p(p-1)\times p} & \mathbf{0}_{p(p-1)\times p(p-1)} \end{bmatrix},$$

which is expressed as $\mathbf{F}_{(1,0)} = n\alpha_g \mathbf{I}_{p^2} + np\beta_g \mathbf{v}\mathbf{v}^T$ with unitary vector $\mathbf{v} = \frac{1}{\sqrt{p}}$ $\left[\, \mathbf{1}_p \mid \mathbf{0}_{p(p-1)} \,\right]$, i.e., $\mathbf{v}^T\mathbf{v} = 1$. Hence, the inverse of the Fisher information matrix can be obtained by the Sherman–Morrison formula. In particular, its vector of eigenvalues can be directly identified as $\frac{1}{n}\left[(\alpha_g + p\beta_g)^{-1}, \alpha_g^{-1}, \ldots, \alpha_g^{-1}\right]$ and summed to obtain its trace. Theorem 2.7 and (2.38) are then applied to conclude. □

Remark 2.9 Contrarily to the Euclidean case of Theorem 2.8, the bound on the natural Riemannian distance in Theorem 2.9 does not depend on the parameter Σ. This is generally a desirable property, as it offers an interpretation grounded solely on intrinsic dimensions of the problem. Additionally, simulation examples in Sect. 2.4.3 show that assessing the error with such criterion (that is more in accordance with the nature of the parameter) can also reveal unexpected properties of the estimates.

2.4.2.3 Fisher–Rao Distance

The Fisher–Rao distance refers to the geodesic distance associated with the Fisher information metric (cf. Sect. 2.2.3). A subtlety is that we voluntarily omit the dependency on n of the Fisher information metric of Theorem 2.2, i.e., the bound will be obtained for using a generic metric in (2.18) with $\alpha = \alpha_g$ and $\beta = \beta_g$. This distinction has two main reasons: (i) it appears more logical to evaluate performance with a distance whose expression does not vary with the sample support of the scenario n; (ii) this allows us to also stress that, though identical, two metrics play a separate role in the derivations: one is inherent to the statistical model and the other is a choice made to measure estimation accuracy. Hence, the elementary tools on $\mathcal{H}_p^{++}$ are

$$\begin{array}{ll} \text{Metric:} & \langle \boldsymbol{\xi}, \boldsymbol{\eta} \rangle_{\Sigma}^{(\alpha_g,\beta_g)} = \mathfrak{Re}(\alpha_g \operatorname{Tr}(\Sigma^{-1}\boldsymbol{\xi}\Sigma^{-1}\boldsymbol{\eta}) + \beta_g \operatorname{Tr}(\Sigma^{-1}\boldsymbol{\xi}) \operatorname{Tr}(\Sigma^{-1}\boldsymbol{\eta})) \\ \text{Logarithm:} & \log_{\Sigma}(\hat{\Sigma}) \;=\; \Sigma \log(\Sigma^{-1}\hat{\Sigma}) \\ \text{Distance:} & \delta^2_{(\alpha_g,\beta_g)}(\Sigma, \hat{\Sigma}) = \alpha_g \| \log(\Sigma^{-1}\hat{\Sigma})\|_2^2 + \beta_g (\log\det(\Sigma^{-1}\hat{\Sigma}))^2. \end{array} \tag{2.43}$$

Recall that full details on this geometry are provided in Sect. 2.2. Contrary to previous geometries, since the considered metric is the Fisher information one, we do not actually need to compute a basis of the tangent space $T_\Sigma \mathcal{H}_p^{++}$ to obtain the bound. However, notice that if needed, such a basis can be obtained using the Gram–Schmidt orthogonalization process. In this case, the Cramér–Rao bound is:

Theorem 2.10 (Cramér–Rao bound on Fisher–Rao distance) *Let $\hat{\Sigma}$ be an unbiased estimator of $\Sigma \in \mathcal{H}_p^{++}$ built from iid data $\{\mathbf{x}_i\}_{i=1}^n$ drawn from $\mathbf{x} \sim$ C-CES$(\mathbf{0}, \Sigma, g)$. The Fisher–Rao distance between $\hat{\Sigma}$ and Σ is bounded in expectation as*

$$E\left[\delta^2_{(\alpha_g,\beta_g)}(\hat{\Sigma}, \Sigma)\right] \geq \frac{p^2}{n}.$$

Proof By definition, we have $\langle\cdot,\cdot\rangle_\Sigma^{\text{FIM}} = n\langle\cdot,\cdot\rangle_\Sigma^{(\alpha_g,\beta_g)}$. Hence, since the basis of interest, denoted $\{\boldsymbol{\xi}_q^{(\alpha_g,\beta_g)}\}_{q=1}^{p^2}$, is orthonormal according to $\langle\cdot,\cdot\rangle_\Sigma^{(\alpha_g,\beta_g)}$, it follows that

$\mathbf{F}_{(\alpha_g,\beta_g)} = n\mathbf{I}_{p^2}$. The trace of its inverse is therefore p^2/n and the proof is concluded by applying Theorem 2.7 and (2.38). □

We notice that Theorems 2.9 and 2.10 coincide in the Gaussian case ($\alpha_g = 1$ and $\beta_g = 0$).

Remark 2.10 Theorem 2.10 actually exemplifies a more of universal result, which illustrates that the Fisher–Rao distance is the most in accordance with the underlying statistical model. Indeed, the proof strategy of Theorem 2.10 holds for any geometry induced by a statistical model (parameter manifold and probability density function). Thus, the Fisher–Rao distance will always be bounded by a ratio between the intrinsic problem dimension and the number of samples.

2.4.3 Simulation Examples

This section illustrates the results of Theorems 2.8–2.10 for the multivariate t-distribution (cf. example of Sect. 2.1.1), and various covariance matrix estimators. In the following, the scatter matrix is built as a $p \times p$ (with $p = 10$) Toeplitz matrix $[\Sigma_T]_{ij} = \rho^{|i-j|}$ with $\rho = 0.9(1+\sqrt{-1})/\sqrt{2}$. For samples distributed as $\mathbf{x} \sim$ C-CES$(\mathbf{0}, \Sigma_T, g_d)$, where g_d is the density generator of the t-distribution with d degrees of freedom, we study the performance of the following estimators of Σ_T:

- SCM: the usual sample covariance matrix, defined as $\hat{\Sigma}_{\text{SCM}} = \frac{1}{n}\sum_{i=1}^{n} \mathbf{x}_i\mathbf{x}_i^H$.
- MLE: The estimator $\hat{\Sigma}_{\text{MLE}}$ defined in (2.26) using the appropriate function $\psi(t) = -\phi(t)$, with ϕ defined in (2.12).
- Mismatched MLE: the M-estimator $\hat{\Sigma}_{\text{m-MLE}}$ constructed as the MLE, except that the parameter d is different from the true parameter. Here, $d = 10$ is set regardless of the underlying distribution.

These performances are evaluated with respect to n (n ranging from 11 to 10^3) through the mean squared distances $\delta^2_{\mathcal{E}}$, $\delta^2_{(1,0)}$ and $\delta^2_{(\alpha_{g_d},\beta_{g_d})}$ (evaluated on 10^4 Monte Carlo simulations) and are compared to the corresponding Cramér–Rao lower bounds from Theorems 2.8–2.10.

The left column of Fig. 2.8 displays the results for a t-distribution with $d = 100$ degrees of freedom. Notice that, in this case, data almost follow a Gaussian distribution (it is usually admitted that $d > 30$ allows to assume Gaussianity of the data). In this setting, $\hat{\Sigma}_{\text{MLE}} \simeq \hat{\Sigma}_{\text{SCM}}$ so these estimators reach similar performances. For all performance measurements (different distances), the mismatched MLE appears not efficient at high sample support, which is due to a bias induced on the scale through the wrong choice of parameter d. Also, $\alpha \simeq 1$ and $\beta \simeq 0$, so $\langle\cdot,\cdot\rangle^{(1,0)}_{\cdot}$ and $\langle\cdot,\cdot\rangle^{(\alpha_{g_d},\beta_{g_d})}_{\cdot}$ generate almost identical distances and corresponding bounds, as observed in Fig. 2.8. Interestingly, as noted in [85], these performance criteria show that the studied estimators are not efficient at low sample support. The natural metric is able to reflect some empirical results in terms of application—the SCM is known to provide an

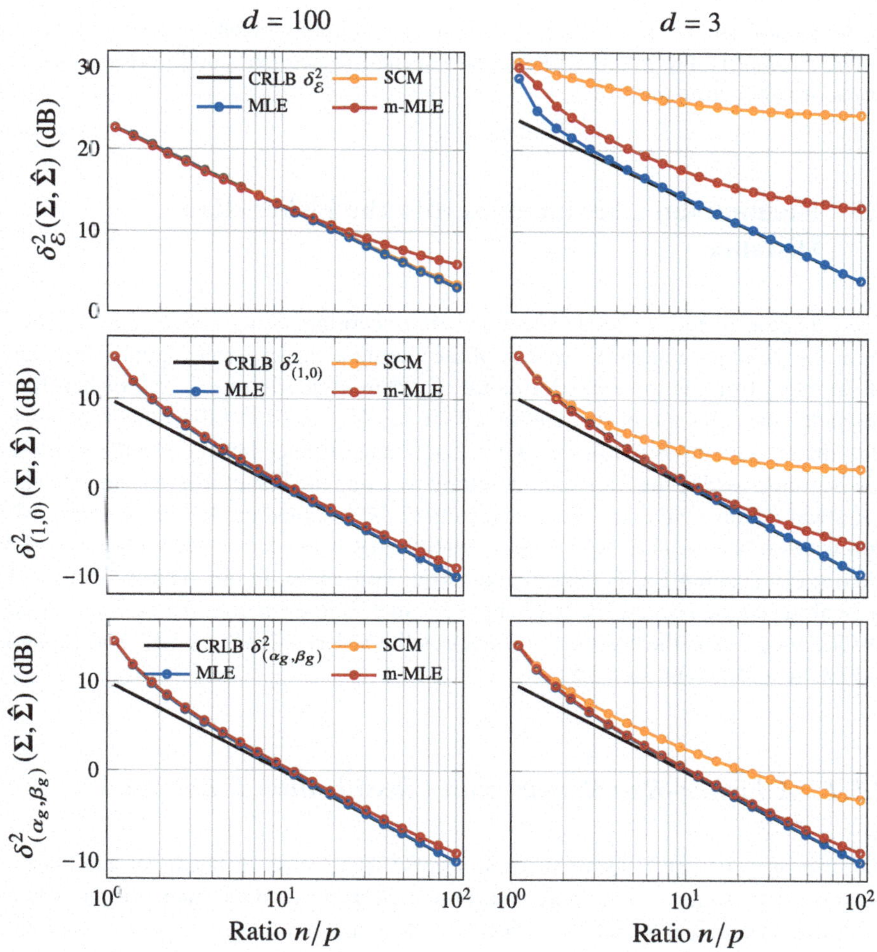

Fig. 2.8 From top to bottom: Euclidean, Natural, CES Fisher–Rao CRLB and mean squared distance scatter matrix for t-distribution with d degrees of freedom versus n/p for $p = 10$. On the left, $d = 100$ (close to Gaussian case) and, on the right, $d = 3$

inaccurate estimation at low sample support—while the Euclidean metric is apparently not, i.e., the Cramér–Rao bound and MSE on the Euclidean metric appear non-informative here.

The right column of Fig. 2.8 displays the same results for a t-distribution with $d = 3$ degrees of freedom. Here, the distribution is heavy-tailed and the SCM, as well as the mismatched MLE, fail to provide an accurate estimator of the scatter matrix. In this case, the study of the Euclidean metric reveals that the MLE is not efficient at low sample support; however, it converges to the bound as n grows. We notice that the convergence toward this regime appears to be slower through the study of the natural and C-CES Fisher–Rao metric, which may be an interesting point in order to quantify the number of samples needed to achieve good performance in terms of application purpose.

2.5 Riemannian Classification with the Fisher–Rao Distance

Classification is a ubiquitous task in machine learning. From a statistical point of view, the problem generally consists of attributing a class to each sample (or batch of samples) from an unlabelled mixture of different distributions. The Fisher–Rao geometry provides a tool that can be efficiently leveraged in this context: as most classification methods are based on the Euclidean distance between samples, these can be transposed to the Riemannian setting by using the Fisher–Rao distance on the statistical feature space (i.e., the parameters of the assumed model). Such transposition is often beneficial as it leverages a metric that is in accordance with the model (e.g., it can account for its natural geometric invariance). In this regard, Sect. 2.5.1 presents a generic framework driven by the Fisher–Rao geometry. An example based on CES models and the nearest centroïd classifier is derived in Sect. 2.5.2 and applied to EEG recordings in Sect. 2.5.3.

2.5.1 A Fisher–Rao Riemannian Classification Framework

The use of statistical features (or descriptors) is common in batch sample classification, as these tend to be more discriminative than raw data. Interestingly, when assuming a statistical model for the batches, the model parameters appear as a natural choice for such statistical features, and the Fisher–Rao distance as a natural tool to compare them. For example, assuming two C-CES models with the same probability density function f, but different parameters Σ_1 and Σ_2, the Fisher–Rao distance (cf. Theorem 2.5 and (2.43)) acts distance between statistical models through the following relation:

$$\underbrace{\delta_{\mathrm{FR}}(f(\mathbf{x}|\Sigma_1), f(\mathbf{x}|\Sigma_2))}_{\text{dist. between models}} \stackrel{\mathrm{def}}{=} \underbrace{\delta_{\mathrm{FR}}(\Sigma_1, \Sigma_2).}_{\text{FR-dist. between parameters}} \tag{2.44}$$

In practice, we handle empirical distributions (i.e., batches of samples), so this distance can be evaluated as

$$\underbrace{\hat{\delta}_{\mathrm{FR}}(\{\mathbf{x}_{i,1}\}_{i=1}^n, \{\mathbf{x}_{i,2}\}_{i=1}^n)}_{\text{dist. between batches}} \stackrel{\mathrm{def}}{=} \underbrace{\delta_{\mathrm{FR}}(\hat{\Sigma}_1, \hat{\Sigma}_2),}_{\text{FR-dist. between estimated covariances}} \tag{2.45}$$

where $\{\mathbf{x}_{i,1}\}_{i=1}^n$ (resp. $\{\mathbf{x}_{i,2}\}_{i=1}^n$) denotes a sample batch, and $\hat{\Sigma}_1$ (resp. $\hat{\Sigma}_2$) denotes an estimate of its covariance matrix, such as the maximum-likelihood estimator presented in Sect. 2.3. From this perspective, a batch classification problem then turns into a problem of classifying covariance matrices on $\mathcal{H}_p^{++}$. Such a task can be achieved by using a standard classification algorithm in which criteria and objects are carefully transposed according to the Fisher–Rao distance (rather than the Euclidean one). For examples related to this setup: the Riemannian nearest centroïd (or minimum distance to mean) classifier [11, 94]; the Riemannian K-means on $\mathcal{H}_p^{++}$ was, e.g., used in [32, 46], Kernel methods based on Riemannian distances were studied in [12, 49, 50], and Riemannian Gaussian mixture models on $\mathcal{H}_p^{++}$ were proposed in [81, 82]. The following section presents the Riemannian counterpart of the nearest centroid classifier for $\mathcal{H}_p^{++}$.

Remark 2.11 Beyond C-CES models, the presented framework generalizes to a generic (model-driven) Riemannian classification methodology, which can be summarized as follows: (i) Model selection: we assume an underlying statistical model, whose parameters should differ between classes; (ii) Statistical Feature extraction: we estimate the corresponding parameters for each batch; (iii) Riemannian classification: the extracted features are classified by leveraging the Fisher–Rao distance.

2.5.2 *Nearest Centroïd Classifier on $\mathcal{H}_p^{++}$ for C-CES Models*

The Riemannian center of mass corresponding to the framework discussed in Sect. 2.5.1 when assuming a Gaussian model has been the reference method to classify electroencephalography (EEG) recordings for the past decade [11]. This section extends this methodology to the C-CES distributions and presents the necessary tools to compute the Riemannian center of masses on $\mathcal{H}_p^{++}$.

Formally, we focus here on the supervised classification of batches of data. Formally, given an unknown batch of a n-sample $\{\mathbf{x}_i\}_{i=1}^n$ and z fixed classes, a classifier $C : (\mathbb{C}^p)^n \to [\![1, \dots, z]\!]$ infers the class label $y \in [\![1, \dots, z]\!]$, i.e.,

$$y = C\left(\{\mathbf{x}_i\}_{i=1}^n\right). \tag{2.46}$$

To provide accurate results, the classifier C is trained on m batches of n samples $\{\{\mathbf{x}_{i,j}\}_{i=1}^n, y_j\}_{j=1}^m$ associated to known class labels $y_j \in [\![1, \dots, z]\!]$. In practice, one usually aims to evaluate the accuracy of a classifier on some dataset $\mathcal{T}$. To do so, the dataset is split into training and test sets, denoted $\mathcal{T}_{\text{train}}$ and $\mathcal{T}_{\text{test}}$, respectively. The classifier C is trained on $\mathcal{T}_{\text{train}}$ and prediction is performed on the testing set $\mathcal{T}_{\text{test}}$. Predicted labels are then compared to actual labels, which yields the accuracy of C on the considered dataset. Notice that there are different ways to build $\mathcal{T}_{\text{train}}$ and $\mathcal{T}_{\text{test}}$ from $\mathcal{T}$, see e.g., the documentation of scikit-learn [76] for more details.

For the model selection step, we consider that each batch $\{\mathbf{x}_{i,j}\}_{i=1}^n$ is distributed according to $\mathbf{x} \sim \text{C-CES}(\mathbf{0}, \Sigma_j, g)$. The statistical parameter extraction step is then performed by maximum-likelihood estimation on each batch (cf. Sect. 2.3). From there, the feature classification problem is set as $\mathcal{T} = \{\hat{\Sigma}_j, y_j\}_{j=1}^m$ on $\mathcal{H}_p^{++}$. We then exploit the Fisher–Rao distance δ of C-CES distribution defined in Theorem 2.5 to generalize the nearest centroïd classifier, also referred to as minimum distance to mean (MDM) classifier, to $\mathcal{H}_p^{++}$. This classification algorithm consists of two steps:

- First, it computes the center of mass of each class, also called class center, from covariance matrices in the training set $\mathcal{T}_{\text{train}}$.
- Then, it assigns the label of the closest class center to each covariance in $\mathcal{T}_{\text{test}}$.

Since the covariance matrices lie on the Riemannian manifold $\mathcal{H}_p^{++}$, the geodesic distance δ from Theorem 2.5 is leveraged in both steps.

We now detail the first step. For every class $y \in [\![1, \ldots, z]\!]$, one must compute the class center $\bar{\boldsymbol{\Sigma}}^{(y)}$ from the training set $\mathcal{T}_{\text{train}}$. It is the center of mass of the set $\{\hat{\Sigma}_j \in \mathcal{T}_{\text{train}} : y_j = y\}$. We thus need to be able to compute the center of mass $\bar{\boldsymbol{\Sigma}}$ of a set $\{\Sigma_j\}_{j=1}^m$ of matrices in $\mathcal{H}_p^{++}$ according to the Fisher–Rao distance in Theorem 2.5. Following [54], the Riemannian center of mass is defined in the following Definition 2.1.

Definition 2.1 (*Riemannian center of mass on* $\mathcal{H}_p^{++}$) The center of mass $\bar{\boldsymbol{\Sigma}}^\star$ of $\{\Sigma_i\}_{i=1}^m$ on $\mathcal{H}_p^{++}$ is defined as the minimizer of the variance computed with the geodesic distance

$$\bar{\boldsymbol{\Sigma}}^\star = \underset{\bar{\boldsymbol{\Sigma}} \in \mathcal{H}_p^{++}}{\arg\min} \quad V(\bar{\boldsymbol{\Sigma}}) \tag{2.47}$$

with $\quad V(\bar{\boldsymbol{\Sigma}}) \overset{\text{def}}{=} \frac{1}{2m} \sum_{j=1}^m \delta^2(\bar{\boldsymbol{\Sigma}}, \Sigma_j)$.

Remark that if the Riemannian distance δ is replaced by its Euclidean counterpart, $\delta_{\mathcal{E}}(\Sigma, \hat{\Sigma}) = \|\Sigma - \hat{\Sigma}\|_2$, then the minimizer of V becomes the arithmetic mean $\bar{\boldsymbol{\Sigma}} = \frac{1}{m} \sum_{j=1}^m \Sigma_j$. Unfortunately, for the Riemannian case, a closed-form solution of (2.47) remains unknown [65] except in very specific cases ($m = 2$, commuting matrices, ...). Hence, one must turn to an iterative optimization procedure. As in [78], we focus here on a Riemannian gradient descent on $\mathcal{H}_p^{++}$. Recall from Sect. 2.3 that, to employ this algorithm, we need to compute the Riemannian gradient of (2.47), choose a retraction and a step size rule. The Riemannian gradient of V was derived in [54, 65], and is provided in Proposition 2.3.

Proposition 2.3 (Riemannian gradient of V) *the Riemannian gradient* grad $V(\bar{\boldsymbol{\Sigma}})$ *of the variance V defined in* (2.47) *at* $\bar{\boldsymbol{\Sigma}} \in \mathcal{H}_p^{++}$ *is*

$$\operatorname{grad} V(\bar{\boldsymbol{\Sigma}}) = -\frac{1}{m} \sum_{j=1}^{m} \log_{\bar{\boldsymbol{\Sigma}}}(\Sigma_j) = -\frac{1}{m} \sum_{j=1}^{m} \bar{\boldsymbol{\Sigma}}^{1/2} \log(\bar{\boldsymbol{\Sigma}}^{-1/2} \Sigma_j \bar{\boldsymbol{\Sigma}}^{-1/2}) \bar{\boldsymbol{\Sigma}}^{1/2}. \tag{2.48}$$

Proof In [65], a technical proof directly deriving the distance of Theorem 2.5 for $\alpha = 1$ and $\beta = 0$ is provided. Here, we propose a more general Riemannian geometry proof, which does not depend on the distance or the manifold. The proved result is well known and can, for instance, be found in [77] without proof. Given $\Sigma \in \mathcal{H}_p^{++}$, we aim to show that the gradient of the function $v(\bar{\boldsymbol{\Sigma}}) = \frac{1}{2}\delta^2(\bar{\boldsymbol{\Sigma}}, \Sigma)$ is grad $v(\bar{\boldsymbol{\Sigma}}) =$

$-\log_{\bar{\Sigma}}(\Sigma)$, where $\log_{\cdot}(\cdot)$ is the Riemannian logarithm mapping corresponding to the Riemannian distance $\delta(\cdot,\cdot)$. Let $\bar{\boldsymbol{\Sigma}}(t)$ the geodesic such that $\bar{\boldsymbol{\Sigma}}(0)=\bar{\boldsymbol{\Sigma}}$ and $\dot{\bar{\boldsymbol{\Sigma}}}(0)=\boldsymbol{\xi}\in\mathcal{H}_p$. It follows that $\mathrm{D}\,v(\bar{\boldsymbol{\Sigma}})[\boldsymbol{\xi}]=\frac{\mathrm{d}}{\mathrm{d}t}v(\bar{\boldsymbol{\Sigma}}(t))\big|_{t=0}$. Let γ_t the geodesic joining $\bar{\boldsymbol{\Sigma}}(t)$ to Σ. By construction, $H(s,t)=\gamma_t(s)$ is a variation of the geodesic γ_0 [41, Definition 3.24]. Furthermore, we have $\frac{\mathrm{d}}{\mathrm{d}t}v(\bar{\boldsymbol{\Sigma}}(t))\big|_{t=0}=\frac{\mathrm{d}}{\mathrm{d}t}E(\gamma_t)\big|_{t=0}$, where $E(\gamma_t)=\frac{1}{2}\int_0^1\langle\dot{\gamma}_t(s),\dot{\gamma}_t(s)\rangle_{\gamma_t(s)}\mathrm{D}$ is the energy of the geodesic γ_t. Let $Y(s)$ such that $H(t,s)=\exp_{\gamma_0(s)}(tY(s))$. From [41, Theorem 3.31], we get the first variation formula of energy

$$\frac{\mathrm{d}}{\mathrm{d}t}E(\gamma_t)=[\langle Y(s),\dot{\gamma}_0(s)\rangle_{\gamma_0(s)}]_0^1-\int_0^1\langle Y(s),\nabla_{\dot{\gamma}_0(s)}\dot{\gamma}_0(s)\rangle_{\gamma_0(s)}\mathrm{D}.$$

Since γ_0 is a geodesic, $\nabla_{\dot{\gamma}_0(s)}\dot{\gamma}_0(s)=0$. Hence, the second term vanishes. Moreover, $Y(1)=\frac{1}{t}\log_{\gamma_0(1)}(\gamma_t(1))$. Since $\gamma_0(1)=\gamma_t(1)=\Sigma$, $Y(1)=0$. We also have $Y(0)=\frac{1}{t}\log_{\gamma_0(0)}(\gamma_t(0))=\frac{1}{t}\log_{\bar{\Sigma}}(\bar{\boldsymbol{\Sigma}}(t))=\frac{1}{t}t\boldsymbol{\xi}=\boldsymbol{\xi}$. It follows that

$$\frac{\mathrm{d}}{\mathrm{d}t}E(\gamma_t)=-\langle\boldsymbol{\xi},\dot{\gamma}_0(0)\rangle_{\gamma_0(0)}=\langle\boldsymbol{\xi},-\log_{\bar{\Sigma}}(\Sigma)\rangle_{\bar{\Sigma}}.$$

We thus get $\mathrm{D}\,v(\bar{\boldsymbol{\Sigma}})[\boldsymbol{\xi}]=\langle\boldsymbol{\xi},-\log_{\bar{\Sigma}}(\Sigma)\rangle_{\bar{\Sigma}}$. The result follows by identification. One can then conclude the proof of the proposition by using the sum property of the gradient operator. □

Then, the most common choice for the retraction is to take the Riemannian exponential mapping (2.21). Furthermore, the stepsize in this case is often simply set to 1. It follows that, given some initialization $\bar{\boldsymbol{\Sigma}}_{(0)}$, the sequence of iterates is

$$\begin{aligned}\bar{\boldsymbol{\Sigma}}_{(k+1)}&=\exp_{\bar{\boldsymbol{\Sigma}}_{(k)}}\left(\tfrac{1}{m}\textstyle\sum_{i=1}^m\log_{\bar{\boldsymbol{\Sigma}}_{(k)}}(\Sigma_j)\right)\\&=\bar{\boldsymbol{\Sigma}}_{(k)}^{1/2}\exp\left(\tfrac{1}{m}\textstyle\sum_{i=1}^m\log(\bar{\boldsymbol{\Sigma}}_{(k)}^{-1/2}\Sigma_j\bar{\boldsymbol{\Sigma}}_{(k)}^{-1/2})\right)\bar{\boldsymbol{\Sigma}}_{(k)}^{1/2}.\end{aligned}\qquad(2.49)$$

The variance V (2.47) is a strictly geodesically convex function over $\mathcal{H}_p^{++}$ [89]. Hence, its minimizer is unique.

Remark 2.12 Notice that there is no dependence on α and β in (2.49). This means that the Riemannian center of mass according to the Fisher–Rao distance in Theorem 2.5 is the same for every C-CES distribution.

The computation of the class centers being solved, we now turn to the second step of the nearest centroïd classifier: the assignment to a class y_j of each estimated covariance matrix $\hat{\Sigma}_j$ belonging to the test set $\mathcal{T}_{\text{test}}$. This is achieved by taking the class that corresponds to the minimal geodesic distance with respect to all class centers, i.e.,

$$y_j=\underset{y\in[\![1,\dots,z]\!]}{\arg\min}\left\{\delta^2(\bar{\boldsymbol{\Sigma}}^{(y)},\hat{\Sigma}_j)\right\}_{y\in[\![1,\dots,z]\!]}.\qquad(2.50)$$

The resulting nearest centroïd classifier on $\mathcal{H}_p^{++}$ is summarized in Algorithm 2.1.

Algorithm 2.1 Nearest centroïd classifier on $\mathcal{H}_p^{++}$

Input: A training set $\mathcal{T}_{\text{train}} = \{(\hat{\Sigma}_j, y_j)\}_{j=1}^{m_{\text{train}}}$ and a test set $\mathcal{T}_{\text{test}} = \{\hat{\Sigma}_j\}_{j=1}^{m_{\text{test}}}$.
Output: Predictions of the test set $\{y_j\}_{j=1}^{m_{\text{test}}}$.
# Training
for $y = 1$ *to* z **do**
 Compute the center of mass $\bar{\boldsymbol{\Sigma}}^{(y)}$ of $\{\hat{\Sigma}_j \in \mathcal{T}_{\text{train}} : y_j = y\}$ with (2.49).
# Testing
for $j = 1$ *to* m_{test} **do**
 Assign $\hat{\Sigma}_j$ to the class with the nearest class center $\bar{\boldsymbol{\Sigma}}^{(y)}$ with (2.50).

Remark 2.13 The Gaussian assumption allows recover the classification algorithm from [11], as in this case: (i) the maximum-likelihood estimator is the sample covariance matrix $\hat{\Sigma}_j = \frac{1}{n}\sum_{i=1}^{n} \mathbf{x}_{i,j}\mathbf{x}_{i,j}^H$; ii) $\alpha = 1$ and $\beta = 0$ in the Fisher–Rao distance δ of Theorem 2.5.

2.5.3 *Application to EEG Classification*

One usually needs to classify EEG recordings in the context of brain–computer interfaces (BCI), where a subject interacts with a computer through brain activity. There are several paradigms for BCI based on EEG. The three main ones are steady-state visually evoked potentials (SSVEP) [53], motor imagery (MI) [91], and event-related potentials (ERP) [5]. This example focuses on ERP data, where subjects are exposed to some stimuli (most often a visual one). These induce a signal response in the brain: the so-called P300, which is a positive wave occurring 300 ms after the stimulus. An ERP dataset consists of a set of trials separated into two classes: a target class (TA), for which the subject is exposed to a stimulus; and a non-target class (NT), for which there is no stimulus. More specifically, we consider the BNCI2014_009 dataset [5], which is available on the MOABB platform.[5] This dataset contains data from 10 subjects, with 3 sessions each. Data were acquired on 16 electrodes at 256 Hz and bandpass filtered between 0.1 and 20 Hz. Recordings were then downsampled to 128 Hz. Each session of each subject contains 1728 trials of 0.8s: 288 target and 1440 non-target ones. Hence, each dataset (one session of one subject) yields $\mathcal{T} = \{\mathbf{X}_j, y_j\}_{j=1}^{m}$ in $\mathbb{R}^{p\times n} \times \{\text{TA}, \text{NT}\}$, where $p = 16$, $n = 102$, $m = 1728$ and $z = 2$.

[5] https://github.com/NeuroTechX/moabb—A standard benchmark platform for BCI.

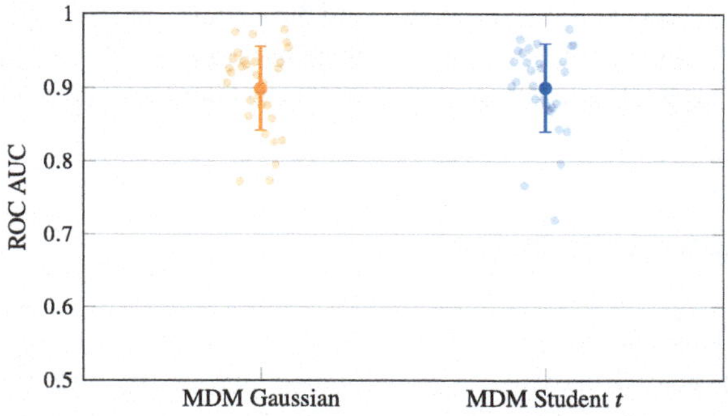

Fig. 2.9 AUC of ROC plots for the nearest centroïd classifiers exploiting the Gaussian distribution (left) and Student t-distribution with $\nu = 2.1$ degrees of freedom (right) applied on the BNCI2014_009 dataset [5] (10 subjects, 3 sessions each)

To perform classification of ERPs, raw data are not directly used. Instead, following [13], augmented data are leveraged. Given the training set $\mathcal{T}_{\text{train}} = \{\mathbf{X}_j\}_{j=1}^{m_{\text{train}}}$, we compute the average target ERP with

$$\mathbf{P}_{\text{TA}} = \frac{1}{m_{\text{TA}}} \sum_{\substack{j=1 \\ y_j=\text{TA}}}^{m_{\text{train}}} \mathbf{X}_j, \tag{2.51}$$

where m_{TA} is the number of target trials in the training set $\mathcal{T}_{\text{train}}$. From there, augmented trials are defined as

$$\tilde{\mathbf{X}}_j = \begin{bmatrix} \mathbf{P}_{\text{TA}} \\ \mathbf{X}_j \end{bmatrix}. \tag{2.52}$$

Covariance matrices $\hat{\Sigma}_j$ are then estimated from these augmented trials both in the training and testing sets. Finally, the nearest centroïd classifier in Algorithm 2.1 is applied on these augmented covariance. We compare two different versions here:

1. **Gaussian version:** covariance matrices estimated through the sample covariance matrix (SCM) and nearest centroïd classifier employed with $\alpha = 1$ and $\beta = 0$.
2. t**-Distribution version:** covariance matrices estimated with the MLE of the t-distribution with $\nu = 2.1$ degrees of freedom and nearest centroïd classifier used with $\alpha = \frac{\nu+p}{\nu+p+2}$ and $\beta = \alpha - 1$.

Achieved accuracies are presented in Fig. 2.9. One can observe that both classifiers feature very good performance on this dataset. One can further notice that they have very similar performance. Indeed, on average, the nearest centroïd classifier with the t-distribution is better by 0.12%. Considering that the SCM is much simpler to compute than the MLE of the t-distribution, one can argue that the nearest centroïd classifier associated with the Gaussian distribution is more advantageous on this dataset. Due to the biological nature of the data, which can be expected to be noisy and contain a non-negligible amount of outliers, one could have expected that a heavy-tail distribution such as the t with $\nu = 2.1$ perform significantly better.

However, the dataset at hand has been curated and the preprocessing has been designed for the Gaussian distribution to work well. Leveraging the t-distribution might be advantageous on real-world non-curated data.

2.6 Conclusion

This chapter presented the Fisher–Rao geometry of C-CES distributions, and its practical uses in statistical signal processing and machine learning. Remark that the methodology that consists in obtaining a Riemannian geometry from the Fisher information metric generalizes to any statistical model (assuming that the parameter space is a smooth manifold). Hence, the approaches presented in this introduction can extend to many other models and applications. Among other examples, such intrinsic analysis has been conducted for the estimation of rotations matrices [27] and for other Lie groups related to tracking problems [56, 57]. In other scopes more directly related to elliptical distributions, we can also mention that geometric tools were used for:

- **Structured covariance matrices** In many applications, the covariance matrix is known to satisfy some form of structural constraint, that can be exploited to reduce the dimension of the estimation problem (see, e.g., [87, 98]). Geometric tools can then be leveraged by expressing the constrained space as a sub-manifold of $\mathcal{H}_p^{++}$. For example: the Fisher information metric was used to obtain structured estimators in [61, 62]; A geometry of Toeplitz matrices was studied in [6]; A framework for probabilistic component analysis (low-rank structured covariance matrices) in C-CES was proposed in [21]; Kronecker products preserve geodesic convexity [96], ans such structure was considered in online covariance matrix estimation in [22]; geometry and structured covariance have also been considered for blind source separation [20].
- **Non-centered models** The geodesics and Fisher–Rao distance of the model $\mathbf{x} \sim \mathcal{CES}(\boldsymbol{\mu}, \Sigma)$ for the mean-and-covariance product manifold $\mathbb{C}^p \times \mathcal{H}_p^{++}$ remains intractable in the general case. Even for the Gaussian distribution $\mathbf{x} \sim \mathcal{N}(\boldsymbol{\mu}, \Sigma)$, only special cases and approximations from geodesic triangles can be obtained [30, 34, 90]. Numerical methods to evaluate these geodesics and corresponding distances were proposed in [68, 69]. Concerning estimation problems, Riemannian optimization was leveraged for non-centered mixture of scaled Gaussian distributions (a sub-family of C-CES distributions) in [33, 36].
- **Mixture models** Mixtures of C-CES can occur within the samples (the observation is the sum of multiple independent contributions) or within batches (the sample set aggregating multiple classes of C-CES). The within-sample mixture is typically used to cast robust models for probabilistic principal component analysis [31, 47, 88]. In this context, geometric tools were developed for low-rank scaled Gaussian signal corrupted by white Gaussian noise in [32]. The within-batch

mixture corresponds to a typical sample-wise classification problem. For this purpose, g-convex relaxations for Gaussian mixture models were studied in [48].

As a final note, we also point out that *information geometry* also refers to a much broader field than the scope covered by this chapter [3, 4]. For comprehensive overviews of the many geometric structures behind families of probability distributions, we refer the readers to [66, 67].

References

1. Absil, P.A., Mahony, R., Sepulchre, R.: Optimization Algorithms on Matrix Manifolds. Princeton University Press (2009)
2. Amari, S.I.: Natural gradient works efficiently in learning. Neural Comput. **10**(2), 251–276 (1998)
3. Amari, S.I.: Information Geometry and Its Applications, vol. 194. Springer (2016)
4. Amari, S.I.: Information geometry. Jpn. J. Math. **16**, 1–48 (2021)
5. Aricò, P., Aloise, F., Schettini, F., Salinari, S., Mattia, D., Cincotti, F.: Influence of P300 latency jitter on event related potential-based brain-computer interface performance. J. Neural Eng. **11**(3), 035008 (2014)
6. Arnaudon, M., Barbaresco, F., Yang, L.: Riemannian medians and means with applications to radar signal processing. IEEE J. Sel. Topics Signal Proc. **7**(4), 595–604 (2013)
7. Arsigny, V., Fillard, P., Pennec, X., Ayache, N.: Log-Euclidean metrics for fast and simple calculus on diffusion tensors. Magn. Reson. Med.: Off. J. Int. Soc. Magn. Reson. Med. **56**(2), 411–421 (2006)
8. Atkinson, C., Mitchell, A.F.: Rao's distance measure. Sankhyā: Indian J. Stat. Ser. A 345–365 (1981)
9. Auderset, C., Mazza, C., Ruh, E.A.: Angular gaussian and cauchy estimation. J. Multivar. Anal. **93**(1), 180–197 (2005)
10. Bangs II, W.J.: Array Processing with Generalized Beam-formers. Yale University (1971)
11. Barachant, A., Bonnet, S., Congedo, M., Jutten, C.: Multiclass brain-computer interface classification by Riemannian geometry. IEEE Trans. Biomed. Eng. **59**(4), 920–928 (2011)
12. Barachant, A., Bonnet, S., Congedo, M., Jutten, C.: Classification of covariance matrices using a Riemannian-based kernel for BCI applications. Neurocomputing **112**, 172–178 (2013)
13. Barachant, A., Congedo, M.: A plug&play P300 BCI using information geometry (2014). arXiv:1409.0107
14. Barrau, A., Bonnabel, S.: A note on the intrinsic Cramér-Rao bound. In: Geometric Science of Information, pp. 377–386. Springer (2013)
15. Berkane, M., Oden, K., Bentler, P.M.: Geodesic estimation in elliptical distributions. J. Multivar. Anal. **63**(1), 35–46 (1997)
16. Besson, O., Abramovich, Y.I.: On the fisher information matrix for multivariate elliptically contoured distributions. IEEE Signal Process. Lett. **20**(11), 1130–1133 (2013)
17. Bhatia, R.: Positive Definite Matrices. Princeton University Press (2009)
18. Bhatia, R., Jain, T., Lim, Y.: On the Bures-Wasserstein distance between positive definite matrices. Expo. Math. **37**(2), 165–191 (2019)
19. Bonnabel, S.: Stochastic gradient descent on Riemannian manifolds. IEEE Trans. Autom. Control **58**(9), 2217–2229 (2013)
20. Bouchard, F., Breloy, A., Ginolhac, G., Renaux, A.: A Riemannian approach to blind separation of t-distributed sources. In: 2020 28th European Signal Processing Conference (EUSIPCO), pp. 965–969. IEEE (2021)

21. Bouchard, F., Breloy, A., Ginolhac, G., Renaux, A., Pascal, F.: A Riemannian framework for low-rank structured elliptical models. IEEE Trans. Signal Process. **69**, 1185–1199 (2021)
22. Bouchard, F., Breloy, A., Mian, A., Ginolhac, G.: On-line Kronecker product structured covariance estimation with Riemannian geometry for t-distributed data. In: 2021 29th European Signal Processing Conference (EUSIPCO), pp. 856–859. IEEE (2021)
23. Bouchard, F., Mian, A., Zhou, J., Said, S., Ginolhac, G., Berthoumieu, Y.: Riemannian geometry for compound gaussian distributions: application to recursive change detection. Signal Process. **176**, 107716 (2020)
24. Boumal, N.: On intrinsic Cramér–Rao bounds for Riemannian submanifolds and quotient manifolds. IEEE Trans. Signal Process. **61**(7), 1809–1821 (2013)
25. Boumal, N.: Optimization and estimation on manifolds. Ph.D. thesis, Université catholique de Louvain (2014)
26. Boumal, N.: An Introduction to Optimization on Smooth Manifolds. Cambridge University Press (2023)
27. Boumal, N., Singer, A., Absil, P.A., Blondel, V.D.: Cramér-Rao bounds for synchronization of rotations. Inf. Inference: J. IMA **3**(1), 1–39 (2014)
28. Breloy, A., Ginolhac, G., Renaux, A., Bouchard, F.: Intrinsic Cramér–Rao bounds for scatter and shape matrices estimation in CES distributions. IEEE Signal Process. Lett. **26**(2), 262–266 (2018)
29. Breloy, A., Ollila, E., Pascal, F.: Spectral shrinkage of Tyler's m-estimator of covariance matrix. In: 2019 IEEE 8th International Workshop on Computational Advances in Multi-Sensor Adaptive Processing (CAMSAP), pp. 535–538. IEEE (2019)
30. Calvo, M., Oller, J.M.: An explicit solution of information geodesic equations for the multivariate normal model. Stat. & Risk Model. **9**(1–2), 119–138 (1991)
31. Chen, T., Martin, E., Montague, G.: Robust probabilistic PCA with missing data and contribution analysis for outlier detection. Comput. Stat. & Data Anal. **53**(10), 3706–3716 (2009)
32. Collas, A., Bouchard, F., Breloy, A., Ginolhac, G., Ren, C., Ovarlez, J.P.: Probabilistic PCA from heteroscedastic signals: geometric framework and application to clustering. IEEE Trans. Signal Process. **69**, 6546–6560 (2021)
33. Collas, A., Bouchard, F., Breloy, A., Ren, C., Ginolhac, G., Ovarlez, J.P.: A Tyler-type estimator of location and scatter leveraging Riemannian optimization. In: ICASSP 2021-2021 IEEE International Conference on Acoustics, Speech and Signal Processing (ICASSP), pp. 5160–5164. IEEE (2021)
34. Collas, A., Bouchard, F., Ginolhac, G., Breloy, A., Ren, C., Ovarlez, J.P.: On the use of geodesic triangles between gaussian distributions for classification problems. In: ICASSP 2022-2022 IEEE International Conference on Acoustics, Speech and Signal Processing (ICASSP), pp. 5697–5701. IEEE (2022)
35. Collas, A., Breloy, A., Ginolhac, G., Ren, C., Ovarlez, J.P.: Robust geometric metric learning. In: 2022 30th European Signal Processing Conference (EUSIPCO), pp. 1447–1451. IEEE (2022)
36. Collas, A., Breloy, A., Ren, C., Ginolhac, G., Ovarlez, J.P.: Riemannian optimization for non-centered mixture of scaled gaussian distributions. IEEE Trans. Signal Process. (2023)
37. Couillet, R., Pascal, F., Silverstein, J.W.: The random matrix regime of Maronna's m-estimator with elliptically distributed samples. J. Multivar. Anal. **139**, 56–78 (2015)
38. Drašković, G., Breloy, A., Pascal, F.: On the asymptotics of maronna's robust PCA. IEEE Trans. Signal Process. **67**(19), 4964–4975 (2019)
39. Drašković, G., Pascal, F.: New insights into the statistical properties of m-estimators. IEEE Trans. Signal Process. **66**(16), 4253–4263 (2018)
40. Duembgen, L., Tyler, D.E.: Geodesic convexity and regularized scatter estimators (2016). arXiv:1607.05455
41. Gallot, S., Hulin, D., Lafontaine, J.: Riemannian Geometry. Springer (1990)
42. Greco, M., Gini, F.: Cramér-Rao lower bounds on covariance matrix estimation for complex elliptically symmetric distributions. IEEE Trans. Signal Process. **61**(24), 6401–6409 (2013)

43. Han, A., Mishra, B., Jawanpuria, P.K., Gao, J.: On Riemannian optimization over positive definite matrices with the Bures-Wasserstein geometry. Adv. Neural. Inf. Process. Syst. **34**, 8940–8953 (2021)
44. Higham, N.J.: Functions of Matrices: Theory and Computation. SIAM (2008)
45. Hippert-Ferrer, A., Bouchard, F., Mian, A., Vayer, T., Breloy, A.: Learning graphical factor models with Riemannian optimization (2022). arXiv:2210.11950
46. Hippert-Ferrer, A., El Korso, M.N., Breloy, A., Ginolhac, G.: Robust low-rank covariance matrix estimation with a general pattern of missing values. Signal Process. **195**, 108460 (2022)
47. Hong, D., Gilman, K., Balzano, L., Fessler, J.A.: HePPCAT: probabilistic PCA for data with heteroscedastic noise. IEEE Trans. Signal Process. **69**, 4819–4834 (2021)
48. Hosseini, R., Sra, S.: Matrix manifold optimization for gaussian mixtures. Advances in Neural Information Processing Systems 28 (2015)
49. Jayasumana, S., Hartley, R., Salzmann, M.: Kernels on Riemannian manifolds. In: Riemannian Computing in Computer Vision, pp. 45–67. Springer (2016)
50. Jayasumana, S., Hartley, R., Salzmann, M., Li, H., Harandi, M.: Kernel methods on the Riemannian manifold of symmetric positive definite matrices. In: Proceedings of the IEEE Conference on Computer Vision and Pattern Recognition, pp. 73–80 (2013)
51. Jeuris, B., Vandebril, R., Vandereycken, B.: A survey and comparison of contemporary algorithms for computing the matrix geometric mean. Electron. Trans. Numer. Anal. **39**, 379–402 (2012)
52. Kai-Tai, F., Yao-Ting, Z.: Generalized Multivariate Analysis. Science Press Beijing and Springer, Berlin (1990)
53. Kalunga, E.K., Chevallier, S., Barthélemy, Q., Djouani, K., Monacelli, E., Hamam, Y.: Online SSVEP-based BCI using Riemannian geometry. Neurocomputing **191**, 55–68 (2016)
54. Karcher, H.: Riemannian center of mass and mollifier smoothing. Commun. Pure Appl. Math. **30**(5), 509–541 (1977)
55. Kay, S.M.: Fundamentals of Statistical Signal Processing. Prentice Hall PTR (1993)
56. Labsir, S., Giremus, A., Yver, B., Benoudiba-Campanini, T.: Joint shape and centroid position tracking of a cluster of space debris by filtering on Lie groups. Signal Process. **183**, 108027 (2021)
57. Labsir, S., Renaux, A., Vilà-Valls, J., Chaumette, E.: Barankin, McAulay–Seidman and Cramér–Rao bounds on matrix lie groups. Automatica **156**, 111199 (2023)
58. Lang, S.: Differential and Riemannian Manifolds. Springer (2012)
59. Lee, J.M.: Riemannian Manifolds: An Introduction to Curvature. Springer (2006)
60. Maronna, R.A., Yohai, V.J.: Robust estimation of multivariate location and scatter. Wiley StatsRef: Statistics Reference Online (1976)
61. Mériaux, B., Ren, C., Breloy, A., El Korso, M.N., Forster, P.: Matched and mismatched estimation of Kronecker product of linearly structured scatter matrices under elliptical distributions. IEEE Trans. Signal Process. **69**, 603–616 (2020)
62. Meriaux, B., Ren, C., El Korso, M.N., Breloy, A., Forster, P.: Robust estimation of structured scatter matrices in (mis) matched models. Signal Process. **165**, 163–174 (2019)
63. Micchelli, C.A., Noakes, L.: Rao distances. J. Multivar. Anal. **92**(1), 97–115 (2005)
64. Mitchell, A.E.: The information matrix, skewness tensor and a-connections for the general multivariate elliptic distribution. Ann. Inst. Stat. Math. **41**, 289–304 (1989)
65. Moakher, M.: A differential geometric approach to the geometric mean of symmetric positive-definite matrices. SIAM J. Matrix Anal. Appl. **26**(3), 735–747 (2005)
66. Nielsen, F.: An elementary introduction to information geometry. Entropy **22**(10), 1100 (2020)
67. Nielsen, F.: The many faces of information geometry. Not. Am. Math. Soc. **69**(1), 36–45 (2022)
68. Nielsen, F.: Fisher-Rao distance and pullback SPD cone distances between multivariate normal distributions (2023). arXiv:2307.10644
69. Nielsen, F.: A simple approximation method for the Fisher-Rao distance between multivariate normal distributions. Entropy **25**(4), 654 (2023)

70. Ollila, E., Eriksson, J., Koivunen, V.: Complex elliptically symmetric random variables - generation, characterization, and circularity tests. IEEE Trans. Signal Process. **59**(1), 58–69 (2011)
71. Ollila, E., Soloveychik, I., Tyler, D.E., Wiesel, A.: Simultaneous penalized M-estimation of covariance matrices using geodesically convex optimization (2016). arXiv:1608.08126
72. Ollila, E., Tyler, D.E.: Regularized m-estimators of scatter matrix. IEEE Trans. Signal Process. **62**(22), 6059–6070 (2014)
73. Ollila, E., Tyler, D.E., Koivunen, V., Poor, H.V.: Complex elliptically symmetric distributions: survey, new results and applications. IEEE Trans. Signal Process. **60**(11), 5597–5625 (2012)
74. Pascal, F., Chitour, Y., Quek, Y.: Generalized robust shrinkage estimator and its application to STAP detection problem. IEEE Trans. Signal Process. **62**(21), 5640–5651 (2014)
75. Pascal, F., Renaux, A.: Statistical analysis of the covariance matrix MLE in K-distributed clutter. Signal Process. **90**(4), 1165–1175 (2010)
76. Pedregosa, F., Varoquaux, G., Gramfort, A., Michel, V., Thirion, B., Grisel, O., Blondel, M., Prettenhofer, P., Weiss, R., Dubourg, V., et al.: Scikit-learn: Machine learning in Python. J. Mach. Learn. Res. **12**, 2825–2830 (2011)
77. Pennec, X.: Hessian of the Riemannian squared distance (2017). https://www-sop.inriafr/members/Xavier.Pennec/AOS-DiffRiemannianLog.pdf
78. Pennec, X., Fillard, P., Ayache, N.: A Riemannian framework for tensor computing. Int. J. Comput. Vision **66**, 41–66 (2006)
79. Rao, C.R.: Information and accuracy attainable in the estimation of statistical parameters. In: S. Kotz, N.L. Johnson (eds.), Breakthroughs in Statistics Volume I: Foundations and Basic Theory, pp. 235–248 (1945)
80. Rao, C.R.: Information and the accuracy attainable in the estimation of statistical parameters. In: Breakthroughs in Statistics: Foundations and Basic Theory, pp. 235–247. Springer (1992)
81. Said, S., Bombrun, L., Berthoumieu, Y., Manton, J.H.: Riemannian gaussian distributions on the space of symmetric positive definite matrices. IEEE Trans. Inf. Theory **63**(4), 2153–2170 (2017)
82. Said, S., Hajri, H., Bombrun, L., Vemuri, B.C.: Gaussian distributions on Riemannian symmetric spaces: statistical learning with structured covariance matrices. IEEE Trans. Inf. Theory **64**(2), 752–772 (2017)
83. Skovgaard, L.T.: A Riemannian geometry of the multivariate normal model. Scandinavian J. Stat. 211–223 (1984)
84. Slepian, D.: Estimation of signal parameters in the presence of noise. Trans. IRE Prof. Group Inf. Theory **3**(3), 68–89 (1954)
85. Smith, S.T.: Covariance, subspace, and intrinsic Cramér–Rao bounds. IEEE Trans. Signal Process. **53**(5), 1610–1630 (2005)
86. Sun, Y., Babu, P., Palomar, D.P.: Regularized Tyler's scatter estimator: existence, uniqueness, and algorithms. IEEE Trans. Signal Process. **62**(19), 5143–5156 (2014)
87. Sun, Y., Babu, P., Palomar, D.P.: Robust estimation of structured covariance matrix for heavy-tailed elliptical distributions. IEEE Trans. Signal Process. **64**(14), 3576–3590 (2016)
88. Sun, Y., Breloy, A., Babu, P., Palomar, D.P., Pascal, F., Ginolhac, G.: Low-complexity algorithms for low rank clutter parameters estimation in radar systems. IEEE Trans. Signal Process. **64**(8), 1986–1998 (2015)
89. Tang, M., Rong, Y., Chen, C.: Riemannian Lp center of mass for scatter matrix estimation in complex elliptically symmetric distributions. In: 2021 IEEE 24th International Conference on Information Fusion (FUSION), pp. 1–8 (2021). https://doi.org/10.23919/FUSION49465.2021.9626967
90. Tang, M., Rong, Y., Zhou, J., Li, X.R.: Information geometric approach to multisensor estimation fusion. IEEE Trans. Signal Process. **67**(2), 279–292 (2018)
91. Tangermann, M., Müller, K.R., Aertsen, A., Birbaumer, N., Braun, C., Brunner, C., Leeb, R., Mehring, C., Miller, K.J., Mueller-Putz, G., et al.: Review of the BCI competition IV. Frontiers in neuroscience p. 55 (2012)

92. Thanwerdas, Y.: Riemannian and stratified geometries on covariance and correlation matrices. Theses, Université Côte d'Azur (2022). https://hal.science/tel-03698752
93. Thanwerdas, Y., Pennec, X.: O (n)-invariant Riemannian metrics on SPD matrices. Linear Algebra Appl. **661**, 163–201 (2023)
94. Tuzel, O., Porikli, F., Meer, P.: Pedestrian detection via classification on Riemannian manifolds. IEEE Trans. Pattern Anal. Mach. Intell. **30**(10), 1713–1727 (2008)
95. Tyler, D.E.: A distribution-free M-estimator of multivariate scatter. Ann. Stat. 234–251 (1987)
96. Wiesel, A.: Geodesic convexity and covariance estimation. IEEE Trans. Signal Process. **60**(12), 6182–6189 (2012). https://doi.org/10.1109/TSP.2012.2218241
97. Wiesel, A.: Unified framework to regularized covariance estimation in scaled gaussian models. IEEE Trans. Signal Process. **60**(1), 29–38 (2012)
98. Wiesel, A., Zhang, T., et al.: Structured robust covariance estimation. Found. Trends® Signal Process. **8**(3), 127–216 (2015)
99. Zhang, H., J Reddi, S., Sra, S.: Riemannian SVRG: fast stochastic optimization on Riemannian manifolds. Advances in Neural Information Processing Systems 29 (2016)
100. Zhang, T., Cheng, X., Singer, A.: Marchenko-Pastur law for Tyler's and Maronna's *m*-estimators (2014). arXiv:1401.3424
101. Zhang, T., Wiesel, A., Greco, M.S.: Multivariate generalized gaussian distribution: convexity and graphical models. IEEE Trans. Signal Process. **61**(16), 4141–4148 (2013)

Chapter 3
Linear Shrinkage of Sample Covariance Matrix or Matrices Under Elliptical Distributions: A Review

Esa Ollila

Abstract This chapter reviews methods for linear shrinkage of the sample covariance matrix (SCM) and matrices (SCM-s) under elliptical distributions in single and multiple populations settings, respectively. In the single sample setting a popular linear shrinkage estimator is defined as a linear combination of the sample covariance matrix (SCM) with a scaled identity matrix. The optimal shrinkage coefficients minimizing the mean-squared error (MSE) under elliptical sampling are shown to be functions of few key parameters only, such as elliptical kurtosis and sphericity parameter. Similar results and estimators are derived for multiple population settings and applications of the studied shrinkage estimators are illustrated in portfolio optimization.

3.1 Introduction

Consider a set of p-dimensional (real-valued) vectors $\mathcal{X} = \{\mathbf{x}_i\}_{i=1}^n$ sampled from a distribution of a random vector $\mathbf{x}$ with unknown mean vector $\boldsymbol{\mu} = \mathsf{E}[\mathbf{x}]$ and unknown positive definite symmetric (PDS) $p \times p$ covariance matrix $\boldsymbol{\Sigma} \equiv \text{cov}(\mathbf{x}) = \mathsf{E}[(\mathbf{x} - \boldsymbol{\mu})(\mathbf{x} - \boldsymbol{\mu})^\top]$. A popular estimate of $\boldsymbol{\Sigma}$ is the *sample covariance matrix (SCM)*, defined by

$$\mathbf{S} = \frac{1}{n-1} \sum_{i=1}^{n} (\mathbf{x}_i - \bar{\mathbf{x}})(\mathbf{x}_i - \bar{\mathbf{x}})^\top, \tag{3.1}$$

where $\bar{\mathbf{x}} = \frac{1}{n} \sum_{i=1}^n \mathbf{x}_i$ denotes the sample mean vector. It has some favourable properties such as being unbiased, i.e., $\mathsf{E}[\mathbf{S}] = \boldsymbol{\Sigma}$, and its scaled version $\mathbf{S}_{\text{ML}} = [(n-1)/n] \cdot \mathbf{S}$ is the maximum-likelihood estimator of the covariance matrix when the samples are independent and identically distributed (i.i.d.) from a multivariate normal (MVN) distribution $\mathcal{N}_p(\boldsymbol{\mu}, \boldsymbol{\Sigma})$.

E. Ollila (✉)
Department of Information and Communications Engineering, Aalto University, Espoo, Finland
e-mail: esa.ollila@aalto.fi

J.-P. Delmas et al. (eds.), *Elliptically Symmetric Distributions in Signal Processing and Machine Learning*, https://doi.org/10.1007/978-3-031-52116-4_3

In many applications, the estimation accuracy (or another performance criterion) can alternatively be improved by using a so-called *tapered* SCM. Such estimate is defined as $\mathbf{W} \circ \mathbf{S}$, where $\circ$ denotes the Hadamard (or Schur) element-wise product, and where $\mathbf{W}$ is a *tapering* matrix (also referred to as covariance matrix taper), i.e., a template that imposes some additional structure to the SCM. Note that above $(\mathbf{W} \circ \mathbf{S})_{ij} = w_{ij}s_{ij}$ for $(\mathbf{W})_{ij} = w_{ij}$ and $(\mathbf{S})_{ij} = s_{ij}$. Covariance matrix tapers have found applications in diverse fields. For example, the true covariance matrix may be known to have a diagonally dominant structure (e.g., in autoregressive models). This means that the variables have a natural order in the sense that $|i - j|$ large implies that the correlation between the ith and the jth variables is close to zero. In this setting, popular estimation approaches use banding-type tapering matrices such as thresholding [1, 2]:

$$(\mathbf{W})_{ij} = \begin{cases} 1, & |i - j| < k \\ 0, & |i - j| \geq k \end{cases} \tag{3.2}$$

for some integer $k \in [\![1, p]\!] = \{1, \ldots, p\}$ called the *bandwidth* parameter. Other types of template matrices are also possible, see [3].

Let $\hat{\boldsymbol{\Sigma}}$ denote an estimator of $\boldsymbol{\Sigma}$ based on a sample $\mathcal{X}$. It is now well known that an estimator that performs better than $\hat{\boldsymbol{\Sigma}}$ can be easily constructed using the concept called regularization or shrinkage which leverages on the concept called *bias-variance tradeoff*. The key idea in shrinkage/regularization is to shift (or shrink) the estimator towards a predetermined target or model. The principle is to decrease the variance of the estimator while introducing some bias, and thus improving the overall performance of the estimation by reducing its *mean-squared error (MSE)*, defined as

$$\mathsf{MSE}(\hat{\boldsymbol{\Sigma}}) = \mathsf{E}\left[\|\hat{\boldsymbol{\Sigma}} - \boldsymbol{\Sigma}\|_{\mathrm{F}}^2\right], \tag{3.3}$$

where $\|\cdot\|_{\mathrm{F}}$ denotes the Frobenius matrix norm, $\|\mathbf{A}\|_{\mathrm{F}} = \sqrt{\mathrm{tr}(\mathbf{A}^\top \mathbf{A})}$ for any matrix $\mathbf{A}$ and $\mathrm{tr}(\cdot)$ denotes the matrix trace, $\mathrm{tr}(\mathbf{A}) = \sum_{i=1}^p a_{ii}$, for any square matrix $\mathbf{A}$. Recall that the bias of $\hat{\boldsymbol{\Sigma}}$ is defined as

$$\mathsf{bias}(\hat{\boldsymbol{\Sigma}}) = \boldsymbol{\Sigma} - \mathsf{E}[\hat{\boldsymbol{\Sigma}}]$$

and an estimator is called unbiased iff $\mathsf{bias}(\hat{\boldsymbol{\Sigma}}) = \mathbf{0}$. This reduction in MSE can be understood via the bias-variance decomposition of the MSE:

$$\mathsf{MSE}(\hat{\boldsymbol{\Sigma}}) = \mathsf{E}\left[\|\hat{\boldsymbol{\Sigma}} - \mathsf{E}[\hat{\boldsymbol{\Sigma}}]\|_{\mathrm{F}}^2\right] + \|\mathsf{bias}(\hat{\boldsymbol{\Sigma}})\|_{\mathrm{F}}^2, \tag{3.4}$$

where the first term on the right-hand side is the *total variance* and the second term is the squared *total bias* of the estimator. If the estimator $\hat{\boldsymbol{\Sigma}}$ is unbiased, then its MSE is equal to its total variance. By using a shrinkage estimator, say $\hat{\boldsymbol{\Sigma}}(\beta)$, where $\beta > 0$ is some tuning parameter that introduces some bias to the estimator $\hat{\boldsymbol{\Sigma}}$, it is possible to reduce its MSE significantly given that the total variance is reduced in larger extent. This will be illustrated in detail in Sect. 3.2.

To derive MSE-optimal shrinkage parameters and their estimates under the assumption that data $\mathcal{X}$ is generated from an elliptically symmetric (ES) distribution [4], one needs to derive the moments of the SCM or tapered SCM, such as its normalized MSE (NMSE). These results as well as some key parameters, the elliptical kurtosis and one particular measure of sphericity, are defined and elaborated in Sect. 3.3.

Shrinkage estimation was introduced by Stein in the context of improved estimation of the mean in his seminal works [5, 6]. These ideas were further studied in [7, 8]. This chapter reviews linear shrinkage estimators of SCM(-s) in single and multiple covariance matrix estimation problems. One of the earliest references studying a linear shrinkage estimator is [9]. A linear shrinkage estimator can often be represented in the form

$$\hat{\mathbf{\Sigma}}(\beta, \alpha) = \beta \mathbf{S} + \alpha \hat{\eta} \mathbf{T}, \tag{3.5}$$

where $\mathbf{T}$ is positive definite symmetric target matrix, α and β are tuning parameters, while $\hat{\eta}$ is a *scale statistics*[1] such as $\hat{\eta} = \mathrm{tr}(\mathbf{S})/p$ or $\hat{\eta} = p/\mathrm{tr}(\mathbf{S}^{-1})$. In (3.5), the SCM is pulled or shrunk towards a predetermined or estimated target structure $\mathbf{T}$, which may be chosen based on prior assumptions about the data. Choosing $\mathbf{T}$ as the identity matrix ($\mathbf{T} = \mathbf{I}$) implies having no a priori knowledge of the shape of the data cloud. One such estimator, defined as $\hat{\mathbf{\Sigma}}(\beta, \alpha) = \beta \mathbf{S} + \alpha \hat{\eta} \mathbf{I}$ with $\hat{\eta} = \mathrm{tr}(\mathbf{S})/p$, was proposed in [11]. This estimator will be described in more detail in Sect. 3.4, where the MSE optimal estimator is considered when $\mathcal{X}$ follows an unspecified ES distribution. Shrinkage estimation of the form $\mathbf{S} + \alpha \mathbf{I}$ (so $\beta = \hat{\eta} = 1, \mathbf{T} = \mathbf{I}$) is often referred to as "diagonal loading" in signal processing literature [12–14].

Different target matrices $\mathbf{T}$ have been considered in the literature. For example, [15] used a target matrix following a single-index market factor model whose motivation stems from portfolio optimization and capital asset pricing model (CAPM), while a constant correlation model was adopted as the target matrix in [16]. It is also possible to shrink toward multiple target matrices simultaneously as proposed in [17–20]. Such multi-target shrinkage covariance matrix estimators are defined by

$$\hat{\mathbf{\Sigma}}(\mathbf{a}) = a_0 \mathbf{S} + \sum_{k=1}^{K} a_k \mathbf{T}_k, \tag{3.6}$$

where $\mathbf{T}_k$, $k = 1, \ldots, K$, are linearly independent target PDS matrices and a_j, $j = 0, \ldots, K$, are the regularization coefficients. It is also common to impose some restrictions on the parameters such as nonnegativity $a_k \geq 0$, and scale constraints, such as $\sum_{k=1}^{K} a_k \leq 1$ for $k = 1, \ldots, K$ and $a_0 = 1 - \sum_{k=1}^{K} a_k$, as in [17, 18].

In the multiple population setting, regularization via pooling the information in the different class samples is also possible. For example, [21] considered covariance

---

[1] Formally, $\eta \equiv \eta(\mathbf{\Sigma})$ is a *scale* parameter if it verifies $\eta(\mathbf{I}) = 1$ and $\eta(a\mathbf{\Sigma}) = a\eta(\mathbf{\Sigma})$ for all $a > 0$ [10]. Then $\hat{\eta}$ is a statistic that estimates this parameter based on data $\mathcal{X}$.

matrix estimation from two independent datasets, whose covariance matrices are different but close to each other. In discriminant analysis classification, the pooled SCM, $\mathbf{S}_{\text{pool}} = \frac{1}{n}\sum_{k=1}^{K} n_k \mathbf{S}_k$, $n = \sum_{k=1}^{K} n_k$, is often used as a shrinkage target and the class covariance matrices are estimated via a convex combination $\hat{\mathbf{\Sigma}}_k = a\mathbf{S}_k + (1-a)\mathbf{S}_{\text{pool}}$, where $a \in [0, 1]$. This was studied in a Bayesian framework in [22, 23], and in the Regularized Discriminant Analysis (RDA) framework in [24]. In this chapter, we consider more general multiple population linear shrinkage settings. First, we consider the coupled linear shrinkage approach [25], where the SCM of kth sample is first linearly shrunk with pooled SCM $\mathbf{S}_{\text{pool}}$, and this estimator is then shrunk toward scaled identity matrix to guarantee positive-definiteness. The optimal coefficients are estimated that minimize the MSE under the assumption that data are sampled from unknown (unspecified) elliptical distributions. Then we consider a more general approach, where the covariance matrix estimator of the kth class is formed as a linear combination of all class SCM-s where coefficients that minimize the MSE are estimated similarly under the elliptical distribution assumption. These developments are discussed in Sect. 3.5. Application to portfolio selection in finance is provided in Sect. 3.6. Finally, Sect. 3.7 concludes.

3.2 Bias-Variance Tradeoff and Shrinkage

To illustrate the idea of shrinkage estimators of covariance matrix, consider the simplest possible shrinkage estimator

$$\hat{\mathbf{\Sigma}}(\beta) = \beta\hat{\mathbf{\Sigma}},$$

where $\beta > 0$ is a shrinkage parameter that can be optimally tuned and $\hat{\mathbf{\Sigma}}$ is some unbiased estimator of $\mathbf{\Sigma}$ such as the SCM, so verifying $\mathsf{E}[\hat{\mathbf{\Sigma}}] = \mathbf{\Sigma}$. First, note that $\hat{\mathbf{\Sigma}}(\beta)$ is obviously biased for any $\beta \neq 1$, the bias being

$$\mathsf{bias}[\hat{\mathbf{\Sigma}}(\beta)] = \mathbf{\Sigma} - \mathsf{E}[\beta\hat{\mathbf{\Sigma}}] = (1-\beta)\mathbf{\Sigma}. \tag{3.7}$$

It is yet possible to improve on the MSE by seeking an optimal constant β_{o} such that $\hat{\mathbf{\Sigma}}_{\text{o}} = \beta_o\hat{\mathbf{\Sigma}}$ attains a smaller MSE than $\hat{\mathbf{\Sigma}}$, i.e.,

$$\mathsf{MSE}(\hat{\mathbf{\Sigma}}_{\text{o}}) < \mathsf{MSE}(\hat{\mathbf{\Sigma}}) \quad \text{for any } \mathbf{\Sigma} \succ 0\,. \tag{3.8}$$

This is equivalent to saying that $\hat{\mathbf{\Sigma}}_{\text{o}}$ is more efficient estimator than $\hat{\mathbf{\Sigma}}$ (regardless of the structure of the true underlying covariance matrix $\mathbf{\Sigma}$). Now consider finding the optimal scaling term as

$$\beta_{\text{o}} = \arg\min_{\beta>0} \mathsf{E}\big[\|\beta\hat{\mathbf{\Sigma}} - \mathbf{\Sigma}\|_{\text{F}}^2\big].$$

Due to (3.4) and (3.7), we have that

$$\mathsf{MSE}(\hat{\boldsymbol{\Sigma}}(\beta)) = \mathsf{E}\big[\|\beta\hat{\boldsymbol{\Sigma}} - \boldsymbol{\Sigma}\|_{\mathrm{F}}^2\big] = \beta^2 \mathsf{MSE}(\hat{\boldsymbol{\Sigma}}) + (1-\beta)^2 \|\boldsymbol{\Sigma}\|_{\mathrm{F}}^2. \tag{3.9}$$

Since $f(\beta) = \mathsf{MSE}(\hat{\boldsymbol{\Sigma}}(\beta))$ is a strictly convex quadratic function, we can easily find the minimum β_o of $f(\beta)$ as solution of $f'(\beta) = 0$, which gives

$$\beta_o = \frac{\|\boldsymbol{\Sigma}\|_{\mathrm{F}}^2}{\mathsf{MSE}(\hat{\boldsymbol{\Sigma}}) + \|\boldsymbol{\Sigma}\|_{\mathrm{F}}^2} = \frac{1}{1 + \mathsf{NMSE}(\hat{\boldsymbol{\Sigma}})}, \tag{3.10}$$

where

$$\mathrm{NMSE}(\hat{\boldsymbol{\Sigma}}) = \frac{\mathsf{E}\big[\|\hat{\boldsymbol{\Sigma}} - \boldsymbol{\Sigma}\|_{\mathrm{F}}^2\big]}{\|\boldsymbol{\Sigma}\|_{\mathrm{F}}^2} \tag{3.11}$$

is the *normalized MSE (NMSE)* of $\hat{\boldsymbol{\Sigma}}$. Equation (3.10) shows that $\beta_o < 1$ since $\mathsf{NMSE}(\hat{\boldsymbol{\Sigma}}) > 0$. It is not yet clear, however, if (3.8) holds. We prove this next.

First, note from (3.9) that

$$\mathsf{MSE}(\hat{\boldsymbol{\Sigma}}_o) = \beta_o^2 \mathsf{MSE}(\hat{\boldsymbol{\Sigma}}) + (1-\beta_o)^2 \|\boldsymbol{\Sigma}\|_{\mathrm{F}}^2. \tag{3.12}$$

Then substituting

$$1 - \beta_o = 1 - \frac{1}{1 + \mathsf{NMSE}(\hat{\boldsymbol{\Sigma}})} = \frac{\mathsf{NMSE}(\hat{\boldsymbol{\Sigma}})}{1 + \mathsf{NMSE}(\hat{\boldsymbol{\Sigma}})} = \beta_o \mathsf{NMSE}(\hat{\boldsymbol{\Sigma}})$$

into (3.12) yields

$$\begin{aligned} \mathsf{MSE}(\hat{\boldsymbol{\Sigma}}_o) &= \beta_o^2 \mathsf{MSE}(\hat{\boldsymbol{\Sigma}}) + \beta_o^2 \{\mathsf{NMSE}(\hat{\boldsymbol{\Sigma}})\}^2 \cdot \|\boldsymbol{\Sigma}\|_{\mathrm{F}}^2 \\ &= \beta_o^2 \mathsf{MSE}(\hat{\boldsymbol{\Sigma}}) + \beta_o^2 \mathsf{NMSE}(\hat{\boldsymbol{\Sigma}}) \cdot \mathsf{MSE}(\hat{\boldsymbol{\Sigma}}) \\ &= \beta_o^2 \mathsf{MSE}(\hat{\boldsymbol{\Sigma}}) \big(1 + \mathsf{NMSE}(\hat{\boldsymbol{\Sigma}})\big) \\ &= \beta_o \mathsf{MSE}(\hat{\boldsymbol{\Sigma}}), \end{aligned} \tag{3.13}$$

where the last identity follows from $1/\beta_o = 1 + \mathsf{NMSE}(\hat{\boldsymbol{\Sigma}})$ due to (3.10). Since $\beta_o < 1$ for all $\boldsymbol{\Sigma} \succ 0$, it thus follows that (3.8) holds, and thus $\beta_o \hat{\boldsymbol{\Sigma}}$ is more efficient estimator than $\hat{\boldsymbol{\Sigma}}$. It is important to observe that this does not only hold for SCM but also for any unbiased estimator $\hat{\boldsymbol{\Sigma}}$ of $\boldsymbol{\Sigma}$.

We now illustrate this fundamental result in the 1-dimensional case ($p = 1$). In this case, the covariance matrix $\boldsymbol{\Sigma}$ is equal to the variance $\sigma^2 = \mathsf{var}(x)$ of a random variable $x \in \mathbb{R}$. Suppose we have a random sample $x_1, \ldots, x_n$ distributed as x. The *sample variance* is defined as

$$s^2 = \frac{1}{n-1}\sum_{i=1}^{n}(x_i - \bar{x})^2, \tag{3.14}$$

where $\bar{x} = \frac{1}{n}\sum_{i=1}^{n} x_i$ denotes the sample mean. Since s^2 is an unbiased estimator of σ^2, we have that

$$\mathsf{MSE}(s^2) = \mathsf{var}(s^2) = \sigma^4\Big(\frac{\mathsf{kurt}(x)}{n} + \frac{2}{n-1}\Big), \tag{3.15}$$

where $\mathsf{kurt}(x)$ denotes the (excess) *kurtosis* of a random variable x, defined as

$$\mathsf{kurt}(x) = \frac{\mathsf{E}[(x-\mu)^4]}{\sigma^4} - 3. \tag{3.16}$$

Let us now consider the shrinkage estimator $\hat{\sigma}^2(\beta) = \beta s^2$. Due to (3.13), we know that $\hat{\sigma}_{\text{o}}^2 = \beta_{\text{o}} s^2$ where $\beta_{\text{o}} < 1$, is always more efficient estimator than the sample variance since

$$\mathsf{MSE}(\hat{\sigma}_{\text{o}}^2) = \beta_{\text{o}}\mathsf{MSE}(s^2) < \mathsf{MSE}(s^2) \quad \text{for any } \sigma^2 > 0.$$

Using (3.10) and (3.15), the optimal scaling constant β_{o} that minimizes $\mathsf{E}[(\beta s^2 - \sigma^2)^2]$ can be expressed compactly as

$$\beta_{\text{o}} = \frac{\sigma^4}{\mathsf{var}(s^2) + \sigma^4} = \frac{n(n-1)}{\mathsf{kurt}(x)(n-1) + n(n+1)}.$$

For example, if the data is from a Gaussian distribution ($x \sim \mathcal{N}(\mu, \sigma^2)$), then $\mathsf{kurt}(x) = 0$, and $\beta_{\text{o}} = (n-1)/(n+1)$, and hence

$$\hat{\sigma}_{\text{o}}^2 = \frac{1}{n+1}\sum_{i=1}^{n}(x_i - \bar{x})^2$$

is always more efficient estimator than the sample variance s^2 for Gaussian samples.

For Gaussian data, $\beta_{\text{o}} \approx 1$, but if the kurtosis is large and positive and n is small, the optimal shrinkage factor β_{o} can be significantly smaller than 1. For example, consider the case that data is from a standard ($\mu = 0, \sigma = 1$) t-distribution with $\nu > 4$ degrees of freedom (d.o.f.) and unit variance. In this case, the kurtosis is $\mathsf{kurt}(x) = 6/(\nu - 4)$. Figure 3.1 depicts the graphs of MSE, squared bias, and variance of $\hat{\sigma}^2(\beta)$ as a function of $\beta \in [0, 1]$ when $n = 10$ and $\nu = 5$. Recall the connections between these quantities through the bias-variance decomposition,

$$\mathsf{MSE}(\hat{\sigma}^2(\beta)) = \mathsf{var}(\hat{\sigma}^2(\beta)) + \mathsf{bias}(\hat{\sigma}^2(\beta))^2.$$

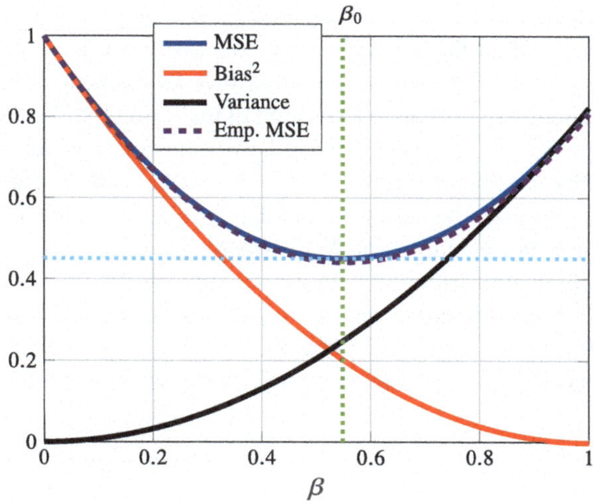

Fig. 3.1 The graphs of MSE, squared bias, and the variance of a shrinkage estimator $\hat{\sigma}^2(\beta) = \beta s^2$ when sampling from a t-distribution of unit variance and d.o.f. $\nu = 5$. Sample size is $n = 10$. The minimum $\mathsf{MSE}(\hat{\sigma}^2(\beta_o))$ is indicated via dotted horizontal line and the value of the optimum β_o via dotted vertical line

The minimum MSE of $\hat{\sigma}_o^2$ is identified as dotted horizontal line and the optimum β_o as a dotted vertical line in the plot. We also computed the empirical MSE averaged over 20000 MC trials. The following conclusions can be drawn. The sample variance s^2 needs to be shrunk nearly by a factor $\beta_o \approx 1/2$ which is substantial scaling. For β close to 1, the bias goes to zero (as expected) while the bias increases when β descends toward 0. The opposite effect is seen in the variance. Optimal tradeoff is obtained by using $\hat{\sigma}_o^2 = \beta_o s^2$. Moreover, one notices that a significant improvement in MSE can be attained by using the MSE-optimal scaled estimator $\beta_o s^2$. One can also notice that the empirical MSE curve has a good match with the theoretical MSE curve.

3.3 NMSE of SCM Under Elliptical Distributions

We remind the reader of some basic properties of elliptically symmetric (ES) distribution reviewed in chapter "Background on real and complex elliptically symmetric distributions" of this book. First, a random vector is said to have an ES distribution if and only if it admits *stochastic representation*,

$$\mathbf{x} = \boldsymbol{\mu} + r\boldsymbol{\Sigma}^{1/2}\mathbf{u}, \tag{3.17}$$

with $\mathbf{u}$ having a uniform distribution on the unit sphere $S^{p-1} = \mathbf{u} \in \{\mathbf{z} \in \mathbb{R}^p : \|\mathbf{z}\| = 1\}$ and $r \geq 0$ being a random variable independent of $\mathbf{u}$. The variable r is called the *modular variate* and, due to (3.17), it verifies

$$r^2 = (\mathbf{x} - \boldsymbol{\mu})^\top \boldsymbol{\Sigma}^{-1}(\mathbf{x} - \boldsymbol{\mu}). \tag{3.18}$$

The parameter $\boldsymbol{\mu} \in \mathbb{R}^p$ is the *symmetry center* and $\boldsymbol{\Sigma}$ is a PDS $p \times p$ matrix parameter, called the *scatter matrix*. We assume $\mathbf{x}$ is an absolutely continuous random vector $\mathbf{x} \in \mathbb{R}^p$ and has finite fourth-order moments. Thus it has a probability density function (p.d.f.) up to a constant of the form $|\boldsymbol{\Sigma}|^{-1/2} g((\mathbf{x}-\boldsymbol{\mu})^\top \boldsymbol{\Sigma}^{-1}(\mathbf{x}-\boldsymbol{\mu}))$, where $g : \mathbb{R}_{\geq 0} \to \mathbb{R}_{>0}$ is called the *density generator* which we without any loss of generality assume to verify $C^{-1} \int_0^\infty t^{p/2} g(t) \mathrm{d}t = p$, where $C = \int_0^\infty t^{p/2-1} g(t) \mathrm{d}t$ which is equivalent to assuming that $\mathsf{E}[r^2] = p$. We write $\mathbf{x} \sim \mathcal{E}_p(\boldsymbol{\mu}, \boldsymbol{\Sigma}, g)$ to denote this case. For example, the MVN distribution $\mathcal{N}_p(\boldsymbol{\mu}, \boldsymbol{\Sigma})$ is a particular instance of the elliptical distribution with $g(t) = \exp(-t/2)$.

The symmetry center $\boldsymbol{\mu}$ is equal to the mean vector $\boldsymbol{\mu} = \mathsf{E}[\mathbf{x}]$ and $\boldsymbol{\Sigma}$ represents the covariance matrix $\boldsymbol{\Sigma} = \mathsf{cov}(\mathbf{x})$. Sometimes, we are only interested in the covariance matrix up to a scaling constant. Hence, we define the *shape matrix* as

$$\boldsymbol{\Lambda} = p \frac{\boldsymbol{\Sigma}}{\operatorname{tr}(\boldsymbol{\Sigma})},$$

which verifies $\operatorname{tr}(\boldsymbol{\Lambda}) = p$.

Two key scalar population parameters in this chapter regarding $\boldsymbol{\Sigma}$ are the *scale* and the *sphericity*. The scale

$$\eta = \frac{\operatorname{tr}(\boldsymbol{\Sigma})}{p} = \frac{1}{p} \sum_{i=1}^{p} \lambda_i \tag{3.19}$$

is the mean of the eigenvalues $\lambda_1, \ldots, \lambda_p$ of $\boldsymbol{\Sigma}$. The sphericity is defined as

$$\gamma = \frac{p \operatorname{tr}(\boldsymbol{\Sigma}^2)}{\operatorname{tr}(\boldsymbol{\Sigma})^2} = \frac{\|\boldsymbol{\Lambda}\|_{\mathrm{F}}^2}{p} = \frac{\frac{1}{p}\sum_{i=1}^p \lambda_i^2}{\left(\frac{1}{p}\sum_{i=1}^p \lambda_i\right)^2}. \tag{3.20}$$

Thus, the sphericity measure (3.20) is the ratio of the mean of the squared eigenvalues of $\boldsymbol{\Sigma}$ relative to the mean of its eigenvalues squared. Letting $s_\lambda^2 = \frac{1}{p}\sum_{i=1}^p (\lambda_i - \eta)^2$ denote the sample variance of the eigenvalues, we may express γ as

$$\gamma = 1 + \frac{s_\lambda^2}{\eta^2} = 1 + \frac{1}{p}\|\boldsymbol{\Lambda} - \mathbf{I}\|_{\mathrm{F}}^2.$$

Thus, the sphericity measures how close $\boldsymbol{\Sigma}$ is to a scaled identity matrix or how concentrated the eigenvalues are around their mean value η. In fact, $\gamma \in [1, p]$, where $\gamma = 1$ if and only if $\boldsymbol{\Sigma} \propto \mathbf{I}$ and $\gamma = p$ if and only if $\boldsymbol{\Sigma}$ has its rank equal to 1. The fact that γ is lower bounded by $\gamma \leq p$ is easiest seen by recalling the submultiplicativity of the matrix trace; namely, for any positive semidefinite matrices $\mathbf{A}$ and $\mathbf{B}$, it holds that $\operatorname{tr}(\mathbf{AB}) \leq \operatorname{tr}(\mathbf{A}) \operatorname{tr}(\mathbf{B})$. Thus $\operatorname{tr}(\boldsymbol{\Sigma}^2) \leq \operatorname{tr}(\boldsymbol{\Sigma})^2$ and consequently $\gamma = p \operatorname{tr}(\boldsymbol{\Sigma}^2)/\operatorname{tr}(\boldsymbol{\Sigma})^2 \leq p$.

A statistical variable describing the heavy-tailedness of the elliptical distribution is *elliptical kurtosis* [26] (and also recall Eq. (45) in chapter "Background on real and complex elliptically symmetric distributions" of the book) which is defined as

$$\kappa = \frac{\mathsf{E}[r^4]}{(\mathsf{E}[r^2])^2}\frac{p}{p+2} - 1 = \frac{\mathsf{E}[r^4]}{p(p+2)} - 1, \tag{3.21}$$

where r^2 is the second-order modular variate defined in (3.18). The latter identity in (3.21) follows due to assumption $\mathsf{E}[r^2] = p$. For kurtosis to exist, we need to assume that the elliptical distribution has finite fourth-order moments. The elliptical kurtosis shares properties similar to the kurtosis of a real random variable. Especially, if $\mathbf{x} \sim \mathcal{N}_p(\boldsymbol{\mu}, \boldsymbol{\Sigma})$, then $\kappa = 0$. This follows by noticing that the quadratic form r^2 has a chi-squared distribution with p degrees of freedom ($r^2 \sim \chi^2_p$) and hence $\mathsf{E}[r^4] = p(p+2)$. This result becomes more obvious when one notices the following relationship of κ with the marginal (excess) kurtosis, $\mathsf{kurt}(x_i)$, of any component of x_i of $\mathbf{x} \sim \mathcal{E}_p(\boldsymbol{\mu}, \boldsymbol{\Sigma}, g)$ [27], [28, Lemma 3]:

$$\kappa = \frac{1}{3} \cdot \mathsf{kurt}(x_i), \text{ for any } i \in \{1, \ldots, p\}. \tag{3.22}$$

3.3.1 NMSE of SCM

We are now ready to derive important results moments of SCM under the elliptical distribution. Before starting the NMSE, we recall the following result.

Lemma 3.1 ([27, Lemma 2]) *Let* $\mathbf{x}_1, \ldots, \mathbf{x}_n \overset{iid}{\sim} \mathcal{E}_p(\boldsymbol{\mu}, \boldsymbol{\Sigma}, g)$ *with* $\boldsymbol{\Sigma} = \mathsf{cov}(\mathbf{x})$ *and assume that finite fourth-order moments exist. Then*

$$\mathsf{E}\left[\|\mathbf{S}\|_F^2\right] = (1 + \tau_1 + \tau_2)\,\|\boldsymbol{\Sigma}\|_F^2 + \tau_1\,\mathrm{tr}(\boldsymbol{\Sigma})^2, \tag{3.23}$$

$$\mathsf{E}\left[\mathrm{tr}(\mathbf{S})^2\right] = 2\tau_1\,\|\boldsymbol{\Sigma}\|_F^2 + \left(1 + \tau_2\right)\mathrm{tr}(\boldsymbol{\Sigma})^2, \tag{3.24}$$

where the scalars are defined by

$$\tau_1 = \frac{1}{n-1} + \frac{\kappa}{n} \quad \text{and} \quad \tau_2 = \frac{\kappa}{n}. \tag{3.25}$$

It is important to notice that these expectations depend on the underlying ES distribution (and hence on the density generator g) only via its kurtosis parameter κ. The NMSE of SCM is given next.

Lemma 3.2 ([27, Lemma 1]) *Let* $\mathbf{x}_1, \ldots, \mathbf{x}_n \overset{iid}{\sim} \mathcal{E}_p(\boldsymbol{\mu}, \boldsymbol{\Sigma}, g)$ *with* $\boldsymbol{\Sigma} = \mathsf{cov}(\mathbf{x})$ *and assume that finite fourth-order moments exist. Then*

$$\mathsf{NMSE}(\mathbf{S}) = \left(1 + \frac{p}{\gamma}\right)\left(\frac{1}{n-1} + \frac{\kappa}{n}\right) + \frac{\kappa}{n}, \tag{3.26}$$

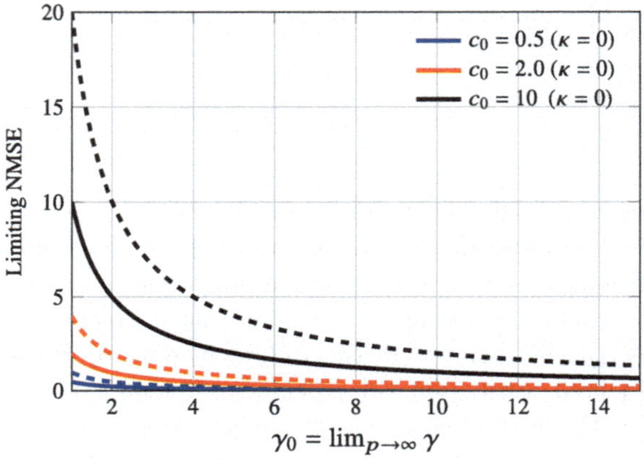

Fig. 3.2 Limiting NMSE in (3.28) as a function of limiting sphericity γ_0 when $p/n \to c_0$ as $p, n \to \infty$. The solid line corresponds to the case $\kappa = 0$ and dotted line $\kappa = 1$

where γ denotes the sphericity parameter.

Sphericity parameter plays crucial role in determining the accuracy of the SCM. Consider the doubly asymptotic regime,

$$c = \frac{p}{n} \to c_0, \quad 0 < c_0 < \infty, \quad \text{as } p, n \to \infty. \tag{3.27}$$

Assume that γ remains bounded, $\gamma \to \gamma_0$ as $p \to \infty$. Then using (3.26), it immediately follows that the limiting NMSE under the doubly asymptotic regime (3.27) is

$$\text{NMSE}(\mathbf{S}) \to \frac{1+\kappa}{\gamma_0} c_0, \tag{3.28}$$

which shows that $\mathbf{S}$ *is not a consistent estimator* of $\boldsymbol{\Sigma}$ unless $c = p/n \to 0$. This is illustrated in Fig. 3.2 which displays the limiting NMSE as a function of γ_0 for different cases of c_0. Again the limiting NMSE is largest when $\boldsymbol{\Sigma}$ is close to being spherical ($\gamma \approx 1$). Moreover, if $c_0 > 1$ (undersampled case), the limiting NMSE can be very large. The solid lines are for case $\kappa = 0$ (which holds for MVN distribution) and the dotted lines for the case $\kappa = 1$. For example, a multivariate t-distribution (MVT) with d.o.f. $\nu = 6$ has $\kappa = 1$. The figure also illustrates that when the distribution is heavy-tailed ($\kappa = 1$) and close to spherical, then the limiting NMSE of SCM can be very large. Finally, we point out that the effect of sphericity in finite sample case is illustrated later in Fig. 3.4a. Since sphericity plays a crucial role in determining the accuracy of the SCM, it is of interest to find an accurate estimator of sphericity. This is the topic of Sect. 3.3.3.

3.3.2 NMSE of Tapered SCM

Let us now derive the MSE of the tapered SCM. For this purpose, assume that template matrix $\mathbf{W} \in \mathcal{W}^+$, where

$$\mathcal{W}^+ = \{\mathbf{W} \in \mathbb{R}^{p\times p}_{\text{Sym}} : w_{ii} = 1, w_{ij} \geq 0 \,\forall i, j \in [\![1, p]\!]\} \tag{3.29}$$

and with $\mathbb{R}^{p\times p}_{\text{Sym}}$ denoting the set of all symmetric $p \times p$ matrices. Write $\text{diag}(\mathbf{A}) \equiv \text{diag}(a_{11}, \ldots, a_{pp})$ for any matrix $\mathbf{A} = (a_{ij})_{p\times p}$. Then we have the following result.

Lemma 3.3 ([3, Lemma 1]) *Let* $\{\mathbf{x}_i\}_{i=1}^n$ *be an i.i.d. random sample from* $\mathcal{E}_p(\boldsymbol{\mu}, \boldsymbol{\Sigma}, g)$ *with finite fourth-order moments. Then for any* $\mathbf{W} \in \mathcal{W}^+$*, it holds that*

$$\mathsf{E}\left[\|\mathbf{W} \circ \mathbf{S}\|_{\mathrm{F}}^2\right] = (1 + \tau_1 + \tau_2)\|\mathbf{W} \circ \boldsymbol{\Sigma}\|_{\mathrm{F}}^2 + \tau_1 \operatorname{tr}((\mathbf{D}_{\boldsymbol{\Sigma}}\mathbf{W})^2),$$

where $\mathbf{D}_{\boldsymbol{\Sigma}} = \text{diag}(\boldsymbol{\Sigma})$ *and* τ_1, τ_2 *are defined in* (3.25).

Notice that the MSE of the tapered SCM is

$$\begin{aligned}\text{MSE}(\mathbf{W} \circ \mathbf{S}) &= \mathsf{E}\left[\|\mathbf{W} \circ \mathbf{S} - \boldsymbol{\Sigma}\|_{\mathrm{F}}^2\right] \\ &= \mathsf{E}\left[\|\mathbf{W} \circ \mathbf{S}\|_{\mathrm{F}}^2\right] + \|\boldsymbol{\Sigma}\|_{\mathrm{F}}^2 - 2\|\mathbf{V} \circ \boldsymbol{\Sigma}\|_{\mathrm{F}}^2,\end{aligned} \tag{3.30}$$

where

$$\mathbf{V} = (v_{ij})_{p\times p} \text{ with } v_{ij} = \sqrt{w_{ij}} \text{ for } \mathbf{W} \in \mathcal{W}^+. \tag{3.31}$$

Thus, plugging in the expression from Lemma 3.3 into (3.30) provides the MSE of the tapered SCM $\mathbf{W} \circ \mathbf{S}$ when sampling from an ES distribution. The NMSE is then obtained from this formula via (3.11).

Figure 3.3 displays the NMSE curve of tapered SCM $\mathbf{W} \circ \mathbf{S}$ when $\mathbf{W}$ is of the form (3.2) and the bandwidth parameter k of $\mathbf{W}$ varies. The data is sampled from an MVN distribution (left panel) and MVT distribution (right panel) with $\nu = 5$ d.o.f., sample size is $n = 100$ and the dimension is $p = 250$. In this example, the true covariance matrix $\boldsymbol{\Sigma}$ has the following structure:

$$(\boldsymbol{\Sigma})_{ij} = \begin{cases} 1 & , i = j \\ \rho|i-j|^{-(\alpha+1)} & , i \neq j, \end{cases} \tag{3.32}$$

where α is a decay parameter and ρ is a correlation parameter which are set to $\alpha = 0.1$ and $\rho = 0.6$, respectively. This model was used in [29]. Figure 3.3 shows the important point. Since the banding template in (3.2) with suitable chosen bandwidth parameter k is well adapted to the true model of $\boldsymbol{\Sigma}$ in (3.32), the NMSE can be significantly reduced with tapered SCM. For MVN data, the best bandwidth $k = 6$ yields the NMSE of 0.089. Note that bandwidth $k = p$ implies $\mathbf{W} = \mathbf{1}\mathbf{1}^\top$ and the tapered SCM reduces to SCM (i.e., $\mathbf{W} \circ \mathbf{S} = \mathbf{S}$). This worst-case bandwidth $k =$

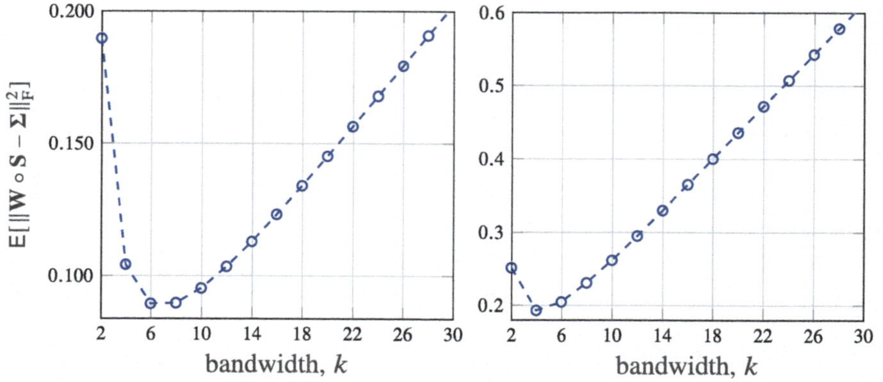

Fig. 3.3 NMSE curve of tapered SCM $\mathbf{W} \circ \mathbf{S}$ as a function of used bandwidth k of $\mathbf{W}$ when sampling from an MVN distribution (left panel) and MVT distribution (right panel) with d.o.f. $\nu = 5$, $\mathbf{\Sigma}$ has structure (3.32) with $\alpha = 0.1$, $n = 100$ and $p = 250$

p gives 1.082 as the NMSE. Thus the tapered SCM improves the MSE of SCM significantly (more than a factor of ten). Performance improvement is even more significant when the data is from a heavy-tailed ES distribution as is illustrated by the more steeply increasing NMSE curve on the right-hand side panel of Fig. 3.3.

3.3.3 Estimator of Sphericity

The spatial sign covariance matrix (SSCM) [30] is an estimate of the shape matrix $\mathbf{\Lambda}$. The scaled[2] SSCM is defined as

$$\hat{\mathbf{\Lambda}} = \frac{p}{n} \sum_{i=1}^{n} \frac{(\mathbf{x}_i - \hat{\boldsymbol{\mu}})(\mathbf{x}_i - \hat{\boldsymbol{\mu}})^\top}{\|\mathbf{x}_i - \hat{\boldsymbol{\mu}}\|^2}, \tag{3.33}$$

where $\hat{\boldsymbol{\mu}} = \arg\min_{\boldsymbol{\mu}} \sum_{i=1}^{n} \|\mathbf{x}_i - \boldsymbol{\mu}\|$ is the sample spatial median [31]. When $\boldsymbol{\mu}$ is known ($\boldsymbol{\mu} = \mathbf{0}$), the SSCM is defined as

$$\hat{\mathbf{\Lambda}} = \frac{p}{n} \sum_{i=1}^{n} \frac{\mathbf{x}_i \mathbf{x}_i^\top}{\|\mathbf{x}_i\|^2}.$$

One of the major selling points of SSCM are its impeccable robustness properties: it possesses the highest possible breakdown point [32] of 1 with fixed location [33] and breakdown point of 1/2 when using the spatial median to estimate the location [34].

[2] The common definition is without the multiplier p.

This can be contrasted to M-estimators of scatter for which the best possible breakdown point is $1/p$ and obtained by Tyler's M-estimator [35].

An estimate of sphericity based on the SSCM, defined by

$$\hat{\gamma} = \frac{n}{n-1}\left(\frac{\|\hat{\mathbf{\Lambda}}\|_F^2}{p} - \frac{p}{n}\right), \tag{3.34}$$

has been studied in many papers (e.g., [20, 36, 37]). In [20], it was shown that (3.34) is asymptotically (as $p \to \infty$) unbiased when sampling from ES distribution under the assumption $\gamma/p \to 0$ as $p \to \infty$. This assumption is sufficiently general and holds for many scatter matrix models [20, Proposition 3]. For example, if $\mathbf{\Sigma}$ has an autoregressive model (**AR(1)**) structure,

$$(\mathbf{\Sigma})_{ij} = \eta \varrho^{|i-j|}, \tag{3.35}$$

where η is the scale (3.19) and ϱ is the correlation parameter, $\varrho \in (-1, 1)$, then

$$\gamma = \frac{p - p\varrho^4 - 2\varrho^2 + 2(\varrho^2)^{p+1}}{p(\varrho^2 - 1)^2}. \tag{3.36}$$

Note that $\gamma = O(1) = o(p)$.

Another estimator proposed in [27, Sect. IV-B] is defined by

$$\hat{\gamma} = \hat{b}_n \left(\frac{p\,\mathrm{tr}(\mathbf{S}^2)}{\mathrm{tr}(\mathbf{S})^2} - \hat{a}_n \frac{p}{n}\right), \tag{3.37}$$

where

$$\hat{a}_n = \left(\frac{n}{n+\hat{\kappa}}\right)\left(\frac{n}{n-1} + \hat{\kappa}\right) \quad \text{and} \quad \hat{b}_n = \frac{(\hat{\kappa}+n)(n-1)^2}{(n-2)(3\hat{\kappa}(n-1) + n(n+1))}.$$

and $\hat{\kappa}$ is an estimate of the elliptical kurtosis. In [28] estimators of γ based on robust M-estimators of scatter were constructed under the assumption that $n > p$ (oversampled case). A comparative study of different estimators of sphericity was recently conducted in [38].

Slightly modified Ell1- or Ell2-estimators of the sphericity parameter of the tapered covariance matrix,

$$\gamma_{\mathbf{W}} \equiv \gamma(\mathbf{W} \circ \mathbf{\Sigma}) = \frac{p\,\mathrm{tr}\left((\mathbf{W} \circ \mathbf{\Sigma})^2\right)}{\mathrm{tr}(\mathbf{\Sigma})^2}, \quad \mathbf{W} \in \mathcal{W}^+ \tag{3.38}$$

can be constructed as shown in [3, Sect. IV].

3.4 Linear Shrinkage of SCM

In this section, we consider the single sample setting and linear shrinkage estimators of the SCM $\mathbf{S}$ or the tapered SCM $\mathbf{W} \circ \mathbf{S}$ in Sects. 3.4.1 and 3.4.2, respectively.

3.4.1 Regularized SCM (RSCM)

The regularized SCM (RSCM) considered in [27] is defined as

$$\hat{\boldsymbol{\Sigma}}(\alpha, \beta) = \beta \mathbf{S} + \alpha \mathbf{I}, \tag{3.39}$$

where $\mathbf{S}$ is the unbiased SCM defined in (3.1), and $\alpha, \beta \geq 0$ are tuning or regularization parameters. The MSE of RSCM can be written as [27, Appendix A]

$$\mathsf{MSE}(\hat{\boldsymbol{\Sigma}}(\alpha, \beta)) = \beta^2 \mathsf{MSE}(\mathbf{S}) + \|\beta \mathbf{S} + \alpha \mathbf{I} - \boldsymbol{\Sigma}\|_{\mathrm{F}}^2. \tag{3.40}$$

Then assuming a sample $\mathbf{x}_1, \ldots, \mathbf{x}_n$ from an arbitrary distribution with finite fourth-order moments, the optimal tuning parameters that minimize the MSE are [27, Theorem 1]

$$\alpha_{\mathrm{o}} = (1 - \beta_{\mathrm{o}})\eta \quad \text{and} \qquad \beta_{\mathrm{o}} = \frac{(\gamma - 1)}{(\gamma - 1) + \gamma \cdot \mathrm{NMSE}(\mathbf{S})}, \tag{3.41}$$

where the scale η and sphericity γ are defined in (3.19) and (3.20), respectively. Note that the NMSE($\mathbf{S}$) for elliptical data is given in Lemma 3.2.

Let $\hat{\boldsymbol{\Sigma}}_{\mathrm{o}} = \hat{\boldsymbol{\Sigma}}(\alpha_{\mathrm{o}}, \beta_{\mathrm{o}})$ denote the optimal or oracle RSCM that has the knowledge of these optimal parameters. Then

$$\mathsf{NMSE}(\hat{\boldsymbol{\Sigma}}_{\mathrm{o}}) = (1 - \beta_{\mathrm{o}}) \frac{\|\boldsymbol{\Sigma} - \eta \mathbf{I}\|_{\mathrm{F}}^2}{\|\boldsymbol{\Sigma}\|_{\mathrm{F}}^2} = (1 - \beta_{\mathrm{o}}) \frac{\gamma - 1}{\gamma}.$$

Next, we give an instructive example illustrating the power of regularization.

Comparing the NMSE of SCM and RSCM

The samples are generated from a $p = 50$-dimensional MVN distribution, $\mathcal{N}_p(\boldsymbol{\mu}, \boldsymbol{\Sigma})$, with AR(1) covariance structure in (3.35). The left panel of Fig. 3.4 displays the NMSE of SCM for varying sample lengths n. As can be noted, the accuracy of SCM $\mathbf{S}$ depends heavily on the value of γ. When $\gamma \approx 1$ (i.e., the distribution is close to being spherical, so $\varrho \approx 0$), the NMSE is largest, and rises steeply when $n < p$. The right panel of Fig. 3.4 displays the NMSE of the optimal RSCM $\hat{\boldsymbol{\Sigma}}_{\mathrm{o}}$. The performance improvement is drastic in the cases when the covariance matrix is close to being spherical (black and red lines) and/or when $n \leq p$.

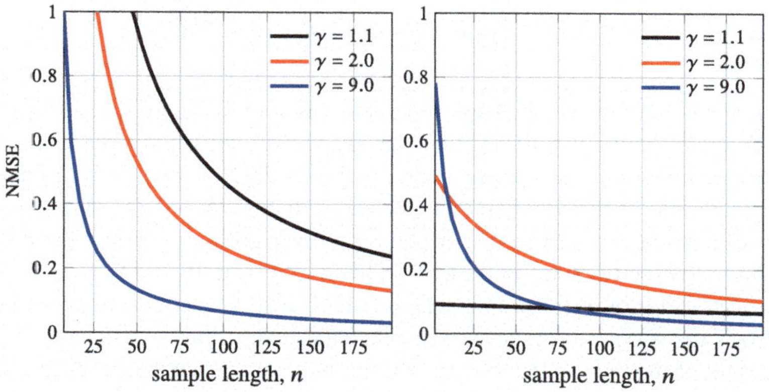

Fig. 3.4 The effect of sphericity γ on the NMSE of SCM $\mathbf{S}$ (left panel) and optimal oracle RSCM $\hat{\boldsymbol{\Sigma}}_o$ (right panel). Samples are from MVN distribution with $\boldsymbol{\Sigma}$ having an AR(1) structure; $p = 50$

In practice, one does not have access to the true α_o or β_o and thus the oracle RSCM is not computable. However, as can be inferred from (3.41) and the NMSE expression in Lemma 3.2, the optimal parameter β_o depends on the sphericity γ and the elliptical kurtosis parameter κ, i.e., $\beta_o \equiv \beta_o(\kappa, \gamma)$. One may compute an estimate $\hat{\kappa}$ using the empirical average of the kurtosis parameters (scaled by $1/3$) due to (3.22) as detailed in [27, Sect. IV], while, for an estimate of sphericity, one may use the estimator defined in (3.34). This gives $\hat{\beta}_o = \beta_o(\hat{\kappa}, \hat{\gamma})$ as the estimate of β_o. To estimate the scale η, one uses $\hat{\eta} = \mathrm{tr}(\mathbf{S})/p$, and then sets $\hat{\alpha}_o = (1 - \hat{\beta}_o)\hat{\eta}$ (recall (3.41)). After estimating these parameters, we can compute the regularized SCM as

$$\hat{\boldsymbol{\Sigma}}_{\mathrm{RSCM}} = \hat{\beta}_o \mathbf{S} + (1 - \hat{\beta}_o)\hat{\eta}\mathbf{I}. \tag{3.42}$$

This estimator was referred to as **RSCM-Ell1**. The estimator using (3.37) as the estimate of sphericity was referred to as **RSCM-Ell2**. MATLAB package is available at http://users.spa.aalto.fi/esollila/regscm/ to compute these estimators.

3.4.2 *Regularized Tapered SCM*

Let $\mathbb{W} = \{\mathbf{W}(k)\}_{k=1}^{K}$ be a finite set of possible template matrices, i.e., matrices satisfying $\mathbf{W}(k) \in \mathcal{W}^+ \ \forall k \in [\![1, K]\!]$, where k is an index that identifies the matrix $\mathbf{W}$ in the set $\mathbb{W}$. For example, the set $\mathbb{W}$ can be the set of all banding matrices $\mathbf{W}(k)$, $k = 1, \ldots, p$ as defined in (3.2) or a union of different types of template matrices. Then, [3] proposed an estimator, referred to as TABASCO (TApered or BAnded Shrinkage COvariance matrix), defined as

$$\hat{\boldsymbol{\Sigma}}(\beta, k) = \beta(\mathbf{W}(k) \circ \mathbf{S}) + (1-\beta)\frac{\mathrm{tr}(\mathbf{S})}{p}\mathbf{I}, \tag{3.43}$$

which benefits both from shrinkage and exploitation of structure via tapering templates $\mathbf{W} \in \mathbb{W}$. Above $\beta \in [0, 1]$ is the shrinkage parameter and $k \in \{1, \dots, K\}$ is the index that identifies the tapering matrix in the set $\mathbb{W}$. Note that $\hat{\boldsymbol{\Sigma}}(\beta, k)$ preserves the original scale of the SCM since $\mathrm{tr}(\mathbf{W} \circ \mathbf{S}) = \mathrm{tr}(\mathbf{S})\ \forall \mathbf{W} \in \mathcal{W}^{+}$. Obviously, the success of banding and/or tapering depends on one's ability to choose the parameters β and k correctly. Since both the RSCM in (3.42) (if $\mathbf{W} = \mathbf{1}\mathbf{1}^{\top} \in \mathbb{W}$ where $\mathbf{1}$ denotes a p-vector of ones) and the tapered SCM (if $\beta = 1$) appear as special cases of (3.43), TABASCO performs never worse than these two estimators in terms of MSE independent of the underlying structure of the true covariance matrix $\boldsymbol{\Sigma}$. Indeed, in the simulation study reported in [3], TABASCO outperformed these estimators as well as many commonly used shrinkage or banding/tapering estimators.

For a given fixed index k, let $\mathbf{W} \equiv \mathbf{W}(k)$ denote the associated template matrix and $\hat{\boldsymbol{\Sigma}}(\beta) \equiv \hat{\boldsymbol{\Sigma}}(\beta, k)$ the associated TABASCO estimator. Then it was shown in [3] that

$$\beta_{\mathrm{o}} = \arg\min_{\beta \in [0,1]} \mathsf{E}\Big[\, \|\hat{\boldsymbol{\Sigma}}(\beta) - \boldsymbol{\Sigma}\|_{\mathrm{F}}^2 \Big] \tag{3.44}$$

$$= \frac{p(\gamma_{\mathbf{V}} - 1)\eta^2}{\mathsf{E}\big[\|\mathbf{W} \circ \mathbf{S}\|_{\mathrm{F}}^2\big] - p^{-1}\mathsf{E}[\mathrm{tr}(\mathbf{S})^2]}, \tag{3.45}$$

where $\mathbf{V} = (v_{ij})$ with $v_{ij} = \sqrt{w_{ij}}$ (as in (3.31)), $\gamma_{\mathbf{V}}$ is the sphericity parameter of $\mathbf{V} \circ \boldsymbol{\Sigma}$, defined via (3.38), and $\eta = \mathrm{tr}(\boldsymbol{\Sigma})/p$ is the scale of $\boldsymbol{\Sigma}$. Under the assumption that data is from an ES distribution, one can derive an explicit analytical expression for β_{o} using expressions for $\mathsf{E}\big[\|\mathbf{W} \circ \mathbf{S}\|_{\mathrm{F}}^2\big]$ and $\mathsf{E}[\mathrm{tr}(\mathbf{S})^2]$ given in Lemmas 3.3 and 3.1, respectively; see [3, Theorem 2] in particularly.

When k is not fixed, then $\beta_{\mathrm{o}} = \beta_{\mathrm{o}}(k)$ depends on k via $\mathbf{W} = \mathbf{W}(k)$ and $\mathbf{V} = \mathbf{V}(k)$. Then, as shown in [3], the MSE optimal index k can be chosen as

$$k_{\mathrm{o}} = \arg\min_{k} \beta_0(k)(1 - \gamma_{\mathbf{V}}(k)), \tag{3.46}$$

where $\gamma_{\mathbf{V}}(k)$ is the sphericity parameter in (3.38) for $\mathbf{V} = \mathbf{V}(k)$.

Naturally, in practice, we need to replace the oracle $\beta_{\mathrm{o}}(k)$ by its estimate $\hat{\beta}_{\mathrm{o}}(k)$. Finally, given $\hat{\beta}_{\mathrm{o}}(k)$ and an estimate of sphericity $\hat{\gamma}_{\mathbf{V}}(k)$, one can choose the best index k (and the associated template $\mathbf{W} = \mathbf{W}(k)$) as $\hat{k}_{\mathrm{o}} = \arg\min_k \hat{\beta}_{\mathrm{o}}(k)(1 - \hat{\gamma}_{\mathbf{V}}(k))$ as in (3.46). These values are then used to obtain the final optimal TABASCO estimator $\hat{\boldsymbol{\Sigma}}_{\text{TABASCO}} = \hat{\boldsymbol{\Sigma}}(\hat{\beta}_{\mathrm{o}}, \hat{k}_{\mathrm{o}})$ via Eq. (3.43), where $\hat{\beta}_{\mathrm{o}} = \hat{\beta}_{\mathrm{o}}(\hat{k}_{\mathrm{o}})$. We refer to [3] for more details of the calculations. Efficient MATLAB toolbox for computing the TABASCO estimator is available at https://github.com/esollila/Tabasco.

3.5 Multiple Class Estimation Problem

In this section, we consider the case where we have K different classes or populations, and we have observed n_k, $k = 1, \dots, K$, i.i.d. p-dimensional samples from these populations. The covariance matrix of class $k \in \{1, \dots, K\}$ is defined as

$$\boldsymbol{\Sigma}_k = \mathsf{E}[(\mathbf{x}_{ik} - \boldsymbol{\mu}_k)(\mathbf{x}_{ik} - \boldsymbol{\mu}_k)^\top],$$

where $\mathbf{x}_{ik}$ denotes the ith sample from class k and $\boldsymbol{\mu}_k = \mathsf{E}[\mathbf{x}_{ik}]$ is the mean of class k. The conventional estimate for the covariance matrix is the unbiased SCM defined for class k by

$$\mathbf{S}_k = \frac{1}{n_k - 1} \sum_{i=1}^{n_k} (\mathbf{x}_{ik} - \bar{\mathbf{x}}_k)(\mathbf{x}_{ik} - \bar{\mathbf{x}}_k)^\top,$$

where $\bar{\mathbf{x}}_k = (1/n_k) \sum_{i=1}^{n_k} \mathbf{x}_{ik}$ is the sample mean of class k.

The estimators that are considered in this section combine or pool the information from the other classes in order to reduce the MSE of the estimator of a given class. The underlying rationale for pooling comes from the often plausible assumption that the class populations share a somewhat similar structure. This is because the same variables that are measured under slightly different population conditions are often positively correlated, and thus, share a similar correlation/covariance structure. Thus, the information available in another class should be used for improving the estimation in the target class.

Since the classes can be assumed to have a similar covariance structure, it is beneficial to shrink the individual class covariance matrix estimates toward the pooled (average) SCM of the classes, using the pooled SCM defined by

$$\mathbf{S}_{\text{pool}} = \sum_{k=1}^{K} \pi_k \mathbf{S}_k, \qquad \pi_k = \frac{n_k}{\sum_{j=1}^{K} n_j}. \tag{3.47}$$

Often better choice is to use a convex combination of the SCM and the pooled SCM; For example, [24] proposed to use the convex combination

$$\hat{\boldsymbol{\Sigma}}_k(\beta) = \beta \mathbf{S}_k + (1 - \beta)\mathbf{S}_{\text{pool}}, \tag{3.48}$$

as an estimate for the class covariance matrix, where $\beta \in [0, 1]$ is the tuning parameter. This partially pooled estimator is then further regularized toward a scaled identity matrix in order to stabilize its eigenvalues and guarantee positive definiteness of the estimator in low sample size settings ($p > n_k$ for some k):

$$\hat{\boldsymbol{\Sigma}}_k(\alpha, \beta) = \alpha \hat{\boldsymbol{\Sigma}}_k(\beta) + (1 - \alpha)\mathbf{I}_{\hat{\boldsymbol{\Sigma}}_k(\beta)}, \tag{3.49}$$

where $\hat{\boldsymbol{\Sigma}}_k(\beta)$ is given in (3.48), $\mathbf{I}_{\mathbf{A}} = (\text{tr}(\mathbf{A})/p)\mathbf{I}$ and $\alpha, \beta \in [0, 1]$ are tuning parameters. The author of [24] then proposed an RDA framework based on this estimator. Similar ideas but from Bayesian perspectives were developed in [22, 23].

3.5.1 Coupled RSCM

We call the estimator in (3.49) as the *coupled RSCM* estimator as it couples two different types of regularization. The task that remains is to determine the optimal tuning parameters $(\alpha_k, \beta_k) \in [0, 1] \times [0, 1]$, for $k = 1, \dots, K$. In RDA [24], one uses $\beta \equiv \beta_k$ and $\alpha \equiv \alpha_k$, i.e., same parameter pair is used *for all* classes $k = 1, \dots, K$, and then one picks up the best pair (α, β) from a grid of values using cross-validation. It is easy to criticize that such an approach is not only suboptimal but also computer-intensive. As a remedy, [25] proposed a data-adaptive approach for choosing class-specific choices (α_k, β_k) that minimize the $\mathsf{MSE}(\hat{\boldsymbol{\Sigma}}_k(\alpha, \beta))$ for each $k = 1, \dots, K$. This method is described in this section in more detail.

Before proceeding, it is worthwhile to point out four special instances of the estimator (3.49):

(C1) *The unpooled regularized SCM estimator* omits the pooled SCM and only shrinks toward the scaled identity matrix:

$$\hat{\boldsymbol{\Sigma}}_k(\alpha_k, \beta_k = 1) = \alpha_k \mathbf{S}_k + (1 - \alpha_k)\mathbf{I}_{\mathbf{S}_k}.$$

This type of shrinkage is typically considered in single-class covariance matrix estimation (see e.g., [11, 27]).

(C2) *The partially pooled estimator* omits regularization toward the scaled identity and only shrinks toward the pooled SCM:

$$\hat{\boldsymbol{\Sigma}}_k(\alpha_k = 1, \beta_k) = \hat{\boldsymbol{\Sigma}}_k(\beta_k) = \beta_k \mathbf{S}_k + (1 - \beta_k)\mathbf{S}.$$

(C3) *The fully pooled estimator* uses the pooled SCM for every class k and shrinks it toward the scaled identity matrix:

$$\hat{\boldsymbol{\Sigma}}_k(\alpha_k, \beta_k = 0) = \alpha_k \mathbf{S}_{\text{pool}} + (1 - \alpha_k)\mathbf{I}_{\mathbf{S}_{\text{pool}}}.$$

Such shrinkage can be considered if all classes have an identical distribution.

(C4) *The scaled identity estimator* uses the partially pooled estimator to scale the identity matrix:

$$\hat{\boldsymbol{\Sigma}}_k(\alpha_k = 0, \beta_k) = \mathbf{I}_{(\beta_k \mathbf{S}_k + (1-\beta_k)\mathbf{S}_{\text{pool}})}.$$

Since it is clear that the tuning parameters are class-specific, we drop the subscripts from α_k and β_k and denote them from now on simply by α and β.

The NMSE of Coupled RSCM and Estimates of Tuning Parameters

We adopt the Setup A from [25] consisting of $K = 4$ classes, which all follow an AR(1) covariance model in (3.35) with correlations $\varrho_k = (0.2, 0.3, 0.4, 0.5)$, sample sizes $n_k = (25, 50, 75, 100)$, and scales $\eta_k \equiv 1\ \forall k$. The data are generated from MVT distribution with d.o.f. $\nu = 8$. The dimension is $p = 200$. Figure 3.5 displays the NMSE of the 4th class $\hat{\mathbf{\Sigma}}_4(\alpha, \beta)$ in (3.49). The gray dots depict the estimated tuning parameters (showing 400 realizations of 4000 Monte Carlo trials) using the estimation method proposed in [25]. The black triangle (▲) identifies the optimal tuning parameter pair, and the blue square (■) depicts the mean of the estimated tuning parameters. One can notice that using the estimator (C3) would be beneficial in this case and using the estimated tuning parameters one obtains an estimator with MSE that is very close to the best possible oracle estimator.

An alternative, streamlined estimator to (3.49) was further proposed in [25] by changing the α-regularization target, and by defining the estimator as

$$\tilde{\mathbf{\Sigma}}_k(\alpha, \beta) = \alpha \hat{\mathbf{\Sigma}}_k(\beta) + (1 - \alpha)\mathbf{I}_{\mathbf{T}}, \tag{3.50}$$

where $\mathbf{T} \in \{\mathbf{S}_k, \mathbf{S}\}$ and $\hat{\mathbf{\Sigma}}_k(\beta)$ is defined in (3.48). This simplifies the expression for the MSE and allows for an analytical solution for the tuning parameters as given below.

Theorem 3.1 ([25, Theorem 3]) *The theoretical MSE of estimator* (3.50) *is a bivariate polynomial of the form*

$$\mathsf{MSE}(\tilde{\mathbf{\Sigma}}_k(\alpha, \beta)) = \alpha^2\beta^2 B_{22} + \alpha^2 \beta B_{21} + \alpha^2 B_{20} + \alpha\beta B_{11} + \alpha B_{10} + B_{00},$$

where the coefficients B_{ij} depend on the scalars $\eta_j = \mathrm{tr}(\mathbf{\Sigma}_j)/p$, $\mathsf{E}[\|\mathbf{S}_j\|_\mathrm{F}^2]$, $\mathsf{E}[\|\mathbf{I}_{\mathbf{S}_j}\|_\mathrm{F}^2]$, and $\langle \mathbf{\Sigma}_i, \mathbf{\Sigma}_j\rangle_\mathrm{F} = \mathrm{tr}(\mathbf{\Sigma}_i\mathbf{\Sigma}_j)$, and are detailed in [25, Appendix B]. If $(\alpha^\star, \beta^\star) \in (0, 1) \times (0, 1)$, the optimal tuning parameters $(\alpha^\star, \beta^\star)$ minimizing the MSE are

$$\alpha^\star = \frac{2B_{10}B_{22} - B_{11}B_{21}}{B_{21}^2 - 4B_{20}B_{22}} \quad \textit{and} \quad \beta^\star = \frac{2B_{11}B_{20} - B_{10}B_{21}}{2B_{10}B_{22} - B_{11}B_{21}}.$$

Otherwise, the optimal parameters are on the boundary of the feasible set $[0, 1] \times [0, 1]$, *and are given by one of the following options:*

(i) $\alpha^\star = \left[-\dfrac{1}{2}\dfrac{B_{10}}{B_{20}}\right]_0^1$ *and* $\beta^\star = 0$,

(ii) $\alpha^\star = \left[-\dfrac{1}{2}\dfrac{B_{10} + B_{11}}{B_{22} + B_{21} + B_{20}}\right]_0^1$ *and* $\beta^\star = 1$,

(iii) $\alpha^\star = 1$ *and* $\beta^\star = \left[-\frac{1}{2} \frac{B_{21} + B_{11}}{B_{22}} \right]_0^1$,

(iv) $\alpha^\star = 0$, *which implies* $\tilde{\boldsymbol{\Sigma}} = \mathbf{I}_\mathbf{T}$ *and that the MSE does not depend on* β.

Above the clip *function* $[c]_a^b = \max\{a, \min\{b, c\}\}$ *projects* c *on to the interval* $[a, b]$.

The unknown constants B_{ij} are replaced by their estimated values when constructing the streamlined estimator, again assuming that data are generated from unspecified elliptical distributions. This provides significant speed-up compared to previous approaches where one uses cross-validation to estimate the tuning parameters involved in the coupled RSCM. The coupled RSCM estimator was adapted and applied to a real data classification problem in the RDA framework in [25, Sects. V-B and VI-B] where the proposed method of estimating the MSE-optimal tuning parameters was compared to different types of cross-validation based methods. The proposed approach performed similarly to cross-validation (CV) in terms of classification accuracy but achieved the same performance with significant computational gain.

It should be emphasized that the main difference between (3.50) to (3.49) is that the trace of (3.50) depends on α, while in (3.49) it does not. However, when $\mathrm{tr}(\mathbf{I}_\mathbf{T}) \approx \mathrm{tr}(\mathbf{I}_{\hat{\boldsymbol{\Sigma}}_k(\beta)})$, the performance of the two estimators is expected to be similar. Simulation results in [25, Table I] illustrate that neither the coupled RSCM in (3.49) nor the streamlined estimator (3.50) was always better than the other. The codes to compute the coupled RSCM or streamlined RSCM with MSE-optimal estimated tuning parameters are available in MATLAB, R, and Python programming languages at https://github.com/EliasRaninen/CoupledRSCM.

3.5.2 Linear Pooling of Sample Covariance Matrices

In [20], a method is proposed to estimate each class covariance matrix as a linear combination of the SCM-s of the classes. For a vector of nonnegative weights $\mathbf{a} \geq \mathbf{0}$, i.e., $\mathbf{a} = (a_i)$, $a_i \geq 0$, $i = 1, \ldots, K$, one defines

$$\mathbf{S}(\mathbf{a}) = \sum_{i=1}^{K} a_i \mathbf{S}_i. \tag{3.51}$$

Restricting the coefficients to be nonnegative ensures that the estimator is positive semidefinite. The goal is to find a $K \times K$ nonnegative coefficient matrix $\mathbf{A}^\star = (\mathbf{a}_1^\star \; \cdots \; \mathbf{a}_K^\star)$ where

$$\mathbf{a}_k^\star = \arg\min_{\mathbf{a} \geq \mathbf{0}} \mathsf{E}\big[\|\mathbf{S}(\mathbf{a}) - \boldsymbol{\Sigma}_k\|_\mathrm{F}^2\big], \; k = 1, \ldots, K. \tag{3.52}$$

Let us define a diagonal matrix consisting of scaled MSE-s of the SCM-s as its diagonal elements as

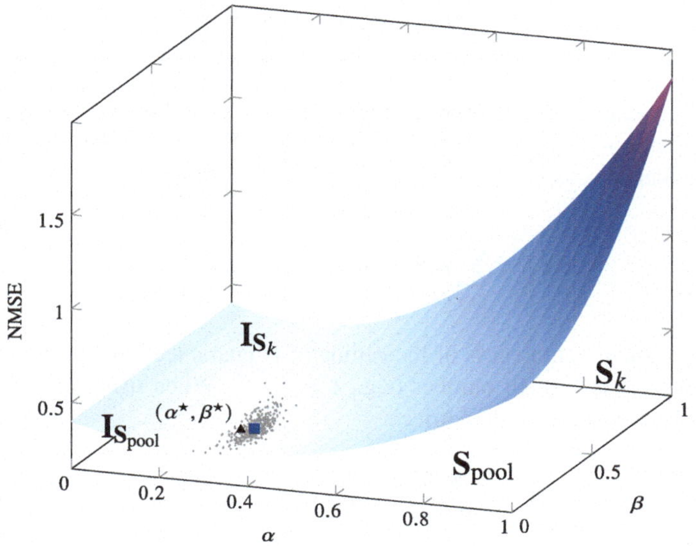

Fig. 3.5 NMSE of $\hat{\boldsymbol{\Sigma}}(\alpha, \beta)$ for the AR(1) covariance model in (3.35) with $\varrho_k = (0.2, 0.3, 0.4, 0.5)$, $n_k = (25, 50, 75, 100)$, dimension $p = 200$, and sampling from MVT distributions with $\nu = 8$ d.o.f

$$\boldsymbol{\Delta} = \text{diag}(\delta_1, \ldots, \delta_K), \quad \delta_k = p^{-1}\mathsf{E}[\|\mathbf{S}_k - \boldsymbol{\Sigma}_k\|_{\text{F}}^2] \tag{3.53}$$

as well as the matrix of scaled inner products of the covariance matrices as

$$\mathbf{C} = \big(\mathbf{c}_1 \cdots \mathbf{c}_K\big) = (c_{ij}) = \big(p^{-1}\,\text{tr}(\boldsymbol{\Sigma}_i\boldsymbol{\Sigma}_j)\big). \tag{3.54}$$

We can then state the following result.

Theorem 3.2 ([20, Propositions 1, 2]) *The scaled MSE in* (3.52) *can be written as*

$$p^{-1}\mathsf{E}[\|\mathbf{S}_k - \boldsymbol{\Sigma}_k\|_{\text{F}}^2] = \mathbf{a}^\top(\boldsymbol{\Delta} + \mathbf{C})\mathbf{a} - 2\mathbf{c}_k^\top\mathbf{a} + c_{kk}, \tag{3.55}$$

where $\boldsymbol{\Delta}$ *and* $\mathbf{C}$ *are defined in* (3.53) *and* (3.54), *respectively. Furthermore,* $\boldsymbol{\Delta} + \mathbf{C}$ *is a positive definite symmetric matrix, and hence the MSE is a strictly convex quadratic function in* $\mathbf{a}$. *The unconstrained solution, which minimizes the MSE in* (3.55) *is*

$$\mathbf{a}_k^\star = (\boldsymbol{\Delta} + \mathbf{C})^{-1}\mathbf{c}_k \Leftrightarrow \mathbf{A}^\star = (\boldsymbol{\Delta} + \mathbf{C})^{-1}\mathbf{C}. \tag{3.56}$$

It is important to notice that if the solution (3.56) to the unconstrained problem is also nonnegative, i.e., verifies $\mathbf{a}_k^\star \geq 0$, then it is solution also to the constrained problem. If this is not the case, then the solution is found by solving the strictly convex quadratic programming (QP) problem

$$\begin{aligned} &\text{minimize } \tfrac{1}{2}\mathbf{a}^\top(\boldsymbol{\Delta}+\mathbf{C})\mathbf{a} - \mathbf{c}_k^\top \mathbf{a} \\ &\text{subject to } \mathbf{a} \geq \mathbf{0}. \end{aligned} \tag{3.57}$$

It is often beneficial to incorporate regularization toward the identity matrix. For example, if $p > n = \sum_k n_k$, then all of the SCM-s $\mathbf{S}_k$ are singular. Regularization toward the identity can easily be added by using the estimator

$$\tilde{\mathbf{S}}(\mathbf{a}) = \sum_{i=1}^{K} a_i \mathbf{S}_j + a_I \mathbf{I}, \quad a_i \geq 0, a_I > \epsilon, \tag{3.58}$$

where the positive definiteness of the estimator is guaranteed due to the constraint $a_I > \epsilon$, where ϵ is a small number (e.g., $\epsilon = 10^{-6}$). When using (3.58), one can simply replace $\boldsymbol{\Delta}$ and $\mathbf{C}$ with matrices

$$\tilde{\boldsymbol{\Delta}} = \begin{pmatrix} \boldsymbol{\Delta} & \mathbf{0} \\ \mathbf{0}^\top & 0 \end{pmatrix} \quad \text{and} \quad \tilde{\mathbf{C}} = \begin{pmatrix} \mathbf{C} & \boldsymbol{\eta} \\ \boldsymbol{\eta}^\top & 1 \end{pmatrix}, \tag{3.59}$$

where $\boldsymbol{\eta} = (\eta_1, \ldots, \eta_K)^\top$ is a vector consisting of scales $\eta_k = \mathrm{tr}(\boldsymbol{\Sigma}_k)/p$ of $\boldsymbol{\Sigma}_k$-s. The coefficient vector $\mathbf{a} = (a_1, \ldots, a_K, a_I)^\top$ that minimize the MSE $\mathsf{E}\big[\|\tilde{\mathbf{S}}(\mathbf{a}) - \boldsymbol{\Sigma}_k\|_{\mathrm{F}}^2\big]$ under the stated constraints in (3.58) can be found by solving the following strictly convex QP problem

$$\begin{aligned} &\text{minimize } \tfrac{1}{2}\mathbf{a}^\top(\tilde{\boldsymbol{\Delta}}+\tilde{\mathbf{C}})\mathbf{a} - \tilde{\mathbf{c}}_k^\top \mathbf{a} \\ &\text{subject to } a_j \geq 0, \ j = 1, \ldots, K, a_I \geq \epsilon. \end{aligned} \tag{3.60}$$

The QP formulation of the problem makes it easy to incorporate additional constraints if needed. For example, in order to find a convex combination of the SCM-s the equality constraint $\mathbf{1}^\top \mathbf{a} = 1$ can be added to the QP (3.60). Such constraint may be preferred in the case that the different population covariance matrices have similar scales, so $\eta_j \approx \eta_k$.

Linearly pooled estimator (3.51) offers more flexibility than the partially pooled estimator (3.48) as it has individual weights for every class SCM. Same holds for their modifications (i.e., (3.58) versus (3.49)). Linearly pooled estimator requires estimation of more coefficients, and thus errors in these estimates may impact its performance. Another benefit of coupled estimator is that it has a similar form as the popular estimator used in RDA and it can thus be easily applied to discriminant analysis classification problems without any modifications. Codes for computing the linear pooled estimator are available at https://github.com/EliasRaninen/LinearPoolingOfSampleCovarianceMatrices.

3.6 Application to Portfolio Selection

Portfolio selection and optimization is one of the most important topics in investment theory. It is a mathematical framework wherein one seeks portfolio allocations

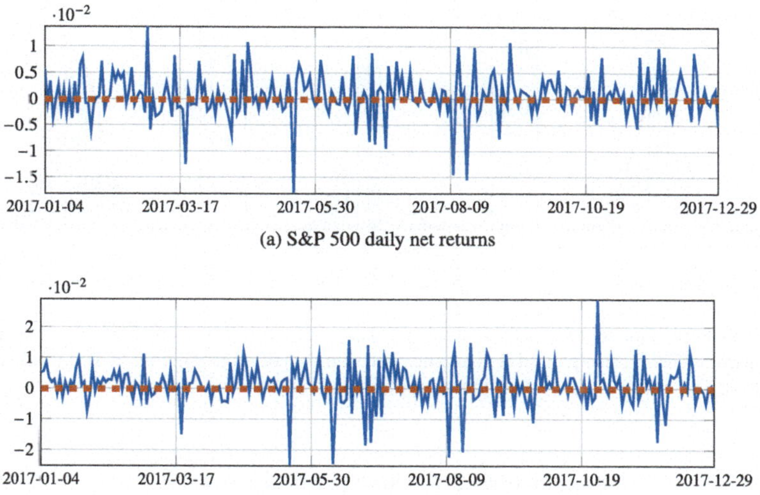

(b) Nasdaq-100 daily net returns

Fig. 3.6 Daily net returns of the stock indices for year 2017

which balance the return-risk tradeoff such that it satisfies the investor's needs. Some historical key references are [39–43].

We consider a portfolio P that consists of p assets which can be stocks, bonds, currencies, exchange-traded funds (ETF-s), etc. We assume that assets are held for a fixed investment period (e.g., 1 month, 1 year). The net return of the ith asset at time t is

$$r_{i,t} = \frac{p_{i,t} - p_{i,t-1}}{p_{i,t-1}} = \frac{p_{i,t}}{p_{i,t-1}} - 1 \in [-1, \infty), \tag{3.61}$$

where $p_{i,t}$ denotes the price of ith asset at time t.

The original time series of stock prices $p_{i,t}$ is not a stationary time series, but it can be argued that a return time series $r_{i,t}$ is close to stationarity within a fixed sufficiently short time period. This is illustrated in Fig. 3.6 which displays daily net returns of Standard & Poor's 500 (S&P 500) and Nasdaq-100 stock indexes for year 2017. Daily net returns are heavy-tailed and non-Gaussian distributed, having occasional large negative or positive returns. Overall, the returns are observed to fluctuate around zero which is displayed by the dotted red line in the figure.

The objective in portfolio optimization is to find optimal portfolio weights which determine the proportion of wealth that is to be invested in each particular asset. That is, a fraction $w_i \in \mathbb{R}$ of the total wealth is invested in the ith asset, $i = 1, \ldots, p$, and the portfolio with p assets is described by the portfolio *weight* or *allocation vector* $\mathbf{w} \in \mathbb{R}^p$ which satisfies the constraint $\mathbf{1}^\top \mathbf{w} = 1$. The *global mean variance portfolio* (GMVP) aims at finding the weight vector that minimizes the portfolio variance (risk or volatility), and hence does not require specifying the mean vector. The GMVP optimization problem is

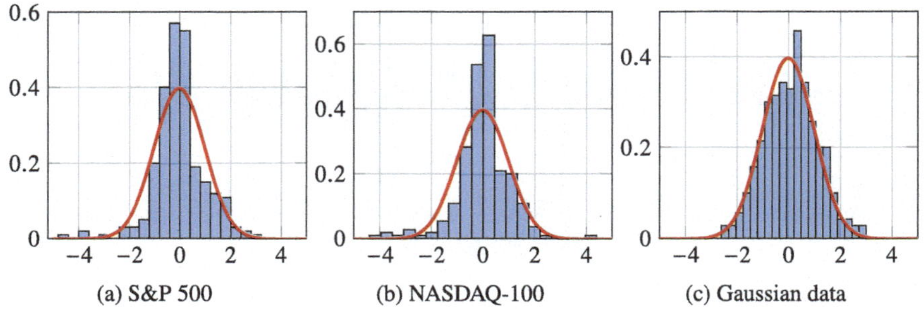

Fig. 3.7 Empirical histograms of standardized daily net returns of S&P 500 and Nasdaq-100 indexes for year 2017. Also plotted is synthetic Gaussian data of same length $n = 100$ from $\mathcal{N}(0, 1)$ distribution. The pdf of $\mathcal{N}(0, 1)$ distribution is plotted in red

$$\underset{\mathbf{w}\in\mathbb{R}^p}{\text{minimize}}\ \mathbf{w}^\top\boldsymbol{\Sigma}\mathbf{w} \quad \text{subject to} \quad \mathbf{1}^\top\mathbf{w} = 1, \tag{3.62}$$

where $\boldsymbol{\Sigma}$ is the covariance matrix of $\mathbf{r}_t = (r_{1,t}, \ldots, r_{p,t})^\top$. The solution to (3.62) is

$$\mathbf{w}_o = \frac{\boldsymbol{\Sigma}^{-1}\mathbf{1}}{\mathbf{1}^\top\boldsymbol{\Sigma}^{-1}\mathbf{1}}. \tag{3.63}$$

Naturally, the covariance matrix is unknown and needs to be estimated from the historical data.

3.6.1 Are Stock Returns Gaussian?

Let us first investigate the hypothesis that the daily net returns of stocks are Gaussian.

Let us start by plotting the histograms of historical daily net returns. These are shown in Fig. 3.7a,b which display the histograms of standardized daily net returns of S&P 500 and Nasdaq-100 indexes. For better comparison of Gaussianity assumption, Fig. 3.7c displays histogram of one realization from a standard Gaussian distribution $\mathcal{N}(0, 1)$ of same length ($n = 100$). Also shown is the p.d.f. of $\mathcal{N}(0, 1)$ distribution plotted in red color. As can be noted, the histograms of daily net returns are not well matched with Gaussian distribution. Instead, we observe that the empirical distribution is more peaked and heavier-tailed. In fact, when Student's t-distribution is fitted to daily log-returns on stocks, it has been observed that the number of degrees of freedom typically lies between 3 and 7 (see e.g., [44, p. 85]).

Figure 3.8 displays the scatter plots of Nasdaq-100 and S&P 500 historical daily net returns for the whole year 2017 and the estimated 99%, 95%, and 50% tolerance ellipses computed using the SCM. Overall, 95.6%, 93.2%, and 65.6% of observations lie inside the 99%, 95%, and 50% tolerance ellipses, respectively. The figure and the obtained numbers further illustrate that the joint distribution of returns is more

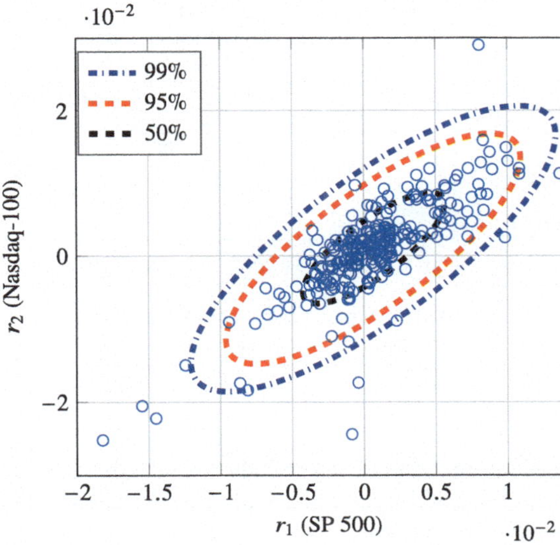

Fig. 3.8 Scatter plots of daily net returns of Nasdaq-100 and S&P 500 over year 2017 and the estimated 50%, 95%, and 99% tolerance ellipses based on the SCM

peaked (concentrated around the mean) and heavier-tailed than bivariate Gaussian distribution as there are many observations that lie outside the 99% tolerance ellipses. Hence, it is fair to say that the joint distribution is not well modeled by the MVN distribution. Instead, an ES distribution that is more peaked and heavier-tailed can provide a better fit.

Although many studies illustrate that for individual stocks or stock index, the value of $\nu > 0$ is often very small, this may not be true when constructing a portfolio over a large set of stocks. Also note that $\nu > 1$, $\nu > 2$, and $\nu > 4$ are required for the existence of first-, second-, and fourth-order moments, respectively. To investigate this, we considered 129 stocks in OMX Helsinki and their daily log returns for each year from 2015 to 2022. This means that each year we have roughly 252 return values on 129 stocks. However, for a given year, we deleted stocks from our analysis that had missing values or several consecutive days of 0 returns. We fitted MVT distribution to the yearly log return data, where the d.o.f. ν was estimated using OPP estimator [28, Algorithm 1] and TWE estimator[3] [46]. As can be noted from Fig. 3.9, the estimated values of ν based on TWE ranges from 5.3 in year 2020 to 13.5 in year 2021, while OPP obtains values from 5.7 in year 2020 to 15.5 in year 2021. Thus, only the year 2020 due to a sudden fall in stock prices, due to the COVID pandemic, indicates a very heavy-tailed MVT distribution. However, the non-Gaussianity is clear from these estimated values.

[3] In the R package `fitHeavyTail` [45], the function `fit_Tyler` implements this method.

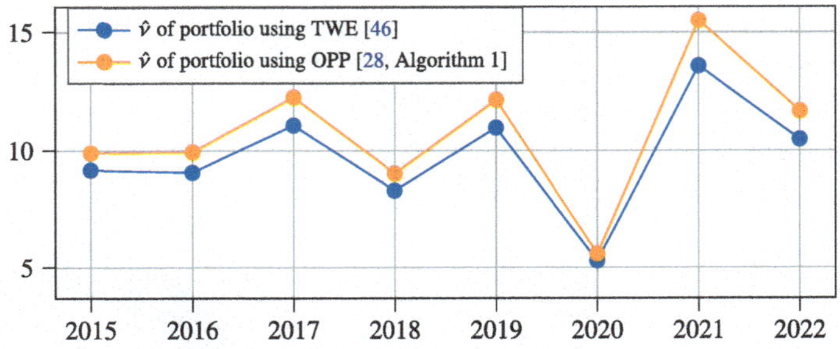

Fig. 3.9 Estimated d.o.f. parameter ν of MVT distribution for OMX Helsinki stock data based on historical daily net returns for each year

The empirical data analysis thus testify that daily return data is not Gaussian but rather better modeled with a heavy-tailed ES distribution. Yet, since the data is not extremely heavy-tailed (as suggested by the obtained estimates of d.o.f. parameter ν), we can anticipate that the SCM $\mathbf{S}$ can be an effective estimator of the covariance matrix for portfolio optimization problems. However, it is important to take into account the fact that the data is non-Gaussian but has higher peakedness and heavier tails. This is the case for linear shrinkage estimators that are reviewed in this chapter since they only assume ellipticity but do not specify the underlying ES distribution.

3.6.2 *Portfolio Analysis*

We now test the performance of RSCM estimators in portfolio optimization using GMVP portfolio selection and historical data. We investigate the out-of-sample portfolio performance of different covariance matrix estimators for three different datasets. The first and second datasets consist of daily net returns of $p = 45$ and $p = 50$ stocks, respectively, that are included in the Hang Seng Index (HSI) from January 4, 2010, to December 24, 2011, and from January 1, 2016, to December 27, 2017, both consisting of $T = 491$ trading days. The third dataset consists of daily net returns of $p = 396$ stocks included in S&P 500 from January 4, 2016, to April 27, 2018, consisting of $T = 583$ trading days.

At a particular day t, we used the previous n days (i.e., from $t - n$ to $t - 1$) as the training window to estimate the covariance matrix, and the portfolio weight vector. The estimated GMVP weight vector $\hat{\mathbf{w}}_o$ was then used to compute the portfolio returns for the following 20 days. (Note that $\hat{\mathbf{w}}_o$ is computed as in (3.63) but unknown $\boldsymbol{\Sigma}$ replaced by its estimate $\hat{\boldsymbol{\Sigma}}$). Next, the window was shifted 20 trading days forward, a new weight vector was computed, and the portfolio returns for another 20 days were

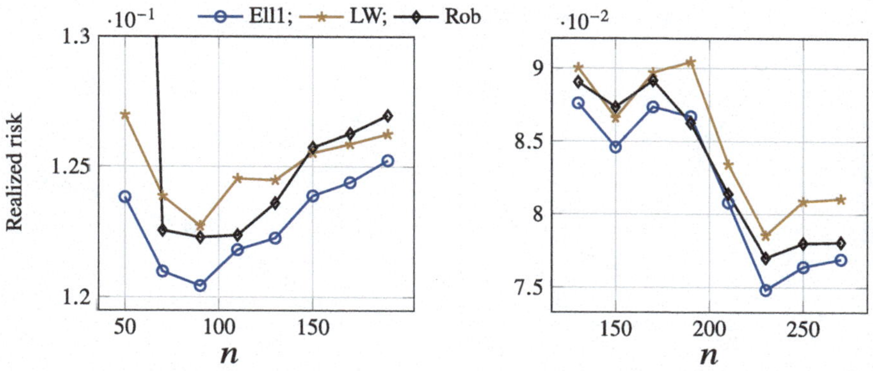

(a) HSI for Jan. 4, 2010 to Dec. 24, 2011. (b) HSI for Jan. 1, 2016 to Dec. 27, 2017)

Fig. 3.10 Annualized realized portfolio risk achieved out-of-sample for the two HSI datasets. The portfolio allocations are obtained using GMVP based on the three different covariance estimators (see text) and different training window lengths n

computed. Hence, this scenario corresponds to the case that the portfolio manager holds the assets for approximately a month (20 trading days), after which they are liquidated and new weights are computed. In this manner, we obtained $T - n$ daily returns from which the realized risk was computed as the sample standard deviation of the obtained portfolio returns. To obtain the annualized realized risk, the sample standard deviations of the daily returns were multiplied by $\sqrt{250}$. In our tests, different training window lengths n were considered.

In our analysis, we compare three different covariance matrix estimators: RSCM-Ell1 [27] described in Sect. 3.4.1 which is compared to RSCM estimator by Ledoit and Wolf [11]. These two estimators both use RSCM in (3.42), defined by

$$\hat{\boldsymbol{\Sigma}}_{\text{RSCM}} = \hat{\beta}_o \mathbf{S} + (1 - \hat{\beta}_o)[\text{tr}(\mathbf{S})/p]\mathbf{I},$$

while they differ only in the approaches to compute $\hat{\beta}_o$. The former utilizes the ellipticity assumption, while the latter builds upon random matrix theory. We also included in our study the robust GMVP weight estimator proposed in [47] that uses a robust regularized Tyler's M-estimator with a tuning parameter selection that is optimized for the GMVP problem. The three estimators are denoted shortly as **Ell1**, **LW**, and **Rob** in the text and figure captions.

Figure 3.10 displays the annualized realized risks for the HSI dataset. Overall, we can notice that RSCM-Ell1 has the best performance for all window lengths and for both periods. For the period 2016–2017, the differences between the estimators were not as large as in the period 2010–2011. Also, note that the optimal training window length which yielded the smallest realized risk was $n = 90$ for the period 2010–2011, but much larger ($n = 230$) for the period 2016–2017. This could be explained by the fact that the stock market were more turbulent in the first period.

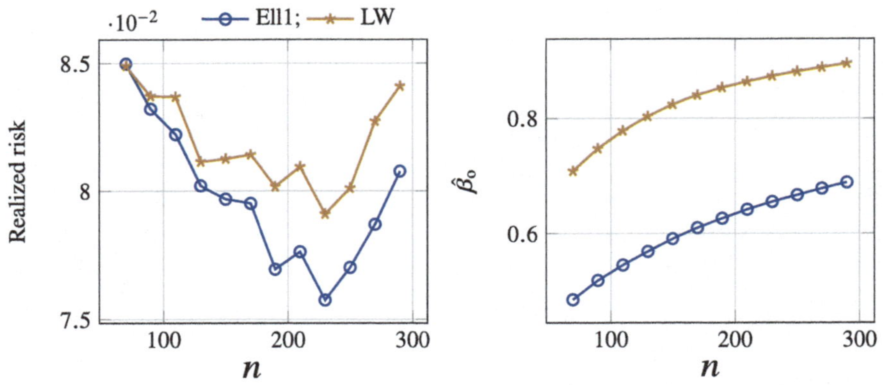

Fig. 3.11 Annualized realized portfolio risk achieved out-of-sample over 583 trading days for a portfolio consisting of $p = 396$ stocks in S&P 500 index for January 4, 2016, to April 27, 2018. The portfolio allocations are obtained using GMVP based on two RSCM estimators (Ell1 and LW) and different training window lengths n. The right panel shows the average $\hat{\beta}_o$ of the RSCM estimators for different training window lengths

The left panel of Fig. 3.11 depicts the annualized realized risks of RSCM-Ell1- and -LW estimators for S&P 500 data. We have excluded the Rob estimator [47] from this study as it is not well suited for very high-dimensional problems. With the S&P 500 data, RSCM-Ell1 achieves the smallest realized risk and outperformed RSCM-LW for all training window lengths n. The optimal training window length which produced the smallest realized risk was $n = 230$ for both methods. Note that, the same result was achieved with HSI data for the period 2016–2017. The right panel of Fig. 3.11 displays the estimated optimal shrinkage parameter $\hat{\beta}_o$ used by the methods. As can be noted, RSCM-LW estimator uses much larger estimate of β_o and thus puts much more weight on the SCM $\mathbf{S}$ than RSCM-Ell1.

3.7 Conclusions

This chapter reviewed methods for linear shrinkage of the SCM(-s) under elliptical distributions in both the single and multiple population settings. Specifically, we considered approaches for choosing the shrinkage parameters that minimize the MSE.

In the single population setting, we reviewed the RSCM estimator proposed in [27, 48] and its generalization called TABASCO [3] that imposes tapering/banding templates to SCM, and thus allows imposing structure to the covariance matrix estimator. In the multiple population setting, we reviewed the coupled RSCM estimator [25] and its generalization, the linearly pooled estimator proposed in [20].

It should be emphasized that only linear shrinkage of SCM was considered in this chapter. Another popular approach is *nonlinear shrinkage* methods which perform nonlinear shrinkage to the eigenvalues $\hat{\lambda}_i, i = 1, \ldots, p$, of SCM $\mathbf{S}$. These estimators are written in the form

$$\hat{\boldsymbol{\Sigma}} = \sum_{i=1}^{p} \phi_i(\hat{\lambda}_i)\mathbf{u}_i\mathbf{u}_i^\top,$$

where $\phi_i : \mathbb{R}_{\geq 0} \to \mathbb{R}_{\geq 0}$ is a nonnegative function and $\mathbf{u}_1, \ldots \mathbf{u}_p$ are the eigenvector of $\mathbf{S}$. Such nonlinear shrinkage approaches often rely upon random matrix theory in their design of the function ϕ_i, see e.g., [49–51].

We also did not cover penalized SCM-s, obtained by adding a penalty term on the covariance matrix to the Gaussian negative log-likelihood function (see e.g., [52–55]). Also note that when a penalty term $\mathrm{tr}(\boldsymbol{\Sigma}^{-1})$ is added to a (scaled) Gaussian negative log-likelihood, one recovers the regularized SCM in (3.39) as the unique solution [53].

Acknowledgements I would like to thank my former doctoral student Elias Raninen whose work has been essential in many of the results presented in this chapter. I would also like to thank my collaborators David E. Tyler, Frédéric Pascal, Daniel P. Palomar, and Arnaud Breloy for many discussions regarding covariance matrix estimation and life.

References

1. Bickel, P.J., Levina, E.: Regularized estimation of large covariance matrices. Ann. Stat. **36**(1), 199–227 (2008)
2. Bickel, P.J., Levina, E.: Covariance regularization by thresholding. Ann. Stat. **36**(6), 2577–2604 (2008)
3. Ollila, E., Breloy, A.: Regularized tapered sample covariance matrix. IEEE Trans. Signal Process. **70**, 2306–2320 (2022)
4. Fang, K.-T., Kotz, S., Ng, K.-W.: Symmetric Multivariate and Related Distributions. Chapman and Hall, London (1990)
5. Stein, C.M.: Estimation of the mean of a multivariate normal distribution. Ann. Stat. **9**(6), 1135–1151 (1981)
6. Stein, C.: Some problems in multivariate analysis. Department of Statistics, Stanford University, Technical Report No. 6 (1956)
7. James, W., Stein, C.: Estimation with quadratic loss. In: Proceedings of the Fourth Berkeley Symposium on Mathematical Statistics and Probability, vol. 1, pp. 361–379 (1961)
8. Efron, B., Morris, C.: Stein's estimation rule and its competitors-an empirical bayes approach. J. Am. Stat. Assoc. **68**(341), 117–130 (1973)
9. Haff, L.: Empirical bayes estimation of the multivariate normal covariance matrix. Ann. Stat. **8**(3), 586–597 (1980)
10. Paindaveine, D.: A canonical definition of shape. Stat. & Probab. Lett. **78**(14), 2240–2247 (2008)
11. Ledoit, O., Wolf, M.: A well-conditioned estimator for large-dimensional covariance matrices. J. Mult. Anal. **88**(2), 365–411 (2004)
12. Carlson, B.D.: Covariance matrix estimation errors and diagonal loading in adaptive arrays. IEEE Trans. Aerosp. Electron. Syst. **24**(4), 397–401 (1988)
13. Li, J., Stoica, P., Wang, Z.: On robust Capon beamforming and diagonal loading. IEEE Trans. Signal Process. **51**(7), 1702–1715 (2003)

14. Du, L., Li, J., Stoica, P.: Fully automatic computation of diagonal loading levels for robust adaptive beamforming. IEEE Trans. Aerosp. Electron. Syst. **46**(1), 449–458 (2010)
15. Ledoit, O., Wolf, M.: Improved estimation of the covariance matrix of stock returns with an application to portfolio selection. J. Empir. Financ. **10**(5), 603–621 (2003)
16. Ledoit, O., Wolf, M.: Honey, i shrunk the sample covariance matrix. J. Portfolio Manag. **30**(4), 110–119 (2004)
17. Lancewicki, T., Aladjem, M.: Multi-target shrinkage estimation for covariance matrices. IEEE Trans. Signal Process. **62**(24), 6380–6390 (2014)
18. Bartz, D., Höhne, J., Müller, K.-R.: Multi-target shrinkage (2014). arXiv:1412.2041
19. Tong, J., Hu, R., Xi, J., Xiao, Z., Guo, Q., Yu, Y.: Linear shrinkage estimation of covariance matrices using low-complexity cross-validation. Signal Process. **148**, 223–233 (2018)
20. Raninen, E., Tyler, D.E., Ollila, E.: Linear pooling of sample covariance matrices. IEEE Trans. Signal Process. **70**, 659–672 (2021)
21. Besson, O.: Maximum likelihood covariance matrix estimation from two possibly mismatched data sets. Signal Process. **167**, 107285 (2020)
22. Greene, T., Rayens, W.S.: Partially pooled covariance matrix estimation in discriminant analysis. Commun. Stat.-Theory Methods **18**(10), 3679–3702 (1989)
23. Rayens, W., Greene, T.: Covariance pooling and stabilization for classification. Comput. Stat. & Data Anal. **11**(1), 17–42 (1991)
24. Friedman, J.H.: Regularized discriminant analysis. J. Amer. Stat. Assoc. **84**(405), 165–175 (1989)
25. Raninen, E., Ollila, E.: Coupled regularized sample covariance matrix estimator for multiple classes. IEEE Trans. Signal Process. **69**, 5681–5692 (2021)
26. Muirhead, R.J.: Aspects of Multivariate Statistical Theory, 704 p. Wiley, New York (1982)
27. Ollila, E., Raninen, E.: Optimal shrinkage covariance matrix estimation under random sampling from elliptical distributions. IEEE Trans. Signal Process. **67**(10), 2707–2719 (2019). May
28. Ollila, E., Palomar, D.P., Pascal, F.: Shrinking the eigenvalues of M-estimators of covariance matrix. IEEE Trans. Signal Process. **69**, 256–269 (2021)
29. Cai, T.T., Zhang, C.-H., Zhou, H.H., et al.: Optimal rates of convergence for covariance matrix estimation. Ann. Stat. **38**(4), 2118–2144 (2010)
30. Visuri, S., Koivunen, V., Oja, H.: Sign and rank covariance matrices. J. Statist. Plann. Inference **91**, 557–575 (2000)
31. Brown, B.: Statistical uses of the spatial median. J. R. Stat. Soc. Ser. B (Methodological) 25–30 (1983)
32. Donoho, D.L., Huber, P.J.: The notion of breakdown point. A festschrift for Erich L. Lehmann, vol. 157184 (1983)
33. Magyar, A.F., Tyler, D.E.: The asymptotic inadmissibility of the spatial sign covariance matrix for elliptically symmetric distributions. Biometrika **101**(3), 673–688 (2014)
34. Croux, C., Dehon, C., Yadine, A.: The k-step spatial sign covariance matrix. Adv. Data Anal. Classif. **4**(2), 137–150 (2010)
35. Dümbgen, L., Tyler, D.E.: On the breakdown properties of some multivariate M-functionals. Scand. J. Stat. **32**(2), 247–264 (2005)
36. Zou, C., Peng, L., Feng, L., Wang, Z.: Multivariate sign-based high-dimensional tests for sphericity. Biometrika **101**(1), 229–236 (2014)
37. Zhang, T., Wiesel, A.: Automatic diagonal loading for Tyler's robust covariance estimator. In: IEEE Statistical Signal Processing Workshop (SSP'16), pp. 1–5 (2016)
38. Ollila, E., Kim, H.-J.: On robust estimators of a sphericity measure in high dimension. In: Robust and Multivariate Statistical Methods: Festschrift in Honor of David E. Tyler, pp. 179–195. Springer (2022)
39. Markowitz, H.: Portfolio selection. J. Financ. **7**(1), 77–91 (1952)
40. Markowitz, H.: Portfolio Selection, Efficient Diversification of Investments. J. Wiley (1959)
41. Tobin, J.: Liquidity preference as behavior towards risk. Rev. Econ. Stud. **25**(2), 65–86 (1958)
42. Sharpe, W.F.: Capital asset prices: A theory of market equilibrium under conditions of risk. J. Financ. **19**(3), 425–442 (1964)

43. Lintner, J.: The valuation of risk assets and the selection of risky investments in stock portfolios and capital budgets. Rev. Econ. Stat. 13–37 (1965)
44. McNeil, A.J., Frey, R., Embrechts, P.: Quantitative Risk Management: Concepts, Techniques and Tools, Revised Edition. Princeton University Press (2005)
45. Palomar, D.P., Zhou, R., Wang, X., Pascal, F., Ollila, E.: fitHeavyTail: mean and covariance matrix estimation under heavy tails (2023). r package version 0.2.0. https://CRAN.R-project.org/package=fitHeavyTail
46. Ollila, E., Palomar, D.P., Pascal, F.: Affine equivariant Tyler's M-estimator applied to tail parameter learning of elliptical distributions. IEEE Signal Process. Lett. **30**, 1017–1021 (2023)
47. Yang, L., Couillet, R., McKay, M.R.: A robust statistics approach to minimum variance portfolio optimization. IEEE Trans. Signal Process. **63**(24), 6684–6697 (2015)
48. Ollila, E.: Optimal high-dimensional shrinkage covariance estimation for elliptical distributions. In: Proceeding of the European Signal Processing Conference (EUSIPCO 2017), Kos, Greece, pp. 1689–1693 (2017)
49. Bun, J., Bouchaud, J.-P., Potters, M.: Cleaning large correlation matrices: tools from random matrix theory. Phys. Rep. **666**, 1–109 (2017)
50. Ledoit, O., Wolf, M.: Analytical nonlinear shrinkage of large-dimensional covariance matrices. Ann. Stat. **48**(5), 3043–3065 (2020)
51. Donoho, D.L., Gavish, M., Johnstone, I.M.: Optimal shrinkage of eigenvalues in the spiked covariance model. Ann. Stat. **46**(4), 1742 (2018)
52. Deng, X., Tsui, K.-W.: Penalized covariance matrix estimation using a matrix-logarithm transformation. J. Comput. Graph. Stat. **22**(2), 494–512 (2013)
53. Ollila, E., Tyler, D.E.: Regularized M-estimators of scatter matrix. IEEE Trans. Signal Process. **62**(22), 6059–6070 (2014)
54. Yi, M., Tyler, D.E.: Shrinking the covariance matrix using convex penalties on the matrix-log transformation. J. Comput. Graph. Stat. **30**(2), 442–451 (2020)
55. Tyler, D.E., Yi, M.: Lassoing eigenvalues. Biometrika **107**(2), 397–414 (2020)

Chapter 4
Robust Estimation with Missing Values for Elliptical Distributions

Alexandre Hippert-Ferrer and Mohammed Nabil El Korso

Abstract In this chapter, we tackle the problem of robust estimation of the mean and the covariance matrix when the data contains missing values. Classical estimation methods either assume a multivariate Gaussian distribution, or suppose an unstructured covariance matrix. However, in many applications, the signal is not well described by a Gaussian model, and very often the data can be efficiently approximated by a low-rank model, inducing a low-rank structure on the covariance matrix which naturally accounts for the underlying signal subspace. By making the most of both (i) robustness to non-Gaussianity and (ii) low-rank structure, this chapter reviews various robust estimation procedures based mainly on the expectation–maximization algorithm which leverages the observed-data likelihood function, where the signal heterogeneity is accounted for through a deterministic scale parameter. Furthermore, the proposed algorithms are designed to handle various patterns of missing values. At the end of the chapter, the performances of the proposed procedures are illustrated on simulated datasets with missing values. We share a link to a code repository for fully reproducible experiments.

4.1 Introduction

The problem of missing data in statistical signal processing has been addressed a long time ago. In 1932, Wilks [81] obtained efficient estimators for the parameters of a normal bivariate distribution when the data contains missing values. This work was extended to the multivariate case by Lord [43] in 1955. Since the early 1970s, the literature in missing data has flourished with the development of computational capacity, leading to major developments in statistical signal processing and

A. Hippert-Ferrer (✉)
Université Gustave Eiffel, IGN-ENSG, Lastig, Champs-sur-Marne, France
e-mail: alexandre.hippert-ferrer@ign.fr

M. N. El Korso
Laboratoire des signaux et systèmes, Université Paris-Saclay, CNRS, CentraleSupélec, 91190 Gif-sur-Yvette, Gif-sur-Yvette, France
e-mail: mohammed.el-korso@universite-paris-saclay.fr

J.-P. Delmas et al. (eds.), *Elliptically Symmetric Distributions in Signal Processing and Machine Learning*, https://doi.org/10.1007/978-3-031-52116-4_4

its related fields, as statistical inference [35, 67] and data analysis [5, 78]. Missing data is everywhere, and it extends to extremely broad applications, as, to name a few, remote sensing [70], longitudinal studies [21], direction-of-arrival estimation [33], ecological statistics [53], chemometrics [79], or political science [32].

Statistical parameter estimation, which is a core subject in signal processing and machine learning, has witnessed particular efforts focusing on the case of incomplete data. One popular approach is to rely on maximum-likelihood (ML) estimation with a prior assumption on the probability distribution of the data. Within this framework, the expectation–maximization (EM) algorithm [14] is a convenient iterative procedure to obtain ML estimates when the ML of the unknown parameters $\boldsymbol{\theta}$ would be easier if the latent (missing) variables $\mathbf{z}$ is assumed known.

Very often, the frameworks using the EM algorithm with missing values assume independent and identically distributed (iid) samples drawn from the Gaussian distribution [4, 27, 40, 44, 68, 72]. Nevertheless, the Gaussian assumption is not realistic in a plethora of signal processing applications, e.g., in high-resolution sensing systems, non-homogeneous environments or in the possible presence of outliers [1, 17, 22, 46, 56, 58, 74, 82]. To overcome these limitations (which also stand for incomplete data), extensive efforts have been put into the development of robust estimation algorithms by considering data drawn from alternative distributions within the real elliptical symmetric (RES) family, as the Student's t−distribution [37, 39, 41], the generalized elliptical symmetric (GES) distribution [18, 19] or the angular Gaussian (AG) distribution [52].

In addition to the data distribution, the structure of the unknown parameter is also of interest and can be exploited to obtain more efficient estimators. For example, models have been considered in which the covariance is assumed to have the structure

$$\boldsymbol{\Sigma} = \sigma^2 \mathbf{I} + \mathbf{H}, \tag{4.1}$$

where $\mathbf{H}$ is a symmetric positive definite (SPD) matrix of rank $q < m$ and σ^2 controls the noise power, a structure known as factor [66] or spiked model [29]. Both models are closely related to signal subspace inference and principal component analysis (PCA), and to its probabilistic counterpart (PPCA) [75], which have recently been proven to be powerful tools to handle missing values [11, 26, 30, 61].

With the previous points in mind, this chapter presents recent EM-based methods and algorithms that handle missing values in the context of (i) robustness to non-Gaussianity and (ii) low-rank structure models. It is organized as follows: in Sect. 4.2, we introduce key concepts and tools of missing-data theory which will be needed along the chapter; Sect. 4.3 reviews existing methods to deal with missing values in the framework of parameter estimation, including the EM algorithm and its variants; Sect. 4.4 presents robust estimation procedures with missing values, starting from a general EM for RES distributions to special cases of RES distributions; Sect. 4.5 proposes a robust formulation of the PPCA model, from which closed-form expression of the statistical parameters are proposed; Sect. 4.6 compares the performances of the estimators of Sect. 4.4 with a robust multiple imputation procedure on simulated data with missing values; in conclusion, we provide in Sect. 4.7 promising perspectives in robust estimation with missing data.

4.2 Tools for Modeling Missing Data

In this section, key tools of missing-data theory are presented, from the concept of missing-data pattern to the one of missing-data mechanism. These concepts will accompany us throughout the chapter. Before that, we need to introduce general and missing-data notation.

General notation In this chapter, x denotes a scalar quantity, $\mathbf{x}$ indicates a vector and $\mathbf{X}$ a matrix. The transpose, trace, and determinant operators are denoted by T, tr, and $|.|$, respectively. The rank of a matrix is indicated by rank(.). The i, j-th element of matrix $\mathbf{X}$ is represented by x_{ij} or $(\mathbf{X})_{i,j}$. The observations (or data) are denoted by matrix $\mathbf{X} \in \mathbb{R}^{m \times n}$, with, on its columns, the random variables (r.v.) $\mathbf{x}_i \in \mathbb{R}^m$, $i = 1, \ldots, n$. The set of $m \times m$ positive definite matrices is $\mathcal{S}_m^{++} = \{\mathbf{\Sigma} \in \mathcal{S}_m : \forall \mathbf{x} \in \mathbb{R}^m \backslash \{0\}, \mathbf{x}^\top \mathbf{\Sigma} \mathbf{x} > 0\}$. $\mathrm{St}_{m,q} = \{\mathbf{U} \in \mathbb{R}^{m \times q} : \mathbf{U}^\top \mathbf{U} = \mathbf{I}_q\}$ is the real Stiefel manifold of $m \times q$ orthogonal matrices. The set of positive semi-definite matrices of rank q is expressed as $\mathcal{S}_{m,q}^{+} = \{\mathbf{\Sigma} \in \mathcal{S}_m, \mathbf{\Sigma} \succeq \mathbf{0}, \mathrm{rank}(\mathbf{\Sigma}) = q\}$.

Missing-data notation As $\mathbf{X}$ might include missing observations, it is necessary to design a model that takes into account data incompleteness. In most studies of missing data, it is common to partition $\mathbf{X}$ into its *observed* and *missing* components, which are denoted $\mathbf{X}^{(0)} \in \mathcal{X}^{(0)}$ and $\mathbf{X}^{(1)} \in \mathcal{X}^{(1)}$, respectively. Here, $\mathcal{X}^{(0)}$ and $\mathcal{X}^{(1)}$ are subsets of the space of observations $\mathcal{X}$. Similarly, the observed and missing elements of each vector $\mathbf{x}_i$ will be noted $\mathbf{x}_i^{(0)}$ and $\mathbf{x}_i^{(1)}$, $i = 1 \ldots, n$. We will use notation O_{cc} for the set of indices i corresponding to fully observed $\mathbf{x}_i$, O_i for the set of indices j for which x_{ij} is observed (similarly O_j for the set of indices i). Finally, O refers to the set of tuples (i, j) for which x_{ij} is observed.

4.2.1 *Missing-Data Model*

Suppose iid samples $(\mathbf{x}_i)_{i=1}^n \in \mathbb{R}^m$ drawn from a multivariate RES_m distribution parameterized by mean (location) vector $\boldsymbol{\mu} \in \mathbb{R}^m$, covariance (scatter) matrix $\mathbf{\Sigma} \in \mathcal{S}_m^{++}$ and density generator g. As we will see further in this chapter, the missing values $\mathbf{x}_i^{(1)}$ in sample $\mathbf{x}_i$ can impede a direct estimation of some unknown parameter $\boldsymbol{\theta}$. For this purpose, it can be useful to extract the observed values $\mathbf{x}_i^{(0)}$ from $\mathbf{x}_i$. For example, in [4], $\mathbf{x}_i^{(0)}$ are simply modeled using selection matrices:

$$\mathbf{x}_i^{(0)} = \mathbf{A}_i \mathbf{x}_i, \quad i = 1, \ldots, n, \tag{4.2}$$

where $\mathbf{A}_i$ is the $p_i \times m$ selection matrix constructed by extracting from the identity matrix the $p_i \leq m$ rows corresponding to the observed variables at the i-th observation.

In this chapter, especially in the first part, it will also be useful to order each sample $\mathbf{x}_i$ by stacking $\mathbf{x}_i^{(0)}$ and $\mathbf{x}_i^{(1)}$ such that

$$\widetilde{\mathbf{x}}_i = \mathbf{P}_i \mathbf{x}_i = \left(\mathbf{x}_i^{(0)} \ \mathbf{x}_i^{(1)}\right)^T, \quad i = 1, \ldots, n, \tag{4.3}$$

where $\mathbf{P}_i$ is a $m \times m$ permutation matrix. This representation allows to clearly identify the observed and missing values and to a better interpretation of the calculus of their statistics. For example, the covariance matrix of $\widetilde{\mathbf{x}}_i$ reads

$$\widetilde{\boldsymbol{\Sigma}}_i = \begin{pmatrix} \widetilde{\boldsymbol{\Sigma}}_i^{(00)} & \widetilde{\boldsymbol{\Sigma}}_i^{(10)} \\ \widetilde{\boldsymbol{\Sigma}}_i^{(01)} & \widetilde{\boldsymbol{\Sigma}}_i^{(11)} \end{pmatrix} = \mathbf{P}_i \boldsymbol{\Sigma} \mathbf{P}_i^T, \quad i = 1, \ldots, n, \tag{4.4}$$

where $\widetilde{\boldsymbol{\Sigma}}_i^{(11)}$, $\widetilde{\boldsymbol{\Sigma}}_i^{(10)}$, $\widetilde{\boldsymbol{\Sigma}}_i^{(00)}$ represent adequate covariance matrices blocks, i.e., of the missing, missing and observed, and observed values, respectively. In the rest of the chapter, we omit the parenthesis in the superscripts of block covariances matrices: for example, $\widetilde{\boldsymbol{\Sigma}}_i^{(00)}$ becomes $\widetilde{\boldsymbol{\Sigma}}_i^{00}$.

4.2.2 *Missing-Data Pattern*

It is also useful to distinguish, for both practical and theoretical reasons, the missing-data *pattern*, i.e., the form of the indicator matrix $\mathbf{R} \in \{0, 1\}^{m \times n}$ which elements are defined by

$$(\mathbf{R})_{ij} = \begin{cases} 0 & \text{if } x_{ij} \text{ is observed,} \\ 1 & \text{otherwise.} \end{cases} \tag{4.5}$$

Figure 4.1 shows possible missing-data patterns in the case of multivariate data. In the following, we give some short examples for each of these patterns (we refer interested readers to the book of Little and Rubin [35], where further developments on the patterns are provided with a rich collection of examples):

- *Univariate* or *multivariate* patterns happen when one or multiple variables are partly missing from a reference and unique time instant. This is the case when one or multiple sensors stop to operate at the *same time* and never record again.
- In contrast, a pattern is named *monotone* if the incomplete data can be sorted in such a way that the jth variable is at least as observed as the $(j-1)$th variable for $j = 2, 3, \ldots, m$ [39]. Similarly to the last example, multiple sensors may stop

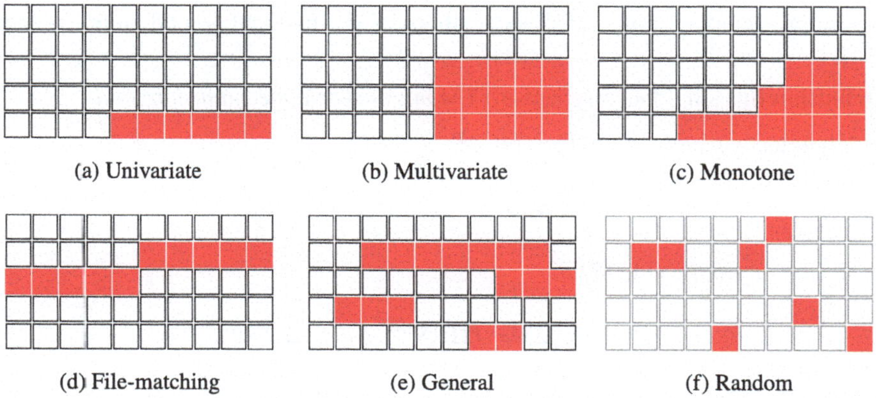

Fig. 4.1 Different types of missing-data patterns in the case of multivariate data (white is observed, red/pink is missing). Note that the random pattern is a special case of the general pattern where each sample a has unique probability of being missing

to function but at *different times*. This pattern has been extensively studied in longitudinal studies [35], wireless communications [33] or finance [41].

- *File matching*: This pattern ensures that all but one variables are at least observed for each observation. For example, when two univariate time series are fusioned to form a bivariate dataset, with variable 1 observed from time t_1 to time t_2 and variable 2 observed from $t_2 + 1$ to t_3.
- *General* or *non-monotone* pattern indicates that multiple blocs of data are missing across multiple variables. This can happen, for example, when multiple sensors *temporary* stop to operate at different times.
- *Random* pattern: This pattern can be considered as a special form of general pattern, but with only individual missing observations, i.e., each observation has a unique random probability of being missing.

4.2.3 Missing-Data Mechanism

Missing-data patterns describe the shape of unobserved data but are ineffective to describe the relation between missing and observed values, i.e., the *cause* of missingness. In his seminal paper [65], D.B. Rubin proposed a classification of the cause of missing data into three different categories, which he called missing-data *mechanisms*.

The concept of missing-data mechanism is defined using the conditional distribution $f(\mathbf{R}|\mathbf{X}, \boldsymbol{\phi})$ of the missing-data pattern $\mathbf{R}$ given the full data $\mathbf{X}$, parametrized by an unknown parameter $\boldsymbol{\phi}$. In the following, we give the classical definition of the missing-data mechanism, being certainly the most popular in the missing-data literature, including in the reference textbooks [35] (first and second editions) and [67].

Definition (missing-data mechanism) *The missing-data mechanism is said*

- *Missing completely at random (MCAR) if*

$$f(\mathbf{R}|\mathbf{X}, \boldsymbol{\phi}) = f(\mathbf{R}|\boldsymbol{\phi}), \quad \forall \boldsymbol{\phi} \in \Phi.$$

- *Missing at random (MAR) if*

$$f(\mathbf{R}|\mathbf{X}, \boldsymbol{\phi}) = f(\mathbf{R}|\mathbf{X}^{(0)}, \boldsymbol{\phi}), \quad \forall \boldsymbol{\phi} \in \Phi.$$

- *Missing not at random (MNAR) if MCAR and MAR assumptions do not hold.*

To put it simply, MCAR happens when the probability of a value being missing is independent of the data values; MAR means that the occurrence of a missing value depends only on the observed values; MNAR when the probability of being missing depends (at least) on the values of missing variables, and possibly on the observed ones.

As pointed out in [69, 71], it is worth noticing that the above definition has been subjected to debate because of the dependence between $\mathbf{R}$ and $\mathbf{X}^{(0)}$. A new definition was proposed by [35], overcoming this ambiguity by using the random variables $\mathbf{x}_i^{(0)}$ and $\mathbf{x}_i^{(1)}$ instead of the data $\mathbf{X}^{(0)}$ and $\mathbf{X}^{(1)}$. We limit ourselves to this brief discussion and leave interested readers in the causes of the debate to [35, 69].

Example 4.1: Real-world scenario of missing-data mechanisms. Consider a sensor that measures atmospheric temperature. As it is placed outdoor, this device can fail due to unpredictable and/or random events (e.g., severe electrical activity, perturbations due to animals, etc.), which implies that the missing temperature values are unrelated to the observed, missing, or any other variable. This setting is clearly MCAR. A MAR scenario would happen when the sensor fails due to an observed variable, such as humidity or precipitation. Finally, a MNAR setting[1] would imply the unavailability of the temperature values because of the temperature itself which do not fall in the sensor range (e.g., too high or too cold).

4.3 A Small Tour of Existing Covariance Estimation Strategies with Missing Data

In this section, we shortly describe important families of methods in statistical estimation with missing values. This framework is one of the many flavors included in what is commonly called missing-data analysis (MDA). When dealing with missing

[1] More precisely, a *self-masked* MNAR setting, i.e., when the unavailability of a missing variable only depends on the variable itself [51].

values, one might want to *impute* the missing values, i.e., replace them with an estimated guess, *infer* some parameters as the mean vector $\boldsymbol{\mu}$ or the covariance matrix $\boldsymbol{\Sigma}$, or *classify* (or *predict*) some outcomes from incomplete data (including labels in a supervised setting).

Indeed, the presence of missing values makes the estimation of an unknown parameter $\boldsymbol{\theta}$ non-trivial. For example, in the case of incomplete data following a multivariate Gaussian distribution, the sample covariance matrix (SCM) can have negative eigenvalues and can contain missing covariates [26], especially with large portions of missing observations. To overcome this issue, many approaches are at hand, and we review some of them below.

4.3.1 Complete-Case Analysis

A ubiquitous strategy when faced with incomplete data is to simply discard the missing samples and analyze the remaining observed ones. As illustrated in Fig. 4.2a, this would imply to remove the missing columns of the data matrix $\mathbf{X}$. In parameter estimation, this is tantamount to estimate an "observed" version of an unknown parameter $\boldsymbol{\theta}$, denoted $\hat{\boldsymbol{\theta}}^{(0)}$, from only *all available* observations, i.e., using $\mathbf{x}_i,\ i \in O_{\text{cc}}$ where O_{cc} is a subset of $\{1, \ldots, n\}$ containing the indices of fully observed samples. For example, the SCM can be computed using fully observed $\mathbf{x}_i$ which yields

$$\widehat{\boldsymbol{\Sigma}}_{\text{cc}} = \frac{1}{|O_{\text{cc}}|} \sum_{i \in O_{\text{cc}}} \left(\mathbf{x}_i - \widehat{\boldsymbol{\mu}}_{\text{cc}}\right)\left(\mathbf{x}_i - \widehat{\boldsymbol{\mu}}_{\text{cc}}\right)^T, \tag{4.6}$$

in which $\widehat{\boldsymbol{\mu}}_{\text{cc}} = (\mu_1, \ldots, \mu_i)^T$ is the sample mean of the observed sample $\mathbf{x}_i$ with $i \in O_{\text{cc}}$. Alternatively, an element-wise estimator can be computed, leading to

$$(\widehat{\boldsymbol{\Sigma}}_{\text{cc}})_{j,k} = \frac{1}{|O|} \sum_{i.\in O} (x_{ij} - (\widehat{\boldsymbol{\mu}}_{\text{cc}})_j)(x_{ik} - (\widehat{\boldsymbol{\mu}}_{\text{cc}})_k)^T, \tag{4.7}$$

where O is the set of indices corresponding to observed values $x_{i.}$ with "." being index j or k, and where $(\widehat{\boldsymbol{\mu}}_{\text{cc}})_j$ refers to the mean of the observed samples in variable j. Unlike estimator (4.6) which uses only fully observed samples, estimator (4.7) uses all available observations such that no missing covariate is created. However, note that the use of this method should be considered carefully. The above estimators will tend to be biased if (*i*) the quantity of missing data increases,[2] leading to a decreasing number of observed samples compared to the number of variables m, (*ii*) the missing-data pattern is not random, and (*iii*) the mechanism is MAR or MNAR as the observed data is not a good representation of the complete data.

---

[2] In [21], authors discourage the use of complete-case analysis when the loss due to missing values exceed 5% of the data.

4.3.2 Imputation Strategies

A second strategy consists first to impute the missing values in $\mathbf{X}$ and then estimate $\hat{\boldsymbol{\theta}}_c$ (letter "c" stands for *complete*) from the infilled (complete) data $\mathbf{X}_c$. There is a wide variety of imputation methods in the literature: most popular methods include mean imputation, multiple imputation, low-rank imputation, and model-based imputation. For an important review of imputation methods, interested readers should consult the book of Van Buuren [78]. In the following, we only discuss the main advantages and drawbacks of mean and multiple imputation methods with regards to ML-based methods.

Firstly, *mean imputation* consists in filling the missing values in variable x_{ij} of $\mathbf{X}$ by the mean of the observed values, $(\widehat{\boldsymbol{\mu}}_{cc})_j$. Despite its simple and computationally efficient implementation, mean imputation cannot be performed in the case of a fully missing variable, as illustrated by Fig. 4.2b. More importantly, it should be mentioned that this method induces important biases in estimators by distorting the data distribution.

A second imputation method is *multiple imputation* [64]. The principle is the following: for a dataset $\mathbf{X}$ with missing values $\mathbf{X}^{(1)}$, S *imputed* (or *completed*) datasets $\mathbf{X}_c^1, \mathbf{X}_c^2, \ldots, \mathbf{X}_c^S$ are generated. For example, the imputed dataset can be drawn S times from a multivariate Gaussian distribution with mean $\widehat{\boldsymbol{\mu}}_{cc}$ and variance Var $(\widehat{\boldsymbol{\mu}}_{cc})$. An estimation of $\boldsymbol{\theta}$ is then computed from the combination (for example, the average) of the q imputed datasets. By taking into account the variability of imputation, this strategy reduces the bias of the estimator $\hat{\boldsymbol{\theta}}$.

Finally, *model-based* methods include parametric and non-parametric approaches: in the former, the imputation can be performed assuming a joint model for the data $(\mathbf{X}^{(0)}, \mathbf{X}^{(1)})$ [78]. In the latter, a common strategy is the impute using a distance between observations as in k-nearest neighbors [76] or hot-decks procedures [3]. More recently, imputation methods based on deep learning and related techniques have been proposed, such as deep convolutional neural networks [83] and variational autoencoders [45].

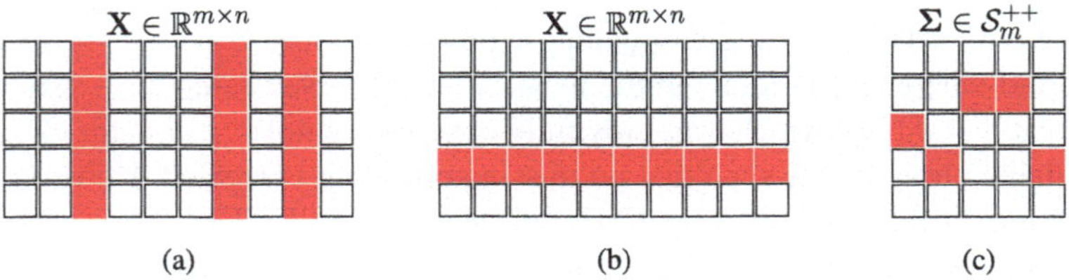

Fig. 4.2 Illustration of different types of missing data: **a** missing samples/observations in the data matrix $\mathbf{X}$, **b** missing variables in $\mathbf{X}$ introducing missing covariates in the covariance matrix $\boldsymbol{\Sigma}$, **c** missing elements in $\boldsymbol{\Sigma}$

4.3.3 Covariance Completion

The principle of covariance completion is fairly simple: from an incomplete covariance matrix $\boldsymbol{\Sigma}$ as in Fig. 4.2c, the goal is to find the best way to reconstruct its missing values [28]. The covariance matrix is possibly corrupted by noise, which often motivates the use of a low-rank model on its structure [8]. Furthermore, some studies assumed a SPD structure on the covariance matrix (see for example, [7] and the references therein). This structure opens many applications, as in the field of brain–computer interfaces, where the state-of-the-art methods for classification employ covariance matrices as statistical features, which are SPD matrices [62].

4.3.4 Likelihood-Based Estimation Under Ignorable Missing-Data Mechanism

4.3.4.1 Likelihood Formulation

Suppose a parameter vector $\boldsymbol{\theta}$ which belongs to a family (or set) of parameters Θ generating the data $\mathbf{X}$. For example, in our case, the unknown parameter is the covariance matrix $\boldsymbol{\Sigma}$ which belongs to $\mathcal{S}_m^{++}$, the set of SPD matrices. Suppose iid samples $\mathbf{x}_1, \ldots, \mathbf{x}_n$ of n observations forming $\mathbf{X}$ distributed according to $f(\mathbf{X}|\boldsymbol{\theta})$ with an unknown parameter $\boldsymbol{\theta}$. The likelihood function associated to $\mathbf{X}$ reads

$$L(\boldsymbol{\theta}|\mathbf{X}) = \prod_{i=1}^{n} f(\mathbf{x}_i|\boldsymbol{\theta}). \tag{4.8}$$

To estimate $\boldsymbol{\theta}$ from the data $\mathbf{X}$, a common strategy relies on maximizing (4.8). The resulting estimator is the well-known ML estimator, which is generally (when it exists and it is unique) consistent in the asymptotic regime, under certain mild conditions.

With missing values, the statistical estimation is conducted on the joint distribution of $(\mathbf{X}, \mathbf{R})$ denoted as $f(., .|\boldsymbol{\theta}, \phi)$, where $\boldsymbol{\theta} \in \boldsymbol{\Theta}$ and $\phi \in \boldsymbol{\Phi}$ are the parameter of the data and the missing-data mechanism distributions, respectively. The *full* likelihood L_{full} under iid assumption is then defined as

$$L_{\text{full}}(\boldsymbol{\theta}, \phi|\mathbf{X}, \mathbf{R}) = \prod_{i=1}^{n} f(\mathbf{x}_i, \mathbf{r}_i|\boldsymbol{\theta}, \phi). \tag{4.9}$$

Since L_{full} involves missing values $\mathbf{X}^{(1)}$, it is not possible to directly compute (4.9). We consider instead the *full observed* likelihood by integrating over the missing values:

$$L_{\text{full,obs}}(\boldsymbol{\theta}, \phi|\mathbf{X}^{(0)}, \mathbf{R}) = \prod_{i=1}^{n} \int_{\mathcal{X}^{(1)}} f(\mathbf{x}_i^{(1)}, \mathbf{x}_i^{(0)}, \mathbf{r}_i|\boldsymbol{\theta}, \phi) d\mathbf{x}_i^{(1)}. \tag{4.10}$$

As noted in [71], the form of (4.10) is rarely closed. However, as we will see in the next section, iterative methods such as the EM algorithm [14] can be used to compute the full observed likelihood.

The missing-data mechanism is said *ignorable* if the statistical estimation of $\boldsymbol{\theta}$ can be conducted by maximizing the *observed* likelihood

$$L_{\text{ign}}(\boldsymbol{\theta}|\mathbf{X}^{(0)}) = \prod_{i=1}^{n} \int_{\mathcal{X}^{(1)}} f(\mathbf{x}_i^{(1)}, \mathbf{x}_i^{(0)}|\boldsymbol{\theta}) d\mathbf{x}_i^{(1)} \tag{4.11}$$

instead of the full observed likelihood (4.10). Specifically, by using the fundamental rule for the joint probability of $(\mathbf{X}, \mathbf{R})$, the full observed likelihood can be rewritten as

$$L_{\text{full,obs}}(\boldsymbol{\theta}, \phi|\mathbf{X}^{(0)}, \mathbf{R}) = \prod_{i=1}^{n} \int_{\mathcal{X}^{(1)}} f(\mathbf{x}_i^{(1)}, \mathbf{x}_i^{(0)}|\boldsymbol{\theta}) f(\mathbf{r}_i|\mathbf{x}_i^{(1)}, \mathbf{x}_i^{(0)}, \phi) d\mathbf{x}_i^{(1)}.$$

In this chapter, we assume that the missing-data mechanism is MAR, which means that data incompleteness depends on the observed values only, so that $f(\mathbf{r}_i|\mathbf{x}_i^{(1)}, \mathbf{x}_i^{(0)}, \phi) = f(\mathbf{r}_i|\mathbf{x}_i^{(0)}, \phi)$ as stated in Definition 1. A handful of computation steps lead to the following expression for the full observed likelihood:

$$\begin{aligned} L_{\text{full, obs}}(\boldsymbol{\theta}, \phi|\mathbf{X}^{(0)}, \mathbf{R}) &= \prod_{i=1}^{n} \int_{\mathcal{X}^{(1)}} f(\mathbf{x}_i^{(0)}, \mathbf{x}_i^{(1)}|\boldsymbol{\theta}) f(\mathbf{r}_i|\mathbf{x}_i^{(0)}, \phi) d\mathbf{x}_i^{(1)} \\ &= \prod_{i=1}^{n} f(\mathbf{r}_i|\mathbf{x}_i^{(0)}, \phi) \prod_{i=1}^{n} \int_{\mathcal{X}^{(1)}} f(\mathbf{x}_i^{(0)}, \mathbf{x}_i^{(1)}|\boldsymbol{\theta}) d\mathbf{x}_i^{(1)} \\ &= \prod_{i=1}^{n} f(\mathbf{r}_i|\mathbf{x}_i^{(0)}, \phi) L_{\text{ign}}(\boldsymbol{\theta}|\mathbf{X}^{(0)}) \\ &\propto L_{\text{ign}}(\boldsymbol{\theta}|\mathbf{X}^{(0)}). \end{aligned} \tag{4.12}$$

This factorization of the full observed likelihood is the second condition for the missing-data mechanism to be considered ignorable, condition one being the distinctness of the parameters $\boldsymbol{\theta}$ and $\boldsymbol{\phi}$ [35]. In this chapter, we assume for simplicity that the missing-data mechanism is ignorable, i.e., the maximum-likelihood estimation of $\boldsymbol{\theta}$ can be conducted by maximizing the observed likelihood (4.11), that is

$$\widehat{\boldsymbol{\theta}}_{\text{ML}} = \underset{\boldsymbol{\theta} \in \boldsymbol{\Theta}}{\arg\max}\, L_{\text{ign}}(\boldsymbol{\theta}|\mathbf{X}^{(0)}). \tag{4.13}$$

The form of the observed likelihood (4.11) is not often closed since it requires the integration over missing data, which motivates the uses of iterative methods such as the EM algorithm. As we shall see in the next section, this technique provides a direct link between the ML estimation of $\boldsymbol{\theta}$ from problem (4.13) and the ML estimation based on the so-called *complete-data*[3] likelihood $L_c(\boldsymbol{\theta}|\mathbf{X})$ [35].

4.3.4.2 The Expectation–Maximization Algorithm as a Way to Handle Missing Data

The EM algorithm [14] is a widely employed iterative method for maximizing the full likelihood in incomplete data problems. The latter algorithm offers a formal and rigorous approach to the intuitive idea of filling-in missing values, except that it is not the missing values themselves that are filled, but the conditional expectation of the sufficient statistics in the (log)likelihood cost function. At step $r = 0$, the algorithm starts with an *initialization* $\boldsymbol{\theta}^{r=0}$, which can be computed using a naive estimator of $\boldsymbol{\theta}$. Then, the following steps are repeated until convergence:

- *E-step (Expectation)*: it consists in computing the so-called Q-function, which is the conditional expectation of the *complete-data* log-likelihood conditioned to the observed data $\mathbf{X}^{(0)}$ and the current estimate of the parameter $\boldsymbol{\theta}^r$. The latter reads (under iid assumption)

$$Q(\boldsymbol{\theta}|\boldsymbol{\theta}^r) = \mathbb{E}_{\mathbf{X}^{(1)}|\mathbf{X}^{(0)},\boldsymbol{\theta}^r}\left[\ell_c(\boldsymbol{\theta}|\mathbf{X})\right] \tag{4.14}$$

$$= \sum_i \int_{\mathcal{X}^{(1)}} \log(f(\mathbf{x}_i^{(1)}, \mathbf{x}_i^{(0)}|\boldsymbol{\theta}))\, f(\mathbf{x}_i^{(1)}|\mathbf{x}_i^{(0)}, \boldsymbol{\theta}^r) d\mathbf{x}^{(1)} \tag{4.15}$$

$$= \sum_i \mathbb{E}_{\mathbf{x}_i^{(1)}|\mathbf{x}_i^{(0)},\boldsymbol{\theta}^r}\left[\ell_c(\boldsymbol{\theta}|\mathbf{x}_i)\right], \tag{4.16}$$

  where $\ell_c(.|.) = \log L_c(.|.)$ is the complete-data *log*-likelihood.
- *M-step (Maximization)*: it updates the parameter by maximizing Q over $\boldsymbol{\theta}$:

$$\boldsymbol{\theta}^{r+1} = \underset{\boldsymbol{\theta}\in\boldsymbol{\Theta}}{\arg\max}\; Q(\boldsymbol{\theta}|\boldsymbol{\theta}^r).$$

The advantages of the EM include its simple implementation, a direct statistical interpretation of each step, and, under general conditions, the guarantee that the sequence $\ell(\boldsymbol{\theta}^r|\mathbf{X}^{(0)})$ converges to stationary values of $\ell(\boldsymbol{\theta}|\mathbf{X}^{(0)})$ if $\ell(\boldsymbol{\theta}|\mathbf{X}^{(0)})$ is bounded. However, an important drawback is that a large fraction of missing information[4] leads to a slow rate of convergence. In fact, [14] found that for $\boldsymbol{\theta}^r$ near a convergence value $\boldsymbol{\theta}^\star$,

---

[3] Recall that the observed-data (or incomplete) is $\mathbf{X}^{(0)}$ and the complete-data is $(\mathbf{X}^{(0)}, \mathbf{X}^{(1)})$.

[4] Missing information is not directly the quantity of missing data, but rather the fraction of information about $\boldsymbol{\theta}$ that is missing in $\ell(\boldsymbol{\theta}|\mathbf{X})$. Strictly speaking, it is defined as the negative of the second derivative of the missing part of the complete-data log-likelihood w.r.t. $\boldsymbol{\theta}$ (see Sect. 8.4 of [36]).

$$|\boldsymbol{\theta}^{r+1} - \boldsymbol{\theta}^\star| = \lambda|\boldsymbol{\theta}^r - \boldsymbol{\theta}^\star|,$$

where λ is the fraction of missing information controlling the speed of convergence of the EM: the greater λ, the slower the speed of convergence.[5]

4.4 Robust Covariance Matrix Estimation Under RES Distribution

As discussed in the introduction, many issues emerge with the Gaussian assumption, especially concerning the robustness of likelihood-based estimators. To overcome these limitations, more general distributions have been considered, which are mainly encompassed by the RES distribution. The aim of this section is to present various robust EM-based estimation procedures of the covariance matrix under the presence of missing data. A general form of the EM for RES distributions is first introduced, along with its stochastic variations. Then, an estimation procedure for a special case of the RES distribution is proposed. Finally, the prior knowledge of a low-rank structure of the covariance matrix is also exploited during the parameter update at the M-step. In addition to this, we provide a simple EM procedure for the widely used Gaussian distribution in the appendix.

4.4.1 A General EM Algorithm for RES Distribution with Missing Data

Using the assumptions and notation from Chap. 1, we consider $\mathbf{x} \sim \mathrm{RES}_m(\boldsymbol{\mu}, \boldsymbol{\Sigma}, g)$, leading to a pdf of the form $f(\mathbf{x}|\boldsymbol{\theta}) = |\boldsymbol{\Sigma}|^{-\frac{1}{2}} g\big((\mathbf{x} - \boldsymbol{\mu})^T \boldsymbol{\Sigma}^{-1} (\mathbf{x} - \boldsymbol{\mu})\big)$.

In the missing-data case, the observed likelihood (under MAR or MCAR) is equivalent to

$$L(\boldsymbol{\theta}|\mathbf{X}^{(0)}) = \prod_{i=1}^{n} f(\mathbf{x}_i^{(0)}|\boldsymbol{\theta}) \tag{4.17}$$

$$= \prod_{i=1}^{n} |\boldsymbol{\Sigma}_i^{(0)}|^{-\frac{1}{2}} g\big((\mathbf{x}_i^{(0)} - \boldsymbol{\mu}_i^{(0)})^T (\boldsymbol{\Sigma}_i^{(0)})^{-1} (\mathbf{x}_i^{(0)} - \boldsymbol{\mu}_i^{(0)})\big). \tag{4.18}$$

By contrast, the complete-data likelihood is defined as

[5] For a clear overview of the rate of convergence of the EM, which notably includes the expression of the fraction of missing information, see Sect. 8.4.3 in [36].

$$L_{\text{full,ign}}(\boldsymbol{\theta}|\mathbf{X}) = \prod_{i=1}^{n} |\boldsymbol{\Sigma}|^{-\frac{1}{2}} g\left((\mathbf{x}-\boldsymbol{\mu})^T \boldsymbol{\Sigma}^{-1}(\mathbf{x}-\boldsymbol{\mu})\right). \tag{4.19}$$

Then, a natural way to design the EM is

- *Initialization*: $\boldsymbol{\theta}^{r=0}$ (using, e.g., an imputation method).

- *E-step*: computation of the surrogate function

$$Q(\boldsymbol{\theta}|\boldsymbol{\theta}^r) = -\frac{n}{2}\log|\boldsymbol{\Sigma}| + \mathbb{E}_{\mathbf{x}_i^{(1)}|\mathbf{x}_i^{(0)},\boldsymbol{\theta}^r}\left[\log g\left((\mathbf{x}-\boldsymbol{\mu})^T \boldsymbol{\Sigma}^{-1}(\mathbf{x}-\boldsymbol{\mu})\right)\right].$$

- *M-step*: $\boldsymbol{\theta}^{r+1} = \underset{\boldsymbol{\theta}\in\boldsymbol{\Theta}}{\arg\max}\; Q(\boldsymbol{\theta}|\boldsymbol{\theta}^r)$.

Unfortunately, the computation of $\mathbb{E}_{\mathbf{x}_i^{(1)}|\mathbf{x}_i^{(0)},\boldsymbol{\theta}^r}\left[\log g\left((\mathbf{x}-\boldsymbol{\mu})^T \boldsymbol{\Sigma}^{-1}(\mathbf{x}-\boldsymbol{\mu})\right)\right]$ is not always tractable.[6] In case of intractable E-step, [9] developed a stochastic EM (SEM), where an approximation of $Q(\boldsymbol{\theta}|\boldsymbol{\theta}^r)$ is used instead of its true intractable value. Specifically, this alternative consists in sampling the missing value $(\mathbf{x}_i^{(1)})_s^r$ via the distribution of the missing random variables conditioned to the observed values, i.e., $(\mathbf{x}_i^{(1)})_s^r \sim f(\mathbf{x}_i^{(1)}|\mathbf{x}_i^{(0)}, \boldsymbol{\theta}^r)$ with $s = 1$. Then, the missing part of $\mathbf{x}_i$ is replaced by $(\mathbf{x}_i^{(1)})_s^r$, which leads to the imputed version $(\mathbf{x}_i)_s^r$. The SEM reads

$$\boldsymbol{\theta}_{\text{SEM}}^{r+1} = \underset{\boldsymbol{\theta}\in\boldsymbol{\Theta}}{\arg\max} -\frac{n}{2}\log|\boldsymbol{\Sigma}| + \log g\left(((\mathbf{x}_i)_s^r-\boldsymbol{\mu})^T \boldsymbol{\Sigma}^{-1}((\mathbf{x}_i)_s^r-\boldsymbol{\mu})\right)$$

with $s = 1$. At the M-step, $\boldsymbol{\theta}^{r+1}$ is computed on the basis of the pseudo-completed data $\mathbf{x}^r$.

In order to enhance the performance of the SEM, [80] proposed the so-called Monte Carlo EM (MCEM). The latter is basically a SEM with several drawings, i.e., $(\mathbf{x}_i^{(1)})_s^r \sim f(\mathbf{x}_i^{(1)}|\mathbf{x}_i^{(0)}, \boldsymbol{\theta}^r)$ with $s = 1, \ldots, S$. In our context, this reads

$$\boldsymbol{\theta}_{\text{MCEM}}^{r+1} = \underset{\boldsymbol{\theta}\in\boldsymbol{\Theta}}{\arg\max} -\frac{n}{2}\log|\boldsymbol{\Sigma}| + \frac{1}{S}\sum_{s}\log g\left(((\mathbf{x}_i)_s^r-\boldsymbol{\mu})^T \boldsymbol{\Sigma}^{-1}((\mathbf{x}_i)_s^r-\boldsymbol{\mu})\right).$$

The MCEM provides an alternative to the analytic computation of the integral in (4.14) by the numerical computation of a Monte Carlo approximation of this integral. This technique is accurate for large S but suffers from a high computational cost.

A third approach, called the stochastic approximation EM (SAEM) [13], consists in replacing the computation of the expectation by a Robbins-Monro-type approxi-

---

[6] Examples of using deterministic EM-type algorithms applied to RES distributions with missing data deal in majority with the Gaussian (see [14] and the appendix), scaled Gaussian [25] and Student t [38]-distributions. In these cases, we can expect to obtain closed-form expressions at the E-step under certain mild conditions.

mation [59], which converges to the ML estimate under some general mild conditions. The E-step of the SAEM reads

$$\theta_{\text{SAEM}}^{r+1} = \underset{\theta \in \Theta}{\arg\max}(1 - \gamma_r) Q(\theta|\theta^{r-1}) - \frac{n}{2} \log |\mathbf{\Sigma}| + \log g\big(((\mathbf{x}_i)_1^r - \boldsymbol{\mu})^T \mathbf{\Sigma}^{-1}((\mathbf{x}_i)_1^r - \boldsymbol{\mu})\big),$$

where γ_r is the step size which should be tuned adequately in order to guarantee the convergence of the SAEM [13]. For an example on the derivation of the SEM, MCEM and SAEM, interested readers can consult [42, 84] in which stochastic EM algorithms have been designed for heavy-tailed autoregressive (AR) and vectors AR (VAR) models for incomplete t-distributed data.

Note that in general, sampling $(\mathbf{x}_i^{(1)})_s^r$ is not trivial and may lead to heavy computational cost. In the following, we focus on a specific case of the RES for which the E-step admits a closed-form expression, leading to a deterministic EM scheme (i.e., avoiding the use of stochastic approximations). Specifically, we focus on the scaled Gaussian distributions.

4.4.2 *Robust Covariance Estimation Under Scaled Gaussian Distributions*

4.4.2.1 Model Setup and Assumptions

As seen in the introduction, most of the covariance matrix estimation methods with missing data use the Gaussian assumption, which can be inaccurate in a heterogeneous context or in the presence of outliers. The mixture of scaled Gaussian (MSG) distributions can tackle this issue by managing heavier tails, which offers a better fit to empirical data [57]. Consider that the m-dimensional iid samples $(\mathbf{x}_i)_{i=1}^n \in \mathbb{R}^m$ follows a zero-mean multivariate MSG distribution. Then it admits the stochastic representation

$$\mathbf{x}_i = \sqrt{\tau_i}\mathbf{n}, \tag{4.20}$$

with:

- $\mathbf{n} \sim \mathbb{R}N(\mathbf{0}, \mathbf{\Sigma}), \quad \mathbf{\Sigma} \in \mathcal{S}_m^{++}$,
- $\mathbf{x}_i|\tau_i \sim \mathbb{R}N(\mathbf{0}, \tau_i \mathbf{\Sigma})$,
- $\tau \in \mathbb{R}_+$: a positive scalar called the *scale* (or texture), which is *deterministic* and unknown.

Model (4.20) is entwined with the (sub)-class of compound Gaussian (CG) distributions, which assumes independence between the scale and $\mathbf{n}$. The name MSG comes from the fact that $(\tau_i)_{i=1}^n$ are seen as scales settings of the Gaussian distribution. Considering deterministic scales instead of assuming a prior f_τ has proven its convenience in terms of robustness to many classes of distributions and its flexibility in the presence of heterogeneous data [49, 57] (see, for example, the second reference for an application to synthetic aperture radar data).

We now focus on an EM formulation to estimate the vector of unknown parameters, which, in the case of the MSG distribution, becomes $\boldsymbol{\theta} = [\text{vec}(\boldsymbol{\Sigma}), \tau_1, \ldots, \tau_n]^T$. The analogy with the Gaussian case using the EM algorithm is presented in the appendix.

4.4.2.2 An EM Algorithm for Mixture of Scaled Gaussian Distributions

In the case of missing data, the complete log-likelihood function of model (4.20) is given by

$$\ell_c(\boldsymbol{\theta}|\widetilde{\mathbf{X}}) = -nm\log(\pi) - n\log|\boldsymbol{\Sigma}| - m\sum_{i=1}^{n}\log\tau_i - \sum_{i=1}^{n}\widetilde{\mathbf{x}}_i^T\left(\tau_i\widetilde{\boldsymbol{\Sigma}}_i\right)^{-1}\widetilde{\mathbf{x}}_i, \tag{4.21}$$

where $\widetilde{\mathbf{x}}_i$ is the permuted version of $\mathbf{x}_i$ as defined in (4.3). Note that the replacement of deterministic textures by 1 for all i in (4.21) leads to the complete log-likelihood function for the Gaussian distribution. In the following, a general procedure using the EM algorithm is derived with reference to the estimation problem (4.13). $\boldsymbol{\theta}^r$ denotes the estimated parameter at iteration r.

- *Initialization*: at step $r = 0$, the estimate $\boldsymbol{\Sigma}^{r=0}$ is initialized with Tyler's M-estimator (see Chap. 1) from available observations in their full dimension m, denoted $\widehat{\boldsymbol{\Sigma}}_{\text{Tyl-obs}}$. In our context, incomplete observations make the direct estimation of τ with the fixed point estimator impracticable: thus, as an initialization, all $\tau_i^{r=0}$ can be set to one.

- *E-step*: compute the conditional expectation of the complete log-likelihood (the so-called Q-function) for each complete vector $\widetilde{\mathbf{x}}_i$, $i = 1, \ldots, n$:

$$\begin{aligned} Q_i(\boldsymbol{\theta}|\boldsymbol{\theta}^r) &= \mathbb{E}_{\mathbf{x}_i^{(1)}|\mathbf{x}_i^{(0)},\boldsymbol{\theta}^r}\left[\ell_c(\boldsymbol{\theta}|\widetilde{\mathbf{x}}_i)\right] \\ &= \mathbb{E}_{\mathbf{x}_i^{(1)}|\mathbf{x}_i^{(0)},\boldsymbol{\theta}^r}\left[-m\log(\pi) - \log|\boldsymbol{\Sigma}| - m\log\tau_i - \widetilde{\mathbf{x}}_i^T\left(\tau_i\widetilde{\boldsymbol{\Sigma}}_i\right)^{-1}\widetilde{\mathbf{x}}_i\right]. \end{aligned} \tag{4.22}$$

Due to the deterministic nature of the first three terms inside the expectation in (4.22), only the expectation of the last term shall be computed. A handful arrangement of the latter leads to

$$\text{tr}\left\{\mathbb{E}_{\mathbf{x}_i^{(1)}|\mathbf{x}_i^{(0)},\boldsymbol{\theta}^r}\left[\begin{pmatrix}\mathbf{x}_i^{(0)}\mathbf{x}_i^{(0)T} & \mathbf{x}_i^{(0)}\mathbf{x}_i^{(1)T} \\ \mathbf{x}_i^{(1)}\mathbf{x}_i^{(0)T} & \mathbf{x}_i^{(1)}\mathbf{x}_i^{(1)T}\end{pmatrix}\right]\left(\tau_i\widetilde{\boldsymbol{\Sigma}}_i\right)^{-1}\right\} = \tau_i^{-1}\text{tr}\{\mathbf{B}_i^r\widetilde{\boldsymbol{\Sigma}}_i^{-1}\},$$

where $\mathbf{B}_i^r = \begin{pmatrix} \mathbf{D}_i^r & \mathbf{E}_i^r \\ \mathbf{E}_i^{rT} & \mathbf{G}_i^r \end{pmatrix}$ is a $m \times m$ matrix containing the expectations of the missing data with blocks given by

$$\mathbf{D}_i^r = \mathbb{E}_{\mathbf{x}_i^{(1)}|\mathbf{x}_i^{(0)},\boldsymbol{\theta}^r}\left[\mathbf{x}_i^{(0)}\mathbf{x}_i^{(0)T}\right] = \mathbf{x}_i^{(0)}\mathbf{x}_i^{(0)T}, \tag{4.23}$$

$$\mathbf{E}_i^r = \mathbb{E}_{\mathbf{x}_i^{(1)}|\mathbf{x}_i^{(0)},\boldsymbol{\theta}^r}\left[\mathbf{x}_i^{(0)}\mathbf{x}_i^{(1)T}\right] = \mathbf{x}_i^{(0)}\mathbb{E}_{\mathbf{x}_i^{(1)}|\mathbf{x}_i^{(0)},\boldsymbol{\theta}^r}\left[\mathbf{x}_i^{(1)T}\right], \tag{4.24}$$

$$\mathbf{G}_i^r = \mathbb{E}_{\mathbf{x}_i^{(1)}|\mathbf{x}_i^{(0)},\boldsymbol{\theta}^r}\left[\mathbf{x}_i^{(1)}\mathbf{x}_i^{(1)T}\right]. \tag{4.25}$$

$\mathbf{B}_i^r$ contains the expectations of the *sufficient* statistics $\mathbf{x}_i^{(1)}$ and $\mathbf{x}_i^{(1)}\mathbf{x}_i^{(1)^T}$ at step r of the EM for the complete data $\widetilde{\mathbf{X}}$, which have a distribution from the regular exponential family [35]. These quantities can be easily computed using a classical result in conditional distributions (see expressions (56)–(57) in Chap. 1 and Theorem 2.5.1 in [2]). In the scaled Gaussian setting, the conditional distribution of $\mathbf{x}_i^{(1)}|\mathbf{x}_i^{(0)}$ is still $\mathbb{R}N(\widetilde{\mu}_i^{1|0}, \widetilde{\boldsymbol{\Sigma}}_i^{11|0})$, but with conditional mean and covariance, respectively, given by

$$\widetilde{\mu}_i^{1|0} = \widetilde{\boldsymbol{\Sigma}}_i^{10}\left(\widetilde{\boldsymbol{\Sigma}}_i^{00}\right)^{-1}\widetilde{\mathbf{x}}_i^{(0)}, \tag{4.26}$$

$$\widetilde{\boldsymbol{\Sigma}}_i^{11|0} = \tau_i\left(\widetilde{\boldsymbol{\Sigma}}_i^{11} - \widetilde{\boldsymbol{\Sigma}}_i^{10}\left(\widetilde{\boldsymbol{\Sigma}}_i^{00}\right)^{-1}\widetilde{\boldsymbol{\Sigma}}_i^{01}\right). \tag{4.27}$$

Compared to the classical Gaussian case, the only change appears in the expression of the conditional covariance $\widetilde{\boldsymbol{\Sigma}}_i^{11|0}$, whereas the expression of $\widetilde{\mu}_i^{1|0}$ does not change because the scale term vanishes due to the application of its inverse. At the r-th iteration, the E-step of the EM algorithm consists in calculating

$$\mathbb{E}_{\mathbf{x}_i^{(1)}|\mathbf{x}_i^{(0)},\boldsymbol{\theta}^r}\left[\mathbf{x}_i^{(1)}\right] = (\widetilde{\mu}_i^{1|0})^r, \tag{4.28}$$

$$\mathbb{E}_{\mathbf{x}_i^{(1)}|\mathbf{x}_i^{(0)},\boldsymbol{\theta}^r}\left[\mathbf{x}_i^{(1)}\mathbf{x}_i^{(1)^T}\right] = (\widetilde{\boldsymbol{\Sigma}}_i^{11|0})^r + (\widetilde{\mu}_i^{1|0})^r(\widetilde{\mu}_i^{1|0^T})^r. \tag{4.29}$$

- *M-step*: finally, $\boldsymbol{\theta}^{r+1}$ is obtained as the solution of the following maximization problem:

$$\begin{aligned} \max_{\boldsymbol{\theta}\in\boldsymbol{\Theta}} \quad & Q(\boldsymbol{\theta}|\boldsymbol{\theta}^r) \\ \text{subject to} \quad & \boldsymbol{\Sigma} \succeq \mathbf{0} \\ & \tau_i > 0, \quad \forall i \end{aligned} \tag{4.30}$$

where $Q(\boldsymbol{\theta}|\boldsymbol{\theta}^r) = \sum_{i=1}^n Q_i(\boldsymbol{\theta}|\boldsymbol{\theta}^r)$ is the total Q-function due to the iid of the observations.

Lemma 1 *The ML estimates $\widehat{\boldsymbol{\Sigma}}$ and $\widehat{\tau}_i$ of problem (4.30) are given by the following closed-form expressions:*

$$\widehat{\tau}_i = \frac{\operatorname{tr}\{\mathbf{B}_i^r \bar{\boldsymbol{\Sigma}}_i^{-1}\}}{p} \quad \textit{for} \quad i \in [1, n], \tag{4.31}$$

$$\widehat{\boldsymbol{\Sigma}} = \frac{m}{n} \sum_{i=1}^{n} \frac{(\mathbf{C}_i^r)^T}{\operatorname{tr}\{\mathbf{C}_i^r \widehat{\boldsymbol{\Sigma}}^{-1}\}} \triangleq \mathcal{H}(\widehat{\boldsymbol{\Sigma}}), \tag{4.32}$$

with $\bar{\boldsymbol{\Sigma}}_i = \mathbf{P}_i \widehat{\boldsymbol{\Sigma}} \mathbf{P}_i^T$, $\mathbf{C}_i^r = \mathbf{P}_i^T \mathbf{B}_i^r \mathbf{P}_i$ *and where* $\mathcal{H}(\cdot)$ *is the fixed point equation.*

We restrain ourselves to the important steps of the proof, which only involves simple calculus and derivation operations, and we left interested readers to check its full version in [25]:

- The first step of the proof consists in properly rewriting the expectation of the complete-data log-likelihood (4.21) for all observations, that is

$$\mathbb{E}_{\mathbf{x}_i^{(1)}|\mathbf{x}_i^{(0)},\boldsymbol{\theta}^r}\left[\ell_c(\boldsymbol{\theta}|\widetilde{\mathbf{X}})\right] \propto -n \log|\boldsymbol{\Sigma}| - m \sum_{i=1}^{n} \log \tau_i - \sum_{i=1}^{n} \tau_i^{-1} \operatorname{tr}\{\mathbf{B}_i^r \widetilde{\boldsymbol{\Sigma}}_i^{-1}\}, \tag{4.33}$$

  where, as described before, the block matrix $\mathbf{B}_i^r$ contains the expectation of the sufficient statistics $\mathbf{x}_i^{(1)}\mathbf{x}_i^{(1)T}$ to be computed at the E-step using (4.23).
- The second step consist in deriving (4.33) with respect to τ_i, which requires trivial calculus and gives $\widehat{\tau}_i$.
- Finally, (4.33) is updated by replacing τ_i by $\widehat{\tau}_i$ and differentiated w.r.t. $\boldsymbol{\Sigma}$. A bit of work in arranging the terms gives the desired closed-form expression of the estimator $\widehat{\boldsymbol{\Sigma}}$.
- As it is clear that the expression of the estimator $\widehat{\boldsymbol{\Sigma}}$ depends on itself, it can be updated using the fixed point algorithm $\boldsymbol{\Sigma}_{k+1} = \mathcal{H}(\boldsymbol{\Sigma}_k)$ where k refers to the iteration index of the fixed point. This rule favorably applies in a $n > m$ regime, which is the case here.

The complete estimation procedure of $\boldsymbol{\theta}$ is given in Algorithm 4.1. The stopping condition of the EM algorithm is ensured by the evaluation of the quantity $||\boldsymbol{\theta}^{r+1} - \boldsymbol{\theta}^r||_F^2$ at each iteration, while the convergence of the fixed point algorithm relies on $||\boldsymbol{\Sigma}_{k+1}^r - \boldsymbol{\Sigma}_k^r||_F^2$.

Note that at step $r = 0$, the estimate $\boldsymbol{\Sigma}^{r=0}$ can be initialized with Tyler's M-estimator from available observations in their full dimension m. As already pointed out, $\tau_i^{r=0}$ cannot be directly computed with the fixed point equation because of missing observations. For this reason, all scales are initialized to 1.

4.4.3 Handling Robustness and Low-Rank Structure

Until now, the structure exhibited by the covariance in the MSG model has not been exploited in the EM estimation procedure. This can be problematic when the sample

Algorithm 4.1 EM-Tyl: robust estimation of $\boldsymbol{\theta}$ under MSG distribution with missing values.

Require: $(\widetilde{\mathbf{x}}_i)_{i=1}^n \sim \mathbb{R}N_m(\mathbf{0}, \tau_i \boldsymbol{\Sigma})$, $(\mathbf{P}_i)_{i=1}^n$
Ensure: $\widehat{\boldsymbol{\Sigma}}$, $(\widehat{\tau}_i)_{i=1}^n$
Initialization:

$$\boldsymbol{\Sigma}^{r=0} = \widehat{\boldsymbol{\Sigma}}_{\text{Tyl-obs}}$$
$$\tau^{r=0} = \mathbf{1}_N^T$$

repeat ▷ *EM loop, r varies*
  E-step:
  Compute $\mathbb{E}[\mathbf{x}_i^{(1)}]$, $\mathbb{E}[\mathbf{x}_i^{(0)}\mathbf{x}_i^{(1)T}]$, $\mathbb{E}[\mathbf{x}_i^{(1)}\mathbf{x}_i^{(1)T}]$ for $i = 1, \ldots, n$
  Compute $\mathbf{C}_i^r = \mathbf{P}_i^T \mathbf{B}_i^r \mathbf{P}_i$ for $i = 1, \ldots, n$
  M-step:
  repeat ▷ *fixed point loop, k varies (optional loop)*
    $\widehat{\boldsymbol{\Sigma}}_{k+1}^r = \mathcal{H}(\widehat{\boldsymbol{\Sigma}}_k^r) = \frac{m}{n} \sum_{i=1}^n \frac{(\mathbf{C}_i^r)^T}{\mathrm{tr}\left\{\mathbf{C}_i^r \widehat{\boldsymbol{\Sigma}}^{-1}\right\}}$
  until $||\boldsymbol{\Sigma}_{k+1}^r - \boldsymbol{\Sigma}_k^r||_F^2$ converges
  Compute $\widehat{\tau}_i^r = \frac{1}{m}\mathrm{tr}\left\{\mathbf{B}_i^r \bar{\boldsymbol{\Sigma}}_i^{-1}\right\}$ for $i = 1, \ldots, n$
  $r \leftarrow r + 1$
until $||\boldsymbol{\theta}^{r+1} - \boldsymbol{\theta}^r||_F^2$ converges

support is low ($n \simeq m$) as the SCM is not a reliable estimate in this setup [34]. In this context, a popular approach consists in imposing a low-rank structure on the covariance matrix. The resulting data is driven by an underlying low-dimensional linear model corrupted by an independent perturbation

$$\mathbf{x}_i = \sqrt{\tau}_i(\mathbf{W}\mathbf{s}_i + \boldsymbol{\epsilon}_i), \tag{4.34}$$

where $\mathbf{W}$ is a rank-q factor loading matrix, $\mathbf{s}_i \in \mathbb{R}^q$ and $\boldsymbol{\epsilon}_i \in \mathbb{R}^m$ are independent r.v. with $\mathbf{s}_i \sim \mathbb{R}N_m(\mathbf{0}, \mathbf{I})$ and $\boldsymbol{\epsilon}_i \sim \mathbb{R}N_m(\mathbf{0}, \sigma^2\mathbf{I}_m)$. Following these probability assumptions, the covariance Σ of $\mathbf{x}_i$ is

$$\mathbb{E}[\mathbf{x}_i\mathbf{x}_i^T] = \tau_i \mathbb{E}[\mathbf{W}\mathbf{s}_i\mathbf{s}_i^T\mathbf{W}^T + \boldsymbol{\epsilon}_i\boldsymbol{\epsilon}_i^T] = \tau_i(\mathbf{H} + \sigma^2\mathbf{I}_m)$$

with $\mathbf{H} = \mathbf{W}\mathbf{W}^T$. This configuration is useful in a large variety of signal applications with a high-dimensional setting (e.g., hyperspectral imagery, genomics or denoising) and is closely related to low-rank factor analysis [60], PPCA [75], and signal subspace estimation. The vector of unknown parameters is now $\boldsymbol{\theta} = [\mathrm{vec}(\mathbf{H}), \sigma^2, \tau_1, \ldots, \tau_n]^T$. At the M-step of the algorithm, the maximization problem (4.30) becomes a low-rank estimation problem:

$$\begin{aligned} \max_{\theta \in \Theta} \quad & Q(\theta|\theta^r) \\ \text{subject to} \quad & \Sigma = \sigma^2 \mathbf{I}_m + \mathbf{H} \\ & \text{rank}(\Sigma) = q \\ & \sigma > 0, \quad \tau_i > 0, \quad \forall i. \end{aligned} \tag{4.35}$$

The seminal work of [75] gives a general solution to this problem, which is recalled hereafter using the formulations of [31, 73]. Let $\Sigma^r \overset{\text{EVD}}{=} \sum_{i=1}^m \lambda_i^r \mathbf{u}_i^r \mathbf{u}_i^{Tr}$ be the eigenvalue decomposition of Σ^r at the r-th iteration of the EM algorithm, where $\lambda_1^r < \cdots < \lambda_m^r$ are the eigenvalues of Σ^r and $\mathbf{u}_1^r, \ldots, \mathbf{u}_m^r$ the corresponding eigenvectors. The closed-form solution of problem (4.35) for Σ and σ^2 are then given by

$$\widehat{\Sigma} = \hat{\sigma}^2 \mathbf{I}_m + \sum_{i=1}^{q} \hat{\lambda}_i \mathbf{u}_i^r \mathbf{u}_i^{Tr}, \tag{4.36}$$

$$\hat{\sigma}^2 = \frac{1}{m-r} \sum_{i=q+1}^{m} \lambda_i^r, \tag{4.37}$$

where $\hat{\lambda}_i = \lambda_i^r - \hat{\sigma}^2$ for $i = 1, \ldots, q$. Since the scales $(\tau_i)_{i=1}^n$ apply on Σ and not on $\mathbf{H}$, their estimation in (4.31) remain the same with problem (4.35). The procedure to estimate all parameters under the low-rank factor model is given in Algorithm 4.2. It uses the same form as Algorithm 4.1 where Eqs. (4.36)–(4.37) are applied just after the fixed point algorithm to each newly estimated $\widehat{\Sigma}_{k+1}^r$. Finally, note that standards convergence results of the majorization–minimization (MM) algorithm ensure the proof of monotonicity of the low-rank estimation algorithm [73].

4.5 Robust Probabilistic Component Analysis with Missing Values

In many datasets, the relevant information often lies in a subspace of lower dimension than the space of observations. As many learning algorithms, principal component analysis (PCA) exploits this underlying structure by projecting the data into a lower-dimensional linear subspace with minimal mean-square error.

4.5.1 Classical Probabilistic Component Analysis

The initial formulations of PCA by Pearson (1901) and Hotelling (1933) were obtained without any assumption on the probabilistic model of the data, whereas

Algorithm 4.2 EM-Tyl-LR: robust low-rank estimation of $\boldsymbol{\theta}$ under MSG distribution with missing values.

Require: $(\widetilde{\mathbf{x}}_i)_{i=1}^n \sim \mathbb{R}N_m(\mathbf{0}, \tau_i \boldsymbol{\Sigma})$, $(\mathbf{P}_i)_{i=1}^n$, rank $q < m$
Ensure: $\widehat{\boldsymbol{\Sigma}}$, $(\widehat{\tau}_i)_{i=1}^n$
Initialization:

$$\boldsymbol{\Sigma}^{r=0} = \widehat{\boldsymbol{\Sigma}}_{\text{Tyl-obs}}$$
$$\tau^{r=0} = \mathbf{1}_N^T$$

repeat ▷ *EM loop, r varies*
  E-step:
  Compute $\mathbb{E}[\mathbf{x}_i^{(1)}]$, $\mathbb{E}[\mathbf{x}_i^{(0)}\mathbf{x}_i^{(1)T}]$, $\mathbb{E}[\mathbf{x}_i^{(1)}\mathbf{x}_i^{(1)T}]$ for $i = 1, \ldots, n$
  Compute $\mathbf{C}_i^r = \mathbf{P}_i^T \mathbf{B}_i^r \mathbf{P}_i$ for $i = 1, \ldots, n$
  M-step:
  repeat ▷ *fixed point loop, k varies (optional loop)*
    $\widehat{\boldsymbol{\Sigma}}_{k+1}^r = \mathcal{H}(\widehat{\boldsymbol{\Sigma}}_k^r)$
    $\widehat{\boldsymbol{\Sigma}}_{k+1}^r \overset{\text{EVD}}{=} \sum_{i=1}^p \lambda_i^r \mathbf{u}_i^r \mathbf{u}_i^{Tr}$
    Update $\widehat{\boldsymbol{\Sigma}}_{k+1}^r$ by computing (4.36) and (4.37)
  until $||\boldsymbol{\Sigma}_{k+1}^r - \boldsymbol{\Sigma}_k^r||_F^2$ converges
  Compute $\widehat{\tau}_i^r$, $i = 1, \ldots, n$
  $r \leftarrow r + 1$
until $||\boldsymbol{\theta}^{r+1} - \boldsymbol{\theta}^r||_F^2$ converges

a *probabilistic* derivation of PCA within a density estimation framework was proposed by [75]:

$$\mathbf{x}_i = \mathbf{W}\mathbf{s}_i + \boldsymbol{\mu} + \mathbf{n}_i, \tag{4.38}$$

where $\mathbf{W}$ is a $m \times q$ factor loading matrix with rank q, $\mathbf{s}_i$ are $q \times 1$ vectors of principal components, $\boldsymbol{\mu}$ is the $m \times 1$ bias (or mean) vector and $\mathbf{n}_i$ is the noise term. Both principal components and noise are assumed normally distributed, i.e., $\mathbf{s}_i \sim \mathbb{R}N(\mathbf{0}, \mathbf{I})$, $\mathbf{n}_i \sim \mathbb{R}N(\mathbf{0}, \sigma^2\mathbf{I})$. The resulting covariance matrix for the observations involves a core that is low-rank, i.e., which lives in a subspace of dimension $q < m$, and is of the form $\boldsymbol{\Sigma} = \mathbf{H} + \sigma^2\mathbf{I}$, where $\mathbf{H} = \mathbf{W}\mathbf{W}^T$ belongs to the set of positive semi-definite matrices of rank q, denoted $\mathcal{S}_{m,q}^{++}$. This has been formulated and studied in the seminal work by Tipping and Bishop in [75]. In this paper, the authors proposed an efficient EM algorithm to find the MLE of the model parameters, namely, $\mathbf{W}$, $\boldsymbol{\mu}$, and σ^2.

However, it is worth mentioning that a rotational ambiguity of the loading matrix remains: if $\mathbf{O}$ denotes an orthogonal rotational matrix and $\mathbf{W}_{\text{ML}}$ the ML estimate of $\mathbf{W}$, then $\mathbf{W}_{\text{ML}}\mathbf{O}$ remains the ML estimate. Generally, either a post-processing is possible [75], or, the spanned subspace by the columns of $\mathbf{W}$ is of interest (i.e., the ML estimate of $\mathbf{W}^T\mathbf{W}$).

The surrogate EM function reads (under some mild conditions)

$$Q(\boldsymbol{\theta}|\boldsymbol{\theta}^r) = \mathbb{E}_{\mathbf{X}^{(1)}|\mathbf{X}^{(0)},\boldsymbol{\theta}^r}\left[\ell_c(\boldsymbol{\theta}|\mathbf{X},\mathbf{S})\right] = \sum_i \mathbb{E}_{\mathbf{X}^{(1)}|\mathbf{X}^{(0)},\boldsymbol{\theta}^r}\left[\ell_c(\boldsymbol{\theta}|\mathbf{x}_i,\mathbf{s}_i)\right], \tag{4.39}$$

where $\mathbf{X}$ and $\mathbf{S}$ are the $m \times n$ matrices having observed and latent variables on their columns, respectively. Almost ten years after the work of Tipping and Bishop, [26] proposed a generalization to the case of MAR data, which appears relevant when applying PCA on sparse observations. In the following, we give the update rules of the EM algorithm as obtained in [26]. Readers who are interested in the original EM formulations should refer to [75]. The E-step mainly consists in finding the conditional expectation of the latent variables $\mathbf{s}_i$ given the data and the current values of the model parameters. During the M-step, the parameters are updated using the following expressions:

$$\begin{aligned}
\boldsymbol{\Sigma} &= \sigma^2\Big(\sigma^2\mathbf{I} + \sum_{j\in O_i}\mathbf{w}_j\mathbf{w}_j^T\Big)^{-1} \qquad (4.40)\\
\bar{\mathbf{s}}_i &= (\sigma^2)^{-1}\boldsymbol{\Sigma}\sum_{j\in O_i}\mathbf{w}_j(x_{ij}-\mu_j), \quad i = 1,\ldots,n\\
\mu_j &= \frac{1}{|O_j|}\sum_{i\in O_j}(x_{ij}-\mathbf{w}_j^T\bar{\mathbf{s}}_i)\\
\mathbf{w}_j &= \Big(\sum_{i\in O_j}[\bar{\mathbf{s}}_i\bar{\mathbf{s}}_i^T + \boldsymbol{\Sigma}]\Big)^{-1}\sum_{i\in O_j}\bar{\mathbf{s}}_i(x_{ij}-\mu_j), \quad j = 1,\ldots,m\\
\sigma^2 &= \frac{1}{n}\sum_{i,j\in O}\Big[(x_{ij}-\mathbf{w}_j^T\bar{\mathbf{s}}_i-\mu_j)^2 + \mathbf{w}_j^T\boldsymbol{\Sigma}\mathbf{w}_j\Big].
\end{aligned}$$

As pointed out in [26], the missing-data case leads to heavier computations since it requires to (*i*) iteratively update μ_j, (*ii*) consider a different $\boldsymbol{\Sigma}_i$ for each $\mathbf{s}_i$ and (*iii*) recompute the rows of $\mathbf{W}$ based only on those columns of $\bar{\mathbf{s}}$ which contribute to the reconstruction of the observed values in the corresponding row of the data matrix. The proposed scheme is efficient in the case of Gaussian measurements but empirical results indicate a dramatic performance degradation in the case of heavy-tailed observations or in the presence of outliers. In the next section, we overcome this by proposing a robust version of PPCA in the presence of missing data.

4.5.2 Robust Probabilistic Component Analysis Under Mixed Effects Model

As explained in Sect. 4.1, modeling the noise component as RES or compound Gaussian leads (except in some specific cases[7]) to an intractable solution in the presence of missing data. In this case, one has to use stochastic approximations, e.g., MCEM, SEM, and SAEM, which might lead to heavy computational cost (see Sects. 4.4.1 and 4.3.4.2 for more details).

An interesting robust alternative modeling reads

$$\mathbf{x}_i = \sqrt{\tau_i}\mathbf{W}\mathbf{s}_i + \boldsymbol{\mu} + \mathbf{n}_i, \tag{4.41}$$

in which $\mathbf{s}_i \sim \mathbb{R}N_m(\mathbf{0}, \mathbf{I})$, $\mathbf{n}_i \sim \mathbb{R}N_m(\mathbf{0}, \sigma^2\mathbf{I})$ and $\sqrt{\tau_i}$ are deterministic and unknown parameters leading to a tractable E-step. Depending on the value of $\sqrt{\tau_i}$, the i-th sample could be downweighted in order to take into account the influence of a possible outlier. A general extension of the above model has been proposed in [24], which reads

$$\mathbf{x}_{ji} = \sqrt{\tau_{ji}}\mathbf{W}_j\mathbf{s}_{ji} + f_j(\mathbf{A}_j, \boldsymbol{\mu}) + \mathbf{n}_{ji} \quad j = 1, \ldots, m, \quad i = 1, \ldots, n_j, \tag{4.42}$$

in which $\mathbf{s}_{ji} \sim \mathbb{R}N_m(\mathbf{0}, \mathbf{I})$, $\mathbf{n}_{ji} \sim \mathbb{R}N_m(\mathbf{0}, \sigma^2\mathbf{I})$. This extension has the advantage to include the possible presence of different groups or different individuals (indexed by j) within the dataset. Groups, or individuals, are described by a same global behavior (denoted by $\boldsymbol{\mu}$) but with different local variations (represented by $\mathbf{s}_{ji}$). The total number of groups is denoted by m, whereas n_j denotes the sample number for the j-th group of dimension m_j, i.e., $\mathbf{x}_{ji} \in \mathbb{R}^{m_j}$. The common behavior $\boldsymbol{\mu}$ (which is also the parameter of interest) is related to the design matrix $\mathbf{A}_j$ and the response function f_j (e.g., the j-th sub-array response in the case of large sensing systems [23]). Consequently, the spanned subspace for the j-th group reads $\mathbf{W}_j\mathbf{W}_j^T$. Such modeling, widely used in the longitudinal data analysis, is referred as *mixed effects model*. In this model, the random and fixed effects are represented, respectively, by $\mathbf{s}_{ji}$ and $\boldsymbol{\mu}$.

From the observation model (4.42), the unknown vector parameter is

$$\boldsymbol{\theta} = \left[\boldsymbol{\mu}^T, \left\{\boldsymbol{\zeta}_j^T\right\}_j, \left\{\tau_{ji}\right\}_{ji}\right]^T,$$

where $\boldsymbol{\zeta}_j$ is the concatenation of the non-redundant elements in[8] $\mathbf{W}_j$.

In order to tackle this problem using an EM-based algorithm, we define the complete data as $\mathbf{Z} = \left\{\mathbf{z}_{ji}\right\}_{ji}$ with $\mathbf{z}_{ji} = \left[\left(\mathbf{x}_{ji}^{(0)}\right)^T, \left(\mathbf{x}_{ji}^{(1)}\right)^T, \mathbf{s}_{ji}^T\right]^T$. Then, the complete

[7] Examples of these exceptions are the Gaussian case studied in [26, 75], the Student's t-distribution [10] or under some relaxation hypothesis, e.g., [24] and [52].

[8] The estimation of $\mathbf{W}_j$ is known to present a rotational ambiguity as discussed above.

log-likelihood reads (under some mild conditions)

$$Q(\boldsymbol{\theta}|\boldsymbol{\theta}^r) = \sum_{ji} \mathbb{E}_{\mathbf{z}_{ji}|\mathbf{x}_{ji}^{(0)},\boldsymbol{\theta}^r}\left[\ell_c(\boldsymbol{\theta}|\mathbf{z}_{ij})\right] = \sum_{ji} \mathbb{E}_{\mathbf{x}_{ji}^{(1)}|\mathbf{x}_{ji}^{(0)},\boldsymbol{\theta}^r}\left[\mathbb{E}_{\mathbf{s}_{ji}|\mathbf{x}_{ji}^{(0)},\mathbf{x}_{ji}^{(1)},\boldsymbol{\theta}^r}\left\{\ell_c(\boldsymbol{\theta}|\mathbf{z}_{ij})\right\}\right]. \tag{4.43}$$

The inner expectation and the outer expectation in (4.43) can be obtained from tower property for conditional Gaussian expectations (see, e.g., [2]). Namely, it is easy to see that

$$\widehat{\mathbf{x}}_{ji} \stackrel{\text{def}}{=} \mathbb{E}_{\mathbf{z}_{ji}|\mathbf{x}_{ji}^{(0)},\boldsymbol{\theta}^r}\left[\mathbf{x}_{ji}\right] = \left[\left(\mathbf{x}_{ji}^{(0)}\right)^T \left(\widehat{\mathbf{x}}_{ji}^{(1)}\right)^T\right]^T \tag{4.44}$$

$$= \begin{bmatrix} \mathbf{x}_{ji}^{(0)} \\ f_j(\mathbf{A}_j,\boldsymbol{\mu})^{(1)} + \boldsymbol{\Sigma}_{ji}^{(1)(0)}\left(\boldsymbol{\Sigma}_{ji}^{(1)(0)}\right)^{-1}\left(\mathbf{x}_{ji}^{(0)} - f_j(\mathbf{A}_j,\boldsymbol{\mu})^{(0)}\right) \end{bmatrix} \tag{4.45}$$

and

$$\widehat{\mathbf{x}_{ji}\mathbf{x}_{ji}^T} \stackrel{\text{def}}{=} \mathbb{E}_{\mathbf{z}_{ji}|\mathbf{x}_{ji}^{(0)},\boldsymbol{\theta}^r}\left\{\mathbf{x}_{ji}\mathbf{x}_{ji}^T\right\}$$

$$= \begin{bmatrix} \mathbf{x}_{ji}^{(0)}\left(\mathbf{x}_{ji}^{(0)}\right)^T & \mathbf{x}_{ji}^{(0)}\left(\widehat{\mathbf{x}}_{ji}^{(1)}\right)^T \\ \widehat{\mathbf{x}}_{ji}^{(1)}\left(\mathbf{x}_{ji}^{(0)}\right)^T & \boldsymbol{\Sigma}_{ji}^{(1)(1)} - \boldsymbol{\Sigma}_{ji}^{(1)(0)}\left(\boldsymbol{\Sigma}_{ji}^{(0)(0)}\right)^{-1}\boldsymbol{\Sigma}_{ji}^{(0)(1)} + \widehat{\mathbf{x}}_{ji}^{(1)}\left(\widehat{\mathbf{x}}_{ji}^{(1)}\right)^T \end{bmatrix}, \tag{4.46}$$

where the mean and the covariance matrix of $\mathbf{x}_{ji}$ are decomposed as $f_j(\mathbf{A}_j,\boldsymbol{\mu}) = \left[\left(f_j(\mathbf{A}_j,\boldsymbol{\mu})^{(0)}\right)^T \left(f_j(\mathbf{A}_j,\boldsymbol{\mu})^{(1)}\right)^T\right]^T$ and $\boldsymbol{\Sigma}_{ji} = \begin{pmatrix} \boldsymbol{\Sigma}_{ji}^{(0)(0)} & \boldsymbol{\Sigma}_{ji}^{(0)(1)} \\ \boldsymbol{\Sigma}_{ji}^{(1)(0)} & \boldsymbol{\Sigma}_{ji}^{(1)(1)} \end{pmatrix}$.

Then, the E-step consists in updating the following expectations (for the sake of clarity, the index related to the EM iteration number is omitted in the following. For example $\boldsymbol{\mu}^r$ and $\mathbf{W}^r$ are denoted by $\boldsymbol{\mu}$ and $\mathbf{W}$, respectively):

$$\widehat{\mathbf{r}}_{ji} \stackrel{\text{def}}{=} \mathbb{E}_{\mathbf{z}_{ji}|\mathbf{x}_{ji}^{(0)},\boldsymbol{\theta}^r}\left\{\mathbf{x}_{ji} - f_j(\mathbf{A}_j,\boldsymbol{\mu})\right\} = \widehat{\mathbf{x}_{ji}} - f_j(\mathbf{A}_j,\boldsymbol{\mu}), \tag{4.47}$$

$$\widehat{\mathbf{r}_{ji}\mathbf{r}_{ji}^T} \stackrel{\text{def}}{=} \mathbb{E}_{\mathbf{z}_{ji}|\mathbf{x}_{ji}^{(0)},\boldsymbol{\theta}^r}\left\{\mathbf{r}_{ji}\mathbf{r}_{ji}^T\right\} \stackrel{\text{def}}{=} \widehat{\mathbf{x}_{ji}\mathbf{x}_{ji}^T} + f_j(\mathbf{A}_j,\boldsymbol{\mu})f_j(\mathbf{A}_j,\boldsymbol{\mu})^T - 2\widehat{\mathbf{x}}_{ji}f_j(\mathbf{A}_j,\boldsymbol{\mu})^T, \tag{4.48}$$

$$\widehat{\mathbf{s}}_{ji} = \mathbb{E}_{\mathbf{z}_{ji}|\mathbf{x}_{ji}^{(0)},\boldsymbol{\theta}^r}\left\{\mathbf{s}_{ji}\right\} \stackrel{\text{def}}{=} \frac{\sqrt{\tau_{ji}}}{\sigma^2}\mathbf{V}_{ji}^{-1}\mathbf{W}_j\widehat{\mathbf{r}}_{ji}, \tag{4.49}$$

$$\widehat{\mathbf{r}_{ji}\mathbf{s}_{ji}^T} \stackrel{\text{def}}{=} \mathbb{E}_{\mathbf{z}_{ji}|\mathbf{x}_{ji}^{(0)},\boldsymbol{\theta}^r}\left\{\mathbf{r}_{ji}\mathbf{s}_{ji}^T\right\} = \frac{\sqrt{\tau_{ji}}}{\sigma^2}\widehat{\mathbf{r}_{ji}\mathbf{r}_{ji}^T}\mathbf{W}_j, \tag{4.50}$$

$$\widehat{\mathbf{s}_{ji}\mathbf{s}_{ji}^T} \stackrel{\text{def}}{=} \mathbb{E}_{\mathbf{z}_{ji}|\mathbf{x}_{ji}^{(0)},\boldsymbol{\theta}^r}\left\{\mathbf{s}_{ji}\mathbf{s}_{ji}^T\right\} = \mathbf{V}_{ji}^{-1} + \frac{\tau_{ji}}{\sigma^4}\mathbf{V}_{ji}^{-1}\mathbf{W}_j^T\widehat{\mathbf{r}_{ji}\mathbf{r}_{ji}^T}\mathbf{W}_j\mathbf{V}_{ji}^{-1}. \tag{4.51}$$

in which $\mathbf{V}_{ji} = \mathbf{I} + \frac{\tau_{ji}}{\sigma^2}\mathbf{W}_j^T\mathbf{W}_j$.

The M-step of the EM algorithm can be costly depending on the form of[9] $f_j(.)$. Nevertheless, in the case of a linear model, i.e., $f_j(\mathbf{A}_j, \boldsymbol{\mu}) = \mathbf{A}_j\boldsymbol{\mu}$, the M-step is tractable.[10] Nevertheless, the EM algorithm is enhanced using one of its variant, the ECME algorithm. Instead of maximizing sequentially the surrogate function $Q(.)$ w.r.t. $\boldsymbol{\theta}$, the ECME maximizes the observed (incomplete) likelihood when it is possible and tractable. Specifically, the loading matrix, τ_{ji} and the mean vector are given as the maximization of the surrogate EM function, whereas the noise power is estimated using the observed likelihood. Consequently, by plugging (4.47)–(4.51) into (4.43), the M-step of the ECME reads

$$\sqrt{\tau_{ji}} = \frac{\operatorname{tr}\left(\widehat{\mathbf{s}_{ji}\mathbf{r}_{ji}^T}\mathbf{W}_j\right)}{\operatorname{tr}(\widehat{\mathbf{s}_{ji}\mathbf{s}_{ji}^T}\mathbf{W}_j^T\mathbf{W}_j)}, \tag{4.52}$$

$$\sigma^2 = \frac{1}{\sum_j n_j}\sum_j\sum_i \operatorname{tr}\left\{\tau_{ji}\widehat{\mathbf{s}_{ji}\mathbf{s}_{ji}^T}\mathbf{W}_j^T\mathbf{W}_j + \widehat{\mathbf{r}_{ji}\mathbf{r}_{ji}^T} - 2\sqrt{\tau_{ji}}\operatorname{tr}\left(\widehat{\mathbf{s}_{ji}\mathbf{r}_{ji}^T}\mathbf{W}_j\right)\right\}, \tag{4.53}$$

$$\mathbf{W}_j = \left(\sum_i \sqrt{\tau_{ji}}\widehat{\mathbf{r}_{ji}\mathbf{s}_{ji}^T}\right)\left(\sum_i \widehat{\mathbf{s}_{ji}\mathbf{s}_{ji}^T}\right)^{-T}, \tag{4.54}$$

$$\boldsymbol{\mu} = \left(\sum_j \mathbf{A}_j^T\left(\sum_i \boldsymbol{\Gamma}_{ji}\right)\mathbf{A}_j\right)^{-1}\left(\sum_j \mathbf{A}_j^T\left(\sum_i \boldsymbol{\Gamma}_{ji}\right)\widehat{\mathbf{x}}_{ji}\right), \tag{4.55}$$

in which $\boldsymbol{\Gamma}_{ji} = \frac{1}{\sigma^2}\left(\mathbf{I} - \mathbf{W}_j\left(\frac{\tau_{ji}}{\sigma^2}\mathbf{I} + \mathbf{W}_j^T\mathbf{W}_j\right)^{-1}\mathbf{W}_j^T\right)$.

A parallel scheme naturally appears from the above equations: the consensus step is enforced via the common mean and homogeneous noise power. It is worth mentioning that the update of the mean involves the inversion of a $m \times m$ matrix in classical robust multivariate covariance and mean estimators (in which $m = \sum_j m_j$). Nevertheless, the above scheme updates the mean locally by the inversion of a smaller matrix of size $m_j \times m_j$ matrix. Then, a consensus is made in the fusion center. This makes the above scheme method computationally efficient by enhancing the computational cost and avoiding a central processor overload (Fig. 4.3).

---

[9] Array signal processing modeling is an example of $f_j(.)$ reading the steering matrix of the j-th sub-array and the mean vector denotes the direction of arrivals [55].

[10] For example, this the case in an imaging context for which the design matrix $\mathbf{A}_j$ is a known linear operator that maps the image from the space domain to the visibility domain and the mean denotes the intensity vector [47].

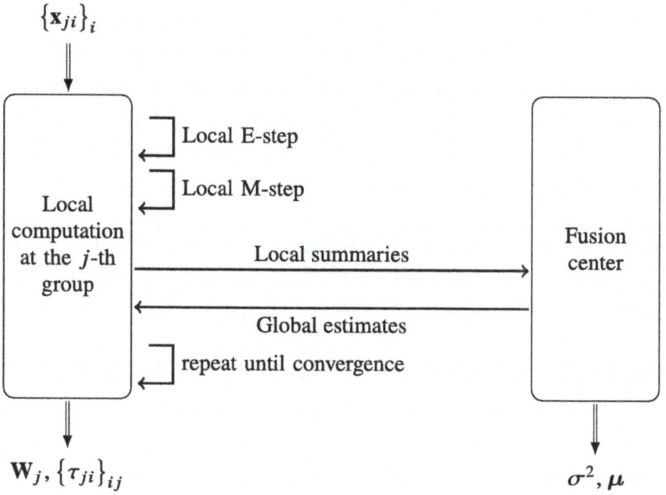

Fig. 4.3 Parallel scheme of the robust PPCA, in which (*i*) Local E-step updates $\widehat{\mathbf{r}_{ji}}, \widehat{\mathbf{x}_{ji}}$, and $\widehat{\mathbf{x}_{ji}\mathbf{x}_{ji}^T}$, whereas, in the (*ii*) Local M-step, we update τ_{ji}, $\mathbf{W}_j$, and $\boldsymbol{\Gamma}_{ji}$. (*iii*) The local summaries refer to $\sum_i \boldsymbol{\Gamma}_{ji}$ and $\kappa_j = \sum_i \mathrm{tr}\{\tau_{ji}\widehat{\mathbf{s}_{ji}\mathbf{s}_{ji}^T}\mathbf{W}_j^H\mathbf{W}_j + \mathbf{r}_{ji}^T\mathbf{r}_{ji} - 2\sqrt{\tau_{ij}}\mathbf{r}_{ji}^T\mathbf{W}_j\widehat{\mathbf{s}_{ji}}$ and finally (iv) the global summaries represent the common parameters, i.e., $\boldsymbol{\mu}$ and σ^2

4.6 Numerical Analysis Under Missing Data

This section is dedicated to the numerical analysis of the aforementioned estimators in the context of scaled Gaussian distributions presented in Sects. 4.4.2 and 4.4.3.[11] The code to reproduce the following experiments is available at: https://github.com/ahippert/em-scaled-gaussian.

First, let us compare, through numerical simulation, the estimation performance of the estimator (4.32) on simulated data $(\mathbf{x}_i)_{i=1}^n$ drawn from the zero-mean multivariate Gaussian and MSG distributions with covariance $\boldsymbol{\Sigma}$. The main aim is to quantify the performance losses in regard to the missing-data ratio and patterns in both full- and low-rank scenarios. To generate a covariance matrix admitting a low-rank structure (4.1), we compute

$$\boldsymbol{\Sigma} = \sigma^2\mathbf{I}_m + \mathbf{U}\mathbf{U}^T,$$

where $\mathbf{U} \in \mathrm{St}_{m,q}$ is the underlying signal subspace basis obtained from the eigenvalue decomposition of the low-rank covariance matrix $\mathbf{H}$ and σ is a free parameter corresponding to the signal-to-noise ratio. $\mathbf{H}$ admits a Toeplitz structure $(\mathbf{H})_{ij} =$

[11] We omit the comparison of the robust PPCA under missing data, since its gain versus classical schemes (discussed below) is similar to the gain of the robust EM algorithm for mixture of scaled Gaussian distributions in the presence of missing data versus classical schemes.

$\rho^{|i-j|}$, $i, j = 1, \ldots, m$, where $\rho = 0.7$ and $m = 15$. Scales $(\tau_i)_{i=1}^n$ are drawn from a Gamma distribution with shape parameter $\alpha = 1$ and scale parameter $\beta = \frac{1}{\alpha}$.

For the sake of the experiment, missing data are also simulated, which allows a full control on their ratio and pattern. Importantly, as the number of observations $n$increases in the comparison, the missing-data ratio decreases. The following covariance estimators are considered for comparison:

EM estimators The covariance matrix for the Gaussian distribution obtained from Algorithm 4.3 (see the appendix), named $\widehat{\boldsymbol{\Sigma}}_{\text{EM-SCM}}$ and the covariance matrix for MSG distributions obtained from Algorithm 4.1, named $\widehat{\boldsymbol{\Sigma}}_{\text{EM-Tyl}}$.

SCM estimators The SCM from the *clairvoyant* data (without missing data) and the SCM $\widehat{\boldsymbol{\Sigma}}_{\text{SCM}}^{(0)}$ estimated from the subset of fully observed vectors as defined in (4.6).

Tyler's M -estimators Tyler's M-estimator from the clairvoyant data as expressed in Eq. (97) of Chap. 1 and Tyler's M-estimator $\widehat{\boldsymbol{\Sigma}}_{\text{Tyl}}^{(0)}$ from the subset of fully observed vectors $(\mathbf{x}_i)_{i \in \mathcal{O}}$.

Robust multiple imputation We propose a robust version of multiple imputation [63, 78]: for each $\mathbf{x}_i$ with missing values $\mathbf{x}_i^{(1)}$, S vectors with *imputed* (or *completed*) missing entries $(\mathbf{x}_{i1,c}, \mathbf{x}_{i2,c}, \ldots, \mathbf{x}_{iS,c})$ (letter "c" stands for *complete*) are generated with missing values drawn from a MSG distribution:

$$\mathbf{x}_{ij,c}^{(1)} \sim \mathbb{R}N(\boldsymbol{\mu}_{ij}^{(0)}, \sqrt{\tau_{ij}}\boldsymbol{\sigma}_{ij}^{(0)}), \quad j = 1, \ldots, S,$$

where $\boldsymbol{\mu}_{ij}^{(0)}$ and $\boldsymbol{\sigma}_{ij}^{(0)}$ are the mean and variance of the observed components of $\mathbf{x}_{ij}$, respectively, and τ_{ij} are the scales parameters drawn from a Gamma distribution $Gamma(\alpha = 1, \beta = 1)$. The estimated covariance is then the mean of the S Tyler's M-estimators computed from the S imputed vectors:

$$\widehat{\boldsymbol{\Sigma}}_{\text{RMI}} = \frac{1}{S}\sum_{j=1}^{S} \widehat{\boldsymbol{\Sigma}}_{\text{Tyl},j} = \frac{p}{nS}\sum_{j=1}^{S}\sum_{i=1}^{n} \frac{\mathbf{x}_{ij,c}\mathbf{x}_{ij,c}^T}{\mathbf{x}_{ij,c}\widehat{\boldsymbol{\Sigma}}_{\text{Tyl},j}^{-1}\mathbf{x}_{ij,c}^T}.$$

All these estimators are compared through the *natural* distance for the set of SPD matrices $\mathcal{S}_m^{++}$, which is $|| \log(\boldsymbol{\Sigma}^{-\frac{1}{2}}\widehat{\boldsymbol{\Sigma}}\boldsymbol{\Sigma}^{-\frac{1}{2}})||$ [6]. Figure 4.4 shows mean estimation errors in functions of the number of observations for three missing-data patterns: monotone, general, and random patterns. Overall, the EM-Tyl estimator outperforms other estimators, except when the number of observations is not sufficient and approaches m. In this case, the RMI estimator is more appropriate. Unsurprisingly, EM-Tyl, RMI, and Tyl-obs reach Tyler's estimator on clairvoyant data as the missing-data ratio decreases in large sample size. In the case of Gaussian estimators, the performances of the EM-SCM estimator are very close to those of the clairvoy-

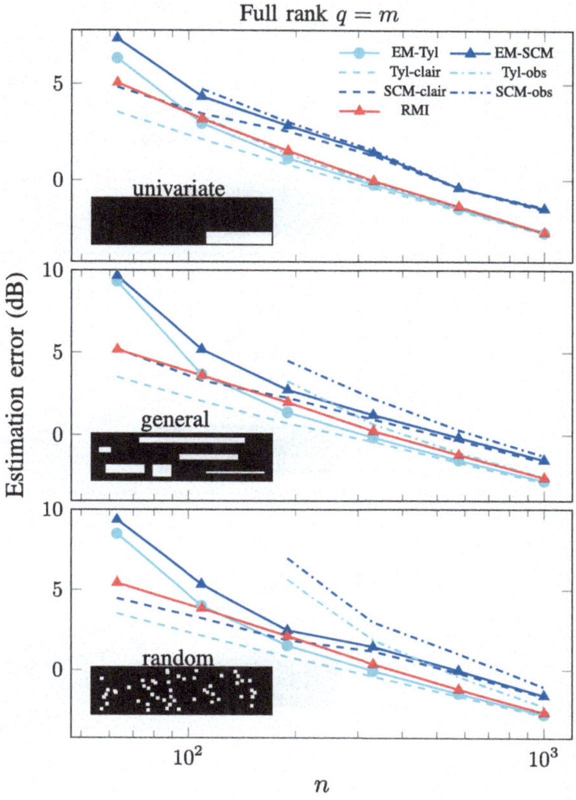

Fig. 4.4 Mean estimation error (500 experiments) as function of the number of samples n for univariate (top), general (middle) and random (bottom) missing-data patterns with $m = 15$

ant SCM. As expected, Tyl-obs and SCM-obs display poor performances when the missing-data ratio increases.

Secondly, in the following, we consider the robust low-rank estimation under scaled Gaussian distributions presented in Sect. 4.4.3. Namely, estimation errors for the low-rank factor model are illustrated in Fig. 4.5 with rank $q = 5$. Interestingly, accounting for the rank in the estimation procedure lowers the error for a small number of observations n, i.e., the gap between the EM-Tyl estimator and robust multiple imputation diminishes compared to the full-rank case. When n increases, it is clear that EM-Tyl and EM-SCM estimators outperform other estimators for MSG and Gaussian distributions, respectively. When the rank is known, the low-rank estimators will produce lower errors compared to full-rank estimators, as illustrated in Fig. 4.6 for $n = 105$.

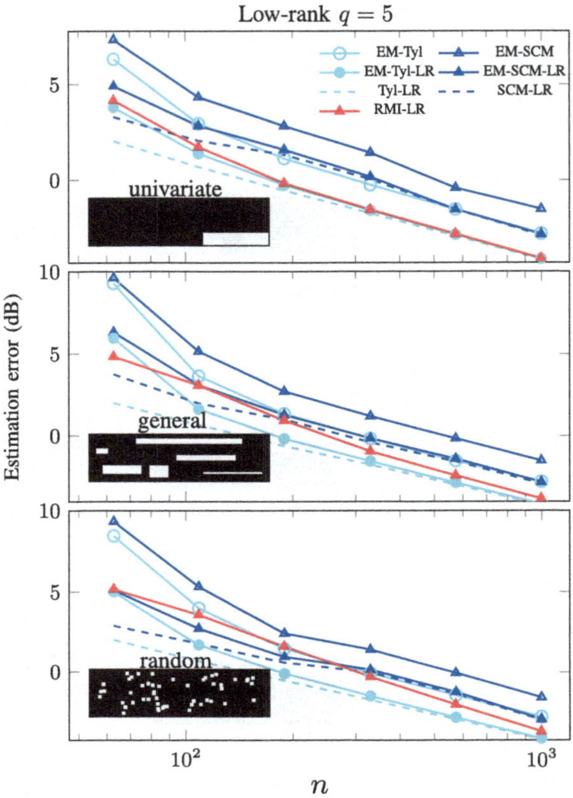

Fig. 4.5 Mean estimation error (500 experiments) as function of the number of samples n for univariate (top), general (middle) and random (bottom) missing-data patterns with a low-rank setting ($q = 5$) and $m = 15$. 'LR' stands for low-rank

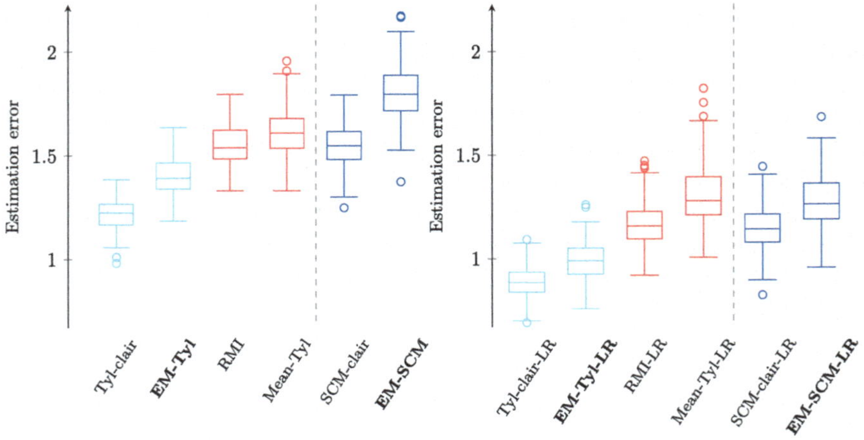

Fig. 4.6 Mean estimation in the full-rank (left) and low-rank (right) covariance model with $n = 105$. Dashed lines separate robust estimators from estimators based on the Gaussian model

4.7 Conclusion and Perspectives

We have reviewed a series of robust methods based on the EM algorithms and its variants to estimate the statistical parameters of data drawn from classes of RES distributions with missing values. In addition to the robustness of non-Gaussian data, these methods take advantage of underlying low-rank structures, which has proven to characterize well many real datasets (see for example, [20] in radar processing or [77] in data science, and the references therein). The performances of the estimators in terms of estimation error have been illustrated[12] under various missing-data patterns, showing the interest in both non-Gaussian and low-rank assumptions compared to robust multiple imputation, especially when the number of samples is large enough compared to m.

In the following, we present three perspectives closely linked to the subjects addressed in this chapter.

1. **Learning the scales with model-based deep neural networks** In Sect. 4.4.2, the relaxation of model (4.20) with a deterministic scale allows to derive statistics which are robust to many classes of distributions [49] and conveniently leads to closed-form expressions for parameters $\boldsymbol{\Sigma}$ and $\boldsymbol{\tau}$ at the M-step. An alternative strategy would be to assume an *unknown* prior distribution f_τ on the scale, and learn the unknown density generator using recent model-based deep neural networks in a supervised fashion [15, 16, 48]. The latter has been applied using fully observed data and extension to the case of the possible presence of missing data is of interest.

2. **Doubly robust probabilistic component analysis** In this chapter, we have considered that the data was MCAR or MAR, so that $f(\mathbf{r}_i|\mathbf{x}_i, \phi) = f(\mathbf{r}_i|\mathbf{x}_i^{(0)}, \phi)$. The consequence is that maximizing the full observed likelihood is equivalent to maximizing the observed likelihood, as seen in (4.12). If the mechanism is MNAR, one has to deal with the expectation of $f_{\mathbf{R}|\mathbf{X}}(\mathbf{r}_i|\mathbf{x}_i, \phi)$ during the E-step, which is clearly not tractable. A challenge would be to design a *doubly robust* algorithm to both data distribution and missing-data mechanism.

3. **Machine learning with structured missingness** Very often, machine learning algorithms need complete data to work [54]. Missing values can create errors at all levels of the analysis pipeline and lead to biased models which are not generalizable. A challenge in statistical estimation with missing data is to efficiently exploit what [50] called *structured missingness*, i.e., all possible combinations of the (possibly unknown) missing-data mechanisms and patterns in the data. To do so, ML algorithms should be designed by (1) exploring the geometry of mechanisms and patterns as both can capture multivariate relationships within the data,

[12] The code to reproduce the experiments is available at: https://github.com/ahippert/em-scaled-gaussian.

(2) explicitly modeling structured missingness using informative statistical priors on the mechanisms and patterns parameters, (3) better characterizing and understanding ethical implications of structured missingness, which can arise from socio-cultural biases during data extraction, analysis and/or interpretation (see [12] for an extensive discussion of these biases in machine learning, especially in ground truth datasets).

Acknowledgements We would like to warmly thank all those who have collaborated with us on this subject. Specifically, in chronological order, Prof. Guillaume Ginolhac (Listic, Université Savoie Mont-Blanc), Prof. Arnaud Breloy (CNAM), Dr. Florent Bouchard (Université Paris-Saclay, CNRS, CentraleSupélec, L2S), and Prof. Frédéric Pascal (Université Paris-Saclay, CNRS, CentraleSupélec, L2S).

Appendix: The Special Case of the Gaussian Distribution

As a particular case, we consider the widely used Gaussian distribution among the RES class. In this scope, the data is modeled as iid real-valued Gaussian random vectors with zero mean and unknown covariance matrix $\boldsymbol{\Sigma}$, which we recall is denoted $\mathbf{x} \sim \mathbb{R}N_m(\mathbf{0}, \boldsymbol{\Sigma})$.

Suppose that $(\text{vec}(\boldsymbol{\Sigma})) = \boldsymbol{\theta}$ is the parameter vector to be estimated from the data $\mathbf{x}$. The ML estimator $\widehat{\boldsymbol{\theta}}_{\text{ML}}$ (see (4.13)) given the observed data $\mathbf{X}^{(0)}$ maximizes the observed-data log-likelihood ℓ_{ign}. This function is given by

$$\ell_{\text{ign}}(\boldsymbol{\theta}|\mathbf{X}^{(0)}) = -\sum_{i=1}^{n} p_i \log(\pi) - \sum_{i=1}^{n} \log|\mathbf{A}_i \boldsymbol{\Sigma} \mathbf{A}_i^T| + \mathbf{x}_i^{(0)} (\mathbf{A}_i \boldsymbol{\Sigma} \mathbf{A}_i^T)^{-1} \mathbf{x}_i^{(0)T}, \tag{4.56}$$

where scalar p_i and matrix $\mathbf{A}_i$ are defined as in (4.2). By contrast, the complete-data log-likelihood ℓ_{c} can be expressed by replacing $\mathbf{x}_i$ by its permuted version $\widetilde{\mathbf{x}}_i$ as defined in (4.3):

$$\ell_{\text{c}}(\boldsymbol{\theta}|\widetilde{\mathbf{X}}) = -nm \log(\pi) - n \log|\boldsymbol{\Sigma}| - \sum_{i=1}^{n} \widetilde{\mathbf{x}}_i^T \widetilde{\boldsymbol{\Sigma}}_i^{-1} \widetilde{\mathbf{x}}_i. \tag{4.57}$$

With reference to the estimation problem (4.13), the EM algorithm is as follows:

- Initialization: At step $r = 0$ of the algorithm, initialize the covariance matrix $\boldsymbol{\Sigma}^{(0)}$ with the SCM computed from fully observed $\mathbf{x}_i$, e.g., using Eqs. (4.6) or (4.7).

- E-step: For each observation i, compute the expectation of $\mathbf{x}_i^{(1)}$ conditioned by the observed data $\mathbf{x}_i^{(0)}$ and the estimation of the parameters at step r:

$$
\begin{aligned}
Q_i(\boldsymbol{\theta}|\boldsymbol{\theta}^r) &= \mathbb{E}_{\mathbf{x}_i^{(1)}|\mathbf{x}_i^{(0)},\boldsymbol{\theta}^r}\left[\ell_c(\boldsymbol{\theta}|\widetilde{\mathbf{x}}_i)\right] \\
&= \mathrm{tr}\left(\mathbf{B}_i^r \widetilde{\boldsymbol{\Sigma}}_i^{-1}\right),
\end{aligned} \tag{4.58}
$$

where $\mathbf{B}_i^r$ contains the expected sufficient statistics as defined in (4.23)–(4.25). These expectations are computed using expressions (4.28) and (4.29). The conditional mean $\widetilde{\mu}_i^{1|0}$ and covariance $\widetilde{\boldsymbol{\Sigma}}_i^{11|0}$ involved in these expressions remain the same, except that the scales need to be dropped from (4.27).

- *M-step*: resolve the following maximization problem:

$$
\boldsymbol{\theta}^{r+1} = \underset{\boldsymbol{\theta}\in\Theta}{\arg\max}\ Q(\boldsymbol{\theta}|\boldsymbol{\theta}^r), \tag{4.59}
$$

where $Q(\boldsymbol{\theta}|\boldsymbol{\theta}^r) = \sum_{i=1}^n Q_i(\boldsymbol{\theta}|\boldsymbol{\theta}^r)$ is the total Q-function due to the iid of the observations. Solving (4.59) is tantamount to solve $\frac{\partial Q(\boldsymbol{\theta}|\boldsymbol{\theta}^r)}{\partial \boldsymbol{\theta}} = 0$. Fortunately, in this case, a closed-form expression for $\boldsymbol{\Sigma}$ is found with simple derivation calculus:

$$
\boldsymbol{\Sigma}^{r+1} = \frac{1}{n}\sum_{i=1}^{n} \mathbf{P}_i (\mathbf{B}_i^r)^T \mathbf{P}_i^T, \tag{4.60}
$$

where $\mathbf{P}_i$, $i = 1, \ldots, n$ are the permutation matrices as defined in (4.3). Notice that in the case of no missing values, $\mathbf{P}_i = \mathbf{I}_m$ and $\mathbf{B}_i^r = \mathbb{E}[\mathbf{x}_i\mathbf{x}_i^T]$, which leads to the classical SCM. The following algorithm sums up this simple procedure.

Algorithm 4.3 EM-SCM: Estimation of $\boldsymbol{\Sigma}$ under the Gaussian distribution with missing values

---

Require: $(\widetilde{\mathbf{x}}_i)_{i=1}^n \sim \mathbb{R}N_m(\mathbf{0}, \boldsymbol{\Sigma})$, $(\mathbf{P}_i)_{i=1}^n$
Ensure: $\widehat{\boldsymbol{\Sigma}}$
1: *Initialization*: compute $\boldsymbol{\Sigma}^{r=0}$ using (4.6) or (4.7)
2: **repeat**
3: *E-step*: compute $\mathbb{E}[\mathbf{x}_i^{(1)}]$, $\mathbb{E}[\mathbf{x}_i^{(0)}\mathbf{x}_i^{(1)T}]$, $\mathbb{E}[\mathbf{x}_i^{(1)}\mathbf{x}_i^{(1)T}]$ for $i = 1, \ldots, n$ using (4.26)–(4.29)
4: *M-step*: compute $\boldsymbol{\Sigma}^{r+1} = \frac{1}{n}\sum_{i=1}^n \mathbf{P}_i(\mathbf{B}_i^r)^T\mathbf{P}_i^T$
5: $r \leftarrow r + 1$
6: **until** $||\boldsymbol{\Sigma}^{r+1} - \boldsymbol{\Sigma}^r||_F^2$ converges

---

References

1. Abdallah, R., Breloy, A., El Korso, M., Lautru, D.: Bayesian signal subspace estimation with compound gaussian sources. Signal Process. Elsevier **167**(3), 107–310 (2020)

2. Anderson, T.: An Introduction to Multivariate Statistical Analysis. Wiley, New York (1965)
3. Andridge, R.R., Little, R.J.: A review of hot deck imputation for survey non-response. Int. Stat. Rev. **78**(1), 40–64 (2010)
4. Aubry, A., De Maio, A., Marano, S., Rosamilia, M.: Structured covariance matrix estimation with missing-(complex) data for radar applications via expectation-maximization. IEEE Trans. Signal Process. **69**, 5920–5934 (2021)
5. Benzécri, J.P., et al.: L'analyse des données, vol. 2. Dunod Paris (1973)
6. Bhatia, R.: Positive definite matrices. In: Positive Definite Matrices. Princeton University Press (2009)
7. Bishop, W.E., Yu, B.M.: Deterministic symmetric positive semidefinite matrix completion. Advances in Neural Information Processing Systems **27** (2014)
8. Candes, E.J., Plan, Y.: Matrix completion with noise. Proc. IEEE **98**(6), 925–936 (2010)
9. Celeux, G.: The SEM algorithm: a probabilistic teacher algorithm derived from the EM algorithm for the mixture problem. Comput. Stat. Q. **2**, 73–82 (1985)
10. Chen, T., Martin, E., Montague, G.: Robust probabilistic PCA with missing data and contribution analysis for outlier detection. Comput. Stati. Data Anal. **53**(1), 3706–3716 (2009)
11. Chen, T., Martin, E., Montague, G.: Robust probabilistic PCA with missing data and contribution analysis for outlier detection. Comput. Stati. Data Anal. **53**(10), 3706–3716 (2009)
12. Crawford, K.: The Atlas of AI: Power, Politics, and the Planetary Costs of Artificial Intelligence. Yale University Press (2021)
13. Delyon, B., Lavielle, M., Moulines, E.: Convergence of a stochastic approximation version of the EM algorithm. Ann. Stat. **63**(3), 94–128 (1999)
14. Dempster, A.P., Laird, N.M., Rubin, D.B.: Maximum likelihood from incomplete data via the EM algorithm. J. R. Stat. Soc. Ser. B (Methodol.) **39**(1), 1–22 (1977)
15. Diskin, T., Draskovic, G., Pascal, F., Wiesel, A.: A robust EM algorithm for radio interferometric imaging in the presence of outliers. In: International Workshop on Computational Advances in Multi-Sensor Adaptive Processing (CAMSAP), pp. 1–5. IEEE Workshop (2017)
16. Diskin, T., Eldar, Y.C., Wiesel, A.: Learning to estimate without bias. IEEE Trans. Signal Process. **71**(3), 2162–2171 (2023)
17. Formont, P., Pascal, F., Vasile, G., Ovarlez, J.P., Ferro-Famil, L.: Statistical classification for heterogeneous polarimetric SAR images. IEEE J. Sel. Top. Signal Process. **5**(3), 567–576 (2010)
18. Frahm, G., Jaekel, U.: A generalization of Tyler's M-estimators to the case of incomplete data. Comput. Stat. Data Anal. **54**(2), 374–393 (2010)
19. Frahm, G., Nordhausen, K., Oja, H.: M-estimation with incomplete and dependent multivariate data. J. Multivar. Anal. **176**, 104569 (2020)
20. Goodman, N.A., Stiles, J.M.: On clutter rank observed by arbitrary arrays. IEEE Trans. Signal Process. **55**(1), 178–186 (2006)
21. Graham, J.W.: Missing data analysis: making it work in the real world. Annu. Rev. Psychol. **60**, 549–576 (2009)
22. Greco, M.S., Gini, F.: Statistical analysis of high-resolution SAR ground clutter data. IEEE Trans. Geosci. Remote Sens. **45**(3), 566–575 (2007)
23. Haardt, M., Pesavento, M.F.R., El Korso, M.N.: A robust and flexible EM algorithm for mixtures of elliptical distributions with missing data. In: Viberg, M. (ed.) Electronic Reference in Signal Processing: Array and Statistical Signal Processing, vol. 3, pp. 651–717 (2014)
24. Hippert-Ferrer, A., El Korso, M.N., Breloy, A., Ginolhac, G.: Robust mean and covariance matrix estimation under heterogeneous mixed-effects model with missing values. Signal Process. **188**, 108195 (2021)
25. Hippert-Ferrer, A., El Korso, M.N., Breloy, A., Ginolhac, G.: Robust low-rank covariance matrix estimation with a general pattern of missing values. Signal Process. **195**, 108460 (2022)
26. Ilin, A., Raiko, T.: Practical approaches to principal component analysis in the presence of missing values. J. Mach. Learn. Res. **11**, 1957–2000 (2010)
27. Jamshidian, M.: An EM algorithm for ML factor analysis with missing data. In: Latent Variable Modeling and Applications to Causality, pp. 247–258. Springer (1997)

28. Johnson, C.R.: Matrix completion problems: a survey. In: Matrix Theory and Applications, vol. 40, pp. 171–198 (1990)
29. Johnstone, I.M.: On the distribution of the largest eigenvalue in principal components analysis. Ann. Stat. **29**(2), 295–327 (2001)
30. Josse, J., Husson, F.: Handling missing values in exploratory multivariate data analysis methods. Journal de la Société Française de Statistique **153**(2), 79–99 (2012)
31. Kang, B., Monga, V., Rangaswamy, M.: Rank-constrained maximum likelihood estimation of structured covariance matrices. IEEE Trans. Aerosp. Electron. Syst. **50**(1), 501–515 (2014)
32. King, G., Honaker, J., Joseph, A., Scheve, K.: Analyzing incomplete political science data: an alternative algorithm for multiple imputation. Am. Polit. Sci. Rev. **95**(1), 49–69 (2001)
33. Larsson, E.G., Stoica, P.: High-resolution direction finding: the missing data case. IEEE Trans. Signal Process. **49**(5), 950–958 (2001)
34. Ledoit, O., Wolf, M.: A well-conditioned estimator for large-dimensional covariance matrices. J. Multivar. Anal. **88**(2), 365–411 (2004)
35. Little, R.J., Rubin, D.B.: Statistical Analysis with Missing Data, vol. 793. John Wiley & Sons (2019)
36. Little, R.J., Rubin, D.B.: Statistical Analysis with Missing Data. John Wiley & Sons (2002)
37. Little, R.J.: Robust estimation of the mean and covariance matrix from data with missing values. J. R. Stat. Soc.: Ser. C (Appl. Stat.) **37**(1), 23–38 (1988)
38. Liu, C., Rubin, D.B.: ML estimation of the t distribution using EM and its extensions, ECM and ECME. Statistica Sinica pp. 19–39 (1995)
39. Liu, C.: Missing data imputation using the multivariate t-distribution. J. Multivar. Anal. **53**(1), 139–158 (1995)
40. Liu, C.: Efficient ML estimation of the multivariate normal distribution from incomplete data. J. Multivar. Anal. **69**, 206–217 (1999)
41. Liu, J., Palomar, D.P.: Regularized robust estimation of mean and covariance matrix for incomplete data. Signal Process. **165**, 278–291 (2019)
42. Liu, J., Kumar, S., Palomar, D.P.: Parameter estimation of heavy-tailed AR model with missing data via stochastic EM. IEEE Trans. Signal Process. **67**(8), 2159–2172 (2019)
43. Lord, F.M.: Estimation of parameters from incomplete data. J. Am. Stat. Assoc. **50**(271), 870–876 (1955)
44. Lounici, K., et al.: High-dimensional covariance matrix estimation with missing observations. Bernoulli **20**(3), 1029–1058 (2014)
45. Mattei, P.A., Frellsen, J.: MIWAE: Deep generative modelling and imputation of incomplete data sets. In: International Conference on Machine Learning, pp. 4413–4423. PMLR (2019)
46. Meriaux, B., Ren, C., El Korso, M., Breloy, A., Forster, P.: Robust estimation of structured scatter matrices in (mis) matched models. Signal Process. Elsevier **165**(3), 163–174 (2019)
47. Mhiri, Y., El Korso, M.N., Breloy, A., Larzabal, P.: A robust EM algorithm for radio interferometric imaging in the presence of outliers. In: SiPS. IEEE Workshop on Signal Processing Systems, pp. 1–5 (2022)
48. Mhiri, Y., El Korso, M.N., Breloy, A., Larzabal, P.: Imagerie radio-interférométrique robuste par dépliement neuronal. In: Gretsi, pp. 1–4 (2023)
49. Mian, A., Ginolhac, G., Ovarlez, J.P., Atto, A.M.: New robust statistics for change detection in time series of multivariate SAR images. IEEE Trans. Signal Process. **67**(2), 520–534 (2018)
50. Mitra, R., McGough, S.F., Chakraborti, T., Holmes, C., Copping, R., Hagenbuch, N., Biedermann, S., Noonan, J., Lehmann, B., Shenvi, A., et al.: Learning from data with structured missingness. Nat. Mach. Intell. **5**(1), 13–23 (2023)
51. Mohan, K.: On handling self-masking and other hard missing data problems. In: AAAI Symposium (2018)
52. Mouret, F., Hippert-Ferrer, A., Pascal, F., Tourneret, J.Y.: A robust and flexible EM algorithm for mixtures of elliptical distributions with missing data. IEEE Trans. Signal Process. **71**, 1669–1682 (2023)
53. Nakagawa, S., Freckleton, R.P.: Missing inaction: the dangers of ignoring missing data. Trends Ecol. Evol. **23**(11), 592–596 (2008)

54. Nijman, S., Leeuwenberg, A., Beekers, I., Verkouter, I., Jacobs, J., Bots, M., Asselbergs, F., Moons, K., Debray, T.: Missing data is poorly handled and reported in prediction model studies using machine learning: a literature review. J. Clin. Epidemiol. **142**, 218–229 (2022)
55. Ollier, V., El Korso, M.N., Boyer, R., Larzabal, P., Pesavento, M.: Joint ML calibration and DOA estimation with separated arrays. In: ICASPP. IEEE International Conference on Acoustics, Speech and Signal Processing, pp. 2996–3000 (2016)
56. Ollier, V., El Korso, M., Ferrari, A., Boyer, R., Larzabal, P.: Robust distributed calibration of radio interferometers with direction dependent distortions. Signal Process. Elsevier **153**(3), 348–354 (2018)
57. Ollila, E., Tyler, D.E., Koivunen, V., Poor, H.V.: Complex elliptically symmetric distributions: survey, new results and applications. IEEE Trans. Signal Process. **60**(11), 5597–5625 (2012)
58. Pascal, F., Chitour, Y., Ovarlez, J.P., Forster, P., Larzabal, P.: Covariance structure maximum-likelihood estimates in compound Gaussian noise: existence and algorithm analysis. IEEE Trans. Signal Process. **56**(1), 34–48 (2008)
59. Robbins, H., Monro, S.: A Stochastic Approximation Method. The Annals of Mathematical Statistics, pp. 400–407 (1951)
60. Robertson, D., Symons, J.: Maximum likelihood factor analysis with rank-deficient sample covariance matrices. J. Multivar. Anal. **98**(4), 813–828 (2007)
61. Robin, G., Josse, J., Moulines, É., Sardy, S.: Low-rank model with covariates for count data with missing values. J. Multivar. Anal. **173**, 416–434 (2019)
62. Rodrigues, P.L.C., Congedo, M., Jutten, C.: A data imputation method for matrices in the symmetric positive definite manifold. In: GRETSI 2019-XXVIIème Colloque francophone de traitement du signal et des images (2019)
63. Royston, P.: Multiple imputation of missing values. Stand Genomic Sci. **4**(3), 227–241 (2004)
64. Rubin, D.B.: Multiple Imputation for Nonresponse in Surveys, vol. 81. John Wiley & Sons (2004)
65. Rubin, D.B.: Inference and missing data. Biometrika **63**(3), 581–592 (1976)
66. Ruppert, D., Matteson, D.S.: Statistics and Data Analysis for Financial Engineering, vol. 13. Springer (2011)
67. Schafer, J.L.: Analysis of Incomplete Multivariate Data. CRC press (1997)
68. Schneider, T.: Analysis of incomplete climate data: Estimation of mean values and covariance matrices and imputation of missing values. J. Climate **14**, 853–871 (2001)
69. Seaman, S., Galati, J., Jackson, D., Carlin, J.: What is meant by "missing at randomâŁž? Stat. Sci. **28**(2), 257–268 (2013)
70. Shen, H., Li, X., Chen, Q., Zeng, C., Yang, G., Li, H., Zhang, L.: Missing information reconstruction of remote sensing data: a technical review. IEEE Geosci. Remote Sens. Mag. **3**, 61–85 (2015)
71. Sportisse, A.: Handling heterogeneous and MNAR missing data in statistical learning frameworks: imputation based on low-rank models, online linear regression with sgd, and model-based clustering. Ph.D. thesis, Sorbonne université (2021)
72. Städler, N., Stekhoven, D.J., Bühlmann, P.: Pattern alternating maximization algorithm for missing data in high-dimensional problems. J. Mach. Learn. Res. **15**(1), 1903–1928 (2014)
73. Sun, Y., Babu, P., Palomar, D.P.: Robust estimation of structured covariance matrix for heavy-tailed elliptical distributions. IEEE Trans. Signal Process. **64**(14), 3576–3590 (2016)
74. Theiler, J., Foy, B., Fraser, A.: Characterizing non-Gaussian clutter and detecting weak gaseous plumes in hyperspectral imagery. Proc. SPIE **5806**, 182–193 (2005)
75. Tipping, M.E., Bishop, C.M.: Probabilistic principal component analysis. J. R. Stat. Soc. Ser. B (Stat. Methodol.) **61**(3), 611–622 (1999)
76. Troyanskaya, O., Cantor, M., Sherlock, G., Brown, P., Hastie, T., Tibshirani, R., Botstein, D., Altman, R.B.: Missing value estimation methods for dna microarrays. Bioinformatics **17**(6), 520–525 (2001)
77. Udell, M., Townsend, A.: Why are big data matrices approximately low rank? SIAM J. Math. Data Sci. **1**(1), 144–160 (2019)
78. Van Buuren, S.: Flexible Imputation of Missing Data. CRC Press (2018)

79. Walczak, B., Massart, D.L.: Dealing with missing data: Part I. Chemom. Intell. Lab. Syst. **58**, 15–27 (2001)
80. Wei, G.C., Tanner, M.A.: A monte carlo implementation of the EM algorithm and the poor man's data augmentation algorithms. J. Am. Stat. Assoc. **85**(411), 699–704 (1990)
81. Wilks, S.S.: Moments and distributions of estimates of population parameters from fragmentary samples. Ann. Math. Stat. **3**(3), 163–195 (1932)
82. Zhang, X., El Korso, M., Pesavento, M.: MIMO radar target localization and performance evaluation under SIRP clutter. Signal Process. Elsevier **130**(3), 217–232 (2017)
83. Zhang, Q., Yuan, Q., Zeng, C., Li, X., Wei, Y.: Missing data reconstruction in remote sensing image with a unified spatial-temporal-spectral deep convolutional neural network. IEEE Trans. Geosci. Remote Sens. **56**(8), 4274–4288 (2018)
84. Zhou, R., Liu, J., Kumar, S., Palomar, D.P.: Student's *t* VAR modeling with missing data via stochastic EM and Gibbs sampling. IEEE Trans. Signal Process. **68**, 6198–6211 (2020)

Part II
Performance Analysis

Chapter 5
Semiparametric Estimation in Elliptical Distributions

Stefano Fortunati

Abstract This chapter has the twofold aim of introducing in an intuitive and accessible manner the general framework of semiparametric inference and then of showing how it can be fruitfully applied to the joint estimation of the *location vector* and the *covariance* (or *scatter*) matrix of a set of elliptically distributed observations in the presence of an unknown density generator. A semiparametric model is a set of probablity density functions (pdfs) parameterized by a finite-dimensional parameter vector of interest and by an infinite-dimensional nuisance parameter, i.e., a function, whose estimation is not strictly required. The presence of this additional functional unknown will clearly lead to some performance losses in the estimation of the finite-dimensional parameter vector of interest. The first goal of this chapter is then to show how the classical estimation theory can be generalized in order to take into account an infinite-dimensional nuisance term. In particular, the three building blocks of the semiparametric theory, that are *the Hilbert space of score vectors*, *the nuisance tangent space* and *the related projection operator*, will be introduced. By means of these abstract concepts, we define the semiparametric counterpart of the Fisher Information Matrix (FIM) and the related semiparametric efficiency bound. After having prepared the theoretical ground, the focus of the second part of the chapter is on the application of the general semiparametric inference framework to the joint estimation of the location vector and of the scatter matrix in Real Elliptically Symmetric (RES) distributed random vectors. A closed-form expression for the semiparametric FIM and the related bound will be provided. We conclude the chapter by presenting the class of the R-estimators of the scatter matrix.

S. Fortunati (✉)
Laboratoire des signaux et systèmes, Université Paris-Saclay, CNRS, CentraleSupélec, 91190 Gif-sur-Yvette, France

DR2I-IPSA,
91190 Ivry sur Seine, France
e-mail: stefano.fortunati@centralesupelec.fr

J.-P. Delmas et al. (eds.), *Elliptically Symmetric Distributions in Signal Processing and Machine Learning*, https://doi.org/10.1007/978-3-031-52116-4_5

5.1 Introduction

Let $\mathbf{x}_1, \ldots, \mathbf{x}_n$ be n observation vectors collected from a random experiment and let $P_0(\mathbf{x}_1, \ldots, \mathbf{x}_n)$ be their (joint) "true" distribution admitting a (joint) "true" pdf $p_0(\mathbf{x}_1, \ldots, \mathbf{x}_n)$. In this chapter, we always assume that the observations are *independent* and *identically distributed* (i.i.d.) random vectors, consequently, the joint pdf can be expressed as $p_0(\mathbf{x}_1, \ldots, \mathbf{x}_n) = \prod_{i=1}^{n} p_0(\mathbf{x}_i)$. In the following, we indicate simply as $\mathbf{x} \sim p_0$ a generic random vector distributed according to the (marginal) "true" pdf p_0. Note that, in practice, the true pdf p_0 is not fully a-priori known, then the main task of any statistical procedure is to infer the missing knowledge on p_0 from the collected data. Inference methods rely then on the definition of a *model*, i.e., a set of pdfs that are able to statistically characterize the data-generating process. The nature of the model to be used depends on the available amount of a-priori information: the more precise our knowledge of the real-world phenomenon is, the narrower the model can be. The most commonly adopted models are the *parametric* ones.

As the name suggests, a parametric model $\mathcal{P}_\theta$ is a set of pdfs parameterized only by a finite-dimensional parameter vector $\boldsymbol{\theta} \in \Theta \subseteq \mathbb{R}^q$:

$$\mathcal{P}_\theta \stackrel{\text{def}}{=} \left\{ p_X(\mathbf{x}|\boldsymbol{\theta}), \boldsymbol{\theta} \in \Theta \subseteq \mathbb{R}^q \right\}.$$

If a parametric model is adopted, the information to be inferred from the observations is summarized in the (finite-dimensional) parameter vector $\boldsymbol{\theta}$ to be estimated. More formally, if the true data pdf $p_0(\mathbf{x})$ belongs to $\mathcal{P}_\theta$, this implies that there exists $\boldsymbol{\theta}_0 \in \Theta$ such that $p_0(\mathbf{x}) = p_X(\mathbf{x}|\boldsymbol{\theta}_0)$. However, in many practical applications, the available a-priori knowledge about the data-generating process may result to be insufficient to validate the chosen parametric model and we could run the risk that the true data pdf p_0 does not belong to the assumed model $\mathcal{P}_\theta$. This *model misspecification* [6] will lead to some performance degradation in the estimation of $\boldsymbol{\theta}$ that may result to be unacceptable. In this case, one may decide to rely on a more general *non-parametric model*. A non-parametric model is a collection of pdfs that possibly satisfy some functional constraints, i.e., the symmetry or the finiteness of their moments up to a given order:

$$\mathcal{P}_p \stackrel{\text{def}}{=} \{ p_X(\mathbf{x}) \in \mathcal{K} \},$$

where $\mathcal{K}$ is some constrained set of pdfs. If, on one hand, the generality of a non-parametric model minimizes the risk of model misspecification, non-parametric inference requires a large amount of homogeneous data that may be difficult (or even impossible) to collect in some applications.

The *semiparametric models* have been then proposed in order to keep the inherent finite-dimensionality of the parameters of interest without sacrificing the generality of the statistical description of a data-generating process [39]. Specifically, a semiparametric model is a set of pdfs characterized by a finite-dimensional parameter $\boldsymbol{\theta} \in \Theta \subseteq \mathbb{R}^q$ along with some infinite-dimensional parameter, i.e., a function, $g \in \mathcal{G}$:

$$\mathcal{P}_{\theta,g} \stackrel{\text{def}}{=} \left\{p_X(\mathbf{x}|\boldsymbol{\theta}, g), \boldsymbol{\theta} \in \Theta \subseteq \mathbb{R}^q, g \in \mathcal{G}\right\}.$$

A huge amount of practical inference problems can be described using this semiparametric formalism (see [2, 29, 39] and the reference therein). Here are some examples of remarcable importance for the signal processing community [5]:

1. **Inference in real elliptically symmetric (RES) distributions** As discussed in Chap. 1, a random vector $\mathbf{x} \in \mathbb{R}^m$ is said to be RES-distributed if it has a relevant pdf of the form:

$$p_X(\mathbf{x}) = |\boldsymbol{\Sigma}|^{-1/2} g((\mathbf{x} - \boldsymbol{\mu})^T \boldsymbol{\Sigma}^{-1}(\mathbf{x} - \boldsymbol{\mu})), \tag{5.1}$$

where $g : \mathbb{R}^+ \to \mathbb{R}^+$ is called *density generator*. In almost all practical applications involving the elliptical model, we are generally interested in the estimation of the *location vector* $\boldsymbol{\mu} \in \mathbb{R}^m$ and/or of the *covariance/scatter* matrix $\boldsymbol{\Sigma} \in \mathcal{M}$ where $\mathcal{M}$ indicates the set of all the positive definite, symmetric $m \times m$ matrices. Therefore, the density generator g can be considered as a nuisance function whose estimation is not required. Note that, as discussed in Sect. 2.1 of Chap. 1, $\boldsymbol{\Sigma} \in \mathcal{M}$ and $g \in \mathcal{G}$ are not jointly indefinable. To avoid the scale ambiguity, we introduce the symmetric and positive definite *shape* matrix as

$$\mathbf{V} = \boldsymbol{\Sigma}/s(\boldsymbol{\Sigma}) \in \mathcal{M}_s \stackrel{\text{def}}{=} \{\mathbf{V} \in \mathcal{M} | s(\mathbf{V}) = 1\} \subset \mathcal{M}, \tag{5.2}$$

where $s : \mathcal{M} \to \mathbb{R}^+$ is a scalar homogeneous functional (see [15, 31], Sect. II, [10]) and $\mathcal{M}_s$ is a constrained smooth sub-manifold of $\mathcal{M}$ whose dimension is $\dim(\mathcal{M}_s) = \dim(\mathcal{M}) - 1$. It is immediate to verify that the RES family can be interpreted as a semiparametric model of the form:

$$\mathcal{P}_{\mu,\mathbf{V},g} \stackrel{\text{def}}{=} \left\{p_X(\mathbf{x}) = |\mathbf{V}|^{-1/2} g((\mathbf{x} - \boldsymbol{\mu})^T \mathbf{V}^{-1}(\mathbf{x} - \boldsymbol{\mu})), (\boldsymbol{\mu}, \mathbf{V}) \in \Theta, g \in \mathcal{G}\right\}, \tag{5.3}$$

where $\Theta \stackrel{\text{def}}{=} \mathbb{R}^m \times \mathcal{M}_s$. We will investigate the semiparametric elliptical model extensively in the second part of the chapter.

2. **Estimation with missing data** Let $\mathbf{z} \triangleq (\mathbf{x}^T, \mathbf{y}^T)^T$ be a *complete* dataset, where:

- $\mathbf{x}$ is the *observed* (available) dataset.
- $\mathbf{y}$ is the *unobservable* (missing) dataset.

Suppose now that we want to estimate the value of a parameter vector $\boldsymbol{\theta} \in \Theta$ from the observed dataset $\mathbf{x}$ when the pdf p_Y of the missing data $\mathbf{y}$ is *unknown*. Direct application of the basic rules of conditional distributions allows one to express the pdf p_X of the observed dataset as:

$$p_X(\mathbf{x}|\boldsymbol{\theta}) = \int_{\mathcal{Y}} p_{X,Y}(\mathbf{x}, \mathbf{y}|\boldsymbol{\theta}) d\mathbf{y} = \int_{\mathcal{Y}} p_{X|Y}(\mathbf{x}|\mathbf{y}, \boldsymbol{\theta}) p_Y(\mathbf{y}) d\mathbf{y}.$$

Consequently, the set of all the pdfs of the observed dataset $\mathbf{x}$ can be framed in a *semiparametric mixture model* of the form [2, Sect. 4.5], [35].

$$\mathcal{P}_{\boldsymbol{\theta},p_Y} \stackrel{\text{def}}{=} \{p_X(\mathbf{x}|\boldsymbol{\theta}, p_Y), \boldsymbol{\theta} \in \Theta, p_Y \in \mathcal{K}\}.$$

3. **Nonlinear regression** According to the amount of a-priori knowledge that we have on the observation model, the nonlinear regression can be conveniently described using the semiparametric formalism:

$$\mathbf{x} = f(\mathbf{z}, \boldsymbol{\theta}) + \boldsymbol{\epsilon},$$

- $\boldsymbol{\theta} \in \Theta$: parameter vector to be estimated,
- $f \in \mathcal{F}$: possibly unknown nonlinear function,
- $\mathbf{z}$: random vector with possibly unknown pdf $p_Z \in \mathcal{K}$,
- $\boldsymbol{\epsilon}$: random noise with possibly unknown pdf $p_\epsilon \in \mathcal{E}$.

The set of all pdfs for $\mathbf{x}$ is a semiparametric model of the form [2, Sect. 4.3]:

$$\mathcal{P}_{\boldsymbol{\theta},f,p_Z,p_\epsilon} \stackrel{\text{def}}{=} \{p_X(\mathbf{x}|\boldsymbol{\theta}, f, p_Z, p_\epsilon), \boldsymbol{\theta} \in \Theta, f \in \mathcal{F}, p_Z \in \mathcal{K}, p_\epsilon \in \mathcal{E}\}.$$

4. **Autoregressive processes (AR)** Let us consider the AR(p) process:

$$x_n = \sum\nolimits_{i=1}^{p} \theta_i x_{n-i} + w_n, \quad n \in (-\infty, \infty),$$

- $\boldsymbol{\theta} \triangleq [\theta_1, \ldots, \theta_p]$: parameter vector to be estimated.
- w_n: i.i.d. innovations with pdf $p_w \in \mathcal{W}$.

The aim is to estimate $\boldsymbol{\theta}$ from a vector $\mathbf{x}$ of autoregressive observations. Usually, a Gaussianity assumption is adopted for the innovation process $\{w_n\}$. This assumption however is generally hardly justifiable. One again, the semiparametric formalism provides us with a suitable framework to statistically describe this problem. In fact, let $\mathbf{x} \in \mathbb{R}^m$ a vector of m observations from the above-mentioned AR(p). The set of all possible pdfs for $\mathbf{x} \in \mathbb{R}^m$ is a semiparametric model [18]:

$$\mathcal{P}_{\boldsymbol{\theta},p_w} \stackrel{\text{def}}{=} \{p_X(\mathbf{x}|\boldsymbol{\theta}, p_w), \boldsymbol{\theta} \in \Theta, p_w \in \mathcal{W}\}.$$

Unlike the classical estimation theory, in semiparametric inference, we need to handle the presence of a nuisance function. The bridge allowing us to go from the classical, "finite-dimensional", estimation theory to the semiparametric one is the general theory of Hilbert spaces.

5.2 Estimation and Hilbert Spaces

The aim of this section is then to introduce the mathematical framework that will allow us to handle the infinite-dimensional (functional) nuisance involved in a semi-parametric model. We will proceed in two steps: firstly, we frame the classical finite-dimensional estimation theory in the geometrical context of Hilbert spaces. Secondly, we will extend these outcomes to the semiparametric framework. Our discussion will be tutorial in nature, but the reader can find the whole formal mathematical derivation in [2, Sects. 2 and 3] and in [35].

5.2.1 Background on Hilbert Spaces, Linear Spans, and Projections

Defining the geometry of a problem means to specify (at least) two elements: the linear vector space in which our problem is embedded and the related inner product. These two concepts can be formalized by the definition of Hilbert space [34, Chap. 4]:

Definition 5.1 A Hilbert space $(\mathcal{H}, \langle\cdot,\cdot\rangle)$ is a real (or complex) inner product space that is also a complete metric space with respect to the distance function $||\cdot|| = \sqrt{\langle\cdot,\cdot\rangle}$ induced by the inner product $\langle\cdot,\cdot\rangle$.

The inner product allows us to define the concept of orthogonality. In particular, two elements $u, h \in \mathcal{H}$ are said to be *orthogonal*, denoted as $u \perp h$, iff $\langle u, h\rangle = 0$. They are called *orthonormal* if, in addition, $||u|| = ||h|| = 1$. The next theorem is a key result for Hilbert spaces since it defines the projection operator.

Theorem 5.1 (The Projection Theorem) *Let $\mathcal{U}$ be a closed subspace of an Hilbert space $\mathcal{H}$. For any $h \in \mathcal{H}$, there exists a unique $u_0 \in \mathcal{U}$ such that (s.t.)*

$$u_0 = \underset{u\in\mathcal{U}}{\operatorname{argmin}}||h-u||, \quad h \in \mathcal{H}. \tag{5.4}$$

Moreover, $u_0 \in \mathcal{U}$ is uniquely determined by the orthogonality constraint

$$\langle h-u_0, u\rangle = 0, \quad \forall u \in \mathcal{U}. \tag{5.5}$$

We indicate $u_0 = \Pi(h|\mathcal{U}) \in \mathcal{U}$ as the projection of $h \in \mathcal{H}$ onto the subspace $\mathcal{U} \subset \mathcal{H}$ and $\Pi(\cdot|\mathcal{U})$ the related projection operator.

Finding the projection of an element $h \in \mathcal{H}$ onto a generic subspace $\mathcal{U} \subset \mathcal{H}$ is, in general, a non-trivial task. However, there is a case of fundamental importance for the subsequent development in which this projection can be explicitly derived. In particular, let us build a vector $\mathbf{v} = (v_1, \ldots, v_r)$ of r arbitrary elements of $\mathcal{H}$. Let us define the *linear span* of $\mathbf{v}$ in $\mathcal{H}$ as

$$\mathcal{U} \stackrel{\text{def}}{=} \left\{u \in \mathcal{H} | u = \mathbf{a}^T \mathbf{v} : \mathbf{a} \text{ is any vector in } \mathbb{R}^r\right\} \subset \mathcal{H}. \tag{5.6}$$

If $\{v_1, \dots, v_r\}$ are linearly independent, the linear span $\mathcal{U}$ is a *finite-dimensional* linear subspace of $\mathcal{H}$ of dimension r. Moreover, it can be shown that the unique orthogonal projection of an arbitrary element $h \in \mathcal{H}$ onto $\mathcal{U}$ in (5.6) can be expressed as [35, Sect. 2]

$$\Pi(h|\mathcal{U}) = \left\langle h, \mathbf{v}^T \right\rangle \left\langle \mathbf{v}, \mathbf{v}^T \right\rangle^{-1} \mathbf{v}, \tag{5.7}$$

where

$$\left\langle h, \mathbf{v}^T \right\rangle \stackrel{\text{def}}{=} \left(\langle h, v_1 \rangle \cdots \langle h, v_r \rangle \right) \in \mathbb{R}^{1 \times r}, \tag{5.8}$$

$$\left\langle \mathbf{v}, \mathbf{v}^T \right\rangle \stackrel{\text{def}}{=} \begin{pmatrix} \langle v_1, v_1 \rangle & \cdots & \langle v_1, v_r \rangle \\ \vdots & \ddots & \vdots \\ \langle v_r, v_1 \rangle & \cdots & \langle u_k, v_r \rangle \end{pmatrix} \in \mathbb{R}^{r \times r}, \tag{5.9}$$

that is assumed to be invertible.

Let us now take it a step further and define the space $\mathcal{H}_q \stackrel{\text{def}}{=} \mathcal{H} \times \cdots \times \mathcal{H}$ as a Hilbert space obtained by the Cartesian product of q copies of $\mathcal{H}$ whose inner product is induced by the one of $\mathcal{H}$. Specifically, let $\mathbf{h} = (h_1, \cdots, h_q)^T$ and $\mathbf{w} = (w_1, \cdots, w_q)^T$ be two vectors in $\mathcal{H}_q$. Then, the inner product of $\mathcal{H}_q$ is given by $\langle \mathbf{h}, \mathbf{w} \rangle = \sum_{i=1}^q \langle h_i, w_i \rangle$. This type of space is generally called *q-replicating* Hilbert space. What about the linear span of $\mathcal{H}_q$? Let $\mathbf{v} = (v_1, \cdots, v_r)^T$ be a column vector of r arbitrary elements of $\mathcal{H}$. The linear span of $\mathbf{v}$ in $\mathcal{H}_q$ is given by [35, Sect. 2]

$$\mathcal{U}_q \stackrel{\text{def}}{=} \left\{\mathbf{u} \in \mathcal{H}_q | \mathbf{u} = \mathbf{A}\mathbf{v} : \mathbf{A} \text{ is any matrix in } \mathbb{R}^{q \times r}\right\} \subset \mathcal{H}_q. \tag{5.10}$$

If the entries of $\mathbf{v}$ are linearly independent, the dimension of $\mathcal{U}_q$ is $q \cdot r$. Moreover, the projection of an arbitrary $\mathbf{h} \in \mathcal{H}_q$ onto the linear span $\mathcal{U}_q$ in (5.10) is given by

$$\Pi(\mathbf{h}|\mathcal{U}_q) = \left\langle \mathbf{h}, \mathbf{v}^T \right\rangle \left\langle \mathbf{v}, \mathbf{v}^T \right\rangle^{-1} \mathbf{v}, \tag{5.11}$$

$$\left\langle \mathbf{h}, \mathbf{v}^T \right\rangle \stackrel{\text{def}}{=} \begin{pmatrix} \langle h_1, v_1 \rangle & \cdots & \langle h_1, v_r \rangle \\ \vdots & \ddots & \vdots \\ \langle h_q, v_1 \rangle & \cdots & \langle h_q, v_r \rangle \end{pmatrix} \in \mathbb{R}^{q \times r}, \tag{5.12}$$

and the matrix $\left\langle \mathbf{v}, \mathbf{v}^T \right\rangle$ is defined as in (5.9) and it is assumed to be invertible.

5.2.2 *The Hilbert Space $\mathcal{H}_q$ of the q-Variate Random Functions*

In this subsection, we introduce a replicating Hilbert space that plays a fundamental role in the geometrical interpretation of the estimation theory. Let $(\mathcal{X}, \mathfrak{F}, P_X)$ be a probability space where the sample space $\mathcal{X}$ is a subset of $\mathbb{R}^m$, $\mathfrak{F}$ is the Borel σ-algebra on $\mathcal{X}$ and P_X is a probability measure. We assume that the related cdf $P_X(\mathbf{x})$ admits a pdf $p_X(\mathbf{x})$ with respect to the Lebesgue measure, s.t. $dP_X(\mathbf{x}) = p_X(\mathbf{x})d\mathbf{x}$.

Let $\mathcal{H}$ be the set of scalar random functions on the sample space $\mathcal{X}$, i.e., $h : \mathcal{X} \to \mathbb{R}$ satisfying the following two properties:

1. *Zero-mean*: $\mathrm{E}(h) = \int_{\mathcal{X}} h(\mathbf{x})dP_X(\mathbf{x}) = 0$.
2. *Finite variance*: $\mathrm{E}(h^2) = \int_{\mathcal{X}} h^2(\mathbf{x})dP_X(\mathbf{x}) = \sigma_h^2 < +\infty$.

The (infinite-dimensional) linear space $\mathcal{H}$ can be endowed with the inner product:

$$\langle h_1, h_2 \rangle \stackrel{\text{def}}{=} \mathrm{E}(h_1 h_2) = \int_{\mathcal{X}} h_1(\mathbf{x})h_2(\mathbf{x})dP_X(\mathbf{x}), \quad \forall h_1, h_2 \in \mathcal{H}. \tag{5.13}$$

Then, $(\mathcal{H}, \langle \cdot, \cdot \rangle)$ is a Hilbert space and the squared norm associated to its inner product in (5.13) is $||h||^2 = \mathrm{E}(h^2) = \sigma_h^2$ that is the variance of $h \in \mathcal{H}$.

We can now define the q-replicating Hilbert space $\mathcal{H}_q = \mathcal{H} \times \cdots \times \mathcal{H}$ as the linear space of the multivariate random functions $\mathbf{h} : \mathcal{X} \to \mathbb{R}^q$ endowed by the following inner product induced by the one of $\mathcal{H}$:

$$\langle \mathbf{h}_1, \mathbf{h}_2 \rangle \stackrel{\text{def}}{=} \mathrm{E}(\mathbf{h}_1^T \mathbf{h}_2) = \sum\nolimits_{i=1}^{q} \mathrm{E}(h_{1,i} h_{2,i}). \tag{5.14}$$

From the replicating nature of $\mathcal{H}_q$, it is immediate to verify that every $\mathbf{h} \in \mathcal{H}_q$ satisfies the two properties:

1. *Zero-mean*: $\mathrm{E}(\mathbf{h}) = \int_{\mathcal{X}} \mathbf{h}(\mathbf{x})dP_X(\mathbf{x}) = \mathbf{0}$.
2. *Finite variance*: $\mathrm{E}(\mathbf{h}^T\mathbf{h}) = \langle \mathbf{h}, \mathbf{h} \rangle = ||\mathbf{h}||^2 < +\infty$.

Furthermore, since $\mathbf{h}$ is a q-variate random function, we can define its $q \times q$ covariance matrix $\mathbf{C}(\mathbf{h})$ in the usual way as

$$\mathbf{C}(\mathbf{h}) \stackrel{\text{def}}{=} \mathrm{E}(\mathbf{h}\mathbf{h}^T) = \begin{pmatrix} \mathrm{E}(h_1 h_1) & \cdots & \mathrm{E}(h_1 h_q) \\ \vdots & \ddots & \vdots \\ \mathrm{E}(h_q h_1) & \cdots & \mathrm{E}(h_q h_q) \end{pmatrix} \in \mathbb{R}^{q \times q}. \tag{5.15}$$

Let us investigate the geometrical structure of $\mathcal{H}_q$. In particular, we want to derive an explicit expression of the orthogonal projection of a generic element $\mathbf{h}$ into a linear span of $\mathcal{H}_q$. Following the discussion given in Sect. 5.2.1, let us define $\mathbf{v} = (v_1, \cdots, v_r)^T$ as a column vector of r arbitrary element of $\mathcal{H}$. Then, its linear span $\mathcal{U}_q$ is defined as in (5.10). Let us consider now the problem of finding the orthogonal projection of an arbitrary element $\mathbf{h} \in \mathcal{H}_q$ onto $\mathcal{U}_q$, i.e., $\Pi(\mathbf{h}|\mathcal{U}_q)$. From the Projection Theorem (Theorem 5.1), we know that this projection is unique and it can be explicitly written using (5.11) as

$$\Pi(\mathbf{h}|\mathcal{U}_q) = \mathrm{E}(\mathbf{h}\mathbf{v}^T)\mathrm{E}(\mathbf{v}\mathbf{v}^T)^{-1}\mathbf{v} = \mathrm{E}(\mathbf{h}\mathbf{v}^T)[\mathbf{C}(\mathbf{v})]^{-1}\mathbf{v}. \tag{5.16}$$

Let us now see how these general concepts on Hilbert space may provide an original look on the classical (finite-dimensional) estimation theory.

5.2.3 *Estimation in Parametric Models: A Geometric Interpretation*

Let $(\mathcal{X}, \mathfrak{F}, P_X)$ be the same probability space introduced in Sect. 5.2.2, and let P_X and p_X the related cdf and pdf, s.t. $dP_X(\mathbf{x}) = p_X(\mathbf{x})d\mathbf{x}$. Let us consider the following *parametric model*:

$$\mathcal{P}_{\boldsymbol{\theta},\boldsymbol{\eta}} \stackrel{\text{def}}{=} \left\{p_X(\mathbf{x}|\boldsymbol{\theta}, \boldsymbol{\eta}), \boldsymbol{\theta} \in \Theta \subseteq \mathbb{R}^q, \boldsymbol{\eta} \in \Gamma \subseteq \mathbb{R}^r\right\}, \tag{5.17}$$

where the q-dimensional vector $\boldsymbol{\theta}$ is the vector of the *parameters of interest*, while the r-dimensional vector $\boldsymbol{\eta}$ is the vector of the *nuisance parameters*. In the rest of this section, we always indicate the *true* parameter vectors as $\boldsymbol{\theta}_0$ and $\boldsymbol{\eta}_0$, and the related *true* pdf and distribution as $p_0(\mathbf{x}) = p_X(\mathbf{x}|\boldsymbol{\theta}_0, \boldsymbol{\eta}_0)$ and $P_0(\mathbf{x})$, respectively. Moreover, the expectation operator and the covariance matrix evaluated w.r.t. the true pdf will be indicated as $\mathrm{E}_0(\cdot)$ and $\mathbf{C}_0(\cdot)$, respectively.

Let us start by defining a key ingredient in estimation theory: the *score vector*.

Definition 5.2 Let $\mathcal{P}_{\boldsymbol{\theta},\boldsymbol{\eta}}$ be a parametric model as in (5.17). The *score vector of the parameters of interest* $\mathbf{s}_{\boldsymbol{\theta}_0}$ and the *nuisance score vector* $\mathbf{s}_{\boldsymbol{\eta}_0}$ are defined, respectively, as the q- and r-variate random functions s.t.:

$$\left[\mathbf{s}_{\boldsymbol{\theta}_0}\right]_i \stackrel{\text{def}}{=} \left[\mathbf{s}_{\boldsymbol{\theta}}(\mathbf{x}; \boldsymbol{\theta}_0, \boldsymbol{\eta}_0)\right]_i = \left[\nabla_{\boldsymbol{\theta}} \ln p_X(\mathbf{x}|\boldsymbol{\theta}_0, \boldsymbol{\eta}_0)\right]_i = \left.\frac{\partial \ln p_X(\mathbf{x}|\boldsymbol{\theta}, \boldsymbol{\eta}_0)}{\partial \theta_i}\right|_{\boldsymbol{\theta}=\boldsymbol{\theta}_0}, \tag{5.18}$$

$$\left[\mathbf{s}_{\boldsymbol{\eta}_0}\right]_j \stackrel{\text{def}}{=} \left[\mathbf{s}_{\boldsymbol{\eta}}(\mathbf{x}; \boldsymbol{\theta}_0, \boldsymbol{\eta}_0)\right]_j = \left[\nabla_{\boldsymbol{\eta}} \ln p_X(\mathbf{x}|\boldsymbol{\theta}_0, \boldsymbol{\eta}_0)\right]_j = \left.\frac{\partial \ln p_X(\mathbf{x}|\boldsymbol{\theta}_0, \boldsymbol{\eta})}{\partial \eta_j}\right|_{\boldsymbol{\eta}=\boldsymbol{\eta}_0}, \tag{5.19}$$

where $i = 1, \ldots, q$ and $j = 1, \ldots, r$. Under the usual regularity conditions (collected in, e.g., [23, Sects. 6.2 and 6.3]) that allow for the order inversion between integral and derivative operators, the score vectors are zero-mean random vectors:

$$\mathrm{E}_0(\mathbf{s}_{\boldsymbol{\theta}_0}) = \mathbf{0}, \qquad \mathrm{E}_0(\mathbf{s}_{\boldsymbol{\eta}_0}) = \mathbf{0}. \tag{5.20}$$

Moreover, if $||\mathbf{s}_{\theta_0}|| < \infty$ and $||\mathbf{s}_{\eta_0}|| < \infty$, we can immediately deduce that the score vectors belongs to two replicating Hilbert spaces of the type defined in Sect. 5.2.2:

$$\mathbf{s}_{\theta_0} \in \mathcal{H}_q, \qquad \mathbf{s}_{\eta_0} \in \mathcal{H}_r. \tag{5.21}$$

This result allows us to use the geometric structure of $\mathcal{H}_q$ and $\mathcal{H}_r$ to our estimation-oriented goal. First of all, we can note that the covariance matrix in (5.15) of $\mathbf{s}_{\theta_0}$ and $\mathbf{s}_{\eta_0}$ is nothing but the Fisher Information Matrices (FIMs) of the parameters of interest and of the nuisance ones:

$$\mathrm{C}_0(\mathbf{s}_{\theta_0}) = \mathrm{E}_0(\mathbf{s}_{\theta_0}\mathbf{s}_{\theta_0}^T) \stackrel{\text{def}}{=} \mathbf{I}_{\theta_0} \quad \mathrm{C}_0(\mathbf{s}_{\eta_0}) = \mathrm{E}_0(\mathbf{s}_{\eta_0}\mathbf{s}_{\eta_0}^T) \stackrel{\text{def}}{=} \mathbf{I}_{\eta_0}. \tag{5.22}$$

For further reference, let us also introduce the cross-information matrices as

$$\mathbf{I}_{\theta_0\eta_0} \stackrel{\text{def}}{=} \mathrm{E}_0(\mathbf{s}_{\theta_0}\mathbf{s}_{\eta_0}^T) = \left[\mathrm{E}_0(\mathbf{s}_{\eta_0}\mathbf{s}_{\theta_0}^T)\right]^T \stackrel{\text{def}}{=} \mathbf{I}_{\eta_0\theta_0}^T. \tag{5.23}$$

Specifically, using the findings related to the projection operator onto the finite-dimensional linear span of $\mathcal{H}_q$ presented in Sect. 5.2.2, we can introduce two fundamental concepts that will be extensively used in the rest of the chapter.

Definition 5.3 The nuisance tangent space. The linear span of $\mathcal{H}_q$ spanned by the entries of the nuisance score vector $\mathbf{s}_{\eta_0}$, i.e.,

$$\mathcal{T}_{\eta_0} \stackrel{\text{def}}{=} \left\{\mathbf{u} \in \mathcal{H}_q | \mathbf{u} = \mathbf{A}\mathbf{s}_{\eta_0} : \mathbf{A} \text{ is any matrix in } \mathbb{R}^{q \times r}\right\} \subset \mathcal{H}_q \tag{5.24}$$

is called *nuisance tangent space of the parametric model* $\mathcal{P}_{\theta,\eta}$ at the true nuisance parameter vector η_0.

Definition 5.4 The efficient score vector $\bar{\mathbf{s}}_0$ is defined as the residual of the score vector of the parameters of interest $\mathbf{s}_{\theta_0}$ after projecting it onto the nuisance tangent space $\mathcal{T}_{\eta_0}$ defined in (5.24) (see e.g., [2, Chap. 2]):

$$\bar{\mathbf{s}}(\mathbf{x}; \boldsymbol{\theta}_0, \eta_0) = \bar{\mathbf{s}}_0 \stackrel{\text{def}}{=} \mathbf{s}_{\theta_0} - \Pi(\mathbf{s}_{\theta_0}|\mathcal{T}_{\eta_0}) = \mathbf{s}_{\theta_0} - \mathbf{I}_{\theta_0\eta_0}\mathbf{I}_{\eta_0}^{-1}\mathbf{s}_{\eta_0}. \tag{5.25}$$

Note that, in the last equality, we used the explicit expression for the projection operator given in (5.16) and the definition of the FIM and of the cross-information matrix in (5.23). Consequently, the **efficient FIM** can be defined as

$$\bar{\mathbf{I}}(\boldsymbol{\theta}_0|\eta_0) \stackrel{\text{def}}{=} \mathrm{E}_0(\bar{\mathbf{s}}_0\bar{\mathbf{s}}_0^T) = \mathbf{I}_{\theta_0} - \mathbf{I}_{\theta_0\eta_0}\mathbf{I}_{\eta_0}^{-1}\mathbf{I}_{\theta_0\eta_0}^T. \tag{5.26}$$

5.2.3.1 Efficiency and Cramér–Rao Inequality in Parametric Models

Score vectors and FIMs are the two building blocs of a well-known result in parametric models: the *Cramér–Rao inequality*. We show that this classical result can be obtained in a different, yet equivalent way, by using the geometrical approach previously introduced. Let $\mathcal{P}_{\theta,\eta}$ be a parametric model as in (5.17) and let $\mathbf{I}(\boldsymbol{\theta}_0, \boldsymbol{\eta}_0)$ be the joint FIM for the parameter vector of interest and for the nuisance one:

$$\mathbf{I}(\boldsymbol{\theta}_0, \boldsymbol{\eta}_0) = \begin{pmatrix} \mathbf{I}_{\boldsymbol{\theta}_0} & \mathbf{I}_{\boldsymbol{\theta}_0\boldsymbol{\eta}_0} \\ \mathbf{I}^T_{\boldsymbol{\theta}_0\boldsymbol{\eta}_0} & \mathbf{I}_{\boldsymbol{\eta}_0} \end{pmatrix}. \tag{5.27}$$

Suppose now to have a set of n i.i.d. observations s.t. $\mathbf{x}_i \sim p_0, \forall i = 1, \dots, n$. We define as $\hat{\boldsymbol{\theta}}_n \stackrel{\text{def}}{=} \hat{\boldsymbol{\theta}}(\mathbf{x}_1, \dots, \mathbf{x}_n)$ an estimator of $\boldsymbol{\theta}_0 \in \Theta$ in the presence of the nuisance vector $\boldsymbol{\eta}_0 \in \Gamma$. Then, if $\hat{\boldsymbol{\theta}}_n$ is unbiased, i.e., $\mathrm{E}_0(\hat{\boldsymbol{\theta}}_n) = \mathbf{0}, \forall n$, the following inequality holds (see e.g., [23, Sects. 2.7 and 6.4]):

$$n\mathrm{E}_0\left((\hat{\boldsymbol{\theta}}_n - \boldsymbol{\theta}_0)(\hat{\boldsymbol{\theta}}_n - \boldsymbol{\theta}_0)^T\right) \geq \mathrm{CRB}(\boldsymbol{\theta}_0|\boldsymbol{\eta}_0), \tag{5.28}$$

where $\mathrm{CRB}(\boldsymbol{\theta}_0|\boldsymbol{\eta}_0)$ can be obtained from the joint FIM using the Matrix Inversion Lemma [32]:

$$\mathrm{CRB}(\boldsymbol{\theta}_0|\boldsymbol{\eta}_0) \stackrel{\text{def}}{=} \left[\mathbf{I}_{\boldsymbol{\theta}_0} - \mathbf{I}_{\boldsymbol{\theta}_0\boldsymbol{\eta}_0}\mathbf{I}^{-1}_{\boldsymbol{\eta}_0}\mathbf{I}^T_{\boldsymbol{\theta}_0\boldsymbol{\eta}_0}\right]^{-1}. \tag{5.29}$$

It is immediate to verify that this classical result can be obtained by using the efficient FIM defined in (5.26), i.e.,

$$\mathrm{CRB}(\boldsymbol{\theta}_0|\boldsymbol{\eta}_0) = \bar{\mathbf{I}}(\boldsymbol{\theta}_0|\boldsymbol{\eta}_0)^{-1}. \tag{5.30}$$

This result allows us to provide an intuitive explanation of the efficient score vector in (5.25): $\bar{\mathbf{s}}_0$ is able to characterize the information that the parametric model $\mathcal{P}_{\theta,\eta}$ has on the parameter vector of interest $\boldsymbol{\theta}_0 \in \Theta$ when a additional nuisance term is present. This fact is of crucial importance since, unlike the classical, "matrix-oriented", derivation of the Cramér–Rao inequality, the one based on the efficient score vector and the related efficient FIM does not require the nuisance term to be finite-dimensional. Then, it may be extended to the semiparametric context, as we will see in the next section. Before moving on, let us answer to the following question: is there any estimator of $\boldsymbol{\theta}_0 \in \Theta$ able to achieve the $\mathrm{CRB}(\boldsymbol{\theta}_0|\boldsymbol{\eta}_0)$, at least asymptotically in n? Under some regularity conditions on the parametric model (see e.g., [23, Sects. 6.1 and 6.2]), the *Maximum Likelihood* (ML) estimator is a good candidate. By indicating with

$$L_n(\boldsymbol{\theta}, \boldsymbol{\eta}) \stackrel{\text{def}}{=} \sum\nolimits_{i=1}^{n} \ln p_X(\mathbf{x}_i|\boldsymbol{\theta}, \boldsymbol{\eta}) \tag{5.31}$$

the *joint likelihood function* of a set of n i.i.d. observations, the joint ML estimator $(\hat{\boldsymbol{\theta}}_{ML,n}, \hat{\boldsymbol{\eta}}_{ML,n})$ can be expressed as [23, Sect. 6.2]

$$L_n(\hat{\boldsymbol{\theta}}_{ML,n}, \hat{\boldsymbol{\eta}}_{ML,n}) = \max\left\{L_n(\boldsymbol{\theta}, \boldsymbol{\eta}); \boldsymbol{\theta} \in \Theta, \boldsymbol{\eta} \in \Gamma\right\}, \tag{5.32}$$

Moreover, under the regularity conditions listed in, e.g., [23, Sects. 6.2 and 6.3], the ML estimator $\hat{\boldsymbol{\theta}}_{ML,n}$ of the parameter vector of interest $\boldsymbol{\theta}_0$ satisfies the two following, well-known, properties:

P1 *$\sqrt{n}$-consistency*: $\sqrt{n}(\hat{\boldsymbol{\theta}}_{ML,n} - \boldsymbol{\theta}_0) = O_{P_X}(1)$.[1]
P2 *Asymptotic normality and efficiency*: $\sqrt{n}(\hat{\boldsymbol{\theta}}_{ML,n} - \boldsymbol{\theta}_0) \underset{n\to\infty}{\sim} \mathcal{N}\left(\mathbf{0}, \mathrm{CRB}(\boldsymbol{\theta}_0|\boldsymbol{\eta}_0)\right)$.

We note in passing that, when the ML estimator in (5.32) exists and it is unique, it can be expressed as the solution of the following nonlinear system of *score-based estimating equations*:

$$\begin{cases} n^{-1/2} \sum_{i=1}^n \mathbf{s}_{\boldsymbol{\theta}}(\mathbf{x}_i; \boldsymbol{\theta}, \boldsymbol{\eta})\big|_{\boldsymbol{\theta}=\hat{\boldsymbol{\theta}}_{S,n}, \boldsymbol{\eta}=\hat{\boldsymbol{\eta}}_{S,n}} = \mathbf{0}, \\ n^{-1/2} \sum_{i=1}^n \mathbf{s}_{\boldsymbol{\eta}}(\mathbf{x}_i; \boldsymbol{\theta}, \boldsymbol{\eta})\big|_{\boldsymbol{\theta}=\hat{\boldsymbol{\theta}}_{S,n}, \boldsymbol{\eta}=\hat{\boldsymbol{\eta}}_{S,n}} = \mathbf{0}, \end{cases} \tag{5.33}$$

where $\mathbf{s}_{\boldsymbol{\vartheta}}$ and $\mathbf{s}_{\boldsymbol{\eta}}$ are the score vectors defined in (5.18) and (5.19), respectively

Even if it is the most popular, the ML estimator in (5.32) is not the only one having the desired properties P1 and P2. Without rigorously listing the required regularity conditions, the following provides two other $\sqrt{n}$-consistent (P1) and efficient (P2) estimators:

- *Efficient score-based estimating equations*: Let $\boldsymbol{\eta}_n^\star = \boldsymbol{\eta}_n^\star(\boldsymbol{\theta}_0)$ be a sub-optimal, *$\sqrt{n}$-consistent but not necessarily efficient*, estimator of the nuisance parameter vector $\boldsymbol{\eta}_0$. Note that $\boldsymbol{\eta}_n^\star$ may depend on the true parameter vector $\boldsymbol{\theta}_0$. For further reference, we indicate $\boldsymbol{\eta}_n^\star$ as a *preliminary estimator*. Then, we can define $\hat{\boldsymbol{\theta}}_{ES,n}$ as the estimator of the parameter vector of interest $\boldsymbol{\theta}_0$ obtained as the solution of the following estimating equations:

$$n^{-1/2} \sum_{i=1}^n \bar{\mathbf{s}}(\mathbf{x}_i; \boldsymbol{\theta}, \boldsymbol{\eta}_n^\star(\boldsymbol{\theta}))\Big|_{\boldsymbol{\theta}=\hat{\boldsymbol{\theta}}_{ES,n}} = \mathbf{0}, \tag{5.34}$$

  where $\bar{\mathbf{s}}(\mathbf{x}; \boldsymbol{\theta}, \boldsymbol{\eta})$ is the efficient score vector given in (5.25). Under the "necessary" regularity conditions [35, Ex. 3.2], it can be proved that $\hat{\boldsymbol{\theta}}_{ES,n}$ satisfies the properties P1 and P2.
- *One-step estimators*: Suppose that, to find an approximate solution of the estimating equations in (5.34), we use the first (one-step) iteration of the Newton–Raphson method as

$$\tilde{\boldsymbol{\theta}}_{ES,n} = \tilde{\boldsymbol{\theta}} - \left[\mathbf{J}(\tilde{\boldsymbol{\theta}}, \boldsymbol{\eta}_n^\star(\tilde{\boldsymbol{\theta}}))\right]^{-1} \overline{\Delta}_n(\tilde{\boldsymbol{\theta}}, \boldsymbol{\eta}_n^\star(\tilde{\boldsymbol{\theta}})), \tag{5.35}$$

  where $\tilde{\boldsymbol{\theta}}$ is the starting point of the Newton–Raphson iteration applied to the function given in (5.34) and called *efficient central sequence*:

$$\overline{\Delta}(\mathbf{x}_1, \ldots, \mathbf{x}_n; \boldsymbol{\theta}, \boldsymbol{\eta}) = \overline{\Delta}_n(\boldsymbol{\theta}, \boldsymbol{\eta}) \stackrel{\text{def}}{=} n^{-1/2} \sum_{i=1}^n \bar{\mathbf{s}}(\mathbf{x}_i; \boldsymbol{\theta}, \boldsymbol{\eta}), \tag{5.36}$$

---

[1] Let x_l be a sequence of random variables. Then $x_l = O_{P_X}(1)$ if for any $\epsilon > 0$, there exists a finite $N > 0$ and a finite $L > 0$, s.t. $\Pr\{|x_l| > N\} < \epsilon, \forall l > L$ (*stochastic boundedness*).

whose Jacobian matrix is given by

$$\mathbf{J}(\boldsymbol{\theta}, \boldsymbol{\eta}_n^\star(\boldsymbol{\theta})) \stackrel{\text{def}}{=} \nabla_{\boldsymbol{\theta}}^T \left[\overline{\Delta}_n(\boldsymbol{\theta}, \boldsymbol{\eta}_n^\star(\boldsymbol{\theta})) \right] = n^{-1/2} \sum\nolimits_{i=1}^{n} \nabla_{\boldsymbol{\theta}}^T \bar{\mathbf{s}}(\mathbf{x}_i; \boldsymbol{\theta}, \boldsymbol{\eta}_n^\star(\boldsymbol{\theta})). \quad (5.37)$$

In his seminal and fundamental works [21], [22, Chap. 6] Le Cam showed that it is possible to construct an efficient one-step estimator $\hat{\boldsymbol{\theta}}_{OP,n}$, inspired by the Newton–Raphson iteration in (5.35), as

$$\hat{\boldsymbol{\theta}}_{OP,n} = \boldsymbol{\theta}_n^\star + n^{-1/2} \left[\mathbf{C}(\boldsymbol{\theta}_n^\star, \boldsymbol{\eta}_n^\star) \right]^{-1} \overline{\Delta}_n(\boldsymbol{\theta}_n^\star, \boldsymbol{\eta}_n^\star), \quad (5.38)$$

where $\boldsymbol{\theta}_n^\star$ is a preliminary, $\sqrt{n}$-consistent but possibly not efficient, estimator of the parameter vector of interest $\boldsymbol{\theta}_0$ and

$$\mathbf{C}(\boldsymbol{\theta}_n^\star, \boldsymbol{\eta}_n^\star) = \frac{1}{n} \left[\sum\nolimits_{i=1}^{n} \bar{\mathbf{s}}(\mathbf{x}_i, \boldsymbol{\theta}_n^\star, \boldsymbol{\eta}_n^\star) \bar{\mathbf{s}}(\mathbf{x}_i, \boldsymbol{\theta}_n^\star, \boldsymbol{\eta}_n^\star)^T \right], \quad (5.39)$$

is the sample estimate of the efficient FIM $\bar{\mathbf{I}}(\boldsymbol{\theta}_0|\boldsymbol{\eta}_0)$ in (5.26). [2] Remarkably, Le Cam showed that the estimator $\hat{\boldsymbol{\theta}}_{OP,n}$ in (5.38) satisfies the properties P1 and P2 (see e.g., [2, Theorem 2 in Sect. 2.5 and Theorem 1 in Sect. 7.8]).

We are now ready to generalize these results to the semiparametric framework.

5.2.4 *Extension to Semiparametric Models*

Let $(\mathcal{X}, \mathfrak{F}, P_X)$ be the same probability space defined in Sect. 5.2.3. We have already defined in Sect. 5.1 a *semiparametric model* as

$$\mathcal{P}_{\boldsymbol{\theta},g} \stackrel{\text{def}}{=} \left\{ p_X(\mathbf{x}|\boldsymbol{\theta}, g), \boldsymbol{\theta} \in \Theta \subseteq \mathbb{R}^q, g \in \mathcal{G} \right\}, \quad (5.40)$$

where $\boldsymbol{\theta} \in \Theta$ is q-dimensional vector of the parameters of interest, while $g \in \mathcal{G}$ is a nuisance function. Following the notation introduced before, we denote the true "semiparametric vector" as $(\boldsymbol{\theta}_0, g_0) \in \Theta \times \mathcal{G}$, and consequently the true pdf is denoted as $p_0(\mathbf{x}) = p_X(\mathbf{x}|\boldsymbol{\theta}_0, g_0)$. From an abstract, "Hilbert space-based" standpoint, there are no conceptual differences between the definition of a parametric model in (5.17) and the one of a semiparametric model in (5.40). In particular, both the parametric and the semiparametric models can be considered as a family of pdfs indexed by a vector belonging to some parameter space. In the case of parametric models, such a parameter space is a finite-dimensional space obtained as the Cartesian product of two finite-dimensional spaces, $(\boldsymbol{\theta}^T, \boldsymbol{\eta}^T)^T \in \Theta \times \Gamma \subseteq \mathbb{R}^q \times \mathbb{R}^r$, i.e., the space of the parameters of interest Θ and the nuisance parameter space Γ. In

[2] To link this result with the Jacobian matrix of the efficient central sequence in (5.37), we can note that, under the necessary regularity conditions, $\bar{\mathbf{I}}(\boldsymbol{\theta}_0|\boldsymbol{\eta}_0) = \mathrm{E}_0(\bar{\mathbf{s}}_0 \bar{\mathbf{s}}_0^T) = -\mathrm{E}_0(\nabla_{\boldsymbol{\theta}_0}^T \bar{\mathbf{s}}_0)$.

the semiparametric case, the parameter space is the Cartesian product of the finite-dimensional space of the parameters of interest Θ and an infinite-dimensional set $\mathcal{G}$ of nuisance functions, i.e., $(\boldsymbol{\theta}, g) \in \Theta \times \mathcal{G}$. Even if, from a conceptual standpoint, there is no difference between parametric and semiparametric models, we cannot directly apply the tools developed in the parametric framework to the semiparametric one because of the infinite-dimensional nature of the nuisance parameter $g \in \mathcal{G}$. To develop an inference theory for semiparametric models, we need to generalize the concepts of efficient score vector in the presence of infinite-dimensional nuisance parameters.

The link between parametric and semiparametric inference is the concept of *parametric sub-model* of a semiparametric model. Formally, the *i-th* parametric sub-model of $\mathcal{P}_{\boldsymbol{\theta},g}$ is defined as [35, Sect. 4.2], [2, Sect. 3.1], [18, Sect. 2.2]:[3]

$$\mathcal{P}_{\boldsymbol{\theta},\nu_i} \stackrel{\text{def}}{=} \left\{ p_X(\mathbf{x}|\boldsymbol{\theta}, \nu_i(\cdot,\cdot,\boldsymbol{\eta})), \boldsymbol{\theta} \in \Theta \subseteq \mathbb{R}^q, \boldsymbol{\eta} \in \Gamma_i \subseteq \mathbb{R}^{r_i} \right\}, \text{ where:} \tag{5.41}$$

$$\begin{aligned} \nu_i : \mathcal{X} \times \Theta \times \Gamma_i &\to \mathcal{G} \\ \boldsymbol{\eta} &\mapsto \nu_i(\mathbf{x}, \boldsymbol{\theta}, \boldsymbol{\eta}) \end{aligned} \tag{5.42}$$

is a *known* function parametrized by a vector of *unknown* parameters. In particular, for every $i \in \mathcal{I}$, $\mathcal{P}_{\boldsymbol{\theta},\nu_i}$ in (5.41) is a parametric model of the form already defined in (5.17) satisfying the following three conditions [35, Sect. 4.2]:

(C0) $\nu_i : \mathcal{X} \times \Theta \times \Gamma_i \to \mathcal{G}$ is a smooth parametric map $\forall i \in \mathcal{I}$,
(C1) $\mathcal{P}_{\boldsymbol{\theta},\nu_i} \subseteq \mathcal{P}_{\boldsymbol{\theta},g}$, $\forall i \in \mathcal{I}$,
(C2) $p_0(\mathbf{x}) \in \mathcal{P}_{\boldsymbol{\theta},\nu_i}$, i.e., $\forall i \in \mathcal{I}$ there exists a vector $(\boldsymbol{\theta}_0^T, \boldsymbol{\eta}_0^T)^T \in \Theta \times \Gamma_i \subseteq \mathbb{R}^q \times \mathbb{R}^{r_i}$ such that $p_X(\mathbf{x}|\boldsymbol{\theta}_0, \nu_i(\cdot,\cdot,\boldsymbol{\eta}_0)) = p_X(\mathbf{x}|\boldsymbol{\theta}_0, g_0)$.

Condition C1 tells us that all the pdfs that compose each possible parametric submodel $\mathcal{P}_{\boldsymbol{\theta},\nu_i}$ must belong to the semiparametric model $\mathcal{P}_{\boldsymbol{\theta},g}$ as well. Moreover, Condition C2 highlights the fact that each parametric submodel $\mathcal{P}_{\boldsymbol{\theta},\nu_i}$ must contain the "true" data pdf, i.e., there must exist a parameter vector $\boldsymbol{\eta}_0 \in \Gamma_i$ such that $p_X(\mathbf{x}|\boldsymbol{\theta}_0, \nu_i(\cdot,\cdot,\boldsymbol{\eta}_0)) = p_0(\mathbf{x})$ for every $i \in \mathcal{I}$. It must be stressed that, unlike a classical parametric model, a parametric sub-model is only an artificial tool that is exploited to develop the semiparametric theory but that cannot be directly used for data analysis. In fact, since we do not know the true data pdf p_0, we cannot explicitly build a set of parametric sub-models satisfying the Condition C2.

Roughly speaking, using a parametric sub-model $\mathcal{P}_{\boldsymbol{\theta},\nu_i}$ in place of $\mathcal{P}_{\boldsymbol{\theta},g}$ we are actually approximating the infinite-dimensional parameter $g \in \mathcal{G}$ with the function $\nu_i(\mathbf{x}, \boldsymbol{\theta}, \boldsymbol{\eta})$, parameterized by the finite-dimensional nuisance parameter vector $\boldsymbol{\eta} \in \Gamma_i \subseteq \mathbb{R}^{r_i}$ whose dimension r_i depends on the particular choice of ν_i. This approximation may be made finer and finer by choosing a sequence of functions $\nu_i(\mathbf{x}, \boldsymbol{\theta}, \boldsymbol{\eta}) \in \mathcal{G}$ that converges to $g \in \mathcal{G}$. The way to generalize the classical theory

[3] Note that the notation $p_X(\mathbf{x}|\boldsymbol{\theta}, \nu_i(\cdot,\cdot,\boldsymbol{\eta}))$ indicates that the pdf p_X is a composition of two functions, i.e., w and ν_i, such that $p_X(\mathbf{x}|\boldsymbol{\theta}, \nu_i(\cdot,\cdot,\boldsymbol{\eta})) = (w \circ \nu_i)(\mathbf{x}, \boldsymbol{\theta}, \boldsymbol{\eta}) = w(\mathbf{x}, \boldsymbol{\theta}, \nu_i(\mathbf{x}, \boldsymbol{\theta}, \boldsymbol{\eta}))$.

developed for parametric models to semiparametric framework should now be more clear: we could exploit the outcomes obtained in a sequence of (artificial) parametric sub-models $\{\mathcal{P}_{\theta,\nu_i}\}_{i\in I}$ to obtain a generalization of the concept of nuisance tangent space and efficient score function.

Let us start by providing a semiparametric generalization of nuisance tangent space. In this chapter, we define the *semiparametric nuisance tangent space* according to the definition given in [29] and [35, Sect. 4.4], while a more general (but more abstract) definition is given in [2, Sect. 3.2], [1] and [37, Chap. 25]. At first, let us recall that the Hilbert space $\mathcal{H}_q$ of the q-variate, zero-mean, random functions is a metric space with (squared) distance given by $||\mathbf{h}_1 - \mathbf{h}_2||^2 = E_0((\mathbf{h}_1 - \mathbf{h}_2)^T(\mathbf{h}_1 - \mathbf{h}_2))$. The notion of convergence adopted in the following is based on this definition of distance.

Definition 5.5 The semiparametric nuisance tangent space $\mathcal{T}_{g_0}$ of $\mathcal{P}_{\theta,g}$ at g_0 is defined as the closure[4] of the union of all the nuisance tangent spaces

$$\mathcal{T}_{\eta_{0,i}} \stackrel{\text{def}}{=} \{\mathbf{u} \in \mathcal{H}_q | \mathbf{u} = \mathbf{C}_i \mathbf{s}_{\eta_{0,i}} : \mathbf{C}_i \text{ is any matrix in } \mathbb{R}^{q\times r_i}\} \subset \mathcal{H}_q \tag{5.43}$$

of the parametric submodels $\{\mathcal{P}_{\theta,\nu_i}\}_{i\in I} \subseteq \mathcal{P}_{\theta,g}$. Specifically, $\mathcal{T}_{g_0} \subseteq \mathcal{H}_q$ is the space of all q-variate, zero-mean, random functions $\mathbf{u} \in \mathcal{H}_q$ for which there exists a sequence $\{\mathbf{C}_i \mathbf{s}_{\eta_{0,i}}\}_{i\in I}$ such that $||\mathbf{u} - \mathbf{C}_i \mathbf{s}_{\eta_{0,i}}|| \to 0$, where, according to (5.19), $\mathbf{s}_{\eta_{0,i}} = \nabla_\eta \ln p_X(\mathbf{x}|\boldsymbol{\theta}_0, \nu_i(\cdot,\cdot,\eta_0)) \in \mathcal{H}_{r_i}$ is the nuisance score vector of the parametric sub-model $\mathcal{P}_{\theta,\nu_i}$ [35, Sect. 4.4], [1, 29]. Using this definition, the semiparametric nuisance tangent space can be simply indicated as

$$\mathcal{T}_{g_0} = \overline{\bigcup_{\{\mathcal{P}_{\theta,\nu_i}\}_{i\in I}} \mathcal{T}_{\eta_{0,i}}} \subseteq \mathcal{H}_q. \tag{5.44}$$

Note that, the closure of a union of linear spaces doesn't need to be linear, in general. However, as discussed in [1, Assumption S] and [35, Sect. 4.4, Remark 5]), $\mathcal{T}_{g_0}$ is a q-replicating Hilbert space in the vast majority of the non-pathological semiparametric models. The closure and the linearity of $\mathcal{T}_{g_0}$ guarantee the existence and the uniqueness of the projection operator $\Pi(\cdot|\mathcal{T}_{g_0})$ through Theorem 5.1. However, unlike its parametric counterpart in (5.24), the semiparametric nuisance tangent space is generally an *infinite-dimensional* linear space. Consequently, a general, closed-form expression for $\Pi(\cdot|\mathcal{T}_{g_0})$ is no longer available.

In the wake of what we have done in Sect. 5.2.3 for a parametric model involving a finite-dimensional nuisance vector, let us try to define the semiparametric counterpart of the efficient score vector in (5.25). To this end, we need three ingredients: the score vector of the parameters of interest $\mathbf{s}_{\theta_0}$, the semiparametric nuisance tangent space $\mathcal{T}_{g_0}$ and the related projection operator $\Pi(\cdot|\mathcal{T}_{g_0})$. We have already defined $\mathcal{T}_{g_0}$ and $\Pi(\cdot|\mathcal{T}_{g_0})$ in Definition 5.5, while $\mathbf{s}_{\theta_0}$ is defined exactly as in (5.18) for the

[4] The closure $\overline{\mathcal{A}}$ of a set $\mathcal{A}$ is defined as the smallest closed set that contains $\mathcal{A}$, or equivalently, as the set of all elements in $\mathcal{A}$ together with all the limit points of $\mathcal{A}$.

parametric case where we need to substitute the finite-dimensional nuisance vector η_0 with the nuisance function g_0, i.e., $\mathbf{s}_{\theta_0} = \nabla_\theta \ln p_X(\mathbf{x}|\boldsymbol{\theta}_0, g_0)$. Consequently, **the semiparametric efficient score vector** $\bar{\mathbf{s}}_0$ is defined as the residual of the score vector of the parameters of interest $\mathbf{s}_{\theta_0}$ after projecting it onto the semiparametric nuisance tangent space $\mathcal{T}_{g_0}$:

$$\bar{\mathbf{s}}(\mathbf{x}; \boldsymbol{\theta}_0, g_0) = \bar{\mathbf{s}}_0 \stackrel{\text{def}}{=} \mathbf{s}_{\theta_0} - \Pi(\mathbf{s}_{\theta_0}|\mathcal{T}_{g_0}). \tag{5.45}$$

Moreover, the related **efficient semiparametric FIM** (SFIM) can be defined as

$$\bar{\mathbf{I}}(\boldsymbol{\theta}_0|g_0) \stackrel{\text{def}}{=} \mathrm{E}_0(\bar{\mathbf{s}}_0\bar{\mathbf{s}}_0^T). \tag{5.46}$$

It is immediate to verify that the previous two definitions are perfectly in line with the one given in Definition 5.4 for the parametric case. However, it is important to stress again here that, unlike the parametric case, a general explicit expression for $\Pi(\cdot|\mathcal{T}_{g_0})$ is not available when $\mathcal{T}_{g_0}$ is infinite-dimensional and it has to be calculated on a case-by-case basis. We will come back on this central point ahead in the section.

5.2.4.1 Semiparametric Efficiency and Related Bound

The definition of the semiparametric efficient score vector and the related efficient SFIM provided in (5.45) and (5.46) lead us directly to the generalization of the Cramér–Rao inequality to the semiparametric framework. The following theorem stating it can be found in [35, Theorem 4.1] and [38, Theorem 4.2]. A more abstract and general formulation can be found in [1], [2, Sect. 3.4] and [37, Chap. 25].

Theorem 5.2 *The Semiparametric Cramér–Rao Bound (SCRB) for the estimation of the finite-dimensional vector $\boldsymbol{\theta}_0 \in \Theta$ in the semiparametric model $\mathcal{P}_{\theta,g}$ is given by*

$$\mathrm{SCRB}(\boldsymbol{\theta}_0|g_0) \stackrel{\text{def}}{=} \sup_{\{\mathcal{P}_{\theta,\nu_i}\}} \left[\mathrm{E}_0(\bar{\mathbf{s}}_{0,i}\bar{\mathbf{s}}_{0,i}^T)\right]^{-1} = \left[\mathrm{E}_0(\bar{\mathbf{s}}_0\bar{\mathbf{s}}_0^T)\right]^{-1} = \left[\bar{\mathbf{I}}(\boldsymbol{\theta}_0|g_0)\right]^{-1}, \tag{5.47}$$

where $\{\mathcal{P}_{\theta,\nu_i}\}_{i\in I}$ is the set of all possible parametric sub-models of $\mathcal{P}_{\theta,g}$, as defined in (5.41). *Moreover, $\bar{\mathbf{s}}_{0,i}$ indicates the efficient score vector of the i-th parametric sub-model that, according to* (5.25), *is given by $\bar{\mathbf{s}}_0 = \mathbf{s}_{\theta_0} - \Pi(\mathbf{s}_{\theta_0}|\mathcal{T}_{\eta_{0,i}})$ where $\mathcal{T}_{\eta_{0,i}}$ is defined in* (5.43). *Finally, $\bar{\mathbf{s}}_0$ and $\bar{\mathbf{I}}(\boldsymbol{\theta}_0|g_0)$ are the semiparametric efficient score vector and the efficient SFIM defined in* (5.45) *and* (5.46), *respectively.*

The proof is long and non-trivial, and we left the interesting reader to check it in, e.g., [35, Theorem 4.1] and [38, Theorem 4.2]. Here, we limit ourselves to two observations:

- The definition of the semiparametric efficient score vector, of the efficient SFIM and of the related SCRB provided in (5.45), (5.46), and (5.47) are formally equivalent to their parametric counterparts given in (5.25), (5.26), and (5.30). Only the definition of the nuisance tangent space and of the related projection operator changes. This is due to the fact that the abstract Hilbert space-based framework introduced in Sect. 5.2.2 is able to handle in a unified way both finite- and infinite-dimensional parameter vectors.

- From (5.47), the SCRB$(\boldsymbol{\theta}_0|g_0)$ in $\mathcal{P}_{\theta,g}$ is defined as the supremum of all the CRB$(\boldsymbol{\theta}_0|\eta_{0,i})$ in each parametric sub-models $\{\mathcal{P}_{\theta,\nu_i}\}_{i\in I}$. This confirms the intuition that a semiparametric model contains less information on the parameter vector of interest $\boldsymbol{\theta}_0 \in \Theta$ then any of its possible parametric sub-models.

The following two points, closely linked to the semiparametric efficiency bound, must now be discussed:

Q1 Which is the class of estimators to which the SCRB$(\boldsymbol{\theta}_0|g_0)$ applies?
Q2 Is there any estimator able to achieve it?

Even if of crucial importance, answering to these two questions is not a trivial task and they are the subject of intensive studies in statistics. Here, without any claim of completeness, we will try to provide some general hints leaving to the interested reader the task of further investigating the topic.

Regarding the first question, we can start by observing that we are looking for a class of estimators able to estimate the finite-dimensional parameter vector $\boldsymbol{\theta}_0 \in \Theta$ without assuming any knowledge on the nuisance function $g_0 \in \mathcal{G}$. In other words, a semiparametric estimator should be *robust* with respect to the lack of a complete specification of the functional form of the pdf of the observed dataset $\{\mathbf{x}_1, \ldots, \mathbf{x}_n\}$. As shown in [2, Sect. 2.2 and Chap. 7], [35, Chap. 3], [33, Chap. 4] and the references therein, this concept can be formalized in the class of *regular and asymptotically linear (RAL) estimators* [2, Sect. 2.2 and Chap. 7]. It is worth mentioning that all the robust M-, S-, L-, and R-estimators belong to this class [33]. Specifically, a semiparametric RAL estimator $\hat{\boldsymbol{\theta}}_n \stackrel{\text{def}}{=} \hat{\boldsymbol{\theta}}(\mathbf{x}_1, \ldots, \mathbf{x}_n)$ of the finite-dimensional parameter vector $\boldsymbol{\theta}_0 \in \Theta$ from a set of n i.i.d. data $\{\mathbf{x}_1, \ldots, \mathbf{x}_n\}$ satisfy the following two properties:

1. *$\sqrt{n}$-consistency*: $\sqrt{n}(\hat{\boldsymbol{\theta}}_n - \boldsymbol{\theta}_0) = O_{P_X}(1)$,
2. *Asymptotically normality*: $\sqrt{n}(\hat{\boldsymbol{\theta}}_n - \boldsymbol{\theta}_0) \underset{n\to\infty}{\sim} \mathcal{N}(\mathbf{0}, \boldsymbol{\Xi}(\boldsymbol{\theta}_0, g_0))$.

Consequently, the following inequality holds [2, Chaps. 2 and 3]:

$$\boldsymbol{\Xi}(\boldsymbol{\theta}_0, g_0) \geq \text{SCRB}(\boldsymbol{\theta}_0|g_0). \tag{5.48}$$

This answer to the first point Q1.

Let us now move to the question Q2: is it possible to characterize the RAL estimator with the lowest error covariance matrix, i.e., the one that achieves the semiparametric lower bound (this means to have en equality in (5.48))? Roughly speaking, here we are looking for the semiparametric counterpart of the ML estimator discussed in Sect. 5.2.3.1. This "one million dollars question" is still largely open and many fundamental aspects remain unresolved. Here, we limit ourselves to provide a very preliminary and informal sketch of answer. Basically, an efficient semiparametric estimator can be built by using the *one-step* procedure that we introduced for parametric models in Sect. 5.2.3.1. We can note in fact that the *one-step* estimator in (5.38) is the only one among the four efficient estimators presented in Sect. 5.2.3.1 that does not require the finite-dimensionality of the nuisance parameter space and then it may be generalized to semiparametric models [2, Chap. 7], [18], [37, Chap. 25].

Suppose to have at our disposal two preliminary, sub-optimal (*$\sqrt{n}$-consistent but not necessarily efficient*) estimators, say $\boldsymbol{\theta}_n^\star$ and $g_n^\star$, of the parameter vector of interest $\boldsymbol{\theta}_0 \in \Theta$ and of the nuisance function $g_0 \in \mathcal{G}$. An asymptotically efficient, one-step estimator $\hat{\boldsymbol{\theta}}_{OP,n}$ of $\boldsymbol{\theta}_0 \in \Theta$ can then be obtained as [2, Theorem 1, Sect. 7.8]:

$$\hat{\boldsymbol{\theta}}_{OP,n} = \boldsymbol{\theta}_n^\star + n^{-1/2}\left[\mathbf{C}(\boldsymbol{\theta}_n^\star, g_n^\star)\right]^{-1}\overline{\Delta}_n(\boldsymbol{\theta}_n^\star, g_n^\star), \tag{5.49}$$

where the *efficient central sequence* is given by

$$\overline{\Delta}(\mathbf{x}_1, \ldots, \mathbf{x}_n; \boldsymbol{\theta}_n^\star, g_n^\star) = \overline{\Delta}_n(\boldsymbol{\theta}_n^\star, g_n^\star) \stackrel{\text{def}}{=} n^{-1/2}\sum\nolimits_{i=1}^{n} \bar{\mathbf{s}}(\mathbf{x}_i; \boldsymbol{\theta}_n^\star, g_n^\star). \tag{5.50}$$

Since $\bar{\mathbf{s}}_0$ depends on the true value of the parameter vector of interest $\boldsymbol{\theta}_0$ and on the true nuisance function g_0, in the definition of the central sequence we exploit the term $\bar{\mathbf{s}}(\cdot, \boldsymbol{\theta}_n^\star, g_n^\star)$ where $\boldsymbol{\theta}_0$ and g_0 are substituted with their two $\sqrt{n}$-consistent preliminary estimates $\boldsymbol{\theta}^\star$ and $g^\star$. Moreover, the matrix $\mathbf{C}(\boldsymbol{\theta}_n^\star, g_n^\star)$ represents the sample estimate of the efficient SFIM $\bar{\mathbf{I}}(\boldsymbol{\theta}_0|g_0) = \mathrm{E}_0(\bar{\mathbf{s}}_0\bar{\mathbf{s}}_0^T)$ in (5.46) and can be expressed as

$$\mathbf{C}(\boldsymbol{\theta}_n^\star, g_n^\star) = \frac{1}{n}\left[\sum\nolimits_{i=1}^{n} \bar{\mathbf{s}}(\mathbf{x}_i, \boldsymbol{\theta}_n^\star, g_n^\star)\bar{\mathbf{s}}(\mathbf{x}_i, \boldsymbol{\theta}_n^\star, g_n^\star)^T\right]. \tag{5.51}$$

According to [2, Theorem 1 and Corollary 1], it can be shown that, "under suitable regularity conditions", $\hat{\boldsymbol{\theta}}_{OP,n}$ in (5.49) satisfy the following properties:

1. *$\sqrt{n}$-consistency:* $\sqrt{n}(\hat{\boldsymbol{\theta}}_{op,n} - \boldsymbol{\theta}_0) = O_{P_X}(1)$,
2. *Asymptotic normality and efficiency:* $\sqrt{n}(\hat{\boldsymbol{\theta}}_{op,n} - \boldsymbol{\theta}_0) \underset{n\to\infty}{\sim} \mathcal{N}(\mathbf{0}, \mathrm{SCRB}(\boldsymbol{\theta}_0|g_0))$.

It is worth stressing that the two preliminary estimators, even if sub-optimal, need to be $\sqrt{n}$-consistent. While it is not difficult to derive $\sqrt{n}$-consistent estimators for the finite-dimensional parameter $\boldsymbol{\theta}_0$, this may be a critical point for the functional estimator $g^\star$ whose rate of convergence is generally slower than $\sqrt{n}$ [37, Sect. 25.8]. This may strongly limit the exploitability of semiparametric one-step estimator of the form in (5.49). As recently pointed out in [18], there may be some *rank-based* alternative procedure allowing us to avoid the need of the non-parametric $\sqrt{n}$-consistent estimator of the nuisance function $g \in \mathcal{G}$. We will further discuss this point in Sect. 5.3.3 for the case of elliptical distributions.

5.2.4.2 Calculation of the Projection Operator $\Pi(\cdot|\mathcal{T}_{g_0})$

Before moving to the second part of the chapter, let us discuss one last fundamental point. All the previous results on semiparametric estimation rely on the definition of the semiparametric efficient score vector $\bar{\mathbf{s}}_0$ in (5.45), which, in turn, is based on the calculation of the projection operator $\Pi(\cdot|\mathcal{T}_{g_0})$. Even if $\Pi(\cdot|\mathcal{T}_{g_0})$ is formally well-defined, i.e., it always exists and it is unique, its direct calculation could be involved and sometime impossible to obtain in closed form. Fortunately, many semiparametric models of interest for practical estimation problems are characterized by some additional regularity or/and invariance properties that could be exploited to derive $\Pi(\cdot|\mathcal{T}_{g_0})$. Specifically, in this subsection, we analyze the case in which the projection operator can be evaluated as a conditional expectation. As we will see in the second part of the chapter, this methodology will be readily applied to obtain the projection operator in the semiparametric model of the elliptical distributions.

In Sect. 5.2.2, we defined $\mathcal{H}_q$ as the Hilbert space of the q-variate, zero-mean, random functions on the probability space $(\mathcal{X}, \mathfrak{F}, P_X)$ with inner product given in (5.14) as $\langle \mathbf{h}_1, \mathbf{h}_2 \rangle = E(\mathbf{h}_1^T \mathbf{h}_2)$. Let us now introduce $\mathfrak{D}(\mathbf{x}) \subseteq \mathfrak{F}$ as the sub-sigma algebra in $\mathfrak{F}$ generated by some (possibly multivariate) transformation d of the random vector $\mathbf{x}$, such that $D = d(\mathbf{x})$. As shown in, e.g., [20, Chap. 23] and [2, Appendix 3], the set of all the q-variate, zero-mean, random functions on the probability space $(\mathcal{X}, \mathfrak{D}(\mathbf{x}), P_X)$ is a closed linear subsapce, say $\mathcal{D}$, of the Hilbert space $\mathcal{H}_q \supseteq \mathcal{D}$. Consequently, according to Theorem 5.1, the projection of a generic element $\mathbf{h} \in \mathcal{H}_q$ onto the linear subspace $\mathcal{D}$, i.e., $\Pi(\mathbf{h}|\mathcal{D})$ exists and it is unique.

The question that we aim at answering here is: is there any relation between the projection operator $\Pi(\cdot|\mathcal{D})$ and the *conditional expectation* $E(\cdot|\mathfrak{D}(\mathbf{x})) = E(\cdot|D)$? To this end, let us provide a geometrical definition of the conditional expectation (see [20, Chap. 23] and [2, Appendix 3] for a more formal discussion).

Definition 5.6 Let $\mathbf{h} \in \mathcal{H}_q$ and $\mathbf{u} \in \mathcal{D} \subseteq \mathcal{H}_q$ be two q-variate, zero-mean, random functions on the probability spaces $(\mathcal{X}, \mathfrak{F}, P_X)$ and $(\mathcal{X}, \mathfrak{D}(\mathbf{x}), P_X)$ with $\mathfrak{D}(\mathbf{x}) \subseteq \mathfrak{F}$. Moreover, let D be a random vector defined on the sub-sigma algebra $\mathfrak{D}(\mathbf{x})$.Then the conditional expectation $E(\mathbf{h}|D)$ is the unique element in $\mathcal{D}$, s.t.:

$$\langle \mathbf{h} - E(\mathbf{h}|\mathfrak{D}(\mathbf{x})), \mathbf{u} \rangle = E((\mathbf{h} - E(\mathbf{h}|D))^T \mathbf{u}) = 0, \quad \forall \mathbf{u} \in \mathcal{D} \subseteq \mathcal{H}_q. \tag{5.52}$$

It is immediate to verify that the definition of conditional expectation given in (5.52) is exactly equal to the one of the orthogonal projection given in (5.5) of Theorem 5.1 that define the projection operator. As a consequence, we have the desired relation:

$$\Pi(\cdot|\mathcal{D}) = E(\cdot|D). \tag{5.53}$$

Why should this relation help us? The answer is that, in some semiparametric model of interest, the semiparametric nuisance tangent space (i) has some invariance structure with respect to a given group of transformations and (ii) it admits a characterization through the sub-sigma algebras generated by the maximal invariant statistic of the relevant group of transformation [24]. As we will see in the next section, the semiparametric model of the elliptical distributions belongs to the class of semiparametric models satisfying these two requirements.

5.3 Semiparametric Estimation in Real Elliptically Symmetric (RES) Distributions

The second part of the chapter is dedicated to the application of the general semiparametric theory previously introduced to the problem of the joint estimation of the location vector and of the scatter matrix of a set of elliptically distributed real observations in the presence of an unknown density generator. An exhaustive presentation of the class of RES distributions can be found in Chap. 1 where a comprehensive list of definitions, properties, and examples (together with the relevant notation and nomenclature) is provided. Here we limit ourselves to recall only some fundamental points that will be extensively used in the rest of this section.

We start by recalling here the definition of *spherically distributed* random vector (see Definition 2 in Chap. 1). Let $\mathbf{x}_s \in \mathbb{R}^m$ be a real-valued random vector and let $\mathcal{O}$ be the set of all the orthogonal transformations such that

$$\begin{aligned} \mathcal{O} \ni O : \mathbb{R}^m &\to \mathbb{R}^m \\ \mathbf{x}_s &\mapsto O(\mathbf{x}_s) = \mathbf{O}\mathbf{x}_s, \end{aligned} \tag{5.54}$$

for any orthogonal matrix $\mathbf{O}$, i.e., for any $\mathbf{O} \in \mathbb{R}^{m\times m}$ such that $\mathbf{O}^T\mathbf{O} = \mathbf{O}\mathbf{O}^T = \mathbf{I}_m$. Then, $\mathbf{x}_s$ is said to be spherically distributed if its distribution is invariant to any orthogonal transformations $\mathbf{O} \in \mathcal{O}$:

$$\mathbf{x}_s =_d \mathbf{O}\mathbf{x}_s. \tag{5.55}$$

As a consequence of the properties already given in Definition 2 of Chap. 1, the set of spherically distributed random vectors admits a *maximal invariant statistic* w.r.t. $\mathcal{O}$. For completeness, we recall here the definition of maximal invariant statistic [24, Chap. 6]. Let $\mathcal{D} = \{d\}$ be a group of one-to-one transformations on a sample space $\mathcal{X}$ and let T an invariant statistic such that $T(\mathbf{x}) =_d T(d(\mathbf{x}))$, $\forall \mathbf{x} \in \mathcal{X}$ and $\forall d \in \mathcal{D}$. Then, T is a maximal invariant statistic on $\mathcal{X}$ w.r.t. $\mathcal{D}$ if $T(\mathbf{x}_1) =_d T(\mathbf{x}_2)$ implies that $\mathbf{x}_1 =_d d(\mathbf{x}_2)$, $\forall \mathbf{x}_1, \mathbf{x}_2 \in \mathcal{X}$ and $\forall d \in \mathcal{D}$. Let us now go back to the spherically distributed random vectors. From the *stochastic representation* given in Sect. 2.2 of Chap. 1, we have

$$\mathbf{x}_s =_d \sqrt{\mathcal{Q}}\mathbf{u} =_d \mathcal{R}\mathbf{u}, \ \mathcal{R} = ||\mathbf{x}_s||, \tag{5.56}$$

where $\mathcal{Q} \sim p_\mathcal{Q}(q)$ (the *second-order modular variate*) and $\mathcal{R} \sim p_\mathcal{R}(r)$ (the *modular variate*), have pdfs given in Eq. (13) of Chap. 1. The random vector $\mathbf{u} \sim \mathcal{U}(\mathbb{R}S^m)$ is uniformly distributed on the real unit sphere of dimension $m-1$ and, consequently,

$||\mathbf{u}|| = 1$, $E(\mathbf{u}) = \mathbf{0}$ and $E(\mathbf{u}\mathbf{u}^T) = m^{-1}\mathbf{I}_m$. Moreover, $\mathcal{Q}$ and $\mathbf{u}$ (or $\mathcal{R}$ and $\mathbf{u}$) are independent. As a direct consequence, we have that $||\mathbf{x}_s||^2$ and $\mathbf{x}_s/||\mathbf{x}_s||(=_d \mathbf{u})$ are independent as well.

The stochastic representation in (5.56) provides us with a one-to-one mapping between a spherically distributed vector $\mathbf{x}_s$ and a couple $(\mathcal{R}, \mathbf{u})$. Moreover, for any couple of spherically distributed random vectors $\mathbf{x}_s$ and $\mathbf{y}_s$, we have that

$$||\mathbf{x}_s|| =_d ||\mathbf{y}_s|| \Rightarrow \mathbf{x}_s =_d \mathbf{O}\mathbf{y}_s, \quad \forall \mathbf{O} \in \mathcal{O}, \tag{5.57}$$

where $\mathcal{O}$ is the group of the orthogonal transformations defined in (5.54). Consequently, according to the definition previously provided, $R = ||\mathbf{x}||$ is a maximal invariant statistic for the set of spherically distributed random vectors.

5.3.1 The RES Distributions as a Semiparametric Group Model

A semiparametric group model is a statistical model generated by the *action* of a group of transformations, say $\mathcal{A}$, on a random vector $\mathbf{x}$ that is distributed according to a given class of distributions [2, Sect. 4.2]. As we will show in the following, this abstract definition fits with the RES distributions class. Note that all the subsequent results can be obtained for the class of the circular Complex Elliptically Symmetric (C-CES) distribution as well. The interested reader can find the details in [7].

Definition 5.7 Let $\boldsymbol{\mu} \in \mathbb{R}^m$ be a real vector and let $\boldsymbol{\Sigma} \in \mathcal{M}$ be a positive definite, symmetric matrix such that $\boldsymbol{\Sigma} = \mathbf{T}\mathbf{T}^T$. Among all the possible $\mathbf{T}$, we always choose $\mathbf{T} = \boldsymbol{\Sigma}^{1/2}$. Let us define the group $\mathcal{A}$ of the affine transformations $\alpha_{\boldsymbol{\mu},\boldsymbol{\Sigma}}(\cdot)$ s.t.:

$$\begin{aligned} \mathcal{A} \ni \alpha_{\boldsymbol{\mu},\boldsymbol{\Sigma}}(\cdot) : \mathbb{R}^m \to \mathbb{R}^m, \ \forall \boldsymbol{\mu} \in \mathbb{R}^m, \forall \boldsymbol{\Sigma} \in \mathcal{M} \\ \mathbb{R}^m \ni \mathbf{y} \mapsto \alpha_{\boldsymbol{\mu},\boldsymbol{\Sigma}}(\mathbf{y}) = \boldsymbol{\mu} + \boldsymbol{\Sigma}^{1/2}\mathbf{y}. \end{aligned} \tag{5.58}$$

Note that the neutral element of the group $\mathcal{A}$ is given by $\alpha_e(\cdot) = \alpha_{\mathbf{0},\mathbf{I}_m}(\cdot)$, while the inversion operator is $\alpha^{-1}_{\boldsymbol{\mu},\boldsymbol{\Sigma}}(\cdot) = \boldsymbol{\Sigma}^{-1/2}(\cdot - \boldsymbol{\mu})$.

From the stochastic representation provided in Eq. (7) of Sect. 2.2 of Chap. 1, we can readily deduce that the set of RES distributions is closed under the action of the group $\mathcal{A}$ on the set of the spherically distributed random vectors. In particular, we have that

$$\mathbf{x} =_d \alpha_{\boldsymbol{\mu},\boldsymbol{\Sigma}}(\mathbf{x}_s) =_d \boldsymbol{\mu} + \boldsymbol{\Sigma}^{1/2}\mathbf{x}_s =_d \boldsymbol{\mu} + \sqrt{\mathcal{Q}}\boldsymbol{\Sigma}^{1/2}\mathbf{u} =_d \boldsymbol{\mu} + \mathcal{R}\boldsymbol{\Sigma}^{1/2}\mathbf{u}. \tag{5.59}$$

Moreover, by noticing that the Jacobian of the inverse operator $\alpha^{-1}_{\boldsymbol{\mu},\boldsymbol{\Sigma}}(\cdot)$ is $|\mathbf{J}(\alpha^{-1}_{\boldsymbol{\mu},\boldsymbol{\Sigma}})| = |\boldsymbol{\Sigma}^{-1/2}| = |\boldsymbol{\Sigma}|^{-1/2}$, in the absolutely continuous case, we can deduce the pdf p_X of any RES-distributed random vector $\mathbf{x}$ from the one of the relevant spherically distributed random vector $\mathbf{x}_s$ given in Eq. (4) of Sect. 2.2 of Chap. 1 as $p(\mathbf{x}_s) = g(||\mathbf{x}_s||^2)$:

$$\mathrm{RES}_m(\boldsymbol{\mu}, \boldsymbol{\Sigma}, g) = p_X(\mathbf{x}; \boldsymbol{\mu}, \boldsymbol{\Sigma}, g) = |\boldsymbol{\Sigma}|^{-1/2} g(||\alpha^{-1}_{\boldsymbol{\mu},\boldsymbol{\Sigma}}(\mathbf{x})||^2), \tag{5.60}$$

that is in line with the expression given in (5.1) in Sect. 5.1. Finally, in order to avoid the scale ambiguity between the scatter matrix $\boldsymbol{\Sigma}$ and the density generator g, we rewrite $p(\mathbf{x}; \boldsymbol{\mu}, \boldsymbol{\Sigma}, g)$ as function of the *shape matrix* $\mathbf{V} = \boldsymbol{\Sigma}/s(\boldsymbol{\Sigma}) \in \mathcal{M}_s$ already defined in (5.2). Finally, the set of RES distribution can be cast as a *semiparametric group model* where the finite-dimensional parameter space is given by[5]

$$\Theta \overset{\text{def}}{=} \{\boldsymbol{\theta} \in \mathbb{R}^{q-1} | \boldsymbol{\theta} = (\boldsymbol{\mu}^T, v(\mathbf{V})^T)^T; \boldsymbol{\mu} \in \mathbb{R}^m, \mathbf{V} \in \mathcal{M}_s\}, \tag{5.61}$$

where $q = m(m+3)/2$ and the "-1" is due to the fact that $\dim(\mathcal{M}_s) = \dim(\mathcal{M}) - 1$ as discussed in Sect. 5.1. For further reference, we introduce the "unconstrained" finite-dimensional parameter space as

$$\Theta_u \overset{\text{def}}{=} \{\boldsymbol{\theta}_u \in \mathbb{R}^{q} | \boldsymbol{\theta} = (\boldsymbol{\mu}^T, v(\mathbf{V})^T)^T; \boldsymbol{\mu} \in \mathbb{R}^m, \mathbf{V} \in \mathcal{M}\}. \tag{5.62}$$

The infinite-dimensional parameter space is given by the set $\mathcal{G} \ni g$ of all the possible density generators. Note that, for technical reasons, we always assume that the set $\mathcal{G}$ is made up of all the density generators satisfying the following two additional properties: (1) the limit $\lim_{t\to+\infty} t^{m/2} g(t)$ exists and (2) $\int_0^\infty t^{m/2} g(t)dt < \infty$ that corresponds to impose $E\{Q\} < \infty$.

The semiparametric model that we are going to consider in the following is

$$\mathcal{P}_{\boldsymbol{\mu},\mathbf{V},g} \overset{\text{def}}{=} \left\{p_X(\mathbf{x}) = |\mathbf{V}|^{-1/2} g(||\alpha^{-1}_{\boldsymbol{\mu},\mathbf{V}}(\mathbf{x})||^2), (\boldsymbol{\mu}^T, v(\mathbf{V})^T) \in \Theta, g \in \mathcal{G}\right\}. \tag{5.63}$$

In the following, we will indicate as $\boldsymbol{\theta}_0 = (\boldsymbol{\mu}_0^T, v(\mathbf{V}_0)^T)^T \in \Theta$ the true finite-dimensional vector of interest that we need to estimate and as $g_0 \in \mathcal{G}$ the true nuisance function. Consequently, the true but unknown, data pdf is $p_0(\mathbf{x}) = |\mathbf{V}_0|^{-1/2} g_0(||\alpha^{-1}_{\boldsymbol{\mu}_0,\mathbf{V}_0}(\mathbf{x})||^2)$. After having formally placed the set of the RES distributions in the semiparametric framework, in the next two sections we will first evaluate the semiparametric Cramér–Rao Bound (SCRB), defined in Theorem 5.2, for the joint estimation of the location vector $\boldsymbol{\mu}_0$ and the shape matrix $\mathbf{V}_0$. Secondly, a class of one-step estimators of $\mathbf{V}_0$ of the type given in (5.49) will be presented.

---

[5] The operator $v(\mathbf{A})$ indicates the $m(m+1)/2$-dimensional vector of the entries of the lower (or upper) sub-matrix of a symmetric $\mathbf{A}$. This notation has been adopted in order to keep the consistency with Chap. 1. Note however that, in our previous publications (see e.g., [7, 8, 10]), it was expressed as vecs(·).

5.3.2 *The Semiparametric Cramér–Rao Bound for RES Distributions*

The derivation of a SCRB for estimation problems involving elliptical data has been firstly investigated by Bickel in [3], where a bound for the estimation of the inverse of the scatter matrix has been derived. Further discussion and analysis related to the joint estimation of $\boldsymbol{\mu}$ and $\mathbf{V}$ have been also presented in [2, Sects. 4.2 and 6.3]. More recently, in a series of papers [13–16, 30, 31], Hallin, Paindaveine and their coauthors exploited an original semiparametric generalization of the Le Cam's theory on Local Asympotical Normal (LAN) families of distributions [18] to derive optimal inference algorithms on elliptically distributed data. However, even if profound and of crucial importance, the work of Hallin and Paindavaine requires a strong knowledge of Le Cam's theory, and this falls outside the scope of this chapter. Our aim here is to show how the geometrical concepts previously introduced can be applied in a concrete problem as the joint estimation of $\boldsymbol{\mu}$ and $\mathbf{V}$ in elliptical data. To this end, we will basically follow the line of reasoning proposed in [2, Sects. 4.2 and 6.3], along with the results in [7, 8].

As stated in Theorem 5.2, the basic ingredient that we need to derive the SCRB if the efficient score vector $\bar{\mathbf{s}}_{u,0} = \mathbf{s}_{\boldsymbol{\theta}_{u,0}} - \Pi(\mathbf{s}_{\boldsymbol{\theta}_{u,0}}|\mathcal{T}_{g_0})$ that, in turn, depends on the score vector of the parameters of interest $\mathbf{s}_{\boldsymbol{\theta}_{u,0}}$, the semiparametric nuisance tangent space $\mathcal{T}_{g_0}$ and the related projection operator $\Pi(\cdot|\mathcal{T}_{g_0})$.

5.3.2.1 Step 1: Calculation of $\mathbf{s}_{\boldsymbol{\theta}_{u,0}}$

According to Definition 5.2, the score vector $\mathbf{s}_{\boldsymbol{\theta}_{u,0}}$ of the "unconstrained" parameters of interest $\boldsymbol{\theta}_{u,0} = (\boldsymbol{\mu}_0^T, v(\mathbf{V}_0)^T)^T \in \Theta_u$ in (5.62) can be expressed as:

$$\mathbf{s}_{\boldsymbol{\theta}_{u,0}} = \nabla_{\boldsymbol{\theta}} \ln p_X(\mathbf{x}|\boldsymbol{\theta}_0, g_0) = \begin{pmatrix} \mathbf{s}_{\boldsymbol{\mu}_0}(\mathbf{x}) \\ \mathbf{s}_{v(\mathbf{V}_0)}(\mathbf{x}) \end{pmatrix}, \quad \boldsymbol{\theta}_0 \in \Theta_u, \tag{5.64}$$

where:

$$\mathbf{s}_{\boldsymbol{\mu}_0} = \nabla_{\boldsymbol{\mu}} \ln p_X(\mathbf{x}|\boldsymbol{\mu}_0, \mathbf{V}_0, g_0), \tag{5.65}$$

$$\mathbf{s}_{v(\mathbf{V}_0)} = \nabla_{v(\mathbf{V})} \ln p_X(\mathbf{x}|\boldsymbol{\mu}_0, \mathbf{V}_0, g_0). \tag{5.66}$$

By using the equality:

$$\mathcal{Q}_0 = ||\alpha^{-1}_{\boldsymbol{\mu}_0,\mathbf{V}_0}(\mathbf{x})||^2 = (\mathbf{x} - \boldsymbol{\mu}_0)^T \mathbf{V}_0^{-1}(\mathbf{x} - \boldsymbol{\mu}_0) =_d \mathcal{Q}, \tag{5.67}$$

through direct calculation [27, Chap. 8], [8], we have

$$\mathbf{s}_{\boldsymbol{\mu}_0} = -2\psi_0(\mathcal{Q}_0)\mathbf{V}_0^{-1}(\mathbf{x} - \boldsymbol{\mu}_0) =_d -2\sqrt{\mathcal{Q}}\psi_0(\mathcal{Q})\mathbf{V}_0^{-1/2}\mathbf{u}, \tag{5.68}$$

where the last two equalities follow directly from the Stochastic Representation of a RES vector given in (5.59) and

$$\psi_0(t) \stackrel{\text{def}}{=} \frac{1}{g_0(t)} \frac{dg_0(t)}{dt}. \tag{5.69}$$

Moreover, after some matrix calculation, we get [8]

$$\begin{aligned}\mathbf{s}_{v(\mathbf{V}_0)} &= -\mathbf{D}_m^T \left(\frac{1}{2}\text{vec}(\mathbf{V}_0^{-1}) + \psi_0(Q_0)\left(\mathbf{V}_0^{-1} \otimes \mathbf{V}_0^{-1}\right)\text{vec}((\mathbf{x}-\boldsymbol{\mu}_0)(\mathbf{x}-\boldsymbol{\mu}_0)^T)\right)\\ &=_d -\mathbf{D}_m^T \left(\frac{1}{2}\text{vec}(\mathbf{V}_0^{-1}) + Q\psi_0(Q)\left(\mathbf{V}_0^{-1/2} \otimes \mathbf{V}_0^{-1/2}\right)\text{vec}(\mathbf{u}\mathbf{u}^T)\right),\end{aligned} \tag{5.70}$$

where, as before, the last two equalities follow from (5.59), while $\mathbf{D}_m$ is the $m^2 \times m(m+1)/2$ *duplication matrix* that is implicitly defined by the equality $\mathbf{D}_m v(\mathbf{A}) = \text{vec}(\mathbf{A})$ for any $m \times m$ symmetric matrix $\mathbf{A}$ [25, 26].

Parametric Cramér–Rao bound If the density generator g_0 is known, the FIM for the estimation of $\boldsymbol{\theta}_{u,0} = (\boldsymbol{\mu}_0^T, v(\mathbf{V}_0)^T)^T \in \Theta_u$, can be obtained as a block matrix of the form:

$$\mathbf{I}(\boldsymbol{\theta}_{u,0}) = \text{E}_0(\mathbf{s}_{\boldsymbol{\theta}_{u,0}}\mathbf{s}_{\boldsymbol{\theta}_{u,0}}^T) = \begin{pmatrix} \text{C}_0(\mathbf{s}_{\boldsymbol{\mu}_0}) & \mathbf{0} \\ \mathbf{0}^T & \text{C}_0(\mathbf{s}_{v(\mathbf{V}_0)}) \end{pmatrix}, \tag{5.71}$$

that is in line with the one provided in Eqs. (88) and (89) of Sect. 6.2 of Chap. 1. Through direct calculation, it is easy to verify that

$$\text{C}_0(\mathbf{s}_{\boldsymbol{\mu}_0}) = 4m^{-1}\text{E}(Q\psi_0^2(Q))\mathbf{V}_0^{-1}, \tag{5.72}$$

$$\text{C}_0(\mathbf{s}_{v(\mathbf{V}_0)}) = \mathbf{D}_m^T\left(a_1 \text{vec}(\mathbf{V}_0^{-1})\text{vec}(\mathbf{V}_0^{-1})^T + a_2\mathbf{V}_0^{-1} \otimes \mathbf{V}_0^{-1}\right)\mathbf{D}_m, \tag{5.73}$$

$$a_1 = \frac{a_2}{2} - \frac{1}{4}, \tag{5.74}$$

$$a_2 = \frac{2\text{E}(Q^2\psi_0^2(Q))}{m(m+2)}. \tag{5.75}$$

Finally, the CRB on the estimation of $\boldsymbol{\theta}_0 = (\boldsymbol{\mu}_0^T, v(\mathbf{V}_0)^T)^T \in \Theta$ in (5.61) with the constraint on the shape $s(\mathbf{V}_0) = 1$ can be obtained by using the same procedure discussed in [28]. Specifically, let us define the Jacobian of the constraint as $\mathbf{J}_s(\mathbf{V}) = \nabla_{v(\mathbf{V}_0)}(s(\mathbf{V}_0) - 1)$ and $\mathbf{U}(\mathbf{V}_0) \equiv \mathbf{U}_{\mathbf{V}_0}$ as the matrix whose columns form a basis for an orthonormal null space of $\mathbf{J}_s(\mathbf{V}_0)$. Then, the Constrained CRB (CCRB) for the estimation of $\boldsymbol{\theta}_0 \in \Theta$ when the density generator in known is given by

$$\text{CCRB}(\boldsymbol{\theta}_0) = \begin{pmatrix} \frac{m}{4\text{E}(Q\psi_0^2(Q))}\mathbf{V}_0 & \mathbf{0} \\ \mathbf{0}^T & \mathbf{U}_{\mathbf{V}_0}\left(\mathbf{U}_{\mathbf{V}_0}^T \text{C}_0(\mathbf{s}_{v(\mathbf{V}_0)})\mathbf{U}_{\mathbf{V}_0}\right)^{-1}\mathbf{U}_{\mathbf{V}_0}^T \end{pmatrix}. \tag{5.76}$$

Note that the block-diagonal structure of $\mathbf{I}(\boldsymbol{\theta}_{u,0})$ in (5.71) implies that not knowing the mean value $\boldsymbol{\mu}_0$ does not have any impact on the asymptotic performance in the estimation of the shape matrix $\mathbf{V}_0$ and vice versa.

5.3.2.2 Step 2: Calculation of the Projection $\Pi(\mathbf{s}_{\boldsymbol{\theta}_{u,0}}|\mathcal{T}_{g_0})$

Let us now move to the crucial step of the evaluation of the projection $\Pi(\mathbf{s}_{\boldsymbol{\theta}_{u,0}}|\mathcal{T}_{g_0})$ of $\mathbf{s}_{\boldsymbol{\theta}_{u,0}}$ in (5.64) onto the semiparametric nuisance tangent space $\mathcal{T}_{g_0}$ at the true, but unknown, density generator g_0. To this end, we apply a fundamental property of the semiparametric group model discussed in [2, Sect. 4.2, Lemma 3]:

Lemma 5.1 *Let $\mathcal{A}$ be the group defined in* (5.58). *Let $\mathcal{T}_{g_0}$ and $\mathcal{S}_{g_0}$ be the semiparametric nuisance tangent spaces of the RES and of the spherically symmetric distributions at the true density generator g_0, respectively. Then, $\mathcal{T}_{g_0}$ can be obtained from $\mathcal{S}_{g_0}$ as*

$$\mathcal{T}_{g_0} = \left\{\mathbf{u} \in \mathcal{H}_q \,\middle|\, \mathbf{u} = \mathbf{v} \circ \alpha^{-1}_{\boldsymbol{\mu}_0,\mathbf{V}_0} : \mathbf{v} \in \mathcal{S}_{g_0} \subseteq \mathcal{H}_q\right\}, \tag{5.77}$$

where $\circ$ indicates the composition of functions. Moreover, the projection of the score vector of the parameters of interest $\mathbf{s}_{\boldsymbol{\theta}_{u,0}} \in \mathcal{H}_q$ on $\mathcal{T}_{g_0}$ can be expressed as

$$\Pi(\mathbf{s}_{\boldsymbol{\theta}_{u,0}}|\mathcal{T}_{g_0}) = \Pi(\mathbf{s}_{\boldsymbol{\theta}_{u,0}} \circ \alpha_{\boldsymbol{\mu}_0,\mathbf{V}_0}|\mathcal{S}_{g_0}) \circ \alpha^{-1}_{\boldsymbol{\mu}_0,\mathbf{V}_0}. \tag{5.78}$$

Intuitively, Lemma 5.1 can be explained as follows. The set of spherically symmetric distributions is characterized by the neutral element $\alpha_e(\cdot) = \alpha_{\mathbf{0},\mathbf{I}_m}(\cdot)$ of the group $\mathcal{A}$ in (5.58). We can derive the semiparametric nuisance tangent space $\mathcal{S}_{g_0}$ and the related projection operator $\Pi(\cdot|\mathcal{S}_{g_0})$ at α_e, then we can translate them in the true RES distribution p_0 characterized by an affine transformation $\alpha_{\boldsymbol{\mu}_0,\mathbf{V}_0}$ by means of the relations (5.77) and (5.78).

The following theorem, whose proof is reported in appendix, provides us with an explicit expression of the semiparametric nuisance tangent space $\mathcal{S}_{g_0}$ of the spherically symmetric distribution at g_0.

Theorem 5.3 *Let $\mathcal{P}_g$ be the non-parametric model of the spherically symmetric pdfs:*

$$\mathcal{P}_g \stackrel{\text{def}}{=} \left\{p_X(\mathbf{x}) = g(||\mathbf{x}||^2), g \in \mathcal{G}\right\}. \tag{5.79}$$

Its semiparametric nuisance tangent space $\mathcal{S}_{g_0} = \mathcal{L} \times \ldots \times \mathcal{L}$ is a q-replicating Hilbert space such that:

$$\mathcal{S}_{g_0} = \{\mathbf{u} \in \mathcal{H}_q | \mathbf{u} = (l_1, l_2, \ldots, l_q)^T, l_j \in \mathcal{L}, j = 1, \ldots, q\}, \tag{5.80}$$

$$\mathcal{L} = \{l \in \mathcal{H} | l(\mathbf{x}) = l(\mathbf{O}\mathbf{x}), \forall \mathbf{O} \in \mathcal{O}\} = \{l \in \mathcal{H} | l(\mathbf{x}) = l(||\mathbf{x}||)\}, \tag{5.81}$$

where $\mathcal{O}$ is the group of orthogonal transformations defined in (5.54).

As shown in (5.57), $||\mathbf{x}|| = R$ is a maximal invariant statistic for the set of spherically distributed random vectors. Then, following the discussion previously given in Sect. 5.2.4.2, the set $\mathcal{L}$ in (5.81) can be equivalently characterized as the set of the univariate random functions defined on the probability space $(\mathcal{X}, \mathfrak{D}(\mathbf{x}), P_X)$, where $\mathfrak{D}(\mathbf{x}) \subseteq \mathfrak{F}$ is the sub-sigma algebra generated by the modular variate $||\mathbf{x}|| = R$. Then, as a consequence of Definition 5.6 (see also [2, Sect. 6.3]), we have that the projection onto $\mathcal{S}_{g_0}$ can be expressed as a conditional expectation w.r.t. the modular variate R:

$$\Pi(\mathbf{h}|\mathcal{S}_{g_0}) = \mathrm{E}(\mathbf{h}|R), \quad \forall \mathbf{h} \in \mathcal{H}_q. \tag{5.82}$$

Finally, the desired $\Pi(\mathbf{s}_{\boldsymbol{\theta}_0}|\mathcal{T}_{g_0})$ can by obtained by combining the results in (5.78) and in (5.82). Specifically, after involved but standard calculation (see [8] for all the details), we get

$$\Pi(\mathbf{s}_{\boldsymbol{\theta}_{u,0}}|\mathcal{T}_{g_0}) = \begin{pmatrix} \Pi(\mathbf{s}_{\boldsymbol{\mu}_0}|\mathcal{T}_{g_0}) \\ \Pi(\mathbf{s}_{v(\mathbf{V}_0)}|\mathcal{T}_{g_0}) \end{pmatrix}, \tag{5.83}$$

where:

$$\Pi(\mathbf{s}_{\boldsymbol{\mu}_0}|\mathcal{T}_{g_0}) =_d -2\sqrt{Q}\psi_0(Q)\mathbf{V}_0^{-1/2}\mathrm{E}(\mathbf{u}) = \mathbf{0}, \tag{5.84}$$

$$\Pi(\mathbf{s}_{v(\mathbf{V}_0)}|\mathcal{T}_{g_0}) =_d -\mathbf{D}_m^T \left(\frac{1}{2} + \frac{1}{m}Q\psi_0(Q)\right) \mathrm{vec}(\mathbf{V}_0^{-1}). \tag{5.85}$$

Note that, Eq. (5.84) tells us that the score vector of the location vector $\mathbf{s}_{\boldsymbol{\mu}_0}$ is orthogonal to the nuisance tangent space $\mathcal{T}_{g_0}$:

$$\Pi(\mathbf{s}_{\boldsymbol{\mu}_0}|\mathcal{T}_{g_0}) = \mathbf{0} \Rightarrow \mathbf{s}_{\boldsymbol{\mu}_0} \perp \mathcal{T}_{g_0}. \tag{5.86}$$

This implies that the missing a-priori knowledge of the true density generator g_0 does not have any (asymptotic) impact on the performance of an estimator of the location vector $\boldsymbol{\mu}_0$ [11].

5.3.2.3 Step 3: Calculation of the Efficient SFIM $\bar{\mathbf{I}}(\boldsymbol{\theta}_{u,0}|g_0)$ and of SCRB$(\boldsymbol{\theta}_0|g_0)$

By subtracting (5.83) from (5.64), the semiparametric efficient score vector in (5.45) can be evaluated as

$$\begin{aligned} \bar{\mathbf{s}}_{u,0} = \mathbf{s}_{\boldsymbol{\theta}_{u,0}} - \Pi(\mathbf{s}_{\boldsymbol{\theta}_{u,0}}|\mathcal{T}_{g_0}) &= \begin{pmatrix} \bar{\mathbf{s}}_{\boldsymbol{\mu}_0} \\ \bar{\mathbf{s}}_{v(\mathbf{V}_0)} \end{pmatrix} \\ &= \begin{pmatrix} -2\sqrt{Q}\psi_0(Q)\mathbf{V}_0^{-1/2}\mathbf{u} \\ -\mathbf{D}_m^T Q\psi_0(Q)\left(\mathbf{V}_0^{-1/2} \otimes \mathbf{V}_0^{-1/2}\mathrm{vec}(\mathbf{u}\mathbf{u}^T) - \frac{1}{m}\mathrm{vec}(\mathbf{V}_0^{-1})\right) \end{pmatrix}. \end{aligned} \tag{5.87}$$

Let us now introduce the matrix

$$\Pi^{\perp}_{\mathrm{vec}(\mathbf{I}_m)} = \mathbf{I}_{m^2} - m^{-1}\mathrm{vec}(\mathbf{I}_m)\mathrm{vec}(\mathbf{I}_m)^T, \tag{5.88}$$

as the orthogonal projection matrix on the orthogonal complement of span(vec($\mathbf{I}_m$)), and the matrix $\mathbf{N}_{\mathbf{V}_0}$ as

$$\mathbf{N}_{\mathbf{V}_0} \stackrel{\text{def}}{=} \left(\mathbf{V}_0^{-1/2} \otimes \mathbf{V}_0^{-1/2}\right) \Pi^{\perp}_{\mathrm{vec}(\mathbf{I}_m)}. \tag{5.89}$$

Then, the semiparametric efficient score vector $\bar{\mathbf{s}}_{v(\mathbf{V}_0)}$ can be expressed as

$$\bar{\mathbf{s}}_{v(\mathbf{V}_0)} =_d -\mathbf{D}_m^T Q\psi_0(Q)\mathbf{N}_{\mathbf{V}_0}\mathrm{vec}(\mathbf{u}\mathbf{u}^T). \tag{5.90}$$

We can now explicitly evaluate the efficient SFIM $\bar{\mathbf{I}}(\boldsymbol{\theta}_0|g_0)$ as [8]:

$$\bar{\mathbf{I}}(\boldsymbol{\theta}_{u,0}|g_0) = \mathrm{E}_0(\bar{\mathbf{s}}_{u,0}\bar{\mathbf{s}}_{u,0}^T) = \begin{pmatrix} \mathbf{C}_0(\bar{\mathbf{s}}_{\boldsymbol{\mu}_0}) & \mathbf{0} \\ \mathbf{0}^T & \mathbf{C}_0(\bar{\mathbf{s}}_{v(\mathbf{V}_0)}) \end{pmatrix}, \tag{5.91}$$

where, as for the FIM in (5.71), the semiparametric efficient cross information matrix $\mathrm{E}_0(\bar{\mathbf{s}}_{\boldsymbol{\mu}_0}\bar{\mathbf{s}}_{v(\mathbf{V}_0)}^T)$ is nil. From the previous results and through some algebra, we get:

$$\mathbf{C}_0(\bar{\mathbf{s}}_{\boldsymbol{\mu}_0}) = \mathbf{C}_0(\mathbf{s}_{\boldsymbol{\mu}_0}) = 4m^{-1}\mathrm{E}(Q\psi_0^2(Q))\mathbf{V}_0^{-1}, \tag{5.92}$$

$$\begin{aligned}
\mathbf{C}_0(\bar{\mathbf{s}}_{v(\mathbf{V}_0)}) &= a_2\mathbf{D}_m^T\left(\mathbf{V}_0^{-1} \otimes \mathbf{V}_0^{-1} - \frac{1}{m}\mathrm{vec}(\mathbf{V}_0^{-1})\mathrm{vec}(\mathbf{V}_0^{-1})^T\right)\mathbf{D}_m \\
&= a_2\mathbf{D}_m^T\left[\left(\mathbf{V}_0^{-1/2} \otimes \mathbf{V}_0^{-1/2}\right)\Pi^{\perp}_{\mathrm{vec}(\mathbf{I}_m)}\left(\mathbf{V}_0^{-1/2} \otimes \mathbf{V}_0^{-1/2}\right)\right]\mathbf{D}_m \\
&= a_2\mathbf{D}_m^T\mathbf{N}_{\mathbf{V}_0}\mathbf{N}_{\mathbf{V}_0}^T\mathbf{D}_m,
\end{aligned} \tag{5.93}$$

where a_2 is given in (5.75). Note also that in the second inequality we use the fact that the projection matrix $\Pi^{\perp}_{\mathrm{vec}(\mathbf{I}_m)}$ is idempotent and that $\mathbf{V}_0$ is symmetric. It is important to highlight that the block-diagonal structure of the efficient SFIM $\bar{\mathbf{I}}(\boldsymbol{\theta}_{u,0}|g_0)$ implies that the estimate of the location vector $\boldsymbol{\mu}_0$ and of the shape matrix $\mathbf{V}_0$ are asymptotically decorrelated. Consequently, for the estimation of $\mathbf{V}_0$ we may center the data by means of any $\sqrt{n}$-consistent estimator of $\boldsymbol{\mu}_0$ without any (asymptotic) impact on the estimation performance of $\mathbf{V}_0$ [11]. Note that this result can also be obtained by using the semiparametric Le Cam theory as shown in [13, 14].

We are finally able to state the desired result: an explicit expression of the Semiparametric Cramér–Rao Bound for the estimation of $\boldsymbol{\theta}_0 = (\boldsymbol{\mu}_0^T, v(\mathbf{V}_0)^T)^T \in \Theta$ in RES-distributed data with the constraint on the shape $s(\mathbf{V}_0) = 1$. By introducing the matrix $\mathbf{U}(\mathbf{V}_0) \equiv \mathbf{U}_{\mathbf{V}_0}$ as we did in (5.76), the *constrained* SCRB$(\boldsymbol{\theta}_0|g_0)$ can be expressed as

$$\text{CSCRB}(\boldsymbol{\theta}_0|g_0) = \begin{pmatrix} \frac{m}{4E(Q\psi_0^2(Q))}\mathbf{V}_0 & \mathbf{0} \\ \mathbf{0}^T & \frac{m(m+2)}{2E(Q^2\psi_0^2(Q))}\mathbf{U}_{\mathbf{V}_0}\left(\mathbf{U}_{\mathbf{V}_0}^T\mathbf{D}_m^T\mathbf{N}_{\mathbf{V}_0}\mathbf{N}_{\mathbf{V}_0}^T\mathbf{D}_m\mathbf{U}_{\mathbf{V}_0}\right)^{-1}\mathbf{U}_{\mathbf{V}_0}^T \end{pmatrix}. \tag{5.94}$$

We note, in passing, that the corresponding bounds for C-CES distributed observation can be found in [7] along with the semiparametric generalization of the celebrated Slepian–Bangs formula.

Let us close the chapter by answering to this fundamental question: *Is it possible to derive a RAL estimator of the shape matrix* $\mathbf{V}_0$ *able to achieve the* $\text{CSCRB}(\boldsymbol{\theta}_0|g_0)$?

5.3.3 *The Rank-Based R-Estimators of the Shape Matrix*

Semiparametric efficient estimators of the finite-dimensional parameter vector $\boldsymbol{\theta}_0 \in \Theta$ in the presence of a nuisace function $g \in \mathcal{G}$ has been already introduced in Sect. 5.2.4.1. In particular, we saw that the *semiparametric one-step estimators* in (5.49) have the remarkable property to be asymptotically semiparametric efficient RAL estimators. In this section, we will specialize the general theory presented in Sect. 5.2.4.1 to the estimation of the shape matrix $\mathbf{V}_0 \in \mathcal{M}_s$ from a set of n i.i.d. observations $\{\mathbf{x}_1, \ldots, \mathbf{x}_n\}$, in the presence of an unknown density generator $g_0 \in \mathcal{G}$. In order to keep the discussion as clear as possible, in the following we assume that the location vector is nil, i.e., $\boldsymbol{\mu} = \mathbf{0}$. Then, the finite-dimensional parameter to be estimated is simply given by $\mathbf{V}_0 \in \mathcal{M}_s$. This reduction of the parameter space can be done without loss of generality, since as already discussed in Sect. 5.3.2.2, the score vector of $\boldsymbol{\mu}$ is orthogonal to the nuisance tangent space, i.e., $\Pi(\mathbf{s}_{\boldsymbol{\mu}_0}|\mathcal{T}_{g_0}) = \mathbf{0}$. Consequently, when $\boldsymbol{\mu} \neq \mathbf{0}$ we could always center the data by means of any $\sqrt{n}$-consistent estimator $\hat{\boldsymbol{\mu}}_n$ without any (asymptotic) impact on the estimation of $\mathbf{V}_0 \in \mathcal{M}_s$ [11]. However, in order to have a good finite-sample performance, we suggest to use a $\sqrt{n}$-consistent and *robust* location estimator as the one proposed in [19]. A further reduction of the parameter space can be done by choosing, among all the possible constraints of the shape matrix, the one that forces its first top-left element to be equal to 1:

$$\mathbf{V}_{0,1} = \boldsymbol{\Sigma}_0/[\boldsymbol{\Sigma}_0]_{11} \Rightarrow s(\mathbf{V}_{0,1}) = [\mathbf{V}_{0,1}]_{11} = 1. \tag{5.95}$$

Note that, in the following we will use the subscript 1, i.e., $\mathbf{A}_1$ to indicate that the generic matrix $\mathbf{A}$ satisfies the constraint in (5.95), that is $[\mathbf{A}]_{11} = 1$. To notice the impact that this choice has on the definition of the parameter space, let us start by introducing some useful notation. Let $\mathbf{A}$ be an $m \times m$ symmetric matrix, then $\underline{v}(\mathbf{A})$ is implicitly defined as $v(\mathbf{A}) = [[\mathbf{A}]_{11}, \underline{v}(\mathbf{A})^\top]^\top$.[6] Moreover, if $[\mathbf{A}]_{11} = 0$, then $\mathbf{M}_n$ is the $(m(m+1)/2 - 1) \times m^2$ matrix s.t. $\mathbf{M}_m^\top \underline{v}(\mathbf{A}) = \text{vec}(\mathbf{A})$. Note that $\mathbf{M}_m^\top$ can be obtained from the *duplication matrix* $\mathbf{D}_m$ by removing its first column. Consequently, the finite-dimensional parameter vector to estimate is given by

[6] Note that, in our previous publications (see e.g., [7, 8, 10]), $\underline{v}(\cdot)$ was expressed as $\underline{\text{vecs}}(\cdot)$

$$\boldsymbol{\theta}_0 = \underline{v}(\mathbf{V}_{1,0}) \in \Theta \subseteq \mathbb{R}^{m(m+1)/2-1}, \ [\mathbf{V}_{0,1}]_{11} = 1. \tag{5.96}$$

For ease of readability, let us recall here the expression of a semiparametric one-step estimator given in (5.49) as

$$\hat{\boldsymbol{\theta}}_{op,n} = \boldsymbol{\theta}_n^\star + n^{-1/2}\left[\mathbf{C}(\boldsymbol{\theta}_n^\star, g_n^\star)\right]^{-1}\overline{\Delta}_n(\boldsymbol{\theta}_n^\star, g_n^\star),$$

and try to list the ingredient that we need to implement it for our estimation problem:

1. *A $\sqrt{n}$ -consistent preliminary estimator $\boldsymbol{\theta}_n^\star$ of the constrained shape matrix.* A good candidate for $\boldsymbol{\theta}_n^\star$ is the celebrated Tyler M-estimator [36] given in Sect. 6.4 of Chap. 1, since it is proven to be $\sqrt{n}$-consistent under any (unknown) density generator. We indicate this estimator as $\mathbf{V}_{1,n}^\star$, where the subscript "1" remember us that the constraint in (5.95) needs to be imposed.
2. *The efficient central sequence* $\overline{\Delta}_n(\mathbf{V}_{1,n}^\star, g_n^\star)$. By specializing the definition of semiparametric efficient score vector for $\mathbf{V}_0$ given in (5.90) to the parameter space (5.96), we have that [10]

$$\bar{\mathbf{s}}_{\underline{v}(\mathbf{V}_{1,0})} =_d -Q\psi_0(Q)\mathbf{K}_{\mathbf{V}_{1,0}}\text{vec}(\mathbf{u}\mathbf{u}^T), \tag{5.97}$$

where, for notation simplicity, we define the matrix

$$\mathbf{K}_{\mathbf{V}_{1,0}} \overset{\text{def}}{=} \mathbf{M}_m\left(\mathbf{V}_{1,0}^{-1/2} \otimes \mathbf{V}_{1,0}^{-1/2}\right)\Pi^{\perp}_{\text{vec}(\mathbf{I}_m)}, \tag{5.98}$$

Consequently, the efficient central sequence can be expressed as

$$\overline{\Delta}_n(\mathbf{V}_{1,n}^\star, g_n^\star) = -n^{-1/2}\mathbf{K}_{\mathbf{V}_{1,n}^\star}\sum_{i=1}^{n} Q_i^\star\psi_n^\star(Q_i^\star)\text{vec}(\mathbf{u}_i^\star(\mathbf{u}_i^\star)^T), \tag{5.99}$$

where, from the stochastic representation in (5.56) and (5.59), we have

$$Q_i^\star = \mathbf{x}_i^T[\mathbf{V}_{1,n}^\star]^{-1}\mathbf{x}_i, \quad \mathbf{u}_i^\star = [Q_i^\star\mathbf{V}_{1,n}^\star]^{-1/2}\mathbf{x}_i, \tag{5.100}$$

and, from (5.69), $\psi_n^\star(t) = \frac{1}{g_n^\star(t)}\frac{dg_n^\star(t)}{dt}$, where $g_n^\star(t)$ is a $\sqrt{n}$-consistent non-parametric estimator of the density generator $g_0 \in \mathcal{G}$.
3. *An estimate* $\mathbf{C}(\mathbf{V}_{1,n}^\star, g_n^\star)$ *of the efficient SFIM* $\bar{\mathbf{I}}(\boldsymbol{\theta}_0|g_0)$ *evaluated in* (5.93). From its definition provided in (5.51) and from the previous results, we have that

$$\mathbf{C}(\mathbf{V}_{1,n}^\star, g_n^\star) = a_n^\star\mathbf{K}_{\mathbf{V}_{1,n}^\star}\mathbf{K}_{\mathbf{V}_{1,n}^\star}^T, \tag{5.101}$$

where $a_n^\star$ is a $\sqrt{n}$-consistent estimator of the term $a_2 = \frac{2\mathrm{E}(Q^2\psi_0^2(Q))}{m(m+2)}$ in (5.75).

By taking a careful look at the previous list of ingredients, it is immediate to realize that the critical point for the implementation of this semiparametric efficient estimator is the derivation of $\sqrt{n}$-consistent estimators for the density generator $g_n^\star(t)$ and for the two related quantities $\psi_n^\star(t)$ and $a_n^\star$. As discussed in Sect. 5.2.4.1, finding a non-parametric estimator able to converge at the $\sqrt{n}$-rate is a difficult, or even impossible, task. What can we do then? Is there any other way out? Fortunately, a positive answer to this question has been recently found by Hallin, Oja and Paindaveine in their seminal work [13]. Even if of great importance, their results strongly rely on an invariance-based extension of the semiparametric Le Cam theory discussed in [18], whose explanation falls outside the scope of this chapter. Here, we limit ourselves to a provides only some hints, without any claim of completeness nor of mathematical rigor. The groundbreaking idea proposed in [13] is the use of a *rank-based approximation* $\tilde{\Delta}_n(\mathbf{V}_{1,n}^\star)$ [12], [37, Chap. 13] (hence the name of R-estimator) of the efficient central sequence $\overline{\Delta}_n(\mathbf{V}_{1,n}^\star, g_n^\star)$ that does not require any non-parametric estimator $g_n^\star$ of the density generator $g_0 \in \mathcal{G}$.

By letting the interested reader to discover the technical details and the proofs in [10, 13], in the following we will provide the necessary tools to implement an R-estimator of the shape matrix. At first, we need to introduce the concept of *ranks*. To this end, let us order the n "preliminary estimated" modular variates $\{Q_1^\star, Q_2^\star, \cdots Q_n^\star, \}$ in (5.100) in an ascending way as $Q_{n(1)}^\star < Q_{n(2)}^\star < \ldots < Q_{n(n)}^\star$. Then, the rank $r_i^\star$ of $Q_i^\star$ is its position index in the ordered sequence [37, Chap. 13]. Let us now introduce the following two rank-based functionals:

1. *The $g_n^\star$-free approximation of the efficient central sequence*. By using, as before, the Tyler M-estimator as the $\sqrt{n}$-consistent preliminary estimator of the constrained shape matrix in (5.95), i.e., $\mathbf{V}_{1,n}^\star$, the semiparametric efficient central sequence $\overline{\Delta}_n(\boldsymbol{\theta}_n^\star, g_n^\star)$ in (5.99) can be approximated as

$$\tilde{\boldsymbol{\Delta}}(\mathbf{V}_1^\star) = n^{-1/2}\mathbf{K}_{\mathbf{V}_1^\star} \sum_{i=1}^{n} K\left(\frac{r_i^\star}{n+1}\right) \operatorname{vec}(\mathbf{u}_i^\star(\mathbf{u}_i^\star)^T), \tag{5.102}$$

where $r_i^\star$ is the rank of $Q_i^\star$, already defined in (5.100) along with $\mathbf{u}_i^\star$. The function $K : (0, 1) \rightarrow \mathbb{R}^+$ belongs to the set $\mathcal{K}$ of continuous, square-integrable functions that can be expressed as the difference of two monotone increasing functions [13]. Here, due to its "optimality property" (see [30] for some additional discussion) we use the *van der Waerden* function:

$$K_{vdW}(t) \stackrel{\text{def}}{=} \Phi_G^{-1}(t), \tag{5.103}$$

where Φ_G^{-1} indicates the inverse cumulative distribution function of a χ^2-distributed random variable with m degrees of freedom.

2. *The $g_n^\star$-free approximation of* a_2 in (5.75). As proved in [13], a $\sqrt{n}$-consistent, $g_n^\star$-free, estimator of $a_2 = \frac{2\mathrm{E}(Q^2\psi_0^2(Q))}{m(m+2)}$ is given by

$$\hat{\alpha}_n = \frac{||\boldsymbol{\Delta}(\mathbf{V}_1^\star + n^{-1/2}\mathbf{H}^0) - \widetilde{\boldsymbol{\Delta}}(\mathbf{V}_1^\star)||}{||\mathbf{K}_{\mathbf{V}_1^\star}\mathbf{K}_{\mathbf{V}_1^\star}^T \underline{v}(\mathbf{H}^0)||}, \tag{5.104}$$

where $\mathbf{H}^0$ is a "small perturbation", symmetric, matrix such that $[\mathbf{H}^0]_{11} = 0$. Following [10], we set $\mathbf{H}^0 = (\mathbf{G} + \mathbf{G}^H)/2$ where $[\mathbf{G}]_{ij} \sim \mathcal{CN}(0, \upsilon^2)$, $[\mathbf{G}]_{11} = 0$ and υ is an hyper-parameter that has to be small enough to guarantee that $\widehat{\mathbf{V}}_1^\star + n^{-1/2}\mathbf{H}^0$ remains a positive-definite matrix. Consequently, we have that a $g_n^\star$-free approximation of the efficient SFIM can be obtained as

$$\widetilde{\mathbf{C}}(\widehat{\mathbf{V}}_{1,n}^\star) = \hat{\alpha}_n \mathbf{K}_{\mathbf{V}_{1,n}^\star}\mathbf{K}_{\mathbf{V}_{1,n}^\star}^T. \tag{5.105}$$

Finally, by putting all the previous results together, we have that a (almost) semiparametric efficient one-step R-estimator of the shape matrix is given by

$$\begin{aligned} \underline{v}(\widehat{\mathbf{V}}_{R,1,n}) &= \underline{v}(\mathbf{V}_{1,n}^\star) + n^{-1/2}[\widetilde{\mathbf{C}}(\mathbf{V}_{1,n}^\star)]^{-1}\widetilde{\boldsymbol{\Delta}}(\mathbf{V}_1^\star) \\ &= \underline{v}(\mathbf{V}_{1,n}^\star) + \frac{1}{n\hat{\alpha}_n}\left[\mathbf{K}_{\mathbf{V}_{1,n}^\star}\mathbf{K}_{\mathbf{V}_{1,n}^\star}^T\right]^{-1}\mathbf{K}_{\mathbf{V}_1^\star}\sum_{i=1}^{n} K\left(\frac{r_i^\star}{n+1}\right)\mathrm{vec}(\mathbf{u}_i^\star(\mathbf{u}_i^\star)^T). \end{aligned} \tag{5.106}$$

As shown in [13], $\underline{v}(\widehat{\mathbf{V}}_{R,1,n})$ is a RAL estimator of $\mathbf{V}_{1,0}$ satisfying the following properties:

1. $\sqrt{n}$-*consistent*: $\sqrt{n}(\underline{v}(\widehat{\mathbf{V}}_{R,1,n} - \mathbf{V}_0)) = O_{P_X}(1)$,
2. *Asymptotically normal*: $\sqrt{n}(\underline{v}(\widehat{\mathbf{V}}_{R,1,n} - \mathbf{V}_0)) \underset{n\to\infty}{\sim} \mathcal{N}(\mathbf{0}, \boldsymbol{\Phi}(\mathbf{V}_0, g_0, K))$.

It is worth highlighting that its asymptotic error covariance matrix $\boldsymbol{\Phi}(\boldsymbol{\theta}_0, g_0, K)$ depends on the choice of the rank-based function K along with the true shape matrix $\mathbf{V}_{1,0}$ and the true density generator $g_0 \in \mathcal{G}$. Moreover, due to the constrained parameter space in (5.96), induced by the specific choice of the constraint in (5.95), the efficiency inequality in (5.48) can be explicitly expressed as

$$\boldsymbol{\Phi}(\mathbf{V}_0, g_0, K) \geq \mathrm{CSCRB}(\mathbf{V}_0|g_0) = \frac{m(m+2)}{2\mathrm{E}(Q^2\psi_0^2(Q))}\left[\mathbf{K}_{\mathbf{V}_{1,0}}\mathbf{K}_{\mathbf{V}_{1,0}}^T\right]^{-1}. \tag{5.107}$$

Before moving to the simulation results, some comments are in order:

- The R-estimator in (5.106) has been generalized to the case of Complex Elliptically Symmetric distributed data in [9–11] where an investigation of its robustness properties has been also performed.
- Building upon the seminal results in [13, Prop. 3.1], in [11] the complex-valued R-estimator has been recast in a matrix form able to be more efficient in terms of numerical calculation.
- The MATLAB and Python codes related to the implementation of the different versions of the R-estimator can be found in [4].

- If we want to constraint the estimated shape matrix with a criterion different from the one in (5.95), we can just re-normalize the estimated shape matrix as:

$$\widehat{\mathbf{V}}_{R,s,n} = \widehat{\mathbf{V}}_{R,1,n}/s(\widehat{\mathbf{V}}_{R,1,n}), \tag{5.108}$$

without any impact on the *asymptotic* properties of [15]. However, it must be noted that generally the non-asymptotic, *finite-sample* estimation performance may be impacted by this re-scaling.

5.3.3.1 Finite-Sample Performance of *R*-Estimators

In this subsection, we compare, through numerical simulation, the estimation performance of the Tyler M-estimator with the one of the R-estimator in (5.106). The main aim is to quantify the efficiency losses described by the inequality (5.107) in a finite-sample regime, i.e., when the number of observations is finite. To this end, we generate a set of *zero-mean* i.i.d. t-distributed data whose distribution is described in Sect. 5.2 of Chap. 1. The scatter matrix is assumed to have a Toeplitz structure $[\mathbf{\Sigma}_0]_{ij} = \rho^{|i-j|}$, $i, j = 1, \ldots, m$ where $\rho = 0.8$ and $m = 8$. The performance index, used to compare the Tyler and the R estimators is the following:

$$\varepsilon_\varphi \stackrel{\text{def}}{=} \left\| \mathrm{E}_0 \left(\underline{v}(\widehat{\mathbf{V}}_{\varphi,1,n} - \mathbf{V}_{1,0}) \underline{v}(\widehat{\mathbf{V}}_{\varphi,1,n} - \mathbf{V}_{1,0})^T \right) \right\|_F, \tag{5.109}$$

where $\varphi = \{Tyler, R\}$ indicates the particular estimator under test, $\mathbf{V}_{1,0} = \mathbf{\Sigma}_0/[\mathbf{\Sigma}_0]_{11}$ and $||\mathbf{A}||_F^2 = \mathrm{tr}(\mathbf{A}^T\mathbf{A})$ is the Frobenius norm of matrix $\mathbf{A}$. This Mean Square Error (MSE) index will be compared with the CCRB in (5.76) and with the CSCRB in (5.107). Note that the scalars a_1 and a_2 (in (5.74) and (5.75) respectively), needed to define the bounds, can be evaluated, for the t-distribution, as [8]:

$$a_1 = -\frac{1}{2(m+2+\nu)}, \quad a_2 = \frac{2\mathrm{E}(Q^2\psi_0^2(Q))}{m(m+2)} = \frac{\nu+m}{2(m+2+\nu)}. \tag{5.110}$$

Finally, for the CCRB in (5.76) and the (5.107), the following indices will be reported:

$$\varepsilon_{CCRB} = ||\mathrm{CCRB}(\mathbf{V}_0)||_F, \; \varepsilon_{CSCRB} = ||\mathrm{CCRB}(\mathbf{V}_0|g_0)||_F. \tag{5.111}$$

We present the behavior of the MSE indices and of the bounds as function of the number of i.i.d. observations n in Fig. 5.1 and as function of the degrees of freedom ν of the t-distribution in Fig. 5.2. We recall, in passing, that when $\nu \to 0^+$, the data tend to be extremely heavy-tailed, while when $\nu \to +\infty$, the t-distribution collapse into the Gaussian one.

The first thing to note is that the distance between ε_{CCRB} and ε_{CSCRB} quantifies the efficiency loss due to the missing knowledge of the density generator (parametric vs. semiparametric estimation). This distance does not goes to zero as n goes to infinity

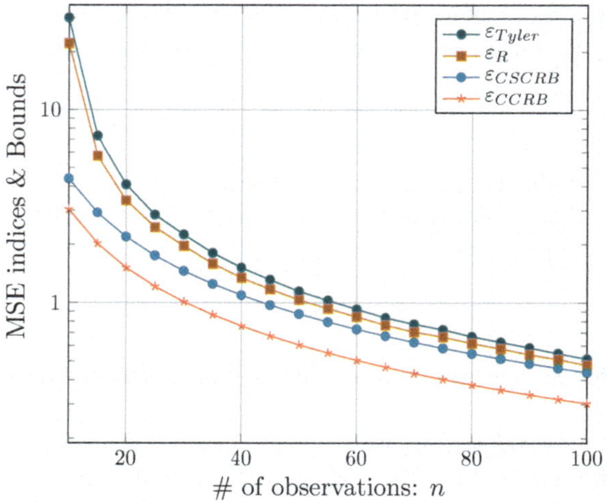

Fig. 5.1 MSE indices and bounds as function of the number of observations n ($\nu = 5, m = 8$)

and it increases as $\nu \to +\infty$, i.e., when the observation are less heavy-tailed. This result remains true for any RES distributions (not only for the t-distribution) [13, 15]. Let us now focus on the performance of the Tyler and R-estimators. Both of them are semiparametric in nature since they does not require the a priori knowledge of the density generator characterizing the data generating process. If we compare their MSE indices, i.e., ε_{Tyler} and ε_R, as function of n as in Fig. 5.1, we can notice a small performance improvement of the R-estimator with respect to the Tyler's one. The benefit of the R-estimator are more evident if we evaluate ε_{Tyler} and ε_R as function of the degrees of freedom ν (the non-Gaussianity parameter) as in Fig. 5.2. As expected, the constrained Tyler M-estimator has a MSE index that is constant with respect to ν. On the other hand, the R-estimator tends to be adaptive with respect to the changing value of ν. Unfortunately, it fails to be semiparametric efficient, i.e., it is not able to achieve the CSCRB in (5.107). There are three main reasons for this:

- As discussed in Sect. 5.3.3, the R-estimator relies on the estimation of the term a_2 in (5.75). The estimator $\hat{\alpha}_n$ given in (5.104) is $\sqrt{n}$-consistent but not optimal in the MSE sense. Then, to improve the global performance of an R-estimator, we should look for a "better" estimator of a_2 in (5.75).
- As shown in (5.107), the asymptotic error covariance matrix $\mathbf{\Phi}(\mathbf{V}_0, g_0, K)$ depends of the choice of the rank function K. A natural question is then: *would it be possible to find the optimal function K that minimize the error covariance under any density generator?* Even if a preliminary answer to this question has been provided in [30], this problem is still open.
- Regarding the Fig. 5.2, it must be noted that ε_R is evaluated in a finite-sample regime ($n = 3\,m$) and not in asymptotic conditions needed to guarantee the optimality of the semiparametric estimator in (5.106). Moreover, it is worth underling that in this non-asymptotic regime, the constraint in (5.108) that we choose to put

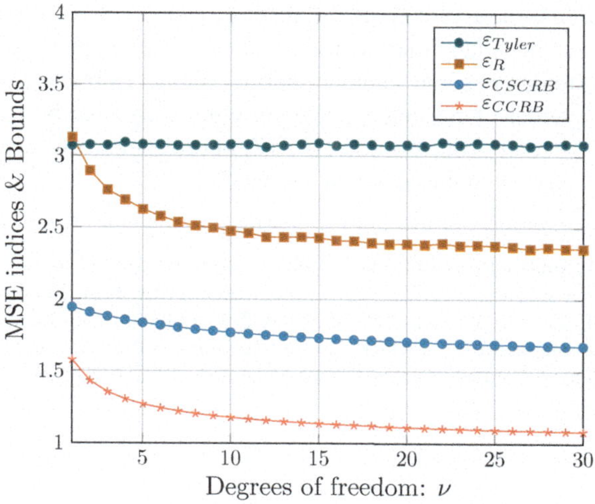

Fig. 5.2 MSE indices and bounds as function of the degrees of freedom ν of the t-distribution ($n = 3m$, $m = 8$)

of the scatter matrix may have an impact on the MSE of the related shape matrix estimator.

5.4 Conclusions

The aim of this chapter was to first provide the reader with a general and tutorial in nature introduction about the estimation problem in semiparametric models. All the necessary mathematical tools have been presented along with some fundamental results as the *Semiparametric Efficiency Bound*. In the second part, we showed how to apply the general framework to the semiparametric estimation of the shape matrix of RES-distributed data in the presence of an unknown density generator. Specifically, the Constrained Semiparametric Cramér–Rao Bound (CSCRB) for the joint estimation of the location vector and of the shape matrix has been proposed together with a rank-based R-estimator of the shape matrix. Here we focused only on the RES case, but the interested reader can find the generalization of all the above-mentioned results to CES-distributed data in [4, 7, 9–11]. We note in passing the semiparametric version of the celebrated Slepian–Bangs formula has also been derived in [7]. Despite the large amount of works already done in the field of the semiparametric statistics, some crucial problems still merit our effort to reach a fully satisfactory solution. Here, we limit ourselves to cite the ones related to the RES distributions:

- As previously discussed, the derivation of a (at least asymptotically) semiparametric efficient estimator is still an open problem. Even if the R-estimator proposed by Hallin, Oja, and Paidaveine in [13] has represented a breakthrough, we still have room for improvement as shown in Figs. 5.1 and 5.2.

- In some applications, we are more interested in a direct estimation of the eigenspace of the shape matrix instead of the shape matrix itself. Would it be possible then to derive some semiparametric efficient estimators of the eigenvectors and eigenvalues of the shape matrix for elliptically distributed data? This problem has been recently addressed in [16, 17] but further works need to be done in order to have an exploitable practical estimator.

Acknowledgements I would like to warmly thank all those who have collaborated with me on this subject. Specifically, in chronological order, Prof. Fulvio Gini, Prof. Maria Sabrina Greco (University of Pisa), Prof. Abdelhak Zoubir (Technische Universität Darmstadt), Prof. Muralidhar Rangaswamy (U.S. AFRL, Sensors Directorate), Prof. Alexandre Renaux, and Prof. Frédéric Pascal (Université Paris-Saclay, CNRS, CentraleSupélec, L2S).

Appendix: Proof of Theorem 5.3

In this appendix, we provide the proof of Theorem 5.3. Before starting the proof, it is worth recalling here that the Hilbert space $\mathcal{H}$ is the set of scalar random functions on the sample space $\mathcal{X}$, i.e., $h : \mathcal{X} \to \mathbb{R}$ satisfying the following two properties:

1. *Zero-mean*: $\mathrm{E}(h) = \int_{\mathcal{X}} h(\mathbf{x}) p_X(\mathbf{x}) d\mathbf{x} = 0$.
2. *Finite variance*: $\mathrm{E}(h^2) = \int_{\mathcal{X}} h^2(\mathbf{x}) p_X(\mathbf{x}) d\mathbf{x} = \sigma_h^2 < +\infty$.

Proof The proof follows the one discussed in [35, Theorem 4.4]. Let us start by defining the generic i-th parametric sub-model of the non-parametric model $\mathcal{P}_g$ in (5.79) as

$$\mathcal{P}_{\nu_i} \stackrel{\text{def}}{=} \left\{ p_X(\mathbf{x}|\boldsymbol{\eta}) = \nu_i(||\mathbf{x}||, \boldsymbol{\eta}), \boldsymbol{\eta} \in \Gamma_i \subseteq \mathbb{R}^{r_i} \right\}, \tag{5.112}$$

where, according to de definition of parametric sub-model in Sect. 5.2.4:

- $\nu_i : \mathcal{X} \times \Gamma_i \to \mathcal{G}$ is a smooth parametric map $\forall i \in \mathcal{I}$.
- Let g_0 be the true density generator, then $\exists \boldsymbol{\eta}_0 \in \Gamma_i$, s.t. $\nu_i(||\mathbf{x}||, \boldsymbol{\eta}_0) = g_0(||\mathbf{x}||^2)$.

According to the definition in (5.43), the i-th nuisance tangent space of $\mathcal{P}_{\nu_i}$ in (5.112) is given by

$$\mathcal{T}_{\boldsymbol{\eta}_{0,i}} \stackrel{\text{def}}{=} \{\mathbf{u} \in \mathcal{H}_q | \mathbf{u} = \mathbf{C}_i \mathbf{s}_{\boldsymbol{\eta}_{0,i}} : \mathbf{C}_i \text{ is any matrix in } \mathbb{R}^{q \times r_i}\} \subset \mathcal{H}_q, \tag{5.113}$$

where the nuisance score vector is expressed as

$$\mathbf{s}_{\boldsymbol{\eta}_{0,i}} \equiv \mathbf{s}_{\boldsymbol{\eta}_{0,i}}(||\mathbf{x}||) = \nabla_{\boldsymbol{\eta}} \nu_i(||\mathbf{x}||, \boldsymbol{\eta}_0). \tag{5.114}$$

From the standard properties of the score vectors, we have that

$$\mathrm{E}_0([\mathbf{s}_{\eta_{0,i}}]_j) = 0 \ \mathrm{E}_0([\mathbf{s}_{\eta_{0,i}}]_j^2) < +\infty \ \Rightarrow [\mathbf{s}_{\eta_{0,i}}]_j \in \mathcal{L}, \tag{5.115}$$

then $\mathbf{s}_{\eta_{0,i}} \in \mathcal{S}_{g_0}$. Consequently:

$$\mathcal{T}_{\eta_{0,i}} \subseteq \mathcal{S}_{g_0}, \ \forall i \in \mathcal{I}, \tag{5.116}$$

that is, all the nuisance tangent spaces $\mathcal{T}_{\eta_{0,i}}$ of any possible parametric sub-model $\mathcal{P}_{\nu_i}$ are contained in $\mathcal{S}_{g_0}$.

To complete the proof, we now need to prove that any element in $\mathcal{S}_{g_0}$ can be expressed as an element of some parametric submodel $\mathcal{T}_{\eta_{0,i}}$. Let start by choosing an arbitrary element $\mathbf{h}(||\mathbf{x}||)$ of $\mathcal{S}_{g_0}$. As parametric sub-model, we choose the following one:

$$\mathcal{P}_\nu = \left\{ p_X(\mathbf{x}|\boldsymbol{\eta}) = \nu(||\mathbf{x}||, \boldsymbol{\eta}), \boldsymbol{\eta} \in \Gamma \subseteq \mathbb{R}^q \right\}, \text{ where:} \tag{5.117}$$

$$\nu(||\mathbf{x}||, \boldsymbol{\eta}) = g_0(||\mathbf{x}||^2) \left(1 + \boldsymbol{\eta}^T \mathbf{h}(||\mathbf{x}||)\right), \tag{5.118}$$

and $\boldsymbol{\eta} \in \Gamma \subseteq \mathbb{R}^q$ is sufficiently small to guarantee that $(1 + \boldsymbol{\eta}^T \mathbf{h}(||\mathbf{x}||)) \geq 0 \ \forall \mathbf{x}$, and then $p_X(\mathbf{x}|\boldsymbol{\eta}) = \nu(||\mathbf{x}||, \boldsymbol{\eta}) \geq 0$. Moreover, from the definition of $\mathbf{h}(||\mathbf{x}||)$ as an element of $\mathcal{S}_{g_0}$, we have that

$$\begin{aligned} \int p_X(\mathbf{x}|\boldsymbol{\eta}) d\mathbf{x} &= \int g_0(||\mathbf{x}||^2) d\mathbf{x} + \boldsymbol{\eta}^T \int \mathbf{h}(||\mathbf{x}||) g_0(||\mathbf{x}||^2) d\mathbf{x} \\ &= 1 + \boldsymbol{\eta}^T \mathrm{E}_0(\mathbf{h}(||\mathbf{x}||)) = 1 + 0 = 1, \end{aligned} \tag{5.119}$$

then $p_X(\mathbf{x}|\boldsymbol{\eta}) = \nu(||\mathbf{x}||, \boldsymbol{\eta})$ represents a proper pdf. For this specific parametric sub-model, we have that the nuisance score vector is given by

$$\mathbf{s}_{\eta_0} = \nabla_{\boldsymbol{\eta}} \nu(||\mathbf{x}||, \boldsymbol{\eta}_0) = \nabla_{\boldsymbol{\eta}} \left[g_0(||\mathbf{x}||^2) \left(1 + \boldsymbol{\eta}^T \mathbf{h}(||\mathbf{x}||)\right) \right] \Big|_{\boldsymbol{\eta}=\boldsymbol{\eta}_0=\mathbf{0}} = \mathbf{h}(||\mathbf{x}||). \tag{5.120}$$

Finally, if we pose $\mathbf{C}_i$ equal to the identity matrix $\mathbf{I}$ in the nuisance tangent space (5.113) (also given in (5.43)), from (5.120), we immediately have that any arbitrary element $\mathbf{h}(||\mathbf{x}||)$ of $\mathcal{S}_{g_0}$ can be expressed as an element of the nuisance tangent space of the parametric model in (5.117). This concludes the proof, i.e., $\mathcal{S}_{g_0}$ is the semiparametric nuisance tangent space of the set of the spherically symmetric distributions.

References

1. Begun, J.M., Hall, W.J., Huang, W.M., Wellner, J.A.: Information and asymptotic efficiency in parametric-nonparametric models. Ann. Stat. **11**(2), 432–452 (1983)
2. Bickel, P., Klaassen, C., Ritov, Y., Wellner, J.: Efficient and Adaptive Estimation for Semiparametric Models. Johns Hopkins University Press (1993)
3. Bickel, P.J.: On adaptive estimation. Ann. Stat. **10**(3), 647–671 (1982)

4. Fortunati, S. (2020). https://github.com/StefanoFor?tab=repositories
5. Fortunati, S., Gini, F., Greco, M., Zoubir, A.M., Rangaswamy, M.: A fresh look at the semiparametric Cramér-Rao bound. In: 2018 26th European Signal Processing Conference (EUSIPCO), pp. 261–265 (2018)
6. Fortunati, S., Gini, F., Greco, M.S., Richmond, C.D.: Performance bounds for parameter estimation under misspecified models: fundamental findings and applications. IEEE Signal Process. Mag. **34**(6), 142–157 (2017)
7. Fortunati, S., Gini, F., Greco, M.S., Zoubir, A.M., Rangaswamy, M.: Semiparametric CRB and Slepian-Bangs formulas for complex elliptically symmetric distributions. IEEE Trans. Signal Process. **67**(20), 5352–5364 (2019)
8. Fortunati, S., Gini, F., Greco, M.S., Zoubir, A.M., Rangaswamy, M.: Semiparametric inference and lower bounds for real elliptically symmetric distributions. IEEE Trans. Signal Process. **67**(1), 164–177 (2019)
9. Fortunati, S., Renaux, A., Pascal, F.: Properties of a new R-estimator of shape matrices. EUSIPCO 2020 (2020)
10. Fortunati, S., Renaux, A., Pascal, F.: Robust semiparametric efficient estimators in complex elliptically symmetric distributions. IEEE Trans. Signal Process. **68**, 5003–5015 (2020)
11. Fortunati, S., Renaux, A., Pascal, F.: Robust semiparametric joint estimators of location and scatter in elliptical distributions. J. Signal Process. Syst. **94**, 133–146 (2022)
12. Hájek, J.: Asymptotic normality of simple linear rank statistics under alternatives. Ann. Math. Statist. **39**(2), 325–346 (1968)
13. Hallin, M., Oja, H., Paindaveine, D.: Semiparametrically efficient rank-based inference for shape II. Optimal R-estimation of shape. Ann. Stat. **34**(6), 2757–2789 (2006)
14. Hallin, M., Paindaveine, D.: Semiparametrically efficient rank-based inference for shape I. Optimal rank-based tests for sphericity. Ann. Stat. **34**(6), 2707–2756 (2006)
15. Hallin, M., Paindaveine, D.: Parametric and semiparametric inference for shape: the role of the scale functional. Stat. Decis. **24**(3), 327–350 (2009)
16. Hallin, M., Paindaveine, D., Verdebout, T.: Optimal rank-based testing for principal components. Ann. Stat. **38**(6), 3245–3299 (2010)
17. Hallin, M., Paindaveine, D., Verdebout, T.: Efficient r-estimation of principal and common principal components. J. Am. Stat. Assoc. **109**(507), 1071–1083 (2014)
18. Hallin, M., Werker, B.J.M.: Semi-parametric efficiency, distribution-freeness and invariance. Bernoulli **9**(1), 137–165 (2003)
19. Hettmansperger, T.P., Randles, R.H.: A practical affine equivariant multivariate median. Biometrika **89**(4), 851–860 (2002)
20. Jacod, J., Protter, P.: Probability Essentials. Springer Series in Statistics (2004)
21. Le Cam, L.: Locally asymptotically normal families of distributions. University of California Publications in Statistics, vol. 3, pp. 37–98 (1960)
22. LeCam, L., L.Yang, G.: Asymptotics in Statistics: Some Basic Concepts, 2 edn. Springer Series in Statistics. Berlin, Germany (2000)
23. Lehmann, E.L.: Theory of Point Estimation. John Willey and Sons Inc., New York (1983)
24. Lehmann, E.L., Romano, J.P.: Testing Statistical Hypotheses. Springer Texts in Statistics (2004)
25. Magnus, J.R., Neudecker, H.: The commutation matrix: some properties and applications. Ann. Stat. **7**(2), 381–394 (1979)
26. Magnus, J.R., Neudecker, H.: The elimination matrix: some lemmas and applications. SIAM J. Algebr. Discret. Methods **1**(4), 422–449 (1980)
27. Magnus, J.R., Neudecker, H.: Matrix Differential Calculus with Applications in Statistics and Econometrics, 3 edn. (1999)
28. Moore, T.J., Kozick, R.J., Sadler, B.M.: The constrained Cramér-Rao bound from the perspective of fitting a model. IEEE Signal Process. Lett. **14**(8), 564–567 (2007). https://doi.org/10.1109/LSP.2006.891316
29. Newey, W.K.: Semiparametric efficiency bounds. J. Appl. Economet. **5**(2), 99–135 (1990)
30. Paindaveine, D.: A Chernoff-Savage result for shape: on the non-admissibility of pseudo-Gaussian methods. J. Multivar. Anal. **97**(10), 2206–2220 (2006)

31. Paindaveine, D.: A canonical definition of shape. Stat. Probab. Lett. **78**(14), 2240–2247 (2008)
32. Petersen, K.B., Pedersen, M.S.: The Matrix Cookbook (2012). http://www2.imm.dtu.dk/pubdb/p.php?3274. (Version 20121115)
33. Rieder, H.: Robust Asymptotic Statistics. Springer Series in Statistics (1994)
34. Rudin, W.: Real and Complex Analysis, 1 edn. McGraw-Hill (1987)
35. Tsiatis, A.: Semiparametric Theory and Missing Data. Springer Series in Statistics (2006)
36. Tyler, D.E.: A distribution-free M-estimator of multivariate scatter. Ann. Stat. **15**(1), 234–251 (1987)
37. van der Vaart, A.W.: Asymptotic Statistics. Cambridge Series in Statistical and Probabilistic Mathematics. Cambridge University Press (1998)
38. Vermeulen, K.: Semiparametric Efficiency. Ph.D. thesis, Universiteit Gent (2011)
39. Wellner, J.A.: Semiparametric models: Progress and problems. In: ISI (ed.) Bulletin of the International Statistical Institute, 4, vol. 51 (1985)

Chapter 6
Estimation and Detection Under Misspecification and Complex Elliptically Symmetric Distributions

Christ D. Richmond and Akshay S. Bondre

Abstract Parameter estimation and detection theory play critical roles in the development of signal and array processing algorithms, especially in applications such as radar, sonar, and wireless communications. Theoretical bounds in detection and estimation are tied to the true underlying data probability distribution. In practice, however, the assumed data model is often imperfect, i.e., it is commonly misspecified at some level. Firstly, we develop the Cramér–Rao bound (CRB) under model misspecification, including a multivariate generalization of Blyth's theorem that guarantees an inequality of the Cramér–Rao type. Secondly, we explore the classic problem of adaptive radar detection under a class of complex multivariate elliptically symmetric (CMES) distributions known as the compound Gaussian. The generalized likelihood ratio test (GLRT) is derived along with its finite sample receiver operating characteristics (ROC). Lastly, the asymptotic (large sample) distribution of the GLRT in general is explored under model misspecification.

6.1 Parameter Estimation

Parameter estimation is a common objective of statistical inference and signal processing [1–3]. Consider observing a series of data observations $x_1, x_2, \ldots, x_m$ that carry information about a scalar parameter θ of interest. A model can be assumed for the data that describes it's relationship to the unknown parameter θ, and this model can be leveraged to draw some inference about the unknown parameter. The fundamental problem of parameter estimation can be stated as follows:

C. D. Richmond (✉)
Department of Electrical and Computer Engineering, Duke University,
Durham, NC, USA
e-mail: christ.richmond@duke.edu

A. S. Bondre
School of Electrical Computer and Energy Engineering, Arizona State University,
Tempe, AZ, USA
e-mail: asbondre@asu.edu

J.-P. Delmas et al. (eds.), *Elliptically Symmetric Distributions in Signal Processing and Machine Learning*, https://doi.org/10.1007/978-3-031-52116-4_6

1. Observe experimental data $\mathbf{x} = [x_1, x_2, \ldots, x_m]^T$.
2. Assume a model for the data, i.e., $\mathbf{x} \sim p(\mathbf{x}|\theta)$ where θ is a scalar parameter.
3. Use data $\mathbf{x}$ and its model to estimate unknown scalar parameter θ, i.e., to find $\widehat{\theta}(\mathbf{x})$.

Note that the estimate $\widehat{\theta}(\mathbf{x})$ is strictly a function of the observed data $\mathbf{x}$, and never depends explicitly on any true unknown parameters such as θ. This qualifies the parameter estimate $\widehat{\theta}(\mathbf{x})$ as a *statistic* [4]. One can treat parameter θ as either nonrandom/deterministic, or one can treat it as a random variable; each treatment, however, can lead to different estimators $\widehat{\theta}(\mathbf{x})$. Strategies for parameter estimation are often categorized based on this assumption. Approaches to parameter estimation are said to be *Bayesian* when θ is treated as a random variable; and approaches are said to be *non-Bayesian* when θ is treated as an unknown deterministic constant. Unless stated otherwise, the discussions in this chapter will be restricted to the latter.

It is always possible that the assumed data distribution $p(\mathbf{x}|\theta)$ differs from the true (and often unknown) data distribution. When the models differ, then the assumed model is said to be *misspecified* or *mismatched.* Model misspecification is arguably an inevitability of engineering practice. Thus, it is useful to develop theory that allows one to determine theoretical bounds on parameter estimation and performance prediction for detection under model misspecification. First, however, we will review some basics of parameter estimation under *perfect model specification*, i.e., when the assumed data model is in fact the true data model.

6.1.1 Mean Squared Error

Perhaps the most commonly used metric for quantifying parameter estimation performance is the *mean squared error (MSE)*. Given the real scalar parameter θ having estimate $\widehat{\theta}(\mathbf{x})$ where data samples $\mathbf{x}$ are continuous, the MSE is defined as

$$\text{MSE}(\theta) \stackrel{\text{def}}{=} \text{E}\{[\widehat{\theta}(\mathbf{x}) - \theta]^2\} = \int_{\Omega} [\widehat{\theta}(\mathbf{x}) - \theta]^2 \cdot p(\mathbf{x}|\theta) d\mathbf{x}, \tag{6.1}$$

i.e., the average squared displacement of the estimator $\widehat{\theta}(\mathbf{x})$ from the true value θ where Ω is the space of all data measurement vectors $\mathbf{x}$. The averaging (integration) is over values of the random data vector $\mathbf{x}$ while conditioned on true parameter value θ. Parameter estimate $\widehat{\theta}(\mathbf{x})$ is itself a random variable (as it is a function of random data) with an average value denoted as $\text{E}[\widehat{\theta}(\mathbf{x})]$. Note that the MSE can be rewritten as

$$\begin{aligned} \text{MSE}(\theta) &= \text{E}[(\widehat{\theta} - \theta)^2] = \text{E}\{(\widehat{\theta} - \text{E}[\widehat{\theta}] + \text{E}[\widehat{\theta}] - \theta)^2\} \\ &= \text{E}\{(\widehat{\theta} - \text{E}[\widehat{\theta}])^2\} + \{\text{E}[\widehat{\theta}] - \theta\}^2 + 2\text{E}\left\{(\widehat{\theta} - \text{E}[\widehat{\theta}])(\text{E}[\widehat{\theta}] - \theta)\right\} \\ &= \text{E}\{(\widehat{\theta} - \text{E}[\widehat{\theta}])^2\} + \{\text{E}[\widehat{\theta}] - \theta\}^2 \\ &\stackrel{\text{def}}{=} \sigma_{\widehat{\theta}}^2 + \{\text{Bias}[\widehat{\theta}]\}^2. \end{aligned} \tag{6.2}$$

The first term is recognized as the variance of $\widehat{\theta}(\mathbf{x})$. The second term is the square of the *bias* of the estimator $\widehat{\theta}(\mathbf{x})$. The estimator MSE is given in general by the sum of its variance and the square of its bias. If $\mathrm{E}[\widehat{\theta}(\mathbf{x})] = \theta$, then the estimator is said to be *unbiased*, and the MSE is equal to the variance of the estimator.

6.1.1.1 MSE Performance: Nonlinear Estimation

Data models in most cases depend nonlinearly on the parameters, i.e., often $p(\mathbf{x}|\theta)$ is a nonlinear function of θ. Thus, nonlinear parameter estimation is a common occurrence in practice. Consider a nonlinear parameter estimation scheme that obtains its estimate $\widehat{\theta}$ by maximizing an objective search function, say $q(\tilde{\theta}, \mathbf{x})$, over scan variable $\tilde{\theta}$, i.e., $\widehat{\theta} = \arg\max_{\tilde{\theta}} q(\tilde{\theta}, \mathbf{x})$. Such nonlinear parameter estimation often exhibits MSE performance with a characteristic behavior as illustrated in Fig. 6.1. The MSE is plotted as a function of signal-to-noise ratio (SNR) where both the MSE and SNR are in decibels (dB). As a function of SNR, there are three distinct regions of MSE performance that can be clearly seen: the no-information region, threshold region, and asymptotic region. These three regions essentially result from varying degrees of stochastic convergence. The asymptotic region corresponds to the high SNR case (i.e., as SNR$\rightarrow \infty$) where the parameter estimate $\widehat{\theta}$ has converged in probability such that it is within a small local neighborhood of the true parameter value θ. The MSE performance tends to decrease smoothly in a linear fashion (in dB space) when in this asymptotic region. As the SNR decreases below a specific value called the *threshold SNR*, the MSE performance of the nonlinear algorithm increases rapidly. This threshold effect has been studied by many researchers and methods for its prediction have been developed based on knowledge of the *ambiguity function* (e.g., see [5] and its references).

6.1.1.2 Parameter Bounds

Regarding the aforementioned typical MSE performance curve, our focus in this chapter will be developing and exploring lower bounds on MSE for the asymptotic region, i.e., finding some number (possibly dependent on parameter θ), say $\sigma^2_{LB}(\theta)$, such that an inequality such as

$$\sigma^2_{LB}(\theta) \leq \mathrm{E}\left\{[\widehat{\theta}(\mathbf{x}) - \theta]^2\right\} \tag{6.3}$$

holds for any estimator $\widehat{\theta}(\mathbf{x})$ perhaps within a specified class. Quantities such as $\sigma^2_{LB}(\theta)$ that lower bound MSE performance in the asymptotic regime are called local error *parameter bounds*. Given an unbiased estimate $\widehat{\theta}(\mathbf{x})$ with variance $\sigma^2_{\widehat{\theta}}$, the bound $\sigma^2_{LB}(\theta)$ is said to be "sharp" if $\sigma^2_{\widehat{\theta}} = \sigma^2_{LB}(\theta)$. Otherwise, when $|\sigma^2_{\widehat{\theta}} - \sigma^2_{LB}(\theta)|$ is small then the bound is said to be "tight." Parameter bounds that can be demonstrated to be at least tight have greater value in practice, because one can rely on

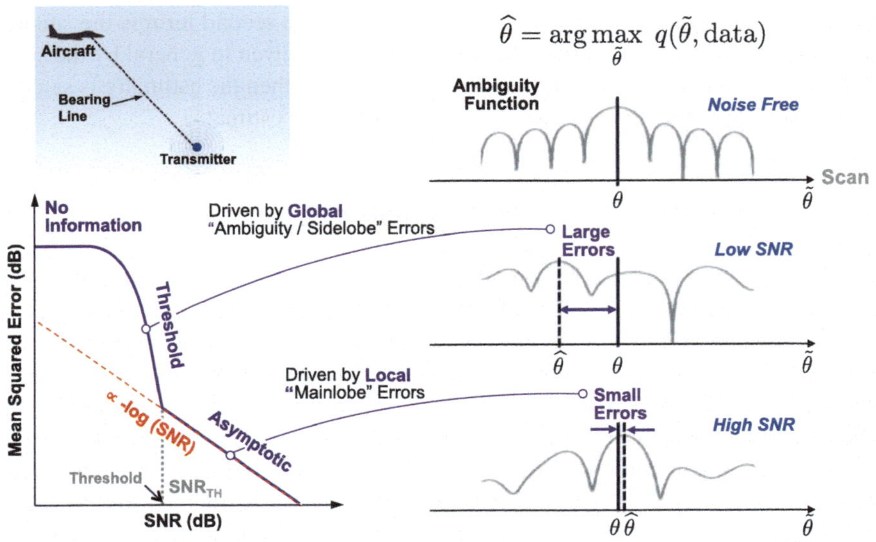

Fig. 6.1 MSE performance of nonlinear estimation

the bound to provide accurate prediction of a "good" algorithm's performance. Evaluating parameter bounds is often easier than designing an optimal algorithm and assessing the algorithm's statistical performance, usually via Monte Carlo simulations. Thus, parameter bounds can be quite valuable for engineering system design and analysis.

6.1.2 Covariance Inequality

A class of local error parameter bounds are derived from the *covariance inequality*. We will first establish this inequality in the classical sense. Later, we will revisit this inequality from the perspective of constrained minimization, as it proves to be an insightful perspective.

6.1.2.1 Scalar Case

Recall from basic probability that the correlation coefficient between two zero mean real scalar random variables, say ζ and η, is a real number whose magnitude is less than or equal to one [6]:

$$\rho_{\zeta,\eta}^2 = \frac{\mathrm{E}^2\{\zeta\eta\}}{\mathrm{E}\{\zeta^2\}\cdot\mathrm{E}\{\eta^2\}} \leq 1 \Longrightarrow \mathrm{E}\{\zeta^2\} \geq \frac{\mathrm{E}^2\{\zeta\eta\}}{\mathrm{E}\{\eta^2\}}, \tag{6.4}$$

where the last inequality is sometimes called the *covariance inequality*. Equality is obtained if and only if $\zeta \propto \eta$, i.e., there's a *linear relationship* between ζ and η [6].

6.1.2.2 Vector Case

Now consider two zero mean real random vectors $\boldsymbol{\zeta}$ and $\boldsymbol{\eta}$, and the deterministic matrix $\mathcal{A}$. Since all covariance matrices are semi-positive definite [7], it follows that

$$\mathrm{E}\{(\boldsymbol{\zeta} - \mathcal{A}\boldsymbol{\eta})(\boldsymbol{\zeta} - \mathcal{A}\boldsymbol{\eta})^T\} \geq \mathbf{0}, \tag{6.5}$$

where it is assumed that the dimensions of matrix $\mathcal{A}$ make sense for the indicated subtraction. Expanding the quadratic, we obtain

$$\begin{gathered}\mathrm{E}\{\boldsymbol{\zeta}\boldsymbol{\zeta}^T - \boldsymbol{\zeta}\boldsymbol{\eta}^T\mathcal{A}^T - \mathcal{A}\boldsymbol{\eta}\boldsymbol{\zeta}^T + \mathcal{A}\boldsymbol{\eta}\boldsymbol{\eta}^T\mathcal{A}^T\} = \\ \mathrm{E}\{\boldsymbol{\zeta}\boldsymbol{\zeta}^T\} - \mathrm{E}\{\boldsymbol{\zeta}\boldsymbol{\eta}^T\}\mathcal{A}^T - \mathcal{A}\mathrm{E}\{\boldsymbol{\eta}\boldsymbol{\zeta}^T\} + \mathcal{A}\mathrm{E}\{\boldsymbol{\eta}\boldsymbol{\eta}^T\}\mathcal{A}^T \geq \mathbf{0}.\end{gathered}$$

Assuming the matrix $\mathrm{E}\{\boldsymbol{\eta}\boldsymbol{\eta}^T\}$ to be nonsingular, we can *complete the square* in $\mathcal{A}$ to obtain

$$\begin{gathered}\mathrm{E}\{\boldsymbol{\zeta}\boldsymbol{\zeta}^T\}+ \\ (\mathcal{A} - \mathrm{E}\{\boldsymbol{\zeta}\boldsymbol{\eta}^T\}\mathrm{E}^{-1}\{\boldsymbol{\eta}\boldsymbol{\eta}^T\})\mathrm{E}\{\boldsymbol{\eta}\boldsymbol{\eta}^T\}(\mathcal{A} - \mathrm{E}\{\boldsymbol{\zeta}\boldsymbol{\eta}^T\}\mathrm{E}^{-1}\{\boldsymbol{\eta}\boldsymbol{\eta}^T\})^T \\ -\mathrm{E}\{\boldsymbol{\zeta}\boldsymbol{\eta}^T\}\mathrm{E}^{-1}\{\boldsymbol{\eta}\boldsymbol{\eta}^T\}\mathrm{E}\{\boldsymbol{\eta}\boldsymbol{\zeta}^T\} \geq \mathbf{0}.\end{gathered} \tag{6.6}$$

Choosing $\mathcal{A} = \mathrm{E}\{\boldsymbol{\zeta}\boldsymbol{\eta}^T\}\mathrm{E}^{-1}\{\boldsymbol{\eta}\boldsymbol{\eta}^T\}$, we obtain the matrix inequality

$$\mathrm{E}\{\boldsymbol{\zeta}\boldsymbol{\zeta}^T\} \geq \mathrm{E}\{\boldsymbol{\zeta}\boldsymbol{\eta}^T\}\mathrm{E}^{-1}\{\boldsymbol{\eta}\boldsymbol{\eta}^T\}\mathrm{E}\{\boldsymbol{\eta}\boldsymbol{\zeta}^T\}. \tag{6.7}$$

This Eq. (6.7) is known as the *multivariate covariance inequality*. Note that (6.7) reduces to the inequality on the right in (6.4) when $\boldsymbol{\zeta}$ and $\boldsymbol{\eta}$ are real scalars.

Next, we shall argue that

1. If $\boldsymbol{\zeta} = \mathcal{A}_0\boldsymbol{\eta}$ for some matrix $\mathcal{A}_0$, then (6.7) obtains equality.
2. If (6.7) obtains equality, then $\boldsymbol{\zeta} = \mathcal{A}_0\boldsymbol{\eta}$ for some matrix $\mathcal{A}_0$.

Regarding 1, note that $\mathrm{E}\{\boldsymbol{\zeta}\boldsymbol{\zeta}^T\} = \mathrm{E}\{\mathcal{A}_0\boldsymbol{\eta}\boldsymbol{\eta}^T\mathcal{A}_0^T\} = \mathcal{A}_0\mathrm{E}\{\boldsymbol{\eta}\boldsymbol{\eta}^T\}\mathcal{A}_0^T$, and $\mathrm{E}\{\boldsymbol{\zeta}\boldsymbol{\eta}^T\} = \mathrm{E}\{\mathcal{A}_0\boldsymbol{\eta}\boldsymbol{\eta}^T\} = \mathcal{A}_0\mathrm{E}\{\boldsymbol{\eta}\boldsymbol{\eta}^T\}$. Thus, (6.7) becomes $\mathcal{A}_0\mathrm{E}\{\boldsymbol{\eta}\boldsymbol{\eta}^T\}\mathcal{A}_0^T \geq \mathcal{A}_0\mathrm{E}\{\boldsymbol{\eta}\boldsymbol{\eta}^T\}\mathrm{E}^{-1}\{\boldsymbol{\eta}\boldsymbol{\eta}^T\}$ $E\{\boldsymbol{\eta}\boldsymbol{\eta}^T\}\mathcal{A}_0^T$, i.e., $\mathcal{A}_0\mathrm{E}\{\boldsymbol{\eta}\boldsymbol{\eta}^T\}\mathcal{A}_0^T = \mathcal{A}_0\mathrm{E}\{\boldsymbol{\eta}\boldsymbol{\eta}^T\}\mathcal{A}_0^T$.

Regarding 2, if $\mathcal{A}_0 = \mathrm{E}\{\boldsymbol{\zeta}\boldsymbol{\eta}^T\}\mathrm{E}^{-1}\{\boldsymbol{\eta}\boldsymbol{\eta}^T\}$, then note by (6.5) and (6.6) that

$$\begin{gathered}E\{(\boldsymbol{\zeta} - \mathcal{A}_0\boldsymbol{\eta})(\boldsymbol{\zeta} - \mathcal{A}_0\boldsymbol{\eta})^T\} = \\ E\{\boldsymbol{\zeta}\boldsymbol{\zeta}^T\} - E\{\boldsymbol{\zeta}\boldsymbol{\eta}^T\}E^{-1}\{\boldsymbol{\eta}\boldsymbol{\eta}^T\}E\{\boldsymbol{\eta}\boldsymbol{\zeta}^T\} \geq \mathbf{0}.\end{gathered} \tag{6.8}$$

Note that if (6.7) obtains equality, i.e., if $\mathrm{E}\{(\boldsymbol{\zeta} - \mathcal{A}_0\boldsymbol{\eta})(\boldsymbol{\zeta} - \mathcal{A}_0\boldsymbol{\eta})^T\} = \mathbf{0}$, then this means that $(\boldsymbol{\zeta} - \mathcal{A}_0\boldsymbol{\eta})$ has zero variation about its mean $\mathrm{E}\{(\boldsymbol{\zeta} - \mathcal{A}_0\boldsymbol{\eta})\} = \mathrm{E}\{\boldsymbol{\zeta}\} - \mathcal{A}_0\mathrm{E}\{\boldsymbol{\eta}\} = \mathbf{0}$, and it must equal its mean deterministically, i.e., $\boldsymbol{\zeta} - \mathcal{A}_0\boldsymbol{\eta} = \mathbf{0}$ almost surely.

Together 1 and 2 establish that the inequality in (6.7) achieves equality if and only if there exists a matrix $\mathcal{A}_0$ such that $\boldsymbol{\zeta} = \mathcal{A}_0\boldsymbol{\eta}$. Note that when $\boldsymbol{\zeta}, \boldsymbol{\eta}$ are real scalars this is equivalent to saying $\zeta \propto \eta$.

It is noteworthy that an analogous multivariate covariance inequality exists for complex vectors; namely, if $\boldsymbol{\zeta}$ and $\boldsymbol{\eta}$ are zero mean complex vectors, then we have the following statement of matrix semi-positive definiteness:

$$\mathrm{E}\{\boldsymbol{\zeta}\boldsymbol{\zeta}^H\} \geq \mathrm{E}\{\boldsymbol{\zeta}\boldsymbol{\eta}^H\}\mathrm{E}^{-1}\{\boldsymbol{\eta}\boldsymbol{\eta}^H\}\mathrm{E}\{\boldsymbol{\eta}\boldsymbol{\zeta}^H\}. \tag{6.9}$$

The proof of this complex version of the multivariate covariance inequality is the same as the real version, but requires replacing matrix transposes with conjugate transposes.

6.1.3 Cramér–Rao Bound: Real Scalar Case

There are several local error parameter bounds that derive from use of the covariance inequality. Perhaps the simplest and most widely known is the *Cramér–Rao bound (CRB)*. Given data $\mathbf{x} = [x_1, x_2, \ldots, x_m]^T \sim p(\mathbf{x}|\theta)$, one can propose a parameter estimate $\widehat{\theta}(\mathbf{x})$ for the unknown real scalar parameter θ. Let the mean value of the proposed parameter estimate be $\mathrm{E}[\widehat{\theta}(\mathbf{x})] = \mu(\theta)$, i.e., some function of the true but unknown parameter θ. Now recall inequality (6.4), and consider the following choices for ζ and η:

$$\zeta(\mathbf{x}, \theta) = \widehat{\theta}(\mathbf{x}) - \mu(\theta), \quad \text{and} \quad \eta(\mathbf{x}, \theta) = \frac{d \ln p(\mathbf{x}|\theta)}{d\theta}. \tag{6.10}$$

Clearly, $\mathrm{E}[\zeta] = 0$; it also follows that $\mathrm{E}[\eta] = 0$ by noting

$$\begin{aligned} \mathrm{E}\left\{\frac{d \ln p(\mathbf{x}|\theta)}{d\theta}\right\} &= \int \frac{d \ln p(\mathbf{x}|\theta)}{d\theta} \cdot p(\mathbf{x}|\theta)d\mathbf{x} = \int \frac{1}{p(\mathbf{x}|\theta)} \frac{dp(\mathbf{x}|\theta)}{d\theta} \cdot p(\mathbf{x}|\theta)d\mathbf{x} \\ &= \int \frac{dp(\mathbf{x}|\theta)}{d\theta} d\mathbf{x} = \frac{d}{d\theta} \int p(\mathbf{x}|\theta)d\mathbf{x} = \frac{d}{d\theta}(1) = 0, \end{aligned} \tag{6.11}$$

where the fourth equality follows if the order of integration and differentiation can be reversed (a formal set of conditions allowing such reversal will be given in Sect. 6.1.6.5). The choice of ζ, η in (6.10), when inserted into inequality (6.4), leads to

$$\mathrm{E}\{(\widehat{\theta}-\mu(\theta))^2\} \geq \frac{\mathrm{E}^2\left\{(\widehat{\theta}-\mu(\theta))\dfrac{d\ln p}{d\theta}\right\}}{\mathrm{E}\left\{\left(\dfrac{d\ln p}{d\theta}\right)^2\right\}}. \tag{6.12}$$

The left-hand side of this inequality we recognize as the variance of the estimator $\widehat{\theta}(\mathbf{x})$. The form of the right-hand side of inequality (6.12) is almost useful, but is somewhat lacking. Indeed, it is lacking because the right-hand side also depends on the estimator, namely via $(\widehat{\theta}(\mathbf{x})-\mu(\theta))$. If the right-hand side can be written such that it does not depend on the estimator, then this could be a very useful expression providing a lower bound on the variance of any estimator with mean $\mu(\theta)$. Specifically, we desire to express (6.12) as a lower bound inequality of the form

$$\mathrm{E}\{(\widehat{\theta}-\mu(\theta))^2\} \geq \frac{b^2(\theta)}{\mathrm{E}\left\{\left(\dfrac{d\ln p}{d\theta}\right)^2\right\}}, \quad \text{where } b(\theta) \text{ is some function that only depends on } \theta \text{ at most.} \tag{6.13}$$

This would be a much more useful form that clearly lower bounds the variance of any estimator $\widehat{\theta}(\mathbf{x})$ with mean $\mu(\theta)$. When the numerator on the right-hand side of (6.12) (sometimes called the square of the *translation*) can be *resolved* into the form shown on the right-hand side of (6.13), the resulting lower bound inequality is said to yield an *inequality of the Cramér–Rao type (ICRT)* [4, 8, 9].

Note that the square root of the numerator in (6.12), i.e., the translation, can be written as

$$\begin{aligned}
\mathrm{E}\left\{(\widehat{\theta}-\mu(\theta))\frac{d\ln p}{d\theta}\right\} &= \int(\widehat{\theta}(\mathbf{x})-\mu(\theta))\frac{d\ln p(\mathbf{x}|\theta)}{d\theta}p(\mathbf{x}|\theta)d\mathbf{x} \\
&= \int\widehat{\theta}(\mathbf{x})\frac{d\ln p(\mathbf{x}|\theta)}{d\theta}p(\mathbf{x}|\theta)d\mathbf{x} - \mu(\theta)\int\frac{d\ln p(\mathbf{x}|\theta)}{d\theta}p(\mathbf{x}|\theta)d\mathbf{x} \\
&= \int\widehat{\theta}(\mathbf{x})\frac{1}{p(\mathbf{x}|\theta)}\frac{dp(\mathbf{x}|\theta)}{d\theta}p(\mathbf{x}|\theta)d\mathbf{x} - \mu(\theta)\int\frac{1}{p(\mathbf{x}|\theta)}\frac{dp(\mathbf{x}|\theta)}{d\theta}p(\mathbf{x}|\theta)d\mathbf{x} \\
&= \int\widehat{\theta}(\mathbf{x})\frac{dp(\mathbf{x}|\theta)}{d\theta}d\mathbf{x} - \mu(\theta)\int\frac{dp(\mathbf{x}|\theta)}{d\theta}d\mathbf{x}.
\end{aligned}$$

If the order of integration and differentiation can be reversed, then the first integral becomes

$$\int\widehat{\theta}(\mathbf{x})\frac{dp(\mathbf{x}|\theta)}{d\theta}d\mathbf{x} = \frac{d}{d\theta}\int\widehat{\theta}(\mathbf{x})p(\mathbf{x}|\theta)d\mathbf{x} = \frac{d}{d\theta}\mu(\theta);$$

and second integral becomes

$$\mu(\theta)\int \frac{dp(\mathbf{x}|\theta)}{d\theta}d\mathbf{x} = \mu(\theta)\cdot\frac{d}{d\theta}\int p(\mathbf{x}|\theta)d\mathbf{x} = \mu(\theta)\cdot\frac{d}{d\theta}(1) = 0.$$

Thus, the translation can be written as

$$\mathrm{E}\left\{(\widehat{\theta} - \mu(\theta))\frac{d\ln p}{d\theta}\right\} = \frac{d\mu(\theta)}{d\theta},$$

and inequality (6.12) can be expressed as

$$\mathrm{E}\{(\widehat{\theta} - \mu(\theta))^2\} \geq \frac{\mathrm{E}^2\left\{(\widehat{\theta} - \mu(\theta))\frac{d\ln p}{d\theta}\right\}}{\mathrm{E}\left\{\left(\frac{d\ln p}{d\theta}\right)^2\right\}} = \frac{\left(\frac{d\mu(\theta)}{d\theta}\right)^2}{\mathrm{E}\left\{\left(\frac{d\ln p}{d\theta}\right)^2\right\}} \overset{\text{def}}{=} \sigma^2_{CRB}(\theta), \quad (6.14)$$

which is clearly of the desired form targeted in (6.13). The expression on the right-hand side of (6.14) is known as Cramér–Rao bound (CRB). It provides a useful lower bound on the variance of all estimators $\widehat{\theta}(\mathbf{x})$ with mean $\mu(\theta)$. The CRB in general is applicable only to estimators sharing the mean $\mu(\theta)$.

The case in which $\mu(\theta) = \theta$, i.e., such that $\mathrm{E}\{\widehat{\theta}(\mathbf{x})\} = \theta$, represents the case for unbiased estimators of parameter θ. Clearly, for this case, $d\mu(\theta)/d\theta = 1$ and (6.14) simplifies to

$$\mathrm{E}\{(\widehat{\theta} - \theta)^2\} \geq \frac{1}{\mathrm{E}\left\{\left(\frac{d\ln p}{d\theta}\right)^2\right\}} = \sigma^2_{CRB}(\theta). \quad (6.15)$$

The denominator is known as the *Fisher information [matrix]* (FIM), i.e.,

$$\mathrm{FIM}(\theta) = \mathrm{E}\left\{\left(\frac{d\ln p(\mathbf{x}|\theta)}{d\theta}\right)^2\right\} = -\mathrm{E}\left\{\frac{d^2\ln p(\mathbf{x}|\theta)}{d\theta^2}\right\}, \quad (6.16)$$

where the last equality is an equivalent form [1, 2]. Thus, the CRB for unbiased estimates of parameter θ is given by the inverse FIM, i.e., $\sigma^2_{CRB}(\theta) = \mathrm{FIM}^{-1}(\theta)$; This inverse FIM provides a lower bound on the variance (and MSE) of all unbiased estimators of θ.

6.1.3.1 Score Functions: On Best Practices and Bound Tightness

Recall that the covariance inequality in (6.4) was used to establish the CRB. Our choice of $\eta(\mathbf{x}, \theta) = \dfrac{d \ln p(\mathbf{x}|\theta)}{d\theta}$, however, appears somewhat arbitrary, but fortuitously resulted in a translation that resolved and an ICRT was obtained. Other choices for $\eta(\mathbf{x}, \theta)$, nonetheless, are possible. The function $\eta(\mathbf{x}, \theta)$ is sometimes called the *score function* in statistics [2, 3]. Varying the choice for the score function $\eta(\mathbf{x}, \theta)$ can result in other valid useful lower bounds on estimator variance, provided one can resolve the translation (e.g., the Barankin bound and Bhattacharyya bound are examples that used different choices for $\eta(\mathbf{x}, \theta)$ [10]).

If we again let $\zeta(\mathbf{x}, \theta) = \widehat{\theta}(\mathbf{x}) - \theta$, then the inequality (6.4) for *arbitrary* $\eta(\mathbf{x}, \theta)$ yields the following:

$$\mathrm{E}\{(\widehat{\theta} - \mu(\theta))^2\} \geq \frac{\mathrm{E}^2\left\{(\widehat{\theta} - \mu(\theta))\eta(\theta)\right\}}{\mathrm{E}\left\{[\eta(\theta)]^2\right\}} = \frac{b^2(\theta)}{\mathrm{E}\left\{[\eta(\theta)]^2\right\}}, \tag{6.17}$$

where the last equality is again the desired form, i.e., an ICRT.

Interestingly, Blyth and Roberts [8] show that if $\eta(\mathbf{x}, \theta)$ is chosen to be a function of the sufficient statistic for estimating θ, then the translation in (6.17) can be resolved to lead to an ICRT. It turns out, however, that a more general theorem regarding the properties of the score function $\eta(\mathbf{x}, \theta)$ exists that guarantees an ICRT. The theorem is called Blyth's theorem and we will discuss it in the next section.

While obtaining an ICRT is a desired goal, it is also desired that the lower bound on the variance obtained is as tight as possible. McWhorter and Scharf [11] showed that choosing the score function $\eta(\mathbf{x}, \theta)$ such that it possesses the following two properties will improve the tightness of the resulting bound: (i) zero mean, i.e., $E\{\eta(\mathbf{x}, \theta)\} = 0$, and (ii) depends on the sufficient statistic for estimating θ.

6.1.3.2 Blyth's Theorem: Score Function and ICRT

The following theorem was established by Blyth that identifies a property of a score function that is guaranteed to resolve the translation and result in an ICRT.

Theorem 6.1 (Blyth's Theorem: univariate under perfect model specification) *If a score function $\eta(\mathbf{x}, \theta)$ has the property that, for every statistic $\widehat{\theta}_Z(\mathbf{x})$ where* $\mathrm{E}\{\widehat{\theta}_Z(\mathbf{x})\} = 0$, *it is orthogonal, i.e.,* $\mathrm{E}\{\widehat{\theta}_Z(\mathbf{x})\eta(\mathbf{x}, \theta)\} = 0$, *then an ICRT is guaranteed, i.e., there exists a function $b(\theta)$ such that*

$$\mathrm{E}\left\{(\widehat{\theta} - \mu(\theta))\eta(\theta)\right\} = b(\theta) \;\; \textit{for all}\; \widehat{\theta}(\mathbf{x}) \;\textit{with mean}\; \mu(\theta).$$

Proof Let $\widehat{\theta}_1(\mathbf{x})$ and $\widehat{\theta}_2(\mathbf{x})$ be any two estimators with a mean of $\mu(\theta)$. Define $\widehat{\theta}_Z(\mathbf{x}) \stackrel{\text{def}}{=} \widehat{\theta}_1(\mathbf{x}) - \widehat{\theta}_2(\mathbf{x})$ and note that $\mathrm{E}\{\widehat{\theta}_Z(\mathbf{x})\} = \mathrm{E}\{\widehat{\theta}_1(\mathbf{x})\} - \mathrm{E}\{\widehat{\theta}_2(\mathbf{x})\} = \mu(\theta) - \mu(\theta) = 0$. Since $\mathrm{E}\{\widehat{\theta}_Z(\mathbf{x})\eta(\mathbf{x}, \theta)\} = 0$ for every zero mean statistic $\widehat{\theta}_Z(\mathbf{x})$, it follows that

$$\begin{aligned}
\mathrm{E}\{\widehat{\theta}_Z(\mathbf{x})\eta(\mathbf{x},\theta)\} &= \mathrm{E}\left\{(\widehat{\theta}_1 - \widehat{\theta}_2)\eta(\theta)\right\} = 0 \\
\mathrm{E}\left\{(\widehat{\theta}_1 - \mu(\theta) - \widehat{\theta}_2 + \mu(\theta))\eta(\theta)\right\} &= 0 \\
\mathrm{E}\left\{(\widehat{\theta}_1 - \mu(\theta))\eta(\theta) - (\widehat{\theta}_2 - \mu(\theta))\eta(\theta)\right\} &= 0 \\
\mathrm{E}\left\{(\widehat{\theta}_1 - \mu(\theta))\eta(\theta)\right\} - \mathrm{E}\left\{(\widehat{\theta}_2 - \mu(\theta))\eta(\theta)\right\} &= 0 \\
\Longrightarrow \mathrm{E}\left\{(\widehat{\theta}_1 - \mu(\theta))\eta(\theta)\right\} &= \mathrm{E}\left\{(\widehat{\theta}_2 - \mu(\theta))\eta(\theta)\right\}.
\end{aligned} \tag{6.18}$$

Since this last equality must be true for *any* two estimators $\widehat{\theta}_1(\mathbf{x}), \widehat{\theta}_2(\mathbf{x})$ with mean $\mu(\theta)$, it must be true for all estimators $\widehat{\theta}(\mathbf{x})$ with mean $\mu(\theta)$. Thus, there must exist a single function $b(\theta) = \mathrm{E}\left\{(\widehat{\theta} - \mu(\theta))\eta(\theta)\right\}$ for all estimators $\widehat{\theta}(\mathbf{x})$ having mean $\mu(\theta)$, i.e., the translation resolves and the resulting inequality obtained will be one of an ICRT. ■

It can be shown that the score function used to obtain the CRB, i.e., $\eta(\mathbf{x}, \theta) = \dfrac{d \ln p(\mathbf{x}|\theta)}{d\theta}$, satisfies the conditions of Blyth's theorem (see [4], pp. 122–123). We will show in Sect. 6.1.6.3 that Blyth's theorem can be generalized to the vector parameter case under model misspecification.

6.1.4 MSE Covariance and Vector Parameters

Multiple parameters occur quite commonly in science and engineering; hence, the desire to jointly estimate them. For example, in radar applications, it is desired to estimate several parameters of a point target scatterer in space; namely, the target's range, Doppler, and angle (azimuth and/or elevation), i.e., joint estimation of the vector $[r, f_d, \phi]^T$. Consider a set of multiple parameters, say $\boldsymbol{\theta} = [\theta_1, \theta_2, \ldots, \theta_J]^T$, that one desires to jointly estimate from the data $\mathbf{x}$. The primary strategy of statistical inference is to assume a model for the data, say that $\mathbf{x} \sim p(\mathbf{x}|\boldsymbol{\theta})$, and to use this model to find a vector estimate $\widehat{\boldsymbol{\theta}}(\mathbf{x}) = [\widehat{\theta}_1(\mathbf{x}), \widehat{\theta}_2(\mathbf{x}), \ldots, \widehat{\theta}_J(\mathbf{x})]^T$. Other strategies to estimate parameters are also possible, but ideally, we'd like to establish fundamental limits on how well any algorithm can estimate parameters $\boldsymbol{\theta}$.

Let the mean value of the proposed vector parameter estimate be denoted

$$\mathrm{E}\{\widehat{\boldsymbol{\theta}}(\mathbf{x})\} = \int \widehat{\boldsymbol{\theta}}(\mathbf{x}) p(\mathbf{x}|\boldsymbol{\theta}) d\mathbf{x} = \boldsymbol{\mu}(\boldsymbol{\theta}) = [\mu_1(\boldsymbol{\theta}), \mu_2(\boldsymbol{\theta}), \ldots, \mu_J(\boldsymbol{\theta})]^T. \tag{6.19}$$

If $\boldsymbol{\mu}(\boldsymbol{\theta}) = \boldsymbol{\theta}$, then the estimator is said to be unbiased. The estimator covariance matrix is therefore $\mathrm{E}\{(\widehat{\boldsymbol{\theta}} - \boldsymbol{\mu}(\boldsymbol{\theta}))(\widehat{\boldsymbol{\theta}} - \boldsymbol{\mu}(\boldsymbol{\theta}))^T\}$. The variance of the i-th element of the vector estimate $\widehat{\boldsymbol{\theta}}(\mathbf{x})$ is given by the diagonals, i.e.,

$$\begin{aligned}
\left[\mathrm{E}\{(\widehat{\boldsymbol{\theta}} - \boldsymbol{\mu}(\boldsymbol{\theta}))(\widehat{\boldsymbol{\theta}} - \boldsymbol{\mu}(\boldsymbol{\theta}))^T\}\right]_{i,i} &= \mathbf{e}_i^T \mathrm{E}\{(\widehat{\boldsymbol{\theta}} - \boldsymbol{\mu}(\boldsymbol{\theta}))(\widehat{\boldsymbol{\theta}} - \boldsymbol{\mu}(\boldsymbol{\theta}))^T\}\mathbf{e}_i \\
&= \mathrm{E}\{(\widehat{\theta}_i - \mu_i(\boldsymbol{\theta}))^2\},
\end{aligned} \tag{6.20}$$

where $\mathbf{e}_i = [0, 0, \ldots, 0, 1, 0, \ldots, 0]^T$ is a standard basis vector; i.e., with i-th element equal to unity and zero otherwise. As in the scalar case, we desire to identify a lower bound on the variance of each element of the vector estimate, i.e., to find $\sigma^2_{LB,i}(\boldsymbol{\theta})$ such that

$$\mathrm{E}\{(\widehat{\theta}_i - \mu_i(\boldsymbol{\theta}))^2\} \geq \sigma^2_{LB,i}(\boldsymbol{\theta}). \tag{6.21}$$

More generally, we seek a $J \times J$ semi-positive matrix $\widetilde{\boldsymbol{\Sigma}}(\boldsymbol{\theta})$ such that for any vector $\mathbf{q} \in \mathbb{R}^J$

$$\mathbf{q}^T \mathrm{E}\{(\widehat{\boldsymbol{\theta}} - \boldsymbol{\mu}(\boldsymbol{\theta}))(\widehat{\boldsymbol{\theta}} - \boldsymbol{\mu}(\boldsymbol{\theta}))^T\}\mathbf{q} \geq \mathbf{q}^T \widetilde{\boldsymbol{\Sigma}}(\boldsymbol{\theta})\mathbf{q}, \tag{6.22}$$

where clearly the choice $\mathbf{q} = \mathbf{e}_i$ results in a lower bound on the variance of the i-th component of any estimate $\widehat{\boldsymbol{\theta}}(\mathbf{x})$ with mean $\boldsymbol{\mu}(\boldsymbol{\theta})$. Equation (6.22) is a more general statement of matrix semi-positive definiteness for the error covariance of estimator $\widehat{\boldsymbol{\theta}}(\mathbf{x})$.

6.1.5 Cramér–Rao Bound: Real Vector Case

A multivariate version of the scalar CRB in (6.14) can be established via use of the multivariate covariance inequality (6.7). Recall that the data is assumed distributed as $\mathbf{x} \sim p(\mathbf{x}|\boldsymbol{\theta})$, and an estimator $\widehat{\boldsymbol{\theta}}(\mathbf{x})$ with mean $\boldsymbol{\mu}(\boldsymbol{\theta})$ is proposed. Consider the choices[1]

$$\boldsymbol{\zeta}(\mathbf{x}, \boldsymbol{\theta}) = \widehat{\boldsymbol{\theta}}(\mathbf{x}) - \boldsymbol{\mu}(\boldsymbol{\theta}), \quad \text{and} \quad \boldsymbol{\eta}(\mathbf{x}, \boldsymbol{\theta}) = \frac{\partial \ln p(\mathbf{x}|\boldsymbol{\theta})}{\partial \boldsymbol{\theta}}. \tag{6.23}$$

Clearly, $\mathrm{E}\{\boldsymbol{\zeta}(\mathbf{x}, \boldsymbol{\theta})\} = \mathrm{E}\{\widehat{\boldsymbol{\theta}}(\mathbf{x})\} - \boldsymbol{\mu}(\boldsymbol{\theta}) = \boldsymbol{\mu}(\boldsymbol{\theta}) - \boldsymbol{\mu}(\boldsymbol{\theta}) = \mathbf{0}$. The steps in (6.11) can be applied element-wise to $\boldsymbol{\eta}(\mathbf{x}, \boldsymbol{\theta})$ to show that $\mathrm{E}\{\boldsymbol{\eta}(\mathbf{x}, \boldsymbol{\theta})\} = \mathbf{0}$. The choices for $\boldsymbol{\zeta}$ and $\boldsymbol{\eta}$ in (6.23), when inserted into (6.7) assuming $\mathrm{E}\{\boldsymbol{\eta}\boldsymbol{\eta}^T\}$ is nonsingular, result in the following matrix inequality:

$$\mathrm{E}\left\{(\widehat{\boldsymbol{\theta}} - \boldsymbol{\mu}(\boldsymbol{\theta}))(\widehat{\boldsymbol{\theta}} - \boldsymbol{\mu}(\boldsymbol{\theta}))^T\right\} \geq$$
$$\mathrm{E}\left\{(\widehat{\boldsymbol{\theta}} - \boldsymbol{\mu}(\boldsymbol{\theta}))\frac{\partial \ln p}{\partial \boldsymbol{\theta}}^T\right\} \mathrm{E}^{-1}\left\{\frac{\partial \ln p}{\partial \boldsymbol{\theta}}\frac{\partial \ln p}{\partial \boldsymbol{\theta}}^T\right\} \mathrm{E}\left\{\frac{\partial \ln p}{\partial \boldsymbol{\theta}}(\widehat{\boldsymbol{\theta}} - \boldsymbol{\mu}(\boldsymbol{\theta}))^T\right\}. \tag{6.24}$$

[1] The gradient of a scalar-valued function $a(\boldsymbol{\theta})$ is defined as the column vector $\frac{\partial a(\boldsymbol{\theta})}{\partial \boldsymbol{\theta}} = \left[\frac{\partial a(\boldsymbol{\theta})}{\partial \theta_1}, \frac{\partial a(\boldsymbol{\theta})}{\partial \theta_2}, \ldots, \frac{\partial a(\boldsymbol{\theta})}{\partial \theta_J}\right]^T$. The gradient of $W \times 1$ vector-valued multivariate function $\mathbf{a}(\boldsymbol{\theta}) = [a_1(\boldsymbol{\theta}), a_2(\boldsymbol{\theta}), \ldots, a_W(\boldsymbol{\theta})]^T$ is defined as the $J \times W$ matrix $\frac{\partial \mathbf{a}(\boldsymbol{\theta})}{\partial \boldsymbol{\theta}} \overset{\text{def}}{=} \left[\frac{\partial a_1(\boldsymbol{\theta})}{\partial \boldsymbol{\theta}} \middle| \frac{\partial a_2(\boldsymbol{\theta})}{\partial \boldsymbol{\theta}} \middle| \cdots \middle| \frac{\partial a_W(\boldsymbol{\theta})}{\partial \boldsymbol{\theta}}\right]$.

As in the scalar case, we desire that the right-hand side of this inequality be independent of the actual estimator $\widehat{\boldsymbol{\theta}}(\mathbf{x})$, i.e., we desire an ICRT. Note that the (i, j)-th component of the translation for the multivariate case is given by

$$\left(\mathrm{E}\left\{(\widehat{\boldsymbol{\theta}} - \boldsymbol{\mu}(\boldsymbol{\theta}))\frac{\partial \ln p}{\partial \boldsymbol{\theta}}^T\right\}\right)_{i,j} = \mathrm{E}\left\{(\widehat{\theta_i} - \mu_i(\boldsymbol{\theta}))\frac{\partial \ln p}{\partial \theta_j}\right\}$$
$$= \int (\widehat{\theta_i}(\mathbf{x}) - \mu_i(\boldsymbol{\theta}))\frac{\partial \ln p(\mathbf{x}|\boldsymbol{\theta})}{\partial \theta_j} p(\mathbf{x}|\boldsymbol{\theta})d\mathbf{x}$$
$$= \int \widehat{\theta_i}(\mathbf{x})\frac{\partial \ln p(\mathbf{x}|\boldsymbol{\theta})}{\partial \theta_j} p(\mathbf{x}|\boldsymbol{\theta})d\mathbf{x} - \mu_i(\boldsymbol{\theta})\int \frac{\partial \ln p(\mathbf{x}|\boldsymbol{\theta})}{\partial \theta_j} p(\mathbf{x}|\boldsymbol{\theta})d\mathbf{x} \tag{6.25}$$
$$= \int \widehat{\theta_i}(\mathbf{x})\frac{1}{p(\mathbf{x}|\boldsymbol{\theta})}\frac{\partial p(\mathbf{x}|\boldsymbol{\theta})}{\partial \theta_j} p(\mathbf{x}|\boldsymbol{\theta})d\mathbf{x} - \mu_i(\boldsymbol{\theta})\int \frac{1}{p(\mathbf{x}|\boldsymbol{\theta})}\frac{\partial p(\mathbf{x}|\boldsymbol{\theta})}{\partial \theta_j} p(\mathbf{x}|\boldsymbol{\theta})d\mathbf{x}$$
$$= \int \widehat{\theta_i}(\mathbf{x})\frac{\partial p(\mathbf{x}|\boldsymbol{\theta})}{\partial \theta_j}d\mathbf{x} - \mu_i(\boldsymbol{\theta})\int \frac{\partial p(\mathbf{x}|\boldsymbol{\theta})}{\partial \theta_j}d\mathbf{x}.$$

If order reversal of integration and differentiation is justified, then we note that

$$\mathrm{E}\left\{(\widehat{\theta_i} - \mu_i(\boldsymbol{\theta}))\frac{\partial \ln p}{\partial \theta_j}\right\} = \frac{\partial}{\partial \theta_j}\int \widehat{\theta_i}(\mathbf{x})p(\mathbf{x}|\boldsymbol{\theta})d\mathbf{x} - \mu_i(\boldsymbol{\theta})\frac{\partial}{\partial \theta_j}\int p(\mathbf{x}|\boldsymbol{\theta})d\mathbf{x}$$
$$= \frac{\partial \mu_i(\boldsymbol{\theta})}{\partial \theta_j} - \mu_i(\boldsymbol{\theta})\frac{\partial}{\partial \theta_j}(1) = \frac{\partial \mu_i(\boldsymbol{\theta})}{\partial \theta_j}, \tag{6.26}$$

i.e., each element of the translation matrix resolves (to be independent of $\widehat{\boldsymbol{\theta}}(\mathbf{x})$). We define the Fisher information matrix (FIM) as

$$\mathbf{FIM}(\boldsymbol{\theta}) \stackrel{\text{def}}{=} \mathrm{E}\left\{\frac{\partial \ln p(\mathbf{x}|\boldsymbol{\theta})}{\partial \boldsymbol{\theta}}\frac{\partial \ln p(\mathbf{x}|\boldsymbol{\theta})}{\partial \boldsymbol{\theta}}^T\right\}. \tag{6.27}$$

This is a multivariate generalization of the scalar parameter case in (6.16). Element-wise this matrix is given by

$$(\mathbf{FIM}(\boldsymbol{\theta}))_{i,j} = \mathrm{E}\left\{\frac{\partial \ln p(\mathbf{x}|\boldsymbol{\theta})}{\partial \theta_i}\frac{\partial \ln p(\mathbf{x}|\boldsymbol{\theta})}{\partial \theta_j}\right\} = -\mathrm{E}\left\{\frac{\partial^2 \ln p(\mathbf{x}|\boldsymbol{\theta})}{\partial \theta_i \partial \theta_j}\right\}, \tag{6.28}$$

where the last equality can be established in a similar way as (6.16) was shown. Thus, the matrix inequality (6.24) can be written as

$$\mathrm{E}\left\{(\widehat{\boldsymbol{\theta}} - \boldsymbol{\mu}(\boldsymbol{\theta}))(\widehat{\boldsymbol{\theta}} - \boldsymbol{\mu}(\boldsymbol{\theta}))^T\right\} \geq \left(\frac{\partial \boldsymbol{\mu}(\boldsymbol{\theta})}{\partial \boldsymbol{\theta}}\right)^T \mathbf{FIM}^{-1}(\boldsymbol{\theta})\left(\frac{\partial \boldsymbol{\mu}(\boldsymbol{\theta})}{\partial \boldsymbol{\theta}}\right). \tag{6.29}$$

This is the multiparameter Cramér–Rao bound for the covariance matrix of all estimators with a mean of $\boldsymbol{\mu}(\boldsymbol{\theta})$. Equation (6.29) is a statement of matrix semi-positive definiteness from which it follows that

$$\begin{aligned}
&\mathbf{e}_i^T \mathrm{E}\left\{(\widehat{\boldsymbol{\theta}} - \boldsymbol{\mu}(\boldsymbol{\theta}))(\widehat{\boldsymbol{\theta}} - \boldsymbol{\mu}(\boldsymbol{\theta}))^T\right\} \mathbf{e}_i \geq \\
&\mathbf{e}_i^T \left(\frac{\partial \boldsymbol{\mu}(\boldsymbol{\theta})}{\partial \boldsymbol{\theta}}\right)^T \mathbf{FIM}^{-1}(\boldsymbol{\theta}) \left(\frac{\partial \boldsymbol{\mu}(\boldsymbol{\theta})}{\partial \boldsymbol{\theta}}\right) \mathbf{e}_i \\
\Longrightarrow \mathrm{E}\{(\widehat{\theta}_i - \mu_i(\boldsymbol{\theta}))^2\} \geq &\left(\frac{\partial \mu_i(\boldsymbol{\theta})}{\partial \boldsymbol{\theta}}\right)^T \mathbf{FIM}^{-1}(\boldsymbol{\theta}) \left(\frac{\partial \mu_i(\boldsymbol{\theta})}{\partial \boldsymbol{\theta}}\right),
\end{aligned} \tag{6.30}$$

where the diagonal elements of the matrices on each side of the inequality yield the desired lower bounds on the variance of each element of the parameter vector estimator $\widehat{\boldsymbol{\theta}}(\mathbf{x})$.

The case when $\boldsymbol{\mu}(\boldsymbol{\theta}) = \boldsymbol{\theta}$ represents the class of unbiased estimators, and it follows that $\dfrac{\partial \boldsymbol{\mu}(\boldsymbol{\theta})}{\partial \boldsymbol{\theta}} = \mathbf{I}_J$, and by (6.29)

$$\mathrm{E}\left\{(\widehat{\boldsymbol{\theta}} - \boldsymbol{\theta})(\widehat{\boldsymbol{\theta}} - \boldsymbol{\theta})^T\right\} \geq \mathbf{FIM}^{-1}(\boldsymbol{\theta}), \tag{6.31}$$

i.e., the CRB lower bound is given by the inverse of the FIM. This is the multiparameter generalization of (6.15). Hence, the lower bound on the variance of unbiased estimates follows from

$$\begin{aligned}
\mathbf{e}_i^T \mathrm{E}\left\{(\widehat{\boldsymbol{\theta}} - \boldsymbol{\theta})(\widehat{\boldsymbol{\theta}} - \boldsymbol{\theta})^T\right\} \mathbf{e}_i &\geq \mathbf{e}_i^T \mathbf{FIM}^{-1}(\boldsymbol{\theta}) \mathbf{e}_i \\
\Longrightarrow \mathrm{E}\{(\widehat{\theta}_i - \theta_i)^2\} &\geq (\mathbf{FIM}^{-1}(\boldsymbol{\theta}))_{i,i},
\end{aligned} \tag{6.32}$$

i.e., the diagonals of the inverse FIM provide lower bounds on the variance (and MSE) of all unbiased estimates of parameter $\boldsymbol{\theta}$.

It is noteworthy that, while we have tacitly assumed the FIM is nonsingular and therefore invertible, all CRB matrix inequalities actually hold when the inverse FIM is replaced with a *generalized inverse* such as the *Moore–Penrose inverse* of the FIM [12]. Thus, bounds on parameter estimation can be established even though the FIM is singular. A singular FIM is usually an indication that there exists some redundancy in the chosen parameter assignment, e.g., for some index k there may exist some function $g(\cdot)$ such that $\theta_k = g(\theta_1, \theta_2, \ldots, \theta_{k-1}, \theta_{k+1}, \theta_{k+2}, \ldots, \theta_J)$; and hence, parameter θ_k provides no new information.

6.1.5.1 Slepian–Bangs Formula for the FIM

When the data $\mathbf{x}$ is multivariate Gaussian distributed, then a simple formula can be established for the elements of the FIM, i.e., for (6.27). Specifically, let data be distributed such that $\mathbf{x} \sim \mathbb{R}N_m(\boldsymbol{\mu}(\boldsymbol{\theta}), \boldsymbol{\Sigma}) = p(\mathbf{x}|\boldsymbol{\theta})$, i.e., parameters $\boldsymbol{\theta}$ only parameterize the mean of the Gaussian distribution via $\boldsymbol{\mu}(\boldsymbol{\theta}) = [\mu_1(\boldsymbol{\theta}), \mu_2(\boldsymbol{\theta}), \ldots, \mu_m(\boldsymbol{\theta})]^T$. The covariance $\mathbf{\Sigma}$ does not depend on $\boldsymbol{\theta}$ in this case. It can be shown that the elements of the FIM are given by the following simple formula:

$$(\mathbf{FIM}(\boldsymbol{\theta}))_{i,j} = \left(\frac{\partial \boldsymbol{\mu}(\boldsymbol{\theta})}{\partial \theta_i}\right)^T \mathbf{\Sigma}^{-1} \left(\frac{\partial \boldsymbol{\mu}(\boldsymbol{\theta})}{\partial \theta_j}\right). \tag{6.33}$$

This formula is attributed to David Slepian [13] (called the Slepian formula).

When $\mathbf{x} \sim \mathbb{R}N_m(\boldsymbol{\mu}, \mathbf{\Sigma}(\boldsymbol{\theta}))$, i.e., parameters $\boldsymbol{\theta}$ only parameterize the covariance $\mathbf{\Sigma}(\boldsymbol{\theta})$, and $\boldsymbol{\mu}$ does not depend on $\boldsymbol{\theta}$, then another simple formula is possible:

$$(\mathbf{FIM}(\boldsymbol{\theta}))_{i,j} = \frac{1}{2}\mathrm{tr}\left[\mathbf{\Sigma}^{-1}(\boldsymbol{\theta})\frac{\partial \mathbf{\Sigma}(\boldsymbol{\theta})}{\partial \theta_i}\mathbf{\Sigma}^{-1}(\boldsymbol{\theta})\frac{\partial \mathbf{\Sigma}(\boldsymbol{\theta})}{\partial \theta_j}\right], \tag{6.34}$$

where the above differentiation of $\mathbf{\Sigma}(\boldsymbol{\theta})$ is done element-wise. This formula is attributed to William Bangs [14] (called Bangs' Formula).

When $\mathbf{x} \sim \mathbb{R}N_m(\boldsymbol{\mu}(\boldsymbol{\theta}), \mathbf{\Sigma}(\boldsymbol{\theta}))$, i.e., parameters $\boldsymbol{\theta}$ parameterize both the mean and covariance, i.e., $\boldsymbol{\mu} = \boldsymbol{\mu}(\boldsymbol{\theta})$, and $\mathbf{\Sigma} = \mathbf{\Sigma}(\boldsymbol{\theta})$, respectively, then a simple formula for the FIM elements results:

$$\begin{aligned}(\mathbf{FIM}(\boldsymbol{\theta}))_{i,j} &= \left(\frac{\partial \boldsymbol{\mu}(\boldsymbol{\theta})}{\partial \theta_i}\right)^T \mathbf{\Sigma}^{-1}(\boldsymbol{\theta}) \left(\frac{\partial \boldsymbol{\mu}(\boldsymbol{\theta})}{\partial \theta_j}\right) \\ &+\frac{1}{2}\mathrm{tr}\left[\mathbf{\Sigma}^{-1}(\boldsymbol{\theta})\frac{\partial \mathbf{\Sigma}(\boldsymbol{\theta})}{\partial \theta_i}\mathbf{\Sigma}^{-1}(\boldsymbol{\theta})\frac{\partial \mathbf{\Sigma}(\boldsymbol{\theta})}{\partial \theta_j}\right],\end{aligned} \tag{6.35}$$

where clearly each element is given by the sum of the Slepian formula and the Bangs formula. This resulting FIM formula is called the Slepian–Bangs formula (see [15] for proof).

6.1.6 Parameter Bounds Under Model Misspecification

When the assumed probability distribution for the measured data differs from the true distribution, the data model is said to be misspecified or mismatched. Researchers have investigated the impact of model misspecification on algorithm performance, and established parameter bounds under such conditions. Most notably, Huber [16] and White [17] have characterized the asymptotic performance of maximum-likelihood estimation under model misspecification, and Vuong [18] has established a generalization of the CRB under model misspecification. These ideas have been revisited more recently in references [19–22]. Our goal in this section is to develop the theory of the misspecified CRB along the lines of more classical treatments seen under perfect model specification, as in [4].

Let the data $\mathbf{x}$ be complex valued in general, i.e., $\mathbf{x} \in \mathbb{C}^m$, and let the assumed data distribution be $p(\mathbf{x}|\boldsymbol{\theta})$, and let the true data distribution be characterized by the probability density function $q(\mathbf{x})$ where $p(\mathbf{x}|\boldsymbol{\theta}) \neq q(\mathbf{x})\ \forall\ \boldsymbol{\theta}$ is allowed. The choice to exclude any parameterization of $q(\cdot)$ by $\boldsymbol{\theta}$ allows for the possibility that the model parameterization could be incorrect; although the results we obtain will allow for $q(\cdot)$ parameterized by $\boldsymbol{\theta}$, but we must assume no knowledge nor access to these parameters of $q(\cdot)$. The parameters $\boldsymbol{\theta} = [\theta_1, \theta_2, \ldots, \theta_J]^T$ used to parameterize the assumed density $p(\mathbf{x}|\boldsymbol{\theta})$ can be either real and/or complex valued in this discussion, but the conjugates of each complex member will also be included to facilitate use of Wirtinger calculus [23] that is the basis of conjugate gradient methods [24, 25]. Specifically, if $\boldsymbol{\theta}_r \in \mathbb{R}^{J_r}$ and $\boldsymbol{\theta}_c \in \mathbb{C}^{J_c}$ are, respectively, the real and complex parameters of interest, then the total parameter set is $\boldsymbol{\theta} = [\boldsymbol{\theta}_r; \boldsymbol{\theta}_c; \boldsymbol{\theta}_c^*]$[2]. and $J = J_r + 2J_c$. Given a data measurement $\mathbf{x}$ and knowledge of the assumed probability density function $p(\mathbf{x}|\boldsymbol{\theta})$, an estimate of the parameters, say $\widehat{\boldsymbol{\theta}}(\mathbf{x}) = [\widehat{\theta}_1(\mathbf{x}), \widehat{\theta}_2(\mathbf{x}), \ldots, \widehat{\theta}_J(\mathbf{x})]^T$, can be proposed. Consider the class of all such estimators sharing a common mean with respect to $q(\mathbf{x})$, i.e., such that $\mathrm{E}_q\{\widehat{\boldsymbol{\theta}}(\mathbf{x})\} = \boldsymbol{\mu}_q = [\mu_{q,1}, \mu_{q,2}, \ldots, \mu_{q,J}]^T$ where the subscript q on the expectation reminds that all statistical averaging will be with respect to the true distribution. The estimator covariance matrix is therefore $\mathrm{E}_q\{(\widehat{\boldsymbol{\theta}} - \boldsymbol{\mu}_q)(\widehat{\boldsymbol{\theta}} - \boldsymbol{\mu}_q)^H\}$. The variance of the estimate for the i-th component of $\boldsymbol{\theta}$ is given by the diagonals, i.e.,

$$\begin{aligned}\left(\mathrm{E}_q\{(\widehat{\boldsymbol{\theta}} - \boldsymbol{\mu}_q)(\widehat{\boldsymbol{\theta}} - \boldsymbol{\mu}_q)^H\}\right)_{i,i} &= \mathbf{e}_i^T \mathrm{E}_q\{(\widehat{\boldsymbol{\theta}} - \boldsymbol{\mu}_q)(\widehat{\boldsymbol{\theta}} - \boldsymbol{\mu}_q)^H\}\mathbf{e}_i \\ &= \mathrm{E}_q\{|\widehat{\theta}_i - \mu_{q,i}|^2\}.\end{aligned} \tag{6.36}$$

Our goal is to identify a lower bound on the variance of the components of any such estimator, i.e., to find $\sigma^2_{LB,i}(\boldsymbol{\theta})$ such that $\mathrm{E}_q\{|\widehat{\theta}_i - \mu_{q,i}|^2\} \geq \sigma^2_{LB,i}(\boldsymbol{\theta})$. More generally, we seek a Hermitian semi-positive definite matrix $\widetilde{\boldsymbol{\Sigma}}(\boldsymbol{\theta})$ such that for any vector $\mathbf{q} \in \mathbb{C}^J$

$$\mathbf{q}^H \mathrm{E}_q\{(\widehat{\boldsymbol{\theta}} - \boldsymbol{\mu}_q)(\widehat{\boldsymbol{\theta}} - \boldsymbol{\mu}_q)^H\}\mathbf{q} \geq \mathbf{q}^H \widetilde{\boldsymbol{\Sigma}}(\boldsymbol{\theta})\mathbf{q}, \tag{6.37}$$

where clearly the choice $\mathbf{q} = \mathbf{e}_i$ identifies lower bounds on the variance of the components. The next section will establish a theorem that will be helpful in obtaining such matrix inequalities.

6.1.6.1 Minimum Norm Theorem

The *minimum norm theorem (MNT)* provides an alternative method to establish lower bounds on parameter estimation [26, 27] and lends insight that is quite useful under model misspecification. It is an attractive framework under model misspecification primarily due to its explicit statement of constraints.

---

[2] We denote $[\mathbf{a}; \mathbf{b}] = [\mathbf{a}^T, \mathbf{b}^T]^T$, where $\mathbf{a}$ and $\mathbf{b}$ are vectors.

The MNT is applicable to any valid *inner product space* [7]. Let $\langle \cdot, \cdot \rangle$ denote the inner product between scalar functions of $\mathbf{x}$. The following is a statement of the MNT.

Theorem 6.2 (Minimum Norm) *Let $\tilde{\zeta}(\mathbf{x})$ and $\tilde{\eta}_1(\mathbf{x}, \boldsymbol{\theta})$, $\tilde{\eta}_2(\mathbf{x}, \boldsymbol{\theta})$, ..., $\tilde{\eta}_K(\mathbf{x}, \boldsymbol{\theta})$ be a set of real and/or complex-valued linearly independent scalar functions defined on the domain of $\mathbf{x}$ and $(\mathbf{x}, \boldsymbol{\theta})$, respectively. Let $K \times 1$ vector $\boldsymbol{\varrho}(\boldsymbol{\theta}) = [\varrho_1(\boldsymbol{\theta}), \varrho_2(\boldsymbol{\theta}), \ldots, \varrho_K(\boldsymbol{\theta})]^T$ be a set of real and/or complex-valued scalar functions defined on the domain of $\boldsymbol{\theta}$. The Gramian matrix $\mathbf{G}$ for function set $\tilde{\eta}_k(\mathbf{x}, \boldsymbol{\theta})$ is given by*

$$(\mathbf{G})_{i,k} \stackrel{def}{=} \langle \tilde{\eta}_i(\mathbf{x}, \boldsymbol{\theta}), \tilde{\eta}_k(\mathbf{x}, \boldsymbol{\theta}) \rangle^*, \quad i, k = 1, 2, \ldots, K. \tag{6.38}$$

The minimum value of the norm squared $||\tilde{\zeta}(\mathbf{x})||^2 = \langle \tilde{\zeta}(\mathbf{x}), \tilde{\zeta}(\mathbf{x}) \rangle$ achievable among all possible functions $\tilde{\zeta}(\mathbf{x})$ when subject to the K inner product constraints

$$\langle \tilde{\zeta}(\mathbf{x}), \tilde{\eta}_k(\mathbf{x}, \boldsymbol{\theta}) \rangle = \varrho_k(\boldsymbol{\theta}), \quad k = 1, 2, \ldots, K \tag{6.39}$$

is given by $\boldsymbol{\varrho}^H(\boldsymbol{\theta})\mathbf{G}^{-1}\boldsymbol{\varrho}(\boldsymbol{\theta})$ (where functions $\tilde{\eta}_k(\cdot, \cdot)$ and $\varrho_k(\cdot)$, $k = 1, 2, \ldots, K$ are fixed).

Since the MNT is applicable to any valid inner product space, the following proof is constructed without defining the inner product space, only assuming that it is valid. The proof is a mild adaptation (to the complex case) of the proof given in Appendix B of [27].

Proof Consider a set $\mathcal{U}$ with a defined inner product $\langle \cdot, \cdot \rangle$ between any two members of set $\mathcal{U}$, and consider subspace $\mathcal{G} \subseteq \mathcal{U}$ given by the span of K linearly independent members $\boldsymbol{\psi}_1, \boldsymbol{\psi}_2, \ldots, \boldsymbol{\psi}_K$ where the goal is to find the member $\mathbf{u} \in \mathcal{U}$ with the smallest norm when subject to K inner product constraints:

$$\min_{\mathbf{u} \in \mathcal{U}} \|\mathbf{u}\|^2 \text{ such that } \langle \mathbf{u}, \boldsymbol{\psi}_k \rangle = \varrho_k, \quad k = 1, 2, \ldots, K. \tag{6.40}$$

Let set $\mathcal{G}_\perp \subseteq \mathcal{U}$ be the orthogonal complement of set $\mathcal{G}$. Thus, $\text{span}(\mathcal{G} \cup \mathcal{G}_\perp) = \mathcal{U}$, or equivalently $\mathcal{G} \oplus \mathcal{G}_\perp = \mathcal{U}$; and if $\mathbf{a} \in \mathcal{G}$ and $\mathbf{b} \in \mathcal{G}_\perp$, then $\langle \mathbf{a}, \mathbf{b} \rangle = 0$ for all such $\mathbf{a}$ and $\mathbf{b}$. Note that $\mathbf{u}$ can be decomposed such that $\mathbf{u} = \mathbf{u}_\mathcal{G} + \mathbf{u}_{\mathcal{G}_\perp}$ where $\mathbf{u}_\mathcal{G} \in \mathcal{G}$ and $\mathbf{u}_{\mathcal{G}_\perp} \in \mathcal{G}_\perp$. The inner product constraints can be written $\langle \mathbf{u}, \boldsymbol{\psi}_k \rangle = \langle \mathbf{u}_\mathcal{G} + \mathbf{u}_{\mathcal{G}_\perp}, \boldsymbol{\psi}_k \rangle = \langle \mathbf{u}_\mathcal{G}, \boldsymbol{\psi}_k \rangle + \langle \mathbf{u}_{\mathcal{G}_\perp}, \boldsymbol{\psi}_k \rangle = \varrho_k$; but $\langle \mathbf{u}_{\mathcal{G}_\perp}, \boldsymbol{\psi}_k \rangle = 0$ due to orthogonality. Thus, these constraints can only be met by the component $\mathbf{u}_\mathcal{G}$. Since $\mathbf{u}_\mathcal{G} \in \mathcal{G}$, it can be represented as $\mathbf{u}_\mathcal{G} = \sum_{i=1}^K \alpha_i \boldsymbol{\psi}_i$ where α_i, $i = 1, 2, \ldots, K$ are the expansion coefficients. Note by properties of inner products that

$$\langle \mathbf{u}_\mathcal{G}, \boldsymbol{\psi}_k \rangle = \left\langle \sum_{i=1}^K \alpha_i \boldsymbol{\psi}_i, \boldsymbol{\psi}_k \right\rangle = \sum_{i=1}^K \alpha_i \langle \boldsymbol{\psi}_i, \boldsymbol{\psi}_k \rangle = \varrho_k, \quad k = 1, 2, \ldots, K. \tag{6.41}$$

Since for any complex member $\mathbf{a}$ belonging to $\mathcal{U}$ the defined inner product must yield $\langle \mathbf{a}, \mathbf{a} \rangle = \|\mathbf{a}\|^2 \geq 0$, the inner product must introduce a complex conjugation to one of its arguments. Here we will assume that for inner product $\langle \mathbf{a}, \mathbf{b} \rangle$ the required conjugation will operate on the elements of $\mathbf{b}$. Hence, for any complex members $\mathbf{a}, \mathbf{b}$ both belonging to $\mathcal{U}$ and inner product $\langle \cdot, \cdot \rangle \in \mathbb{C}$, we have that $\langle \mathbf{a}, \mathbf{b} \rangle = \langle \mathbf{b}, \mathbf{a} \rangle^*$. Similarly, we will have that $\langle \gamma \mathbf{a}, \beta \mathbf{b} \rangle = \gamma \beta^* \langle \mathbf{a}, \mathbf{b} \rangle$ for a valid inner product space. Define $K \times K$ matrix $\mathbf{G}$ such that $(\mathbf{G})_{k,i} \stackrel{\text{def}}{=} \langle \boldsymbol{\psi}_i, \boldsymbol{\psi}_k \rangle = \langle \boldsymbol{\psi}_k, \boldsymbol{\psi}_i \rangle^*$, $i, k = 1, 2, \ldots, K$, and vectors $\boldsymbol{\varrho} \stackrel{\text{def}}{=} [\varrho_1, \varrho_2, \ldots, \varrho_K]^T$, and $\boldsymbol{\alpha} \stackrel{\text{def}}{=} [\alpha_1, \alpha_2, \ldots, \alpha_K]^T$. Clearly, $\mathbf{G} = \mathbf{G}^H$, i.e., it is a conjugate symmetric matrix. Constraints in (6.41) can be written as the linear matrix equation $\mathbf{G}\boldsymbol{\alpha} = \boldsymbol{\varrho}$ where the expansion coefficients are given by the solution $\boldsymbol{\alpha}_\star \stackrel{\text{def}}{=} \mathbf{G}^{-1}\boldsymbol{\varrho}$. Thus, the component of $\mathbf{u}$ in $\mathcal{G}$ is given by $\mathbf{u}_\mathcal{G} = \sum_{i=1}^K (\boldsymbol{\alpha}_\star)_i \cdot \boldsymbol{\psi}_i$; note that this completely determines $\mathbf{u}_\mathcal{G}$, i.e., both its magnitude and direction. The magnitude will satisfy the following:

$$
\begin{aligned}
\|\mathbf{u}_\mathcal{G}\|^2 = \langle \mathbf{u}_\mathcal{G}, \mathbf{u}_\mathcal{G} \rangle &= \left\langle \sum_{i=1}^K (\boldsymbol{\alpha}_\star)_i \cdot \boldsymbol{\psi}_i, \sum_{i=1}^K (\boldsymbol{\alpha}_\star)_i \cdot \boldsymbol{\psi}_i \right\rangle \\
= \sum_{i=1}^K \sum_{k=1}^K (\boldsymbol{\alpha}_\star)_i \cdot (\boldsymbol{\alpha}_\star)_k^* \cdot \langle \boldsymbol{\psi}_i, \boldsymbol{\psi}_k \rangle &= \sum_{i=1}^K \sum_{k=1}^K (\boldsymbol{\alpha}_\star)_i \cdot (\boldsymbol{\alpha}_\star)_k^* \cdot (\mathbf{G})_{k,i} \\
&= \sum_{i=1}^K \sum_{k=1}^K (\boldsymbol{\alpha}_\star)_i \cdot (\boldsymbol{\alpha}_\star)_k^* \cdot \left(\mathbf{G}^*\right)_{i,k} = \boldsymbol{\alpha}_\star^T \mathbf{G}^* \boldsymbol{\alpha}_\star^* \\
&= \boldsymbol{\alpha}_\star^H \mathbf{G} \boldsymbol{\alpha}_\star = \boldsymbol{\varrho}^H \mathbf{G}^{-1} \mathbf{G} \mathbf{G}^{-1} \boldsymbol{\varrho} = \boldsymbol{\varrho}^H \mathbf{G}^{-1} \boldsymbol{\varrho},
\end{aligned}
\tag{6.42}
$$

where third equality follows from properties of a valid inner product space; the fifth equality follows from properties of the Gramian; and the seventh equality follows because $\|\cdot\|^2$ is always real and non-negative, hence, it equals to it's conjugate. Note that

$$
\begin{aligned}
\|\mathbf{u}\|^2 &= \|\mathbf{u}_\mathcal{G} + \mathbf{u}_{\mathcal{G}_\perp}\|^2 = \langle \mathbf{u}_\mathcal{G} + \mathbf{u}_{\mathcal{G}_\perp}, \mathbf{u}_\mathcal{G} + \mathbf{u}_{\mathcal{G}_\perp} \rangle \\
&= \langle \mathbf{u}_\mathcal{G}, \mathbf{u}_\mathcal{G} \rangle + \langle \mathbf{u}_\mathcal{G}, \mathbf{u}_{\mathcal{G}_\perp} \rangle + \langle \mathbf{u}_{\mathcal{G}_\perp}, \mathbf{u}_\mathcal{G} \rangle + \langle \mathbf{u}_{\mathcal{G}_\perp}, \mathbf{u}_{\mathcal{G}_\perp} \rangle \\
&= \langle \mathbf{u}_\mathcal{G}, \mathbf{u}_\mathcal{G} \rangle + \langle \mathbf{u}_{\mathcal{G}_\perp}, \mathbf{u}_{\mathcal{G}_\perp} \rangle \\
&= \|\mathbf{u}_\mathcal{G}\|^2 + \|\mathbf{u}_{\mathcal{G}_\perp}\|^2,
\end{aligned}
\tag{6.43}
$$

where fourth equality follows from $\langle \mathbf{u}_{\mathcal{G}_\perp}, \mathbf{u}_\mathcal{G} \rangle = \langle \mathbf{u}_\mathcal{G}, \mathbf{u}_{\mathcal{G}_\perp} \rangle = 0$. Recall that the objective is to find $\mathbf{u}$ with the smallest norm. Since component $\mathbf{u}_\mathcal{G}$ is completely determined, and therefore fixed by the constraints, the only degree of freedom available to minimize the norm of $\mathbf{u}$ is the choice of component $\mathbf{u}_{\mathcal{G}_\perp}$. The best one can do is choose $\mathbf{u}_{\mathcal{G}_\perp} = \mathbf{0}$ such that $\|\mathbf{u}_{\mathcal{G}_\perp}\|^2 = 0$. Hence, it follows that

$$
\min_{\mathbf{u} \in \{\mathcal{U} \mid \langle \mathbf{u}, \boldsymbol{\psi}_k \rangle = \varrho_k,\ k = 1, 2, \ldots, K\}} \|\mathbf{u}\|^2 \geq \|\mathbf{u}_\mathcal{G}\|^2 = \boldsymbol{\varrho}^H \mathbf{G}^{-1} \boldsymbol{\varrho}. \tag{6.44}
$$

The geometry of this argument is illustrated conceptually in Fig. 6.2 where $\mathcal{G}$ is the red table top "x-y" plane; $\mathcal{G}_\perp$ is the vertical "z" axis; and $\mathcal{U}$ is the entire "x-y-z" space. ■

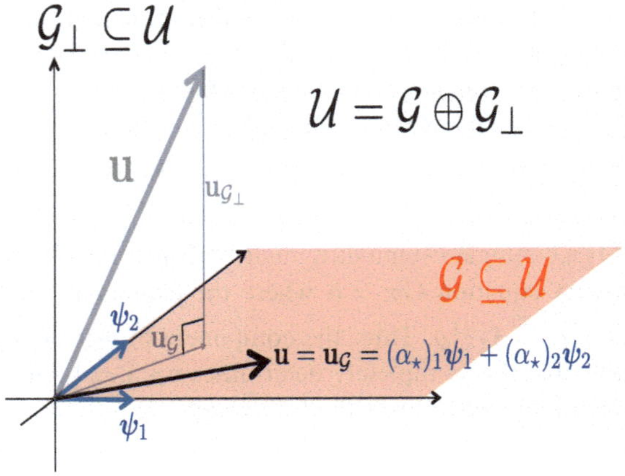

Fig. 6.2 Geometry of minimum norm theorem

6.1.6.2 Multivariate Covariance Inequality via MNT

The goal of this section is to use the MNT to establish lower bounds of the form of (6.37) for the multivariate parameter case, i.e., $J \geq 1$, under model misspecification, i.e., when $p(\mathbf{x}|\boldsymbol{\theta}) \neq q(\mathbf{x})\ \forall\, \boldsymbol{\theta}$. We will first use the MNT to establish the multivariate covariance inequality (6.9), and then explore implications and meanings of the chosen constraints.

Consider the set of all complex scalar-valued statistics $\mathcal{U} = \{u(\mathbf{x}) \in \mathbb{C}\}$. For $a(\mathbf{x}), b(\mathbf{x}) \in \mathcal{U}$ define our inner product as

$$\langle a(\mathbf{x}), b(\mathbf{x})\rangle \stackrel{\text{def}}{=} \mathrm{E}_q\{a(\mathbf{x})b^*(\mathbf{x})\} = \int a(\mathbf{x})b^*(\mathbf{x})q(\mathbf{x})d\mathbf{x}. \tag{6.45}$$

Clearly, $\|u(\mathbf{x})\|^2 = \langle u(\mathbf{x}), u(\mathbf{x})\rangle = \mathrm{E}_q\{|u(\mathbf{x})|^2\}$, i.e., the length of objects in $\mathcal{U}$ is given by the average squared magnitude. Recall from Sect. 6.1.3 that in our effort to develop parameter bounds for the real scalar case, i.e., $J = 1$, we chose $\zeta(\mathbf{x}) = \widehat{\theta}(\mathbf{x}) - \mu_\theta$ and $\eta(\mathbf{x}, \theta)$ as the score function. Considering the vector case (with real and/or complex elements), i.e., with $J \geq 1$, it is useful to define vectors $\boldsymbol{\zeta}(\mathbf{x}) = \widehat{\boldsymbol{\theta}}(\mathbf{x}) - \boldsymbol{\mu}_q$, and $\boldsymbol{\eta}(\mathbf{x}, \boldsymbol{\theta}) = [\eta_1(\mathbf{x}, \boldsymbol{\theta}), \eta_2(\mathbf{x}, \boldsymbol{\theta}), \ldots, \eta_L(\mathbf{x}, \boldsymbol{\theta})]^T$, as well as their arbitrary linear combinations, i.e., the complex scalar-valued functions $\tilde{\zeta}(\mathbf{x}) = \mathbf{q}_1^H\boldsymbol{\zeta}(\mathbf{x})$, and $\tilde{\eta}(\mathbf{x}, \boldsymbol{\theta}) = \mathbf{q}_2^H\boldsymbol{\eta}(\mathbf{x}, \boldsymbol{\theta})$ where $\mathbf{q}_1 \in \mathbb{C}^J$ and $\mathbf{q}_2 \in \mathbb{C}^L$. Note that we are free to choose vectors $\mathbf{q}_1$ and $\mathbf{q}_2$ as we like; it will become clear what choices are most useful. Since the tightest possible bounds are desired, we focus here only on zero mean score functions, i.e., such that $\mathrm{E}_q\{\boldsymbol{\eta}(\mathbf{x}, \boldsymbol{\theta})\} = \mathbf{0}$. Consider the following constrained minimization problem defined on this set $\mathcal{U}$:

$$\min_{\tilde{\zeta} \in \mathcal{U}} \|\tilde{\zeta}(\mathbf{x})\|^2 \ \text{ such that } \begin{array}{c} \langle \tilde{\zeta}(\mathbf{x}), 1\rangle = 0, \\ \langle \tilde{\zeta}(\mathbf{x}), \tilde{\eta}(\mathbf{x})\rangle = \varrho. \end{array} \tag{6.46}$$

This formulation is motivated by observing that the left-hand side of inequality (6.37) is in fact synonymous with $\|\tilde{\zeta}(\mathbf{x})\|^2$ when $\mathbf{q} = \mathbf{q}_1$. The two constraints may appear a bit mysterious until we unwrap them, but we will do that momentarily.

By the MNT, it follows that

$$\|\tilde{\zeta}\|^2 \geq [0, \varrho^*] \begin{bmatrix} \langle 1, 1 \rangle & \langle 1, \tilde{\eta} \rangle^* \\ \langle 1, \tilde{\eta} \rangle & \langle \tilde{\eta}, \tilde{\eta} \rangle \end{bmatrix}^{-1} \begin{bmatrix} 0 \\ \varrho \end{bmatrix} = \frac{|\varrho|^2}{\|\tilde{\eta}\|^2} = \frac{\left|\mathbf{q}_1^H \mathrm{E}_q\{\boldsymbol{\zeta}\boldsymbol{\eta}^H\}\mathbf{q}_2\right|^2}{\mathbf{q}_2^H \mathrm{E}_q\{\boldsymbol{\eta}\boldsymbol{\eta}^H\}\mathbf{q}_2}, \quad (6.47)$$

where the equalities follow from:

$$\begin{aligned} \langle 1, \tilde{\eta} \rangle^* &= [\mathrm{E}_q\{1 \cdot (\mathbf{q}_2^H \boldsymbol{\eta})^*\}]^* = (\mathrm{E}_q\{\boldsymbol{\eta}^H \mathbf{q}_2\})^* \\ &= (\mathrm{E}_q\{\boldsymbol{\eta}\}^H \mathbf{q}_2)^* = \mathbf{q}_2^H \mathrm{E}_q\{\boldsymbol{\eta}\} = \mathbf{q}_2^H \mathbf{0} = 0. \end{aligned} \quad (6.48)$$

Of course, this implies that $\langle 1, \tilde{\eta} \rangle = 0$, and in (6.47), we are inverting a diagonal matrix;

$$\begin{aligned} \langle \tilde{\eta}, \tilde{\eta} \rangle &= \mathrm{E}_q\{|\tilde{\eta}|^2\} = \mathrm{E}_q\{\mathbf{q}_2^H \boldsymbol{\eta} \cdot (\mathbf{q}_2^H \boldsymbol{\eta})^*\} = \mathrm{E}_q\{\mathbf{q}_2^H \boldsymbol{\eta}\boldsymbol{\eta}^H \mathbf{q}_2\} \\ &= \mathbf{q}_2^H \mathrm{E}_q\{\boldsymbol{\eta}\boldsymbol{\eta}^H\}\mathbf{q}_2; \end{aligned} \quad (6.49)$$

$$\begin{aligned} \langle \tilde{\zeta}, \tilde{\eta} \rangle &= \mathrm{E}_q\{\tilde{\zeta}\tilde{\eta}^*\} = \mathrm{E}_q\{\mathbf{q}_1^H \boldsymbol{\zeta} \cdot (\mathbf{q}_2^H \boldsymbol{\eta})^*\} = \mathrm{E}_q\{\mathbf{q}_1^H \boldsymbol{\zeta}\boldsymbol{\eta}^H \mathbf{q}_2\} \\ &= \mathbf{q}_1^H \mathrm{E}_q\{\boldsymbol{\zeta}\boldsymbol{\eta}^H\}\mathbf{q}_2 = \varrho, \end{aligned} \quad (6.50)$$

where the last equality follows from the second constraint in (6.46). Since we are allowing for arbitrary choices of $\mathbf{q}_1, \mathbf{q}_2$ and $\boldsymbol{\theta}, \boldsymbol{\mu}_q$, any constraint must likewise depend on these vectors; i.e., $\varrho = \varrho(\boldsymbol{\theta}, \boldsymbol{\mu}_q, \mathbf{q}_1, \mathbf{q}_2)$. Since we can choose $\mathbf{q}_2$ arbitrarily, we will choose it to maximize the right side of inequality (6.47); this strategy should lead to tighter lower bounds. Let $\mathbf{b} = \mathrm{E}_q^{1/2}\{\boldsymbol{\eta}\boldsymbol{\eta}^H\}\mathbf{q}_2$ and note that

$$\frac{\left|\mathbf{q}_1^H \mathrm{E}_q\{\boldsymbol{\zeta}\boldsymbol{\eta}^H\}\mathbf{q}_2\right|^2}{\mathbf{q}_2^H \mathrm{E}_q\{\boldsymbol{\eta}\boldsymbol{\eta}^H\}\mathbf{q}_2} = \frac{\left|\mathbf{q}_0^H \mathbf{b}\right|^2}{\mathbf{b}^H \mathbf{b}} = \left|\mathbf{q}_0^H \frac{\mathbf{b}}{\|\mathbf{b}\|}\right|^2 \leq \|\mathbf{q}_0\|^2, \quad (6.51)$$

where $\mathbf{q}_0 = \mathrm{E}_q^{-1/2}\{\boldsymbol{\eta}\boldsymbol{\eta}^H\}\mathrm{E}_q\{\boldsymbol{\eta}\boldsymbol{\zeta}^H\}\mathbf{q}_1$. The first and second equalities in (6.51) recognize the form as an Euclidean dot product between complex vector $\mathbf{q}_0$ and a complex unit vector in the direction of $\mathbf{b}$. Here we have tacitly assumed that matrix $\mathrm{E}_q\{\boldsymbol{\eta}\boldsymbol{\eta}^H\}$ is nonsingular. Thus, there's a one-to-one mapping between $\mathbf{q}_2$ and $\mathbf{b}$, i.e., maximizing over $\mathbf{b}$ is equivalent to maximizing over $\mathbf{q}_2$. It is clear that, to maximize, we must choose $\mathbf{b}$ such that $\mathbf{b} \propto \mathbf{q}_0$; hence, the desired lower bound is maximized and given by $\|\mathbf{q}_0\|^2 = \mathbf{q}_0^H \mathbf{q}_0$. This leads to the optimal choice for $\mathbf{q}_2$ as

$$\mathbf{q}_2 = \mathrm{E}_q^{-1/2}\{\boldsymbol{\eta}\boldsymbol{\eta}^H\}\mathbf{q}_0 = \mathrm{E}_q^{-1}\{\boldsymbol{\eta}\boldsymbol{\eta}^H\}\mathrm{E}_q\{\boldsymbol{\eta}\boldsymbol{\zeta}^H\}\mathbf{q}_1 \quad (6.52)$$

for which the right side of inequality (6.47) becomes $\mathbf{q}_1^H \mathrm{E}_q\{\boldsymbol{\zeta}\boldsymbol{\eta}^H\}\mathrm{E}_q^{-1}\{\boldsymbol{\eta}\boldsymbol{\eta}^H\}\mathrm{E}_q\{\boldsymbol{\eta}\boldsymbol{\zeta}^H\}\mathbf{q}_1$. Regarding the left side of inequality (6.47), note that

$$\begin{aligned}\|\tilde{\zeta}\|^2 = \langle\tilde{\zeta},\tilde{\zeta}\rangle = \mathrm{E}_q\{|\tilde{\zeta}|^2\} = \mathrm{E}_q\{\mathbf{q}_1^H\boldsymbol{\zeta}\cdot(\mathbf{q}_1^H\boldsymbol{\zeta})^*\} = \mathrm{E}_q\{\mathbf{q}_1^H\boldsymbol{\zeta}\boldsymbol{\zeta}^H\mathbf{q}_1\}\\ = \mathbf{q}_1^H\mathrm{E}_q\{\boldsymbol{\zeta}\boldsymbol{\zeta}^H\}\mathbf{q}_1.\end{aligned} \tag{6.53}$$

Thus, MNT inequality (6.47) can be written as

$$\begin{aligned}\|\tilde{\zeta}\|^2 = \mathbf{q}_1^H\mathrm{E}_q\{\boldsymbol{\zeta}\boldsymbol{\zeta}^H\}\mathbf{q}_1 \geq \mathbf{q}_1^H\mathrm{E}_q\{\boldsymbol{\zeta}\boldsymbol{\eta}^H\}\mathrm{E}_q^{-1}\{\boldsymbol{\eta}\boldsymbol{\eta}^H\}\mathrm{E}_q\{\boldsymbol{\eta}\boldsymbol{\zeta}^H\}\mathbf{q}_1,\\ \text{for all } \mathbf{q}_1 \in \mathbb{C}^J;\end{aligned} \tag{6.54}$$

this is recognized as a statement of matrix semi-positive definiteness and has the desired form of (6.37). This relationship is usually written simply as

$$\mathrm{E}_q\{\boldsymbol{\zeta}\boldsymbol{\zeta}^H\} \geq \mathrm{E}_q\{\boldsymbol{\zeta}\boldsymbol{\eta}^H\}\mathrm{E}_q^{-1}\{\boldsymbol{\eta}\boldsymbol{\eta}^H\}\mathrm{E}_q\{\boldsymbol{\eta}\boldsymbol{\zeta}^H\} \tag{6.55}$$

that we recognize as the *multivariate covariance inequality*, but under possible model misspecification. Recalling our specific choice $\boldsymbol{\zeta} = \widehat{\boldsymbol{\theta}}(\mathbf{x}) - \boldsymbol{\mu}_q$, we have the inequality

$$\mathrm{E}_q\{(\widehat{\boldsymbol{\theta}} - \boldsymbol{\mu}_q)(\widehat{\boldsymbol{\theta}} - \boldsymbol{\mu}_q)^H\} \geq \mathrm{E}_q\{(\widehat{\boldsymbol{\theta}} - \boldsymbol{\mu}_q)\boldsymbol{\eta}^H\}\mathrm{E}_q^{-1}\{\boldsymbol{\eta}\boldsymbol{\eta}^H\}\mathrm{E}_q\{\boldsymbol{\eta}(\widehat{\boldsymbol{\theta}} - \boldsymbol{\mu}_q)^H\} \tag{6.56}$$

that provides a lower bound on the covariance of any estimator $\widehat{\boldsymbol{\theta}}(\mathbf{x})$ having mean $\boldsymbol{\mu}_q$. Although from (6.9) we already knew that relationship (6.55) was valid and therefore (6.56), by obtaining the same inequality via the MNT proves that this covariance inequality is the solution to a constrained minimization problem, where the constraints can be stated explicitly; namely, the inner product constraints $\langle\tilde{\zeta}, 1\rangle = 0$ and $\langle\tilde{\zeta}, \tilde{\eta}\rangle = \varrho$. We will interpret these constraints in the next section.

6.1.6.3 MNT Constraints and Multivariate Covariance Inequality

The first constraint used in the MNT formulation (6.46) is intended to enforce the assumption that estimator $\widehat{\boldsymbol{\theta}}(\mathbf{x})$ has mean $\boldsymbol{\mu}_q$. Note that the inner product is given by

$$\begin{aligned}\langle\tilde{\zeta}, 1\rangle = \mathrm{E}_q\{\mathbf{q}_1^H\boldsymbol{\zeta}\cdot 1\} = \mathbf{q}_1^H\mathrm{E}_q\{\widehat{\boldsymbol{\theta}} - \boldsymbol{\mu}_q\} = \mathbf{q}_1^H(\mathrm{E}_q\{\widehat{\boldsymbol{\theta}}\} - \boldsymbol{\mu}_q) = 0\\ \text{for all } \mathbf{q}_1 \in \mathbb{C}^J.\end{aligned} \tag{6.57}$$

Clearly, this constraint is satisfied for all $\mathbf{q}_1 \in \mathbb{C}^J$ only if $\mathrm{E}_q\{\widehat{\boldsymbol{\theta}}\} = \boldsymbol{\mu}_q$.

The inner product defining the second constraint was shown in (6.50) to be given by $\langle\tilde{\zeta}, \tilde{\eta}\rangle = \mathrm{E}_q\{\tilde{\zeta}\tilde{\eta}^*\} = \mathbf{q}_1^H\mathrm{E}_q\{\boldsymbol{\zeta}\boldsymbol{\eta}^H\}\mathbf{q}_2$. Recall that we've already chosen a value for $\mathbf{q}_2$ in (6.52); using this value for $\mathbf{q}_2$, the second constraint takes on the form

$$\begin{aligned}\mathbf{q}_1^H\mathrm{E}_q\{\boldsymbol{\zeta}\boldsymbol{\eta}^H\}\mathrm{E}_q^{-1}\{\boldsymbol{\eta}\boldsymbol{\eta}^H\}\mathrm{E}_q\{\boldsymbol{\eta}\boldsymbol{\zeta}^H\}\mathbf{q}_1 = \varrho(\boldsymbol{\theta}, \boldsymbol{\mu}_q, \mathbf{q}_1),\\ \text{for all } \mathbf{q}_1 \in \mathbb{C}^J,\ \mathbf{q}_1 \neq \mathbf{0}.\end{aligned} \tag{6.58}$$

The interpretation of this second constraint can be better understood by considering *any* two estimators $\widehat{\boldsymbol{\theta}}_1(\mathbf{x})$ and $\widehat{\boldsymbol{\theta}}_2(\mathbf{x})$ both having mean $\boldsymbol{\mu}_q$, i.e., such that $\mathrm{E}_q\{\widehat{\boldsymbol{\theta}}_1(\mathbf{x})\} = \mathrm{E}_q\{\widehat{\boldsymbol{\theta}}_2(\mathbf{x})\} = \boldsymbol{\mu}_q$. As a shorthand notation, let the *translation matrix* for each estimator be denoted as $\mathrm{E}_q\{(\widehat{\boldsymbol{\theta}}_k - \boldsymbol{\mu}_q)\boldsymbol{\eta}^H\} \overset{\text{def}}{=} \boldsymbol{\mathcal{M}}_k, k = 1, 2$. The second constraint requires that for all nonzero $\mathbf{q}_1 \in \mathbb{C}^J$ that

$$\mathbf{q}_1^H \boldsymbol{\mathcal{M}}_1 \mathrm{E}_q^{-1}\{\boldsymbol{\eta}\boldsymbol{\eta}^H\}\boldsymbol{\mathcal{M}}_1^H \mathbf{q}_1 = \mathbf{q}_1^H \boldsymbol{\mathcal{M}}_2 \mathrm{E}_q^{-1}\{\boldsymbol{\eta}\boldsymbol{\eta}^H\}\boldsymbol{\mathcal{M}}_2^H \mathbf{q}_1 = \varrho(\boldsymbol{\theta}, \boldsymbol{\mu}_q, \mathbf{q}_1). \tag{6.59}$$

Clearly, this constraint will be satisfied if $\boldsymbol{\mathcal{M}}_1 = \boldsymbol{\mathcal{M}}_2 = \boldsymbol{\mathcal{M}}$ for some fixed matrix $\boldsymbol{\mathcal{M}}$. When this is true, then it follows that

$$\begin{aligned} \boldsymbol{\mathcal{M}}_1 - \boldsymbol{\mathcal{M}}_2 &= \mathrm{E}_q\{(\widehat{\boldsymbol{\theta}}_1 - \boldsymbol{\mu}_q)\boldsymbol{\eta}^H\} - \mathrm{E}_q\{(\widehat{\boldsymbol{\theta}}_2 - \boldsymbol{\mu}_q)\boldsymbol{\eta}^H\} \\ &= \mathrm{E}_q\{[(\widehat{\boldsymbol{\theta}}_1 - \boldsymbol{\mu}_q) - (\widehat{\boldsymbol{\theta}}_2 - \boldsymbol{\mu}_q)]\boldsymbol{\eta}^H\} = \mathrm{E}_q\{(\widehat{\boldsymbol{\theta}}_1 - \widehat{\boldsymbol{\theta}}_2)\boldsymbol{\eta}^H\} = \mathbf{0}. \end{aligned} \tag{6.60}$$

Since Eq. (6.60) must hold for any arbitrary choice of two estimators with mean $\boldsymbol{\mu}_q$, we are motivated to interpret (6.60) as a characteristic property of the score function vector $\boldsymbol{\eta}(\mathbf{x}, \boldsymbol{\theta})$. Note first that $\widehat{\boldsymbol{\theta}}_Z(\mathbf{x}) = \widehat{\boldsymbol{\theta}}_1(\mathbf{x}) - \widehat{\boldsymbol{\theta}}_2(\mathbf{x})$ is a statistic with zero mean, i.e., $\mathrm{E}_q\{\widehat{\boldsymbol{\theta}}_Z(\mathbf{x})\} = \mathrm{E}_q\{\widehat{\boldsymbol{\theta}}_1(\mathbf{x})\} - \mathrm{E}_q\{\widehat{\boldsymbol{\theta}}_2(\mathbf{x})\} = \mathbf{0}$. Now suppose that we could choose score function vector $\boldsymbol{\eta}(\mathbf{x}, \boldsymbol{\theta})$ such that $\mathrm{E}_q\{\widehat{\boldsymbol{\theta}}_Z\boldsymbol{\eta}^H\} = \mathbf{0}$ for all statistics $\widehat{\boldsymbol{\theta}}_Z(\mathbf{x})$ where $\mathrm{E}_q\{\widehat{\boldsymbol{\theta}}_Z(\mathbf{x})\} = \mathbf{0}$. If we can identify such a score function $\boldsymbol{\eta}(\mathbf{x}, \boldsymbol{\theta})$, then, by setting $\widehat{\boldsymbol{\theta}}_Z(\mathbf{x}) = \widehat{\boldsymbol{\theta}}_1(\mathbf{x}) - \widehat{\boldsymbol{\theta}}_2(\mathbf{x})$, it follows that

$$\begin{aligned} \mathrm{E}_q\{\widehat{\boldsymbol{\theta}}_Z\boldsymbol{\eta}^H\} &= \mathrm{E}_q\{(\widehat{\boldsymbol{\theta}}_1 - \widehat{\boldsymbol{\theta}}_2)\boldsymbol{\eta}^H\} = \mathrm{E}_q\{(\widehat{\boldsymbol{\theta}}_1 - \boldsymbol{\mu}_q + \boldsymbol{\mu}_q - \widehat{\boldsymbol{\theta}}_2)\boldsymbol{\eta}^H\} \\ &= \mathrm{E}_q\{[(\widehat{\boldsymbol{\theta}}_1 - \boldsymbol{\mu}_q) - (\widehat{\boldsymbol{\theta}}_2 - \boldsymbol{\mu}_q)]\boldsymbol{\eta}^H\} \\ &= \mathrm{E}_q\{(\widehat{\boldsymbol{\theta}}_1 - \boldsymbol{\mu}_q)\boldsymbol{\eta}^H\} - \mathrm{E}_q\{(\widehat{\boldsymbol{\theta}}_2 - \boldsymbol{\mu}_q)\boldsymbol{\eta}^H\} = \mathbf{0}. \end{aligned} \tag{6.61}$$

Thus, for such score functions, it will be true that $\mathrm{E}_q\{(\widehat{\boldsymbol{\theta}}_1 - \boldsymbol{\mu}_q)\boldsymbol{\eta}^H\} = \mathrm{E}_q\{(\widehat{\boldsymbol{\theta}}_2 - \boldsymbol{\mu}_q)\boldsymbol{\eta}^H\} = \mathbf{B}(\boldsymbol{\theta}, \boldsymbol{\mu}_q)$, i.e., the translation matrices for any two estimators having mean $\boldsymbol{\mu}_q$ will be equal; and whatever that translation matrix value is, we call it the matrix function $\mathbf{B}(\boldsymbol{\theta}, \boldsymbol{\mu}_q)$. Since this equality will be true for *any* two estimators sharing mean $\boldsymbol{\mu}_q$, it must be true for *every* estimator with mean $\boldsymbol{\mu}_q$.

To summarize thus far, the second constraint in the MNT formulation (6.46) that is detailed in (6.58) serves to enforce an ICRT as the resulting solution inequality, i.e., one for which the translation matrices resolve to be invariant to the actual estimator $\widehat{\boldsymbol{\theta}}(\mathbf{x})$ with mean $\boldsymbol{\mu}_q$. The development above suggests that we may seek to satisfy this second constraint with a judicious choice of our score function $\boldsymbol{\eta}(\mathbf{x}, \boldsymbol{\theta})$.

The prior discussion interpreting the second constraint essentially proves the following multivariate statement of Blyth's theorem applicable under model misspecification.

Theorem 6.3 (Blyth's Theorem: multivariate with possible misspecification) *If for every statistic $\widehat{\boldsymbol{\theta}}_Z(\mathbf{x})$ where $\mathrm{E}_q\{\widehat{\boldsymbol{\theta}}_Z(\mathbf{x})\} = \mathbf{0}$, a score function vector $\boldsymbol{\eta}(\mathbf{x}, \boldsymbol{\theta})$ has the property that it is orthogonal, i.e., $\mathrm{E}_q\{\widehat{\boldsymbol{\theta}}_Z(\mathbf{x})\boldsymbol{\eta}^H(\mathbf{x}, \boldsymbol{\theta})\} = \mathbf{0}$, then an ICRT is guaranteed, i.e., there exists a matrix function $\mathbf{B}(\boldsymbol{\theta}, \boldsymbol{\mu}_q)$ such that*

$$\mathrm{E}_q\left\{(\widehat{\boldsymbol{\theta}}-\boldsymbol{\mu}_q)\boldsymbol{\eta}^H\right\}=\mathbf{B}(\boldsymbol{\theta},\boldsymbol{\mu}_q)\ \textit{for all}\ \widehat{\boldsymbol{\theta}}(\mathbf{x})\ \textit{with mean}\ \boldsymbol{\mu}_q.$$

This theorem holds under model misspecification because the parameter estimator $\widehat{\boldsymbol{\theta}}(\mathbf{x})$ with mean $\boldsymbol{\mu}_q$, and the score function vector $\boldsymbol{\eta}(\mathbf{x},\boldsymbol{\theta})$ can be chosen arbitrarily, including those choices based on the assumed data distribution $p(\mathbf{x}|\boldsymbol{\theta})$, whereas all statistical averaging is determined by the true distribution $q(\mathbf{x})$.

Interestingly, the constraint condition (6.59) can be satisfied via a weaker and slightly more general condition than $\mathcal{M}_1=\mathcal{M}_2=\mathcal{M}$ for some fixed $\mathcal{M}$. Indeed, it can be shown that the more general condition is the following

$$\mathcal{M}_2=\mathcal{M}_1\mathrm{E}_q^{-1/2}\{\boldsymbol{\eta}\boldsymbol{\eta}^H\}\mathbf{Q}_0\mathrm{E}_q^{1/2}\{\boldsymbol{\eta}\boldsymbol{\eta}^H\},\tag{6.62}$$

where $\mathbf{Q}_0$ is any $L\times L$ unitary matrix, i.e., such that $\mathbf{Q}_0^H\mathbf{Q}_0=\mathbf{Q}_0\mathbf{Q}_0^H=\mathbf{I}_L$. Note that, for this choice, we have

$$\begin{aligned}\mathcal{M}_2\mathrm{E}_q^{-1}\{\boldsymbol{\eta}\boldsymbol{\eta}^H\}\mathcal{M}_2^H&=\\ \mathcal{M}_1\mathrm{E}_q^{-1/2}\{\boldsymbol{\eta}\boldsymbol{\eta}^H\}\mathbf{Q}_0\mathrm{E}_q^{1/2}\{\boldsymbol{\eta}\boldsymbol{\eta}^H\}&\mathrm{E}_q^{-1}\{\boldsymbol{\eta}\boldsymbol{\eta}^H\}\\ \times\mathrm{E}_q^{1/2}\{\boldsymbol{\eta}\boldsymbol{\eta}^H\}\mathbf{Q}_0^H&\mathrm{E}_q^{-1/2}\{\boldsymbol{\eta}\boldsymbol{\eta}^H\}\mathcal{M}_1^H\\ =\mathcal{M}_1&\mathrm{E}_q^{-1}\{\boldsymbol{\eta}\boldsymbol{\eta}^H\}\mathcal{M}_1^H.\end{aligned}\tag{6.63}$$

Thus, the translation matrices for different estimators sharing the same mean $\boldsymbol{\mu}_q$ need not be algebraically equal (nor equal but opposite in sign) in order to satisfy the second constraint in (6.59). This subject is discussed in more detail in [28] for the interested reader.

6.1.6.4 Translation Matrix Invariance Theorem

The following theorem is an obvious consequence of Blyth's theorem 6.3, but it can be a very useful one, as we will demonstrate.

Theorem 6.4 (Translation Matrix Invariance) *If for every finite variance unbiased estimator of the zero vector* $\widehat{\boldsymbol{\theta}}_Z(\mathbf{x})$*, i.e., where* $\mathrm{E}_q\{\widehat{\boldsymbol{\theta}}_Z(\mathbf{x})\}=\mathbf{0}$ *and* $\mathrm{E}_q\{|(\widehat{\boldsymbol{\theta}}_Z)_i|^2\}<\infty$ *for* $i=1,2,\ldots,J$*, the chosen score function vector* $\boldsymbol{\eta}(\mathbf{x},\boldsymbol{\theta})$ *(perhaps with additional regularity conditions) is orthogonal, i.e.,* $\mathrm{E}_q\{\widehat{\boldsymbol{\theta}}_Z\boldsymbol{\eta}^H\}=\mathbf{0}$*, then the translation matrix* $\mathrm{E}_q\{(\widehat{\boldsymbol{\theta}}-\boldsymbol{\mu}_q)\boldsymbol{\eta}^H\}$ *is invariant to any unbiased estimator* $\widehat{\boldsymbol{\theta}}(\mathbf{x})$ *of* $\boldsymbol{\mu}_q$.

By "invariant", we mean that if we are able to evaluate the expectation $\mathrm{E}_q\{(\widehat{\boldsymbol{\theta}}-\boldsymbol{\mu}_q)\boldsymbol{\eta}^H\}$ by assuming $\widehat{\boldsymbol{\theta}}(\mathbf{x})$ to be a specific estimator with mean $\boldsymbol{\mu}_q$ (e.g., a maximum-likelihood estimator), then the result obtained will be universal, i.e., the exact same expression for the translation matrix would be obtained for every other estimator with mean $\boldsymbol{\mu}_q$. Recall that Eq. (6.56) serves as the basis for establishing lower bounds on the covariance of parameter estimators, including under model misspecification. After choosing a specific score function $\boldsymbol{\eta}(\mathbf{x},\boldsymbol{\theta})$, the challenge remains to resolve the translation matrix $\mathrm{E}_q\{(\widehat{\boldsymbol{\theta}} - \boldsymbol{\mu}_q)\boldsymbol{\eta}^H\}$, i.e., to find a closed-form expression that is invariant to the actual estimator $\widehat{\boldsymbol{\theta}}(\mathbf{x})$. The important implication of Theorem 6.4 is that it suggests a general strategy for resolving the translation and obtaining a useful expression for the desired lower bound when the score function vector satisfies the stated orthogonality, i.e., $\mathrm{E}_q\{\widehat{\boldsymbol{\theta}}_Z\boldsymbol{\eta}^H\} = \mathbf{0}$. The strategy is to pick an estimator $\widehat{\boldsymbol{\theta}}(\mathbf{x})$ for which the evaluation of expectation $\mathrm{E}_q\{(\widehat{\boldsymbol{\theta}} - \boldsymbol{\mu}_q)\boldsymbol{\eta}^H\}$ is possible and perhaps easiest.

6.1.6.5 The Misspecified Cramér–Rao Bound (MCRB)

Since an estimator $\widehat{\boldsymbol{\theta}}(\mathbf{x})$ will be based on the assumed data distribution $p(\mathbf{x}|\boldsymbol{\theta})$, we will denote this connection with a subscript p; namely, via $\widehat{\boldsymbol{\theta}}_p(\mathbf{x})$. The mean of the estimator is by definition given by

$$\mathrm{E}_q\{\widehat{\boldsymbol{\theta}}_p\} = \int \widehat{\boldsymbol{\theta}}_p(\mathbf{x}) \cdot q(\mathbf{x})d\mathbf{x} \stackrel{\text{def}}{=} \boldsymbol{\mu}_q \tag{6.64}$$

that in general is independent of $\boldsymbol{\theta}$ under misspecification. Recall the multivariate covariance inequality under misspecification in (6.55) and choose $\boldsymbol{\zeta}(\mathbf{x}) = \widehat{\boldsymbol{\theta}}_p(\mathbf{x}) - \boldsymbol{\mu}_q$, and consider the following score function[3]

$$\boldsymbol{\eta}(\mathbf{x}, \boldsymbol{\theta}) = \frac{\partial \ln p(\mathbf{x}|\boldsymbol{\theta})}{\partial \boldsymbol{\theta}^*} - \mathrm{E}_q\left\{\frac{\partial \ln p(\mathbf{x}|\boldsymbol{\theta})}{\partial \boldsymbol{\theta}^*}\right\}, \tag{6.65}$$

i.e., the usual CRB score function (when the assumed data distribution is $p(\mathbf{x}|\boldsymbol{\theta})$) minus a possibly nonzero mean value under the true distribution $q(\mathbf{x})$. We subtract its mean value under $q(\mathbf{x})$, because we desire a score function with zero mean, i.e., such that $\mathrm{E}_q\{\boldsymbol{\eta}\} = \mathbf{0}$. This will lead to a tighter lower bound [11].

We will prove that the chosen score function (6.65) upon which the MCRB will be based possesses the desired orthogonality property that $\mathrm{E}_q\{\widehat{\boldsymbol{\theta}}_Z\boldsymbol{\eta}^H\} = \mathbf{0}$ when $E_q\{\widehat{\boldsymbol{\theta}}_Z(\mathbf{x})\} = \mathbf{0}$. First, however, we state an important well-known theorem from analysis that details the *regularity conditions* allowing order reversal of integration and differentiation alluded to several times in deriving the usual CRB.

Consider an arbitrary function f approached by a sequence of functions f_n. Specifically, suppose that $\lim_{n\to\infty} f_n(x) = f(x)$; it then follows that

$$\int f(x)dx = \int \lim_{n\to\infty} f_n(x)dx. \tag{6.66}$$

It is, however, sometimes convenient and easier to reverse the limit and the integration and evaluate instead $\lim_{n\to\infty} \int f_n(x)dx$, but doing so only makes sense if

[3] Anticipating the required Wirtinger calculus [23], the gradient with respect to the conjugate variable is used.

$$\int \lim_{n\to\infty} f_n(x)dx \stackrel{?}{=} \lim_{n\to\infty} \int f_n(x)dx. \tag{6.67}$$

The following theorem provides conditions that guarantee that such a reversal is valid, i.e., equivalence is maintained.

Theorem 6.5 (Dominated Convergence Theorem) *If* $\lim_{n\to\infty} f_n(x) = f(x)$ *and there exists an integrable function* $g(x)$ *that dominates* $f_n(x)$, *i.e., such that* $|f_n(x)| \le g(x)$ *for all* x, *then* $f_n(x)$ *and* $f(x)$ *are integrable and*

$$\int \lim_{n\to\infty} f_n(x)dx = \lim_{n\to\infty} \int f_n(x)dx. \tag{6.68}$$

A proof of this theorem can be found in Rudin [29].

Recalling Theorem 6.4, we will now prove that the MCRB score function (6.65) possesses the aforementioned desired orthogonality property by making use of the Dominated Convergence Theorem.

Theorem 6.6 (MCRB Score Function Orthogonality) *The score function defined in (6.65) (to be used as the basis for the MCRB) possesses the desired orthogonality property* $\mathrm{E}_q\{\widehat{\boldsymbol{\theta}}_Z \boldsymbol{\eta}^H\} = \mathbf{0}$ *for every finite variance statistic* $\widehat{\boldsymbol{\theta}}_Z(\mathbf{x})$ *such that* $\mathrm{E}_q\{\widehat{\boldsymbol{\theta}}_Z(\mathbf{x})\} = \mathbf{0}$.

Proof It will be assumed that there exists a non-negative real number $\tilde{\sigma}^2$ such that $\mathrm{E}_q\{|(\widehat{\boldsymbol{\theta}}_Z)_i|^2\} \le \tilde{\sigma}^2 < \infty$ for $i = 1, 2, \ldots, J$, and that for $k = 1, 2, \ldots, J$, the function $|(1/p(\mathbf{x}|\boldsymbol{\theta})) \cdot (\partial p(\mathbf{x}|\boldsymbol{\theta})/\partial\theta_k^*)|$ is dominated by a square-integrable function under measure $q(\mathbf{x})d\mathbf{x}$, call it $g_\theta(\cdot)$, at least in a small neighborhood of $\boldsymbol{\theta} = \boldsymbol{\mu}_q$; i.e., for $\|\boldsymbol{\theta} - \boldsymbol{\mu}_q\| < \epsilon$, $\epsilon > 0$ we will assume for $k = 1, 2, \ldots, J$ that

$$\left| \frac{1}{p(\mathbf{x}|\boldsymbol{\theta})} \frac{\partial p(\mathbf{x}|\boldsymbol{\theta})}{\partial \theta_k^*} \right| \le g_\theta(\mathbf{x}) \text{ for all } \mathbf{x}, \quad \text{where} \quad \int g_\theta^2(\mathbf{x}) \cdot q(\mathbf{x})d\mathbf{x} < \infty. \tag{6.69}$$

Note the following correlation[4]

$$\begin{aligned}
\left(\mathrm{E}_q\{\widehat{\boldsymbol{\theta}}_Z \boldsymbol{\eta}^H\}\right)_{i,k} &= \mathrm{E}_q\left\{(\widehat{\boldsymbol{\theta}}_Z)_i \frac{\partial \ln p(\mathbf{x}|\boldsymbol{\theta})}{\partial\theta_k^*} - (\widehat{\boldsymbol{\theta}}_Z)_i \mathrm{E}_q\left\{\frac{\partial \ln p(\mathbf{x}|\boldsymbol{\theta})}{\partial\theta_k^*}\right\}\right\} \\
&= \mathrm{E}_q\left\{(\widehat{\boldsymbol{\theta}}_Z)_i \frac{\partial \ln p(\mathbf{x}|\boldsymbol{\theta})}{\partial\theta_k^*}\right\} - \mathrm{E}_q\left\{(\widehat{\boldsymbol{\theta}}_Z)_i\right\} \mathrm{E}_q\left\{\frac{\partial \ln p(\mathbf{x}|\boldsymbol{\theta})}{\partial\theta_k^*}\right\} \\
&= \int (\widehat{\boldsymbol{\theta}}_Z(\mathbf{x}))_i \frac{\partial \ln p(\mathbf{x}|\boldsymbol{\theta})}{\partial\theta_k^*} q(\mathbf{x})d\mathbf{x},
\end{aligned} \tag{6.70}$$

where last equality follows from $\mathrm{E}_q\left\{\widehat{\boldsymbol{\theta}}_Z(\mathbf{x})\right\} = \mathbf{0}$. Since the individual elements of parameter vector $\boldsymbol{\theta} = [\theta_1, \theta_2, \ldots, \theta_J]^T$ can be either real or complex, we will first

[4] $p(\mathbf{x}|\boldsymbol{\theta})$ is a density, and therefore it is real-valued, i.e., $p(\mathbf{x}|\boldsymbol{\theta}) = p^*(\mathbf{x}|\boldsymbol{\theta})$. Wirtinger calculus informs that $(\partial p(\mathbf{x}|\boldsymbol{\theta})/\partial\theta_k^*)^* = \partial p^*(\mathbf{x}|\boldsymbol{\theta})/\partial\theta_k = \partial p(\mathbf{x}|\boldsymbol{\theta})/\partial\theta_k$; similarly, $(\partial p(\mathbf{x}|\boldsymbol{\theta})/\partial\theta_k)^* = \partial p^*(\mathbf{x}|\boldsymbol{\theta})/\partial\theta_k^* = \partial p(\mathbf{x}|\boldsymbol{\theta})/\partial\theta_k^*$ [23].

focus on those that are real, and then discuss those that are complex; for k such that θ_k is a real parameter, i.e., $\theta_k^* = \theta_k$, note that

$$\begin{aligned}
\left(\mathrm{E}_q\{\widehat{\boldsymbol{\theta}}_{\mathcal{Z}}\boldsymbol{\eta}^H\}\right)_{i,k} &= \int (\widehat{\boldsymbol{\theta}}_{\mathcal{Z}}(\mathbf{x}))_i \frac{\partial \ln p(\mathbf{x}|\boldsymbol{\theta})}{\partial \theta_k} q(\mathbf{x})d\mathbf{x} \\
&= \int (\widehat{\boldsymbol{\theta}}_{\mathcal{Z}}(\mathbf{x}))_i \cdot \frac{1}{p(\mathbf{x}|\boldsymbol{\theta})} \cdot \frac{\partial p(\mathbf{x}|\boldsymbol{\theta})}{\partial \theta_k} q(\mathbf{x})d\mathbf{x} \\
&= \int (\widehat{\boldsymbol{\theta}}_{\mathcal{Z}}(\mathbf{x}))_i \cdot \frac{1}{p(\mathbf{x}|\boldsymbol{\theta})} \cdot \lim_{\Delta \to 0} \frac{p(\mathbf{x}|\boldsymbol{\theta} + \Delta \mathbf{e}_k) - p(\mathbf{x}|\boldsymbol{\theta})}{\Delta} q(\mathbf{x})d\mathbf{x} \\
&= \int \lim_{\Delta \to 0} (\widehat{\boldsymbol{\theta}}_{\mathcal{Z}}(\mathbf{x}))_i \cdot \frac{1}{p(\mathbf{x}|\boldsymbol{\theta})} \cdot \frac{p(\mathbf{x}|\boldsymbol{\theta} + \Delta \mathbf{e}_k) - p(\mathbf{x}|\boldsymbol{\theta})}{\Delta} q(\mathbf{x})d\mathbf{x},
\end{aligned} \tag{6.71}$$

where the third equality employs the limit definition of the partial derivative where $\mathbf{e}_k$ is a $J \times 1$ standard basis vector with unity in the k-th position and zero elsewhere; and the last equality follows from properties (algebra) of limits.[5] By (6.69), it follows for $\boldsymbol{\theta}$ near $\boldsymbol{\mu}_q$ that

$$\left| (\widehat{\boldsymbol{\theta}}_{\mathcal{Z}}(\mathbf{x}))_i \cdot \frac{1}{p(\mathbf{x}|\boldsymbol{\theta})} \cdot \frac{p(\mathbf{x}|\boldsymbol{\theta} + \Delta \mathbf{e}_k) - p(\mathbf{x}|\boldsymbol{\theta})}{\Delta} \right| \le |(\widehat{\boldsymbol{\theta}}_{\mathcal{Z}}(\mathbf{x}))_i| \cdot g_\theta(\mathbf{x}) \quad \text{for all } \mathbf{x} \text{ and } |\Delta| \le \epsilon, \quad \epsilon > 0. \tag{6.72}$$

Thus, function $|(\widehat{\boldsymbol{\theta}}_{\mathcal{Z}}(\mathbf{x}))_i| \cdot g_\theta(\mathbf{x})$ dominates the integrand in (6.71). If we can now show that this function is integrable, then the dominated convergence theorem can be invoked. Note by the Cauchy–Schwarz inequality that

$$\begin{aligned}
&\int |(\widehat{\boldsymbol{\theta}}_{\mathcal{Z}}(\mathbf{x}))_i| \cdot g_\theta(\mathbf{x}) \cdot q(\mathbf{x})d\mathbf{x} \\
&\le \left(\int |(\widehat{\boldsymbol{\theta}}_{\mathcal{Z}}(\mathbf{x}))_i|^2 \cdot q(\mathbf{x})d\mathbf{x} \right)^{1/2} \times \left(\int g_\theta^2(\mathbf{x}) \cdot q(\mathbf{x})d\mathbf{x} \right)^{1/2} \\
&\le \tilde{\sigma} \times \left(\int g_\theta^2(\mathbf{x}) \cdot q(\mathbf{x})d\mathbf{x} \right)^{1/2} < \infty.
\end{aligned} \tag{6.73}$$

Since $g_\theta(\mathbf{x})$ is square integrable under measure $q(\mathbf{x})d\mathbf{x}$, it follows that $|(\widehat{\boldsymbol{\theta}}_{\mathcal{Z}}(\mathbf{x}))_i| \cdot g_\theta(\mathbf{x})$ is integrable. Hence, by the dominated convergence theorem the order of integration and differentiation in (6.71) may be reversed. It follows that

[5] If $\lim_{n\to\infty} s_n = S$ and $\lim_{n\to\infty} t_n = T$, then $\lim_{n\to\infty} s_n t_n = ST$ [29].

$$
\begin{aligned}
&\left(\mathrm{E}_q\{\widehat{\boldsymbol{\theta}}_{\mathcal{Z}}\boldsymbol{\eta}^H\}\right)_{i,k} = \\
&\lim_{\Delta\to 0}\int (\widehat{\boldsymbol{\theta}}_{\mathcal{Z}}(\mathbf{x}))_i \cdot \frac{1}{p(\mathbf{x}|\boldsymbol{\theta})}\cdot\frac{p(\mathbf{x}|\boldsymbol{\theta}+\Delta\mathbf{e}_k)-p(\mathbf{x}|\boldsymbol{\theta})}{\Delta}q(\mathbf{x})d\mathbf{x} \\
&= \lim_{\Delta\to 0}\left[\frac{1}{\Delta}\int (\widehat{\boldsymbol{\theta}}_{\mathcal{Z}}(\mathbf{x}))_i \cdot \frac{p(\mathbf{x}|\boldsymbol{\theta}+\Delta\mathbf{e}_k)}{p(\mathbf{x}|\boldsymbol{\theta})}\cdot q(\mathbf{x})d\mathbf{x}\right. \\
&\left. -\frac{1}{\Delta}\int (\widehat{\boldsymbol{\theta}}_{\mathcal{Z}}(\mathbf{x}))_i \cdot \frac{p(\mathbf{x}|\boldsymbol{\theta})}{p(\mathbf{x}|\boldsymbol{\theta})}\cdot q(\mathbf{x})d\mathbf{x}\right]
\end{aligned}
\tag{6.74}
$$

since integration is a linear operation. Now define function α (for fixed $\tilde{\boldsymbol{\theta}}$) such that

$$
\alpha(\boldsymbol{\theta};\tilde{\boldsymbol{\theta}}) \stackrel{\text{def}}{=} \int (\widehat{\boldsymbol{\theta}}_{\mathcal{Z}}(\mathbf{x}))_i \cdot \frac{p(\mathbf{x}|\boldsymbol{\theta})}{p(\mathbf{x}|\tilde{\boldsymbol{\theta}})}\cdot q(\mathbf{x})d\mathbf{x}. \tag{6.75}
$$

Note that (6.74) can be written in terms of $\alpha(\boldsymbol{\theta};\tilde{\boldsymbol{\theta}})$:

$$
\left(\mathrm{E}_q\{\widehat{\boldsymbol{\theta}}_{\mathcal{Z}}\boldsymbol{\eta}^H\}\right)_{i,k} = \lim_{\Delta\to 0}\left[\frac{1}{\Delta}\alpha(\boldsymbol{\theta}+\Delta\mathbf{e}_k;\boldsymbol{\theta}) - \frac{1}{\Delta}\alpha(\boldsymbol{\theta};\boldsymbol{\theta})\right]. \tag{6.76}
$$

By definition of the partial derivative

$$
\frac{\partial\alpha(\boldsymbol{\theta};\tilde{\boldsymbol{\theta}})}{\partial\theta_k} = \lim_{\Delta\to 0}\frac{\alpha(\boldsymbol{\theta}+\Delta\mathbf{e}_k;\tilde{\boldsymbol{\theta}})-\alpha(\boldsymbol{\theta};\tilde{\boldsymbol{\theta}})}{\Delta}. \tag{6.77}
$$

Thus, it follows that

$$
\begin{aligned}
\left(\mathrm{E}_q\{\widehat{\boldsymbol{\theta}}_{\mathcal{Z}}\boldsymbol{\eta}^H\}\right)_{i,k} &= \left.\frac{\partial\alpha(\boldsymbol{\theta};\tilde{\boldsymbol{\theta}})}{\partial\theta_k}\right|_{\tilde{\boldsymbol{\theta}}=\boldsymbol{\theta}} = \frac{\partial}{\partial\theta_k}\int (\widehat{\boldsymbol{\theta}}_{\mathcal{Z}}(\mathbf{x}))_i \cdot \frac{p(\mathbf{x}|\boldsymbol{\theta})}{p(\mathbf{x}|\boldsymbol{\theta})}\cdot q(\mathbf{x})d\mathbf{x} \\
&= \frac{\partial}{\partial\theta_k}\int (\widehat{\boldsymbol{\theta}}_{\mathcal{Z}}(\mathbf{x}))_i \cdot q(\mathbf{x})d\mathbf{x} = \frac{\partial}{\partial\theta_k}\mathrm{E}_q\{(\widehat{\boldsymbol{\theta}}_{\mathcal{Z}}(\mathbf{x}))_i\} = 0,
\end{aligned}
\tag{6.78}
$$

where last equality follows since $\mathrm{E}_q\{\widehat{\boldsymbol{\theta}}_{\mathcal{Z}}(\mathbf{x})\} = \mathbf{0}$.

Regarding k such that θ_k is a complex parameter, let $\theta_k \stackrel{\text{def}}{=} \theta_{R,k} + j\theta_{I,k}$ and note from Wirtinger calculus [23] the definition of the conjugate derivative leads to correlation

$$
\begin{aligned}
\left(\mathrm{E}_q\{\widehat{\boldsymbol{\theta}}_{\mathcal{Z}}\boldsymbol{\eta}^H\}\right)_{i,k} &= \int (\widehat{\boldsymbol{\theta}}_{\mathcal{Z}}(\mathbf{x}))_i \frac{\partial \ln p(\mathbf{x}|\boldsymbol{\theta})}{\partial\theta_k^*}q(\mathbf{x})d\mathbf{x} \\
&= \int (\widehat{\boldsymbol{\theta}}_{\mathcal{Z}}(\mathbf{x}))_i \cdot \frac{1}{2}\left(\frac{\partial \ln p(\mathbf{x}|\boldsymbol{\theta})}{\partial\theta_{R,k}} + j\frac{\partial \ln p(\mathbf{x}|\boldsymbol{\theta})}{\partial\theta_{I,k}}\right)\cdot q(\mathbf{x})d\mathbf{x} \\
&= \frac{1}{2}\int (\widehat{\boldsymbol{\theta}}_{\mathcal{Z}}(\mathbf{x}))_i \frac{\partial \ln p(\mathbf{x}|\boldsymbol{\theta})}{\partial\theta_{R,k}}\cdot q(\mathbf{x})d\mathbf{x} + j\frac{1}{2}\int (\widehat{\boldsymbol{\theta}}_{\mathcal{Z}}(\mathbf{x}))_i \frac{\partial \ln p(\mathbf{x}|\boldsymbol{\theta})}{\partial\theta_{I,k}}\cdot q(\mathbf{x})d\mathbf{x} = 0.
\end{aligned}
\tag{6.79}
$$

The real and imaginary parts of $(\mathrm{E}_q\{\widehat{\boldsymbol{\theta}}_{\mathcal{Z}}\boldsymbol{\eta}^H\})_{i,k}$ for complex θ_k are real integrals with the same form as seen in the case of purely real θ_k discussed in (6.71)–(6.78). Hence, they are also zero for $\boldsymbol{\theta}$ near $\boldsymbol{\mu}_q$. ■

Covariance inequality (6.56) establishes a lower bound on the covariance of all unbiased estimators of $\boldsymbol{\mu}_q$. Having chosen score function (6.65), the corresponding *misspecified Fisher Information Matrix (MFIM)* $\mathrm{E}_q\{\boldsymbol{\eta}\boldsymbol{\eta}^H\}$ must be determined. First note under regularity conditions that

$$\begin{aligned}\mathrm{E}_q\left\{\frac{\partial \ln p(\mathbf{x}|\boldsymbol{\theta})}{\partial \theta_k^*}\right\} &= \int q(\mathbf{x})\cdot\frac{\partial \ln p(\mathbf{x}|\boldsymbol{\theta})}{\partial \theta_k^*}d\mathbf{x} = \frac{\partial}{\partial \theta_k^*}\int q(\mathbf{x})\ln p(\mathbf{x}|\boldsymbol{\theta})d\mathbf{x} \\ &= \frac{\partial}{\partial \theta_k^*}\mathrm{E}_q\{\ln p(\mathbf{x}|\boldsymbol{\theta})\}.\end{aligned} \tag{6.80}$$

Recalling that the Kullback–Leibler (KL) divergence [30] between the assumed density $p(\mathbf{x}|\boldsymbol{\theta})$ and the true density $q(\mathbf{x})$ is given by

$$D(q_{\mathbf{x}}||p_{\mathbf{x}|\boldsymbol{\theta}}) \stackrel{\text{def}}{=} \int q(\mathbf{x})\ln\left(\frac{q(\mathbf{x})}{p(\mathbf{x}|\boldsymbol{\theta})}\right)d\mathbf{x} = \mathrm{E}_q\{\ln q(\mathbf{x})\} - \mathrm{E}_q\{\ln p(\mathbf{x}|\boldsymbol{\theta})\}, \tag{6.81}$$

the mean subtracted in (6.65) can be expressed as

$$\mathrm{E}_q\left\{\frac{\partial \ln p(\mathbf{x}|\boldsymbol{\theta})}{\partial \boldsymbol{\theta}^*}\right\} = -\frac{\partial D(q_{\mathbf{x}}||p_{\mathbf{x}|\boldsymbol{\theta}})}{\partial \boldsymbol{\theta}^*}, \tag{6.82}$$

i.e., the negative gradient with respect to $\boldsymbol{\theta}^*$ of the divergence. Thus, the MFIM is given by

$$\begin{aligned}\mathrm{E}_q\{\boldsymbol{\eta}\boldsymbol{\eta}^H\} &= \mathrm{E}_q\left\{\frac{\partial \ln p(\mathbf{x}|\boldsymbol{\theta})}{\partial \boldsymbol{\theta}^*}\frac{\partial \ln p(\mathbf{x}|\boldsymbol{\theta})}{\partial \boldsymbol{\theta}^*}^H\right\} - \frac{\partial D(q_{\mathbf{x}}||p_{\mathbf{x}|\boldsymbol{\theta}})}{\partial \boldsymbol{\theta}^*}\frac{\partial D(q_{\mathbf{x}}||p_{\mathbf{x}|\boldsymbol{\theta}})}{\partial \boldsymbol{\theta}^*}^H \\ &\stackrel{\text{def}}{=} \mathbf{MFIM}(\boldsymbol{\theta}).\end{aligned} \tag{6.83}$$

The translation matrix $\mathrm{E}_q\{(\widehat{\boldsymbol{\theta}} - \boldsymbol{\mu}_q)\boldsymbol{\eta}^H\}$, by virtue of Theorem 6.4, can be found by evaluating this expectation for a specific estimator. The specific estimator we will focus on is the maximum-likelihood (ML) estimator for $\boldsymbol{\theta}$, i.e.,

$$\widehat{\boldsymbol{\theta}}_{ML}(\mathbf{x}) = \arg\max_{\boldsymbol{\theta}} \ln p(\mathbf{x}\,|\boldsymbol{\theta}) \rightarrow \boldsymbol{\theta}_A \stackrel{\text{def}}{=} \arg\min_{\boldsymbol{\theta}} D(q_{\mathbf{x}}\,\|\,p_{\mathbf{x}|\boldsymbol{\theta}}). \tag{6.84}$$

Under model misspecification, the ML estimator is known to converge in probability (in the limit of large sample support) to point $\boldsymbol{\theta}_A$, i.e., the value of $\boldsymbol{\theta}$ that minimizes the KL divergence between the true and assumed distributions [16, 17]. The value $\boldsymbol{\theta}_A$

is often referred to as the *pseudo-true parameter vector*, and it can be shown asymptotically that ML has mean[6] $\boldsymbol{\mu}_q = \mathrm{E}_q\{\widehat{\boldsymbol{\theta}}_{ML}(\mathbf{x})\} \simeq \boldsymbol{\theta}_A$ [17–19]. Define the average Hessian matrix under $q(\mathbf{x})$ as

$$(\mathbf{C}(\boldsymbol{\theta}))_{i,k} \stackrel{\text{def}}{=} \mathrm{E}_q\left\{\frac{\partial^2 \ln p(\mathbf{x}|\boldsymbol{\theta})}{\partial\theta_i^*\partial\theta_k}\right\}, \quad i = 1, 2, \ldots, J; \; k = 1, 2, \ldots, J. \tag{6.85}$$

The translation matrix can be evaluated for the ML estimator via a Taylor series approximation around point $\boldsymbol{\theta}_A$ and shown to be (see Appendix B of [19]):

$$\mathrm{E}_q\{(\widehat{\boldsymbol{\theta}}_{ML} - \boldsymbol{\theta}_A)\boldsymbol{\eta}^H\} \simeq -\mathbf{C}^{-1}(\boldsymbol{\theta}_A) \cdot \mathbf{MFIM}(\boldsymbol{\theta}_A). \tag{6.86}$$

By Theorem 6.4, this translation matrix is valid for every estimator having the same mean as ML, i.e., for every unbiased estimator of the pseudo-true parameters $\boldsymbol{\theta}_A$. The resulting lower bound on the covariance of all unbiased estimators of pseudo-true parameters $\boldsymbol{\theta}_A$ is, by (6.56),

$$\mathrm{E}_q\{(\widehat{\boldsymbol{\theta}}_p - \boldsymbol{\theta}_A)(\widehat{\boldsymbol{\theta}}_p - \boldsymbol{\theta}_A)^H\} \geq \mathbf{C}^{-1}(\boldsymbol{\theta}_A)\mathbf{MFIM}(\boldsymbol{\theta}_A)\mathbf{C}^{-1}(\boldsymbol{\theta}_A). \tag{6.87}$$

This is the MCRB inequality and the right-hand side is synonymous with *Huber's sandwich covariance* [16, 17]. We will explore this MCRB for specific cases of misspecification in the next two sections.

6.1.6.6 Generalized Slepian and Bangs Formulae: Complex Circular Gaussian

Consider the case where the true data distribution and the assumed distribution are both complex circular Gaussian, but perhaps with incorrect parameterization. Specifically, let the complex data be distributed as $\mathbf{x} \sim q(\mathbf{x}) = \mathbb{C}N_m(\mathbf{d}, \mathbf{B})$, but the assumed data distribution is $\mathbf{x} \sim p(\mathbf{x}|\boldsymbol{\theta}) = \mathbb{C}N_m(\boldsymbol{\mu}(\boldsymbol{\theta}), \boldsymbol{\Sigma}^{-1})$. Note that the parameterization of the assumed distribution via $\boldsymbol{\theta}$ is exclusively in the mean. It is shown in [19] that the MFIM is given by

$$\begin{aligned}\mathbf{MFIM}_D(\boldsymbol{\theta}) &\stackrel{\text{def}}{=} \left(\frac{\partial\boldsymbol{\mu}^*(\boldsymbol{\theta})}{\partial\boldsymbol{\theta}^*}\right)\boldsymbol{\Sigma}^{-1}\mathbf{B}\boldsymbol{\Sigma}^{-1}\left(\frac{\partial\boldsymbol{\mu}(\boldsymbol{\theta})}{\partial\boldsymbol{\theta}}\right)^T \\ &+ \left(\frac{\partial\boldsymbol{\mu}(\boldsymbol{\theta})}{\partial\boldsymbol{\theta}^*}\right)\boldsymbol{\Sigma}^{-*}\mathbf{B}^*\boldsymbol{\Sigma}^{-*}\left(\frac{\partial\boldsymbol{\mu}^*(\boldsymbol{\theta})}{\partial\boldsymbol{\theta}}\right)^T,\end{aligned} \tag{6.88}$$

and the average Hessian matrix is given by

[6] The notation "$A \simeq B$" is used to denote that A is asymptotically equal to B.

$$(\mathbf{C}_D(\boldsymbol{\theta}))_{i,k} \stackrel{\text{def}}{=} -\left[\frac{\partial \boldsymbol{\mu}^*(\boldsymbol{\theta})}{\partial \theta_i^*}\right]^T \boldsymbol{\Sigma}^{-1}\frac{\partial \boldsymbol{\mu}(\boldsymbol{\theta})}{\partial \theta_k} - \left[\frac{\partial \boldsymbol{\mu}^*(\boldsymbol{\theta})}{\partial \theta_k}\right]^T \boldsymbol{\Sigma}^{-1}\frac{\partial \boldsymbol{\mu}(\boldsymbol{\theta})}{\partial \theta_i^*}$$
$$+\left[\frac{\partial^2 \boldsymbol{\mu}^*(\boldsymbol{\theta})}{\partial \theta_i^* \partial \theta_k}\right]^T \boldsymbol{\Sigma}^{-1}\delta\boldsymbol{\mu}(\boldsymbol{\theta}) + \delta\boldsymbol{\mu}^H(\boldsymbol{\theta})\boldsymbol{\Sigma}^{-1}\frac{\partial^2 \boldsymbol{\mu}(\boldsymbol{\theta})}{\partial \theta_i^* \partial \theta_k}, \tag{6.89}$$

where $\delta\boldsymbol{\mu}(\boldsymbol{\theta}) \stackrel{\text{def}}{=} \mathbf{d} - \boldsymbol{\mu}(\boldsymbol{\theta})$. The MCRB inequality in this case is

$$\mathrm{E}_q\{(\widehat{\boldsymbol{\theta}}_p - \boldsymbol{\theta}_A)(\widehat{\boldsymbol{\theta}}_p - \boldsymbol{\theta}_A)^H\} \geq \mathbf{C}_D^{-1}(\boldsymbol{\theta}_A)\mathbf{MFIM}_D(\boldsymbol{\theta}_A)\mathbf{C}_D^{-1}(\boldsymbol{\theta}_A). \tag{6.90}$$

Equations (6.88) and (6.89) represent a generalization of the Slepian formula in (6.33) to the complex data case under model misspecification with possible complex parameters.

Again let the complex data true distribution be $\mathbf{x} \sim q(\mathbf{x}) = \mathbb{C}N_m(\mathbf{d}, \mathbf{B})$, but the assumed data distribution is $\mathbf{x} \sim p(\mathbf{x}|\boldsymbol{\theta}) = \mathbb{C}N_m(\boldsymbol{\mu}, \boldsymbol{\Sigma}(\boldsymbol{\theta}))$. The parameterization of the assumed distribution via $\boldsymbol{\theta}$ in this case is exclusively in the covariance. It is shown in [19] that the MFIM is given by

$$(\mathbf{MFIM}_S(\boldsymbol{\theta}))_{i,k} =$$
$$\mathrm{tr}\left[\boldsymbol{\Sigma}^{-1}\frac{\partial \boldsymbol{\Sigma}(\boldsymbol{\theta})}{\partial \theta_i^*}\boldsymbol{\Sigma}^{-1}\mathbf{B}\boldsymbol{\Sigma}^{-1}\frac{\partial \boldsymbol{\Sigma}(\boldsymbol{\theta})}{\partial \theta_k}\boldsymbol{\Sigma}^{-1}(\mathbf{B} + \delta\boldsymbol{\mu}\delta\boldsymbol{\mu}^H)\right]$$
$$+\delta\boldsymbol{\mu}^H\boldsymbol{\Sigma}^{-1}\frac{\partial \boldsymbol{\Sigma}(\boldsymbol{\theta})}{\partial \theta_k}\boldsymbol{\Sigma}^{-1}\mathbf{B}\boldsymbol{\Sigma}^{-1}\frac{\partial \boldsymbol{\Sigma}(\boldsymbol{\theta})}{\partial \theta_i^*}\boldsymbol{\Sigma}^{-1}\delta\boldsymbol{\mu}, \tag{6.91}$$

where $\delta\boldsymbol{\mu} = \mathbf{d} - \boldsymbol{\mu}$; and that the average Hessian is given by

$$(\mathbf{C}_S(\boldsymbol{\theta}))_{i,k} = \mathrm{tr}\left\{\boldsymbol{\Sigma}^{-1}\frac{\partial^2 \boldsymbol{\Sigma}(\boldsymbol{\theta})}{\partial \theta_i^* \partial \theta_k}\left[\boldsymbol{\Sigma}^{-1}(\mathbf{B} + \delta\boldsymbol{\mu}\delta\boldsymbol{\mu}^H) - \mathbf{I}_m\right]\right\}$$
$$-\mathrm{tr}\left\{\boldsymbol{\Sigma}^{-1}\frac{\partial \boldsymbol{\Sigma}(\boldsymbol{\theta})}{\partial \theta_k}\boldsymbol{\Sigma}^{-1}\frac{\partial \boldsymbol{\Sigma}(\boldsymbol{\theta})}{\partial \theta_i^*}\left[\boldsymbol{\Sigma}^{-1}(\mathbf{B} + \delta\boldsymbol{\mu}\delta\boldsymbol{\mu}^H) - \mathbf{I}_m\right]\right\}$$
$$-\mathrm{tr}\left[\boldsymbol{\Sigma}^{-1}\frac{\partial \boldsymbol{\Sigma}(\boldsymbol{\theta})}{\partial \theta_i^*}\boldsymbol{\Sigma}^{-1}\frac{\partial \boldsymbol{\Sigma}(\boldsymbol{\theta})}{\partial \theta_k}\boldsymbol{\Sigma}^{-1}(\mathbf{B} + \delta\boldsymbol{\mu}\delta\boldsymbol{\mu}^H)\right]. \tag{6.92}$$

The MCRB inequality in this case is

$$\mathrm{E}_q\{(\widehat{\boldsymbol{\theta}}_p - \boldsymbol{\theta}_A)(\widehat{\boldsymbol{\theta}}_p - \boldsymbol{\theta}_A)^H\} \geq \mathbf{C}_S^{-1}(\boldsymbol{\theta}_A)\mathbf{MFIM}_S(\boldsymbol{\theta}_A)\mathbf{C}_S^{-1}(\boldsymbol{\theta}_A). \tag{6.93}$$

Equations (6.91) and (6.92) represent a generalization of the Bangs formula in (6.34) to the complex data case under model misspecification with possible complex parameters.

Although not discussed here, a generalization of the full Slepian–Bangs formula (6.35) to the complex data case under model misspecification with possible complex parameters can be found in [31].

6.1.6.7 Generalized Slepian and Bangs Formulae: Circular Complex Compound Gaussian

Let random vector $\mathbf{x}$ have a true distribution that is circular complex compound Gaussian (C-CCG), and recall that it has a stochastic representation $\mathbf{x} =_d \mathbf{d} + \sqrt{\tau}\mathbf{n}$ where $\mathbf{n} \sim \mathbb{C}N_m(\mathbf{0}, \mathbf{B})$, and τ is a real positive random variable with probability density function q_τ where $\mathbf{n}$ and τ are independent. If the assumed distribution is $\mathbf{x} \sim \mathbb{C}N_m(\boldsymbol{\mu}(\boldsymbol{\theta}), \boldsymbol{\Sigma}) = p(\mathbf{x}|\boldsymbol{\theta})$, then we can determine the MCRB easily from previous results. Note that if we condition on τ, then the true distribution is $\mathbf{x} \sim \mathbb{C}N_m(\mathbf{d}, \tau \cdot \mathbf{B}) = q(\mathbf{x}|\tau)$. Recall that the MCRB inequality is given by (6.87) where the MFIM and the average Hessian are defined by expectations (6.83) and (6.85), respectively. By the law of total expectation, it follows that $\mathrm{E}_q\{\boldsymbol{\eta}\boldsymbol{\eta}^H\} = \mathrm{E}_{q_\tau}\{\mathrm{E}_{q_{\mathbf{x}|\tau}}\{\boldsymbol{\eta}\boldsymbol{\eta}^H|\tau\}\}$. Evaluation of $\mathrm{E}_{q_{\mathbf{x}|\tau}}\{\boldsymbol{\eta}\boldsymbol{\eta}^H|\tau\}$ follows from the results of Sect. 6.1.6.6 because both the assumed data distribution $p(\mathbf{x}|\boldsymbol{\theta})$ and the conditional true distribution $q(\mathbf{x}|\tau)$ are complex circular Gaussian. The MFIM conditioned on τ must be

$$\mathbf{MFIM}_D(\boldsymbol{\theta}|\tau) \stackrel{\text{def}}{=} \tau \cdot \mathbf{MFIM}_D(\boldsymbol{\theta}), \tag{6.94}$$

where $\mathbf{MFIM}_D(\boldsymbol{\theta})$ is defined as in (6.88). This follows by simply replacing $\mathbf{B}$ with $\tau\mathbf{B}$ in (6.88). The final unconditional MFIM is obtained by averaging over τ, i.e., $\mathbf{MFIM}_D(\boldsymbol{\theta}) = \mathrm{E}_{q_\tau}\{\mathbf{MFIM}_D(\boldsymbol{\theta}|\tau)\} = \mathrm{E}_{q_\tau}\{\tau\} \cdot \mathbf{MFIM}_D(\boldsymbol{\theta})$.

Similarly, the law of total expectation can be applied to obtain the average Hessian. When conditioned on τ, however, note from (6.89) that $\mathbf{C}_D(\boldsymbol{\theta}|\tau) = \mathbf{C}_D(\boldsymbol{\theta})$ where $\mathbf{C}_D(\boldsymbol{\theta})$ is defined as in (6.89), because it does not dependent on $\mathbf{B}$; namely, there's no $\mathbf{B}$ to replace with $\tau\mathbf{B}$. Hence, the final unconditional average Hessian is given by (6.89). The final MCRB inequality in this case is given by

$$\mathrm{E}_q\{(\widehat{\boldsymbol{\theta}}_p - \boldsymbol{\theta}_A)(\widehat{\boldsymbol{\theta}}_p - \boldsymbol{\theta}_A)^H\} \geq \mathrm{E}_{q_\tau}\{\tau\} \cdot \mathbf{C}_D^{-1}(\boldsymbol{\theta}_A)\mathbf{MFIM}_D(\boldsymbol{\theta}_A)\mathbf{C}_D^{-1}(\boldsymbol{\theta}_A). \tag{6.95}$$

This same approach can be used to determine the MCRB under a C-CCG for the parameterized covariance case in (6.91)–(6.93). The expressions are more involved and omitted here, but easily computed.

6.1.7 *Numerical Results*

In this section, we compare the root-mean-squared error (RMSE) achieved by the ML estimator to the MCRB for the direction-of-arrival (DoA) estimation problem under varying degrees of model misspecification. We begin with the problem setup considered in the numerical results Sect. VIII of [19], where the true antenna array response differs from the assumed array response due to errors present in the assumed array element positions. We extend the simulation scenario of [19] to the case when the interference follows a complex t-distribution but is assumed to be complex circular Gaussian.

Consider a planewave signal with DoA ϕ_{T} and complex amplitude S_{T} impinging on an $m = 18$ element uniform linear array (ULA) with $\lambda/2.25$ element spacing and a 3 dB beamwidth ≈ 7.2 degs. We also consider the presence of a jammer at a known angle ϕ_J with jammer power σ_J^2. Let $\mathbf{p}_l$ represent the assumed position vector of the lth antenna element. In this section, we focus on the case where the standard radar data model is assumed with the vector of unknown target parameters $\boldsymbol{\theta} = [\phi, S, S^*]^T$ where S and ϕ denote the unknown target return signal complex amplitude and the angle of arrival, respectively. The assumed steering vector $\mathbf{v}(\phi)$ is given by $(\mathbf{v}(\phi))_l = \exp(j\mathbf{k}_\phi^T \mathbf{p}_l)$, where $\mathbf{u}(\phi)$ and $\mathbf{k}_\phi = (2\pi/\lambda)\mathbf{u}(\phi)$ represent the direction unit vector and the wavenumber vector respectively. Thus, the assumed distribution of the received signal is given by $\mathbf{x} \sim p(\mathbf{x}|\boldsymbol{\theta}) = \mathbb{C}N_m(S \cdot \mathbf{v}(\phi), \boldsymbol{\Sigma})$, where the covariance matrix is given by $\boldsymbol{\Sigma} = \mathbf{I}_m + \sigma_J^2 \mathbf{v}(\phi_J)\mathbf{v}^H(\phi_J)$ that represents a jamming signal at angle ϕ_J in a background of white noise. Note that the unknown parameters appear only in the mean of the assumed distribution, thus the generalized Slepian formulae (6.90) and (6.95) are applicable. The ML estimator for ϕ_{T} based on the above-assumed model can be easily derived and is given by

$$\hat{\phi}_{ML} = \arg\max_{\phi} \frac{|\mathbf{v}^H(\phi)\boldsymbol{\Sigma}^{-1}\mathbf{x}|^2}{\mathbf{v}^H(\phi)\boldsymbol{\Sigma}^{-1}\mathbf{v}(\phi)}. \tag{6.96}$$

6.1.7.1 Array Position Error with Complex Circular Gaussian-Distributed Data

The model misspecification scenario in [19] considers the case where the true array element positions are given by $\mathbf{p}_l + \mathbf{e}_l$, where $\mathbf{e}_l$ represents the array position errors. The position errors are drawn from a distribution $\mathbf{e}_l \sim \mathbb{R}N_3(\mathbf{0}, (0.01\lambda)^2 \cdot \mathbf{I}_3)$. We consider a fixed realization drawn from this distribution for the numerical results in this section. Considering the case that the true distribution is a complex circular Gaussian, the received signal can be expressed as

$$\mathbf{x} = \mathbf{d}(\phi_{\mathrm{T}}) + \mathbf{n}, \qquad \text{where } \mathbf{n} \sim \mathbb{C}N_m(\mathbf{0}, \mathbf{B}), \tag{6.97}$$

and the true target response is $\mathbf{d}(\phi_{\mathrm{T}}) = S_{\mathrm{T}} \cdot \check{\mathbf{v}}(\phi_{\mathrm{T}})$, where $(\check{\mathbf{v}}(\phi))_l = \exp(j\mathbf{k}_\phi^T(\mathbf{p}_l + \mathbf{e}_l))$, and $\mathbf{B} = \mathbf{I}_m + \sigma_J^2 \check{\mathbf{v}}(\phi_J)\check{\mathbf{v}}^H(\phi_J)$. Thus, the true distribution of the received signal is given by $\mathbf{x} \sim q(\mathbf{x}) = \mathbb{C}N_m(\mathbf{d}(\phi_{\mathrm{T}}), \mathbf{B})$ where we set $\phi_{\mathrm{T}} = 90$ degs and $\phi_J = 87$ degs. For such a case where both the assumed and true models follow a complex circular Gaussian distribution, the closed form expression for the MFIM and the average Hessian matrix are given by the generalized Slepian formulae (6.88) and (6.89). For this example, the 3×3 MFIM is given by

$$\mathbf{MFIM}_D(\boldsymbol{\theta}) = \begin{bmatrix} \mathbf{J}_1 & \mathbf{J}_2 \\ \mathbf{J}_2^H & 2|S|^2 \cdot \mathring{\mathbf{v}}^H(\phi)\boldsymbol{\Sigma}^{-1}\mathbf{B}\boldsymbol{\Sigma}^{-1}\mathring{\mathbf{v}}(\phi) \end{bmatrix}, \tag{6.98}$$

where $\mathbf{J}_1$ is a 2×2 identity matrix multiplied by $\mathbf{v}^H(\phi)\boldsymbol{\Sigma}^{-1}\mathbf{B}\boldsymbol{\Sigma}^{-1}\mathbf{v}(\phi)$, $\mathbf{J}_2 = [S\mathbf{v}^H(\phi)\boldsymbol{\Sigma}^{-1}\mathbf{B}\boldsymbol{\Sigma}^{-1}\mathring{\mathbf{v}}(\phi), S^*\mathbf{v}^H(\phi)\boldsymbol{\Sigma}^{-1}\mathbf{B}\boldsymbol{\Sigma}^{-1}\mathring{\mathbf{v}}(\phi)]^T$, where $\mathring{\mathbf{v}}(\phi) = d\mathbf{v}(\phi)/d\phi$. The average Hessian matrix is given by

$$\mathbf{C}_D(\boldsymbol{\theta}) = -\mathbf{J}_p(\boldsymbol{\theta}) + \begin{bmatrix} \mathbf{C}_1 & \mathbf{C}_2 \\ \mathbf{C}_2^H & 2\text{Re}\{S^*\mathring{\mathring{\mathbf{v}}}^H(\phi)\boldsymbol{\Sigma}^{-1}\delta\boldsymbol{\mu}(\boldsymbol{\theta})\} \end{bmatrix}, \tag{6.99}$$

where $\mathbf{C}_1$ is 2×2 zero matrix, $\mathbf{C}_2 = [\mathring{\mathbf{v}}^H(\phi)\boldsymbol{\Sigma}^{-1}\delta\boldsymbol{\mu}(\boldsymbol{\theta}), \delta\boldsymbol{\mu}(\boldsymbol{\theta})^H\boldsymbol{\Sigma}^{-1}\mathring{\mathbf{v}}(\phi)]^T$, $\delta\boldsymbol{\mu}(\boldsymbol{\theta}) = \mathbf{d}(\phi_{\mathbb{T}}) - S\mathbf{v}(\phi)$, and $\mathring{\mathring{\mathbf{v}}}(\phi) = d^2\mathbf{v}(\phi)/d\phi^2$. The matrix $\mathbf{J}_p(\boldsymbol{\theta})$ denotes the FIM under perfect model specification, which can be obtained by substituting $\mathbf{B} = \boldsymbol{\Sigma}$ in (6.98). It can be shown that the pseudo-true parameter $\boldsymbol{\theta}_A$ is given by [19]

$$\phi_A = \arg\max_{\phi} \frac{|\mathbf{v}^H(\phi)\boldsymbol{\Sigma}^{-1}\mathbf{d}(\phi_{\mathbb{T}})|^2}{\mathbf{v}^H(\phi)\boldsymbol{\Sigma}^{-1}\mathbf{v}(\phi)} \quad \text{and} \quad S_A = \frac{\mathbf{v}^H(\phi_A)\boldsymbol{\Sigma}^{-1}\mathbf{d}(\phi_{\mathbb{T}})}{\mathbf{v}^H(\phi_A)\boldsymbol{\Sigma}^{-1}\mathbf{v}(\phi_A)}. \tag{6.100}$$

Thus, the MCRB can be obtained by using (6.98)–(6.100) in (6.90). The closed form expression for the MCRB for $\phi_{\mathbb{T}}$ has been derived in [19].

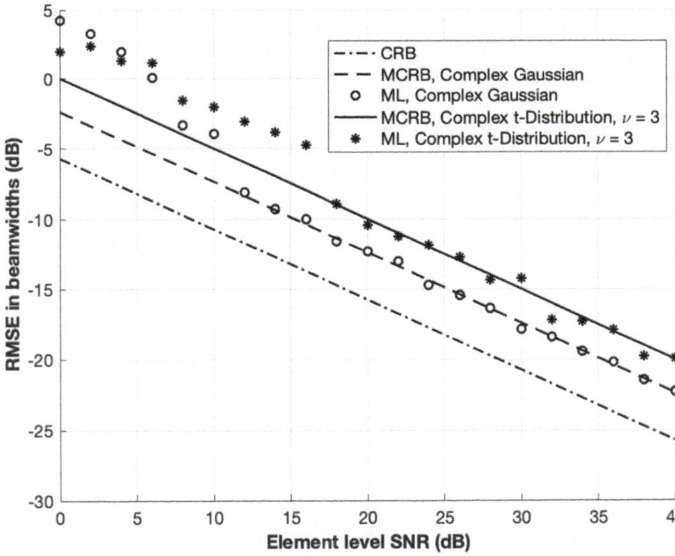

Fig. 6.3 RMSE performance of the ML estimator compared to the MCRB for the DoA $\phi_{\mathbb{T}}$ for the cases when the true distribution is a complex circular Gaussian and a complex t-distribution with parameter $\nu = 3$

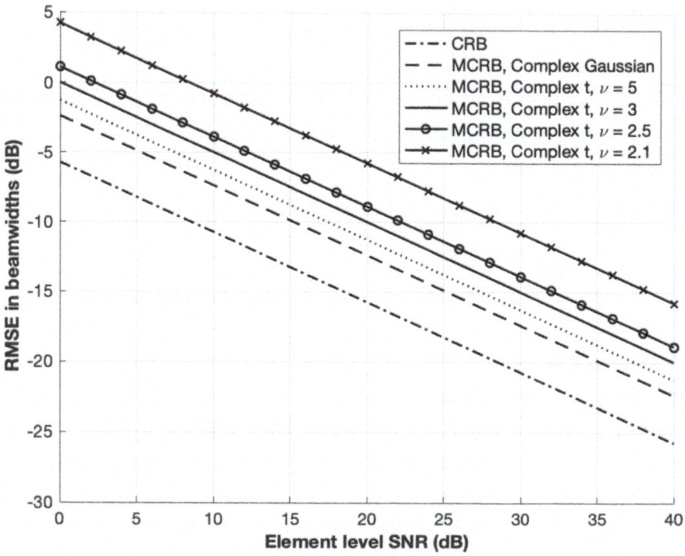

Fig. 6.4 MCRB for the array position error example as the parameter ν of the complex t-distribution is varied

6.1.7.2 Array Position Error with Complex t-Distributed Data

The complex t-distribution belongs to the family of C-CCG distributions, where the texture variable τ follows $\tau^{-1} \sim \text{Gamma}(\nu/2, 2/\nu)$, where $\text{E}(\tau) = \nu/(\nu - 2)$ for $\nu > 2$. When the true distribution of the interference is given by a complex circular t-distribution instead of the complex circular Gaussian, the received signal is given by

$$\mathbf{x} = \mathbf{d}(\phi_{\mathbb{T}}) + \mathbf{n}\sqrt{\tau}, \qquad \text{where } \mathbf{n} \sim \mathbb{C}N_m(\mathbf{0}, \mathbf{B}). \tag{6.101}$$

The MCRB for the case when the assumed distribution is complex circular Gaussian but the true distribution is a C-CCG can be obtained from (6.95). Figure 6.3 demonstrates the RMSE performance (in units of the beamwidth) of the ML estimator and the MCRB for the parameter ϕ for the two cases of true distributions discussed above. The element level signal-to-noise ratio (SNR) is defined as $|S_{\mathbb{T}}|^2$. The CRB for the case when there is no model misspecification is also shown for reference. We can see that the MCRB is higher for the case when the true distribution is a complex t, as compared to the case when the true distribution is a complex circular Gaussian. Figure 6.4 shows the MCRB as the parameter ν for the true complex t-distribution is varied. We can see that as ν increases, the MCRB approaches the MCRB for the case when the true distribution is a complex circular Gaussian since as $\nu \to \infty$, the complex circular t-distribution approaches the complex circular Gaussian distribution.

6.2 Detection

The problem of detection is a classical one that is an exercise in *binary hypothesis testing* [1, 3, 32]. The detection problem can be stated as follows:

1. Observe measured data $\mathbf{x}$.
2. Use data $\mathbf{x}$ along with a model to decide between two possible hypotheses:

$$\begin{aligned} H_0 &: \mathbf{x} \sim p(\mathbf{x}|H_0), \quad \text{Null Hypothesis} \\ H_1 &: \mathbf{x} \sim p(\mathbf{x}|H_1), \quad \text{Alternative Hypothesis.} \end{aligned}$$

Each hypothesis $H_i, i = 0, 1$ assumes a model for the data in the form of a probability density function $p(\mathbf{x}|H_i)$. Only one of these two hypotheses can be true for the observed data $\mathbf{x}$. The decision as to which hypothesis is actually true will be made by leveraging the evidentiary measured data $\mathbf{x}$ and the assumed data models. Although not a requirement, traditionally the *null hypothesis* is chosen to be the hypothesis that is believed to be true in most cases. The *alternative hypothesis*, on the other hand, usually represents the hypothesis that is rarely true.

6.2.1 *Adaptive Radar Detection*

Our discussion of detection in this section will focus primarily on the problem of *adaptive radar detection* [33–37]. Consider the set of data measurements consisting of (i) *primary data* observations $\mathbf{x}(k), k = 1, 2, \ldots, L_s$ in which a radar target return signal's presence is sought; and (ii) *secondary data* observations $\mathbf{x}(k)$, $k = L_s + 1, L_s + 2, \ldots, L_s + L_n$ that are target free but contain noise plus interference. Under each hypothesis, the data measurements may be represented as

$$\begin{aligned} &H_0 : \begin{cases} \mathbf{x}(k) = \mathbf{n}(k), \; k = 1, 2, \ldots, L_s \\ \mathbf{x}(k) = \mathbf{n}(k), \; k = L_s + 1, L_s + 2, \ldots, L_s + L_n \end{cases} \\ &H_1 : \begin{cases} \mathbf{x}(k) = w(k) \cdot \mathbf{V}\mathbf{s} + \mathbf{n}(k), \; k = 1, 2, \ldots, L_s \\ \mathbf{x}(k) = \mathbf{n}(k), \qquad\qquad\qquad k = L_s + 1, L_s + 2, \ldots, L_s + L_n, \end{cases} \end{aligned} \tag{6.102}$$

where for $k = 1, 2, \ldots, L_s + L_n$ the $\mathbf{n}(k)$ is colored noise with unknown covariance proportional to $\mathbf{R}$; $w(k)$ is a known waveform (e.g., known Doppler phase response); $\mathbf{V}$ is a $m \times G$ matrix representing a known low rank signal subspace (e.g., a set of spatial array or space–time steering vectors [38]); and $\mathbf{s}$ is a $G \times 1$ vector representing a set of unknown complex amplitudes. The signal component $w(k)\mathbf{V}\mathbf{s}$ represents the target point scatterer response of the radar return signal whose presence we desire to detect against a background of noise plus interference. Our decision about target presence will be based on the totality of data received. Specifically, define the primary and secondary data matrices, respectively, as

$$\begin{aligned} \mathbf{X}_s &\stackrel{\text{def}}{=} [\mathbf{x}(1)|\mathbf{x}(2)|\cdots|\mathbf{x}(L_s)], \\ \mathbf{X}_n &\stackrel{\text{def}}{=} [\mathbf{x}(L_s+1)|\mathbf{x}(L_s+2)|\cdots|\mathbf{x}(L_s+L_n)]. \end{aligned} \tag{6.103}$$

The totality of data is given by the $m \times (L_s + L_n)$ matrix $\mathbf{X} \stackrel{\text{def}}{=} [\mathbf{X}_s|\mathbf{X}_n]$. If each data sample vector $\mathbf{x}(k)$ (sometimes called a snapshot) is statistically independent identically distributed as complex elliptically symmetric (CES), then the model for the data takes the form

$$\mathbf{X} \sim \prod_{i=1}^{L_s+L_n} |\mathbf{R}|^{-1} g[(\mathbf{x}(i) - \mathbf{m}(i))^H \mathbf{R}^{-1} (\mathbf{x}(i) - \mathbf{m}(i))], \tag{6.104}$$

where $\mathbf{m}(i) = \mathbf{0}$ or $w(i)\mathbf{Vs}$ for $i = 1, 2, \ldots, L_s$ (depending on which hypothesis is true), and $\mathbf{m}(i) = \mathbf{0}$ for $i = L_s + 1, L_s + 2, \ldots, L_s + L_n$. Several authors have explored this type of model that when a C-CCG is considered allows the texture random variable to vary independently for each snapshot [39–44]. The model that we will adopt, however, for the total data in this discussion is the following complex multivariate elliptically symmetric (CMES) distribution

$$\mathbf{X} \sim |\mathbf{R}|^{-(L_s+L_n)} g\left[\mathrm{tr}\mathbf{R}^{-1}(\mathbf{X} - \mathbf{M})(\mathbf{X} - \mathbf{M})^H\right], \tag{6.105}$$

where $\mathbf{M} = [\mathbf{m}(1)|\mathbf{m}(2)|\cdots|\mathbf{m}(L_s + L_n)]$ and for the C-CCG case a single texture r.v. is used for all $L_s + L_n$ snapshots. Each snapshot is mutually uncorrelated, but they share the same covariance, i.e.,

$$E\{(\mathbf{x}(i) - \mathbf{m}(i))(\mathbf{x}(k) - \mathbf{m}(k))^H\} \propto \begin{cases} \mathbf{0}, & i \neq k \\ \mathbf{R}, & i = k. \end{cases} \tag{6.106}$$

The models (6.104) and (6.105) coincide if and only if $\mathbf{X}$ is circular complex Gaussian. Several previous authors have explored non-Gaussian data models similar to (6.105) [45–49].

It is noteworthy that the CMES distribution in (6.105) is expressed as a function of the data matrix $\mathbf{X}$. If, however, we define $\tilde{\mathbf{x}} = \mathrm{vec}(\mathbf{X})$ and $\tilde{\boldsymbol{\mu}} = \mathrm{vec}(\mathbf{M})$, then (6.105) can be expressed as a function of vector $\tilde{\mathbf{x}}$:

$$\tilde{\mathbf{x}} \sim |\mathbf{I}_{L_s+L_n} \otimes \mathbf{R}|^{-1} g[(\tilde{\mathbf{x}} - \tilde{\boldsymbol{\mu}})^H (\mathbf{I}_{L_s+L_n} \otimes \mathbf{R})^{-1} (\tilde{\mathbf{x}} - \tilde{\boldsymbol{\mu}})], \tag{6.107}$$

i.e., the CMES distribution (6.105) is a C-CES distribution. Thus, (6.106) is valid for all such C-CES distributions, and is not limited to C-CCG assumption. The following Sect. 6.2.2 will derive the GLRT under the CMES distribution (6.105), or equivalently the C-CES distribution (6.107). The ROC analysis provided in Sect. 6.2.3 for the GLRT, however, will be based on the special subclass of CMES distributions given by the C-CCG.

6.2.2 GLRT Under CMES Data

Let the data distribution under each hypothesis be given by

$$g_{H_i} = |\mathbf{R}|^{-(L_s+L_n)} g^{(i)} \left[\mathrm{tr}\mathbf{R}^{-1}(\mathbf{X}-\mathbf{M}_i)(\mathbf{X}-\mathbf{M}_i)^H\right],\ i=0,1. \quad (6.108)$$

Define the known waveform vector $\mathbf{w} \stackrel{\text{def}}{=} [w(1), w(2), \ldots, w(L_s)]^T$ and the mean matrices given by

$$\mathbf{M}_0 \stackrel{\text{def}}{=} \mathbf{0}_{m\times(L_s+L_n)}, \quad \mathbf{M}_1 \stackrel{\text{def}}{=} [\mathbf{Vsw}^T|\mathbf{0}_{m\times L_n}]. \quad (6.109)$$

The following assumptions about the functions $g^{(i)}(t)$, $i=0,1$ are made:

1. $g^{(i)}(t)$, $i=0,1$ on $t\geq 0$ are (i) continuous, and (ii) at least have tails which decrease monotonically.
2. Assumptions (i)–(ii) guarantee that there exists $t^{(i)}_{MAX} > 0$ such that $h^{(i)}(t) = t^{m(L_s+L_n)} g^{(i)}(m\cdot t)$ attains a maximum at $t^{(i)}_{MAX}$, $i=0,1$, respectively (see Lemma 1 in [50] and its proof; similar proofs found in [51, 52]); call them $h^{(i)}_{MAX} \stackrel{\text{def}}{=} h^{(i)}(t^{(i)}_{MAX})$.

It is noteworthy that this model includes the complex circular Gaussian considered by Kelly [34] when we choose $g^{(i)}(t) = e^{-t}\pi^{-m(L_n+L_s)}$, $i=0,1$, and $L_s=1$. The adopted model (6.108) also allows for multimodal behavior, and allows the functional form of each distribution to differ under each hypothesis, i.e., $g^{(0)}(t) \neq g^{(1)}(t)$ for $t\geq 0$ is allowed. The *generalized likelihood ratio test (GLRT)* procedure forms the asymptotically optimal test. The GLRT is obtained by forming the likelihood ratio test (LRT), and then replacing all unknown parameters with their ML estimates under each hypothesis [1, 3, 32]:

$$\frac{\max_{\mathbf{s},\mathbf{R}} g_{H_1}(\cdot)}{\max_{\mathbf{R}} g_{H_0}(\cdot)} \mathop{\gtrless}_{H_0}^{H_1} \eta. \quad (6.110)$$

Let $\mathbf{D}_i \stackrel{\text{def}}{=} (\mathbf{X}-\mathbf{M}_i)(\mathbf{X}-\mathbf{M}_i)^H$, $\mathbf{A}_i \stackrel{\text{def}}{=} \mathbf{R}^{-1/2}\mathbf{D}_i\mathbf{R}^{-1/2}$, and note that the maximization of the data distributions over $\mathbf{R}$ can be written as

$$\begin{aligned}\max_{\mathbf{R}} g_{H_i} &= \max_{\mathbf{R}} |\mathbf{R}|^{-(L_s+L_n)} g^{(i)}\left[\mathrm{tr}\mathbf{R}^{-1}(\mathbf{X}-\mathbf{M}_i)(\mathbf{X}-\mathbf{M}_i)^H\right] \\ &= \max_{\mathbf{A}_i} |\mathbf{D}_i|^{-(L_s+L_n)} |\mathbf{A}_i|^{(L_s+L_n)} g^{(i)}\left[\mathrm{tr}\mathbf{A}_i\right],\end{aligned} \quad (6.111)$$

where the last equality follows due to the one-to-one mapping between matrices $\mathbf{R}$ and $\mathbf{A}_i$. Recalling the relationship of the matrix trace and determinant to its eigenvalues [53], note that

$$
\begin{aligned}
g_{H_i} &= |\mathbf{D}_i|^{-(L_s+L_n)} \left[\prod_{k=1}^{m} \lambda_k^{\frac{1}{m}}(\mathbf{A}_i) \right]^{m(L_s+L_n)} g^{(i)} \left[\sum_{k=1}^{m} \lambda_k(\mathbf{A}_i) \right] \\
&\leq |\mathbf{D}_i|^{-(L_s+L_n)} \left[\frac{1}{m} \sum_{k=1}^{m} \lambda_k(\mathbf{A}_i) \right]^{m(L_s+L_n)} g^{(i)} \left[\sum_{k=1}^{m} \lambda_k(\mathbf{A}_i) \right],
\end{aligned}
\tag{6.112}
$$

where $\lambda_k(\mathbf{A}_i)$ is the k-th eigenvalue of matrix $\mathbf{A}_i$. The upper bound follows from the relationship between the geometric mean and arithmetic mean with equality obtained if and only if $\lambda_k(\mathbf{A}_i) = \lambda_0(i)$ for $k = 1, 2, \ldots, m$. This implies that the upper bound is obtained only by the choice $\mathbf{A}_i = \lambda_0(i)\mathbf{I}_m$, or equivalently $\mathbf{R} = (1/\lambda_0(i))\mathbf{D}_i$. Because $\mathbf{A}_i$ is hermitian positive definite, it follows that $\lambda_0(i)$, $i = 0, 1$ are real and positive. Thus, we have

$$
\begin{aligned}
\max_{\mathbf{R}} g_{H_i} &= |\mathbf{D}_i|^{-(L_s+L_n)} \max_{\lambda_0(i)} [\lambda_0(i)]^{m(L_s+L_n)} g^{(i)}[m\lambda_0(i)] \\
&= |\mathbf{D}_i|^{-(L_s+L_n)} h_{MAX}^{(i)}, \quad i = 0, 1,
\end{aligned}
\tag{6.113}
$$

where the second equality follows from the aforementioned assumptions about functions $g^{(i)}(\cdot)$, $i = 0, 1$. Thus, the GLRT is given by

$$
\frac{\max_{\mathbf{s}} |\mathbf{D}_1|^{-(L_s+L_n)} h_{MAX}^{(1)}}{|\mathbf{D}_0|^{-(L_s+L_n)} h_{MAX}^{(0)}} = \left[\frac{|\mathbf{D}_0|}{\min_{\mathbf{s}} |\mathbf{D}_1|} \right]^{(L_s+L_n)} \cdot \frac{h_{MAX}^{(1)}}{h_{MAX}^{(0)}} \underset{H_0}{\overset{H_1}{\gtrless}} \eta.
\tag{6.114}
$$

Regarding minimization of $|\mathbf{D}_1|$ over $\mathbf{s}$, note that

$$
\begin{aligned}
\mathbf{D}_1 &= ([\mathbf{X}_s|\mathbf{X}_n] - [\mathbf{V}\mathbf{s}\mathbf{w}^T|\mathbf{0}])([\mathbf{X}_s|\mathbf{X}_n] - [\mathbf{V}\mathbf{s}\mathbf{w}^T|\mathbf{0}])^H \\
&= ([\mathbf{X}_s - \mathbf{V}\mathbf{s}\mathbf{w}^T|\mathbf{X}_n])([\mathbf{X}_s - \mathbf{V}\mathbf{s}\mathbf{w}^T|\mathbf{X}_n])^H \\
&= (\mathbf{X}_s - \mathbf{V}\mathbf{s}\mathbf{w}^T)(\mathbf{X}_s - \mathbf{V}\mathbf{s}\mathbf{w}^T)^H + \mathbf{X}_n\mathbf{X}_n^H \\
&\overset{\text{def}}{=} \mathbf{D}_s + \mathbf{X}_n\mathbf{X}_n^H,
\end{aligned}
\tag{6.115}
$$

where matrix $\mathbf{D}_s = (\mathbf{X}_s - \mathbf{V}\mathbf{s}\mathbf{w}^T)(\mathbf{X}_s - \mathbf{V}\mathbf{s}\mathbf{w}^T)^H$ has been defined. This matrix can be written as

$$
\begin{aligned}
\mathbf{D}_s &= \mathbf{X}_s\mathbf{X}_s^H - \mathbf{V}\mathbf{s}\mathbf{w}^T\mathbf{X}_s^H - \mathbf{X}_s\mathbf{w}^*\mathbf{s}^H\mathbf{V}^H + \mathbf{V}\mathbf{s}\mathbf{w}^T\mathbf{w}^*\mathbf{s}^H\mathbf{V}^H \\
&= \mathbf{X}_s\mathbf{X}_s^H - \mathbf{V}\mathbf{s}\mathbf{w}^T\mathbf{X}_s^H - \mathbf{X}_s\mathbf{w}^*\mathbf{s}^H\mathbf{V}^H + \mathbf{V}\mathbf{s}\mathbf{s}^H\mathbf{V}^H \cdot \|\mathbf{w}\|^2 \\
&= \|\mathbf{w}\|^2 \left[\frac{1}{\|\mathbf{w}\|^2}\mathbf{X}_s\mathbf{X}_s^H - \frac{1}{\|\mathbf{w}\|^2}\mathbf{V}\mathbf{s}\mathbf{w}^T\mathbf{X}_s^H \right. \\
&\quad \left. - \frac{1}{\|\mathbf{w}\|^2}\mathbf{X}_s\mathbf{w}^*\mathbf{s}^H\mathbf{V}^H + \mathbf{V}\mathbf{s}\mathbf{s}^H\mathbf{V}^H \right].
\end{aligned}
\tag{6.116}
$$

If we define vectors $\tilde{\boldsymbol{\mu}}_s \stackrel{\text{def}}{=} \mathbf{X}_s\mathbf{w}^*/\|\mathbf{w}\|^2$, i.e., a generalized *sample mean vector* for the primary snapshots; and define $\mathbf{v}_s \stackrel{\text{def}}{=} \mathbf{V}\mathbf{s}$, then the matrix $\mathbf{D}_s$ can be expressed as

$$\begin{aligned}\mathbf{D}_s &= \|\mathbf{w}\|^2\left[\frac{1}{\|\mathbf{w}\|^2}\mathbf{X}_s\mathbf{X}_s^H - \mathbf{v}_s\tilde{\boldsymbol{\mu}}_s^H - \tilde{\boldsymbol{\mu}}_s\mathbf{v}_s^H + \mathbf{v}_s\mathbf{v}_s^H\right]\\ &= \|\mathbf{w}\|^2\left[\frac{1}{\|\mathbf{w}\|^2}\mathbf{X}_s\mathbf{X}_s^H - \tilde{\boldsymbol{\mu}}_s\tilde{\boldsymbol{\mu}}_s^H + \tilde{\boldsymbol{\mu}}_s\tilde{\boldsymbol{\mu}}_s^H - \mathbf{v}_s\tilde{\boldsymbol{\mu}}_s^H - \tilde{\boldsymbol{\mu}}_s\mathbf{v}_s^H + \mathbf{v}_s\mathbf{v}_s^H\right]\\ &\stackrel{\text{def}}{=} \|\mathbf{w}\|^2\left[\mathbf{D}_{s1} + (\tilde{\boldsymbol{\mu}}_s - \mathbf{v}_s)(\tilde{\boldsymbol{\mu}}_s - \mathbf{v}_s)^H\right],\end{aligned} \tag{6.117}$$

where matrix $\mathbf{D}_{s1} = (1/\|\mathbf{w}\|^2)\mathbf{X}_s\mathbf{X}_s^H - \tilde{\boldsymbol{\mu}}_s\tilde{\boldsymbol{\mu}}_s^H = (1/\|\mathbf{w}\|^2)\mathbf{X}_s\mathfrak{P}(\mathbf{w}^*|\mathbf{I}_{L_s})\mathbf{X}_s^H$ has been defined and the weighted projection operator is defined as

$$\mathfrak{P}(\mathbf{V}|\mathbf{H}) \stackrel{\text{def}}{=} \mathbf{H} - \mathbf{H}\mathbf{V}(\mathbf{V}^H\mathbf{H}\mathbf{V})^{-1}\mathbf{V}^H\mathbf{H}. \tag{6.118}$$

Inserting (6.117) into (6.115), the following form for $\mathbf{D}_1$ is obtained

$$\begin{aligned}\mathbf{D}_1 &= \|\mathbf{w}\|^2(\tilde{\boldsymbol{\mu}}_s - \mathbf{v}_s)(\tilde{\boldsymbol{\mu}}_s - \mathbf{v}_s)^H + \|\mathbf{w}\|^2\mathbf{D}_{s1} + \mathbf{X}_n\mathbf{X}_n^H\\ &\stackrel{\text{def}}{=} \|\mathbf{w}\|^2(\tilde{\boldsymbol{\mu}}_s - \mathbf{v}_s)(\tilde{\boldsymbol{\mu}}_s - \mathbf{v}_s)^H + \widehat{\mathbf{R}}\\ &= \widehat{\mathbf{R}}^{1/2}\left[\|\mathbf{w}\|^2\widehat{\mathbf{R}}^{-1/2}(\tilde{\boldsymbol{\mu}}_s - \mathbf{v}_s)(\tilde{\boldsymbol{\mu}}_s - \mathbf{v}_s)^H\widehat{\mathbf{R}}^{-1/2} + \mathbf{I}_m\right]\widehat{\mathbf{R}}^{1/2},\end{aligned} \tag{6.119}$$

where we've defined the matrix

$$\widehat{\mathbf{R}} \stackrel{\text{def}}{=} \|\mathbf{w}\|^2\mathbf{D}_{s1} + \mathbf{X}_n\mathbf{X}_n^H = \mathbf{X}_s\mathfrak{P}(\mathbf{w}^*|\mathbf{I}_{L_s})\mathbf{X}_s^H + \mathbf{X}_n\mathbf{X}_n^H, \tag{6.120}$$

i.e., the unnormalized *sample covariance matrix estimate* of $\mathbf{R}$ that leverages the totality of data given by $\mathbf{X}_s$ and $\mathbf{X}_n$. Thus, the determinant of $\mathbf{D}_1$ can be written as

$$\begin{aligned}|\mathbf{D}_1| &= |\widehat{\mathbf{R}}|\cdot\left|\|\mathbf{w}\|^2\widehat{\mathbf{R}}^{-1/2}(\tilde{\boldsymbol{\mu}}_s - \mathbf{v}_s)(\tilde{\boldsymbol{\mu}}_s - \mathbf{v}_s)^H\widehat{\mathbf{R}}^{-1/2} + \mathbf{I}_m\right|\\ &= |\widehat{\mathbf{R}}|\cdot\left[\|\mathbf{w}\|^2(\tilde{\boldsymbol{\mu}}_s - \mathbf{v}_s)^H\widehat{\mathbf{R}}^{-1}(\tilde{\boldsymbol{\mu}}_s - \mathbf{v}_s) + 1\right]\\ &= |\widehat{\mathbf{R}}|\cdot\left[\|\mathbf{w}\|^2(\tilde{\boldsymbol{\mu}}_s - \mathbf{V}\mathbf{s})^H\widehat{\mathbf{R}}^{-1}(\tilde{\boldsymbol{\mu}}_s - \mathbf{V}\mathbf{s}) + 1\right],\end{aligned} \tag{6.121}$$

where second equality follows from the determinant property $|\mathbf{I} + \mathbf{AB}| = |\mathbf{I} + \mathbf{BA}|$, or simply noting that a matrix determinant is given by the product of its eigenvalues [53]; the last equality follows from our definition of $\mathbf{v}_s$. This form for the determinant of $\mathbf{D}_1$ facilitates minimization over vector $\mathbf{s}$. Indeed, the determinant depends on $\mathbf{s}$ exclusively through the stated quadratic in (6.121). If we define the filter weight matrix $\widehat{\mathbf{W}}_{ML} \stackrel{\text{def}}{=} \widehat{\mathbf{R}}^{-1}\mathbf{V}(\mathbf{V}^H\widehat{\mathbf{R}}^{-1}\mathbf{V})^{-1}$, then note that we can complete the square in $\mathbf{s}$ to obtain the form

$$\begin{aligned}&(\tilde{\boldsymbol{\mu}}_s - \mathbf{V}\mathbf{s})^H\widehat{\mathbf{R}}^{-1}(\tilde{\boldsymbol{\mu}}_s - \mathbf{V}\mathbf{s}) =\\ &(\mathbf{s} - \widehat{\mathbf{W}}_{ML}^H\tilde{\boldsymbol{\mu}}_s)^H(\mathbf{V}^H\widehat{\mathbf{R}}^{-1}\mathbf{V})(\mathbf{s} - \widehat{\mathbf{W}}_{ML}^H\tilde{\boldsymbol{\mu}}_s) + \tilde{\boldsymbol{\mu}}_s^H\mathfrak{P}(\mathbf{V}|\widehat{\mathbf{R}}^{-1})\tilde{\boldsymbol{\mu}}_s.\end{aligned} \tag{6.122}$$

The determinant of $\mathbf{D}_1$ is clearly minimized by the following value of $\mathbf{s}$:

$$\arg\min_{\mathbf{s}} |\mathbf{D}_1| = \widehat{\mathbf{W}}_{ML}^{H} \tilde{\boldsymbol{\mu}}_s \stackrel{\text{def}}{=} \widehat{\mathbf{s}}_{ML}; \tag{6.123}$$

Thus,

$$\min_{\mathbf{s}} |\mathbf{D}_1| = |\widehat{\mathbf{R}}| \cdot \left[1 + \|\mathbf{w}\|^2 \tilde{\boldsymbol{\mu}}_s^H \mathfrak{P}(\mathbf{V}|\widehat{\mathbf{R}}^{-1}) \tilde{\boldsymbol{\mu}}_s\right]. \tag{6.124}$$

Note that matrix $\mathbf{D}_0$ can be written as

$$\begin{aligned}
\mathbf{D}_0 = \mathbf{X}\mathbf{X}^H &= \mathbf{X}_s\mathbf{X}_s^H + \mathbf{X}_n\mathbf{X}_n^H \\
&= \|\mathbf{w}\|^2 \left[\frac{1}{\|\mathbf{w}\|^2}\mathbf{X}_s\mathbf{X}_s^H - \tilde{\boldsymbol{\mu}}_s\tilde{\boldsymbol{\mu}}_s^H + \tilde{\boldsymbol{\mu}}_s\tilde{\boldsymbol{\mu}}_s^H\right] + \mathbf{X}_n\mathbf{X}_n^H \\
&= \|\mathbf{w}\|^2 \left[\mathbf{D}_{s1} + \tilde{\boldsymbol{\mu}}_s\tilde{\boldsymbol{\mu}}_s^H\right] + \mathbf{X}_n\mathbf{X}_n^H \\
&= \|\mathbf{w}\|^2 \tilde{\boldsymbol{\mu}}_s\tilde{\boldsymbol{\mu}}_s^H + \|\mathbf{w}\|^2\mathbf{D}_{s1} + \mathbf{X}_n\mathbf{X}_n^H = \|\mathbf{w}\|^2 \tilde{\boldsymbol{\mu}}_s\tilde{\boldsymbol{\mu}}_s^H + \widehat{\mathbf{R}} \\
&= \widehat{\mathbf{R}}^{1/2}\left[\|\mathbf{w}\|^2\widehat{\mathbf{R}}^{-1/2}\tilde{\boldsymbol{\mu}}_s\tilde{\boldsymbol{\mu}}_s^H\widehat{\mathbf{R}}^{-1/2} + \mathbf{I}_m\right]\widehat{\mathbf{R}}^{1/2}.
\end{aligned} \tag{6.125}$$

Thus, the determinant of $\mathbf{D}_0$ can be written as

$$\begin{aligned}
|\mathbf{D}_0| &= |\widehat{\mathbf{R}}| \cdot \left|\|\mathbf{w}\|^2\widehat{\mathbf{R}}^{-1/2}\tilde{\boldsymbol{\mu}}_s\tilde{\boldsymbol{\mu}}_s^H\widehat{\mathbf{R}}^{-1/2} + \mathbf{I}_m\right| \\
&= |\widehat{\mathbf{R}}|\left[1 + \|\mathbf{w}\|^2\tilde{\boldsymbol{\mu}}_s^H\widehat{\mathbf{R}}^{-1}\tilde{\boldsymbol{\mu}}_s\right].
\end{aligned} \tag{6.126}$$

Using (6.126) and (6.124), we can write the GLRT in (6.114) as

$$\tilde{t}_{GLRT}(\mathbf{X}) \stackrel{\text{def}}{=} \frac{1 + \|\mathbf{w}\|^2\tilde{\boldsymbol{\mu}}_s^H\widehat{\mathbf{R}}^{-1}\tilde{\boldsymbol{\mu}}_s}{1 + \|\mathbf{w}\|^2\tilde{\boldsymbol{\mu}}_s^H\mathfrak{P}(\mathbf{V}|\widehat{\mathbf{R}}^{-1})\tilde{\boldsymbol{\mu}}_s} \underset{H_0}{\overset{H_1}{\gtrless}} \left[\frac{h_{MAX}^{(0)}}{h_{MAX}^{(1)}}\eta\right]^{\frac{1}{(L_s+L_n)}} \stackrel{\text{def}}{=} \tilde{\eta}, \tag{6.127}$$

where constants have been moved to the right-hand side of the inequality, and we've taken the $(L_s + L_n)$-th root of the statistic. The GLRT in (6.127) generalizes the test obtained in [48] in two ways: (i) it allows the functional form of the probability densities to differ under each hypothesis, i.e., $g^{(0)}(t) \neq g^{(1)}(t)$ for $t \geq 0$ is allowed; and (ii) it allows for multiple primary data samples, i.e., $L_s \geq 1$ is allowed.

If we define the *adaptive matched filter (AMF)* statistic as

$$t_{AMF}(\mathbf{X}) \stackrel{\text{def}}{=} \tilde{\boldsymbol{\mu}}_s^H\widehat{\mathbf{R}}^{-1}\mathbf{V}(\mathbf{V}^H\widehat{\mathbf{R}}^{-1}\mathbf{V})^{-1}\mathbf{V}^H\widehat{\mathbf{R}}^{-1}\tilde{\boldsymbol{\mu}}_s, \tag{6.128}$$

then note the GLRT can be written as

$$\tilde{t}_{GLRT}(\mathbf{X}) = \frac{1 + \|\mathbf{w}\|^2 \tilde{\boldsymbol{\mu}}_s^H \widehat{\mathbf{R}}^{-1} \tilde{\boldsymbol{\mu}}_s}{1 + \|\mathbf{w}\|^2 \tilde{\boldsymbol{\mu}}_s^H \widehat{\mathbf{R}}^{-1} \tilde{\boldsymbol{\mu}}_s - \|\mathbf{w}\|^2 t_{AMF}(\mathbf{X})}$$
$$= \left[1 - \frac{\|\mathbf{w}\|^2 t_{AMF}(\mathbf{X})}{1 + \|\mathbf{w}\|^2 \tilde{\boldsymbol{\mu}}_s^H \widehat{\mathbf{R}}^{-1} \tilde{\boldsymbol{\mu}}_s}\right]^{-1} \underset{H_0}{\overset{H_1}{\gtrless}} \tilde{\eta}. \tag{6.129}$$

Thus, it is convenient to define the follow equivalent GLRT statistic

$$t_{GLRT}(\mathbf{X}) \stackrel{\text{def}}{=} \frac{\|\mathbf{w}\|^2 t_{AMF}(\mathbf{X})}{1 + \|\mathbf{w}\|^2 \tilde{\boldsymbol{\mu}}_s^H \widehat{\mathbf{R}}^{-1} \tilde{\boldsymbol{\mu}}_s} \underset{H_0}{\overset{H_1}{\gtrless}} \eta. \tag{6.130}$$

A version of this GLRT is mentioned in Eq. (8) of Ref. [5] that was obtained under the circular complex Gaussian data assumption. The test in (6.130), however, is more general as it holds for a class of non-Gaussian distributions with signal subspaces having signal matrices $\mathbf{V}$ with ranks greater than unity.

6.2.3 On ROC Analysis of GLRT for CMES Data

The exact closed-form analysis for the *receiver operating characteristic (ROC)* curves for the GLRT (6.130) under the general CMES assumption in (6.108) is provided in [47] for the $L_s = 1$ case. The present discussion, however, will restrict attention to the subclass of CMES distributions given by C-CCG for $L_s \geq 1$. Under hypothesis H_1, the total data matrix $\mathbf{X}$ in the C-CCG case can be represented stochastically as

$$\mathbf{X} =_d [\mathbf{V}\mathbf{s}\mathbf{w}^T | \mathbf{0}_{m \times L_n}] + \mathbf{N}\sqrt{\tau}, \tag{6.131}$$

where $\text{vec}(\mathbf{N}) \sim \mathbb{C}N_{m(L_s+L_n)}(\mathbf{0}, \mathbf{I}_{L_s+L_n} \otimes \mathbf{R})$, and τ is the texture random variable with probability density function (pdf) $p(\tau)$ defined on $\tau \geq 0$; τ and $\mathbf{N}$ are statistically independent. If we define the complex central beta distribution as

$$p(\beta) = \frac{(L_s + L_n - 1)!}{(L_s + L_n - (m-G) - 1)!(m - G - 1)!}$$
$$\times \beta^{L_s+L_n-(m-G)-1}(1-\beta)^{(m-G)-1}, \quad 0 \leq \beta \leq 1, \tag{6.132}$$

then the probability of detection (PD) for the GLRT conditioned on β and τ is shown in the appendix to be given by

$$\text{PD}_{GLRT|\beta,\tau} = 1 - \frac{(\tilde{\tilde{\eta}})^G}{(1+\tilde{\tilde{\eta}})^{L_s+L_n-m+G-1}} \sum_{k=0}^{L_s+L_n-m-1} \binom{L_s+L_n-m+G-1}{k+G} \times (\tilde{\tilde{\eta}})^k \cdot IG_{k+1}\left(\frac{\delta^2\beta/\tau}{1+\tilde{\tilde{\eta}}}\right), \tag{6.133}$$

where $\tilde{\tilde{\eta}}$ is the threshold; $IG_n(a) = e^{-a}\sum_{k=0}^{n-1} a^k/k!$ is the incomplete Gamma function; δ^2 is the integrated signal-to-interference plus noise ratio (SINR) defined as

$$\delta^2 \stackrel{\text{def}}{=} \|\mathbf{w}\|^2 \mathbf{s}^H \mathbf{V}^H \mathbf{R}^{-1} \mathbf{V}\mathbf{s}. \tag{6.134}$$

The unconditional PD for the GLRT is given by the double integral

$$\text{PD}_{GLRT} = \int_0^\infty p(\tau)\left[\int_0^1 p(\beta)\cdot \text{PD}_{GLRT|\beta,\tau} d\beta\right] d\tau. \tag{6.135}$$

The probability of false alarm (PFA) can be obtained from the PD by setting $\mathbf{s} = \mathbf{0}$, or equivalently setting $\delta = 0$. Specifically, it can be shown that the GLRT PFA conditioned on β and τ is given by

$$\text{PFA}_{GLRT|\beta,\tau} = \frac{1}{(\tilde{\tilde{\eta}}+1)^{L_s+L_n-m+G-1}} \sum_{k=0}^{G-1} \binom{L_s+L_n-m+G-1}{k} \left(\tilde{\tilde{\eta}}\right)^k, \tag{6.136}$$

which is independent of β and τ. Thus, $\text{PFA}_{GLRT} = \text{PFA}_{GLRT|\beta,\tau}$.

6.2.3.1 Numerical Results

Consider a uniform linear array of $m = 18$ elements spaced at $\lambda/2.25$ (i.e., the same array example from Sect. 6.1.7, but with no array position errors, no mismatch, nor misspecification of any kind). Here we focus simply on illustrating the impact of the non-Gaussian nature of the CMES data assumption. The radar target signal return is a rank one signal (i.e., $G = 1$) at array broadside, and the colored noise consists of a 0 dB white noise component and an interfering jammer three degrees from array broadside and 30 dB above the white noise. The GLRT threshold is chosen to satisfy a false alarm constraint of PFA$= 10^{-6}$. The GLRT PD is obtained via (6.135), but this can be simplified by noting that the texture variable only multiplies the noncentrality parameter δ^2 in (6.133). If we consider the texture variable such that $\tau^{-1} \sim \text{Gamma}(\nu/2, 2/\nu)$, then note it can be shown that

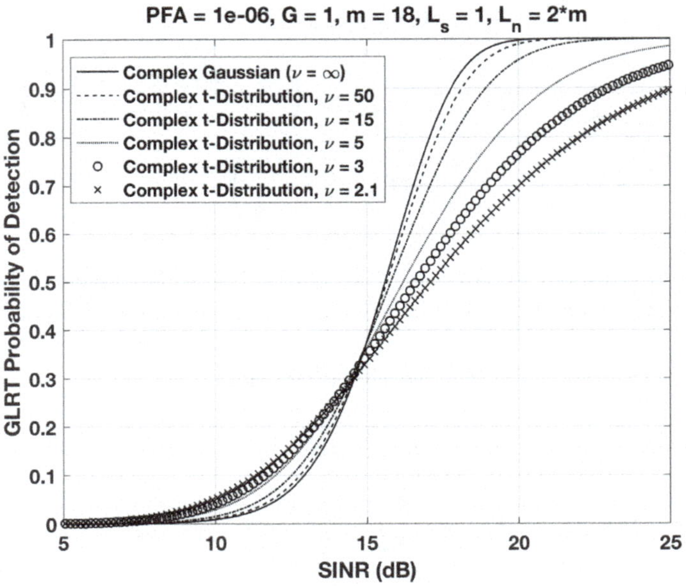

Fig. 6.5 GLRT ROC curves under CMES t-distribution where texture variable is distributed as $\tau^{-1} \sim \text{Gamma}(\nu/2, 2/\nu)$. PD versus integrated SINR for $G = 1$, $m = 18$, $L_s = 1$, $L_n = 2 * m$ shows detection loss as tails of t-distribution grow, i.e., as $\nu \to 2$

$$\widetilde{IG}_n(a) \stackrel{\text{def}}{=} \text{E}_{1/\tau}\left\{IG_n(a/\tau)\right\} = \sum_{i=0}^{n-1} \frac{\Gamma\left(\frac{\nu}{2}+i\right)}{!i \cdot \Gamma(\nu/2) \cdot (2/\nu)^{\nu/2}} \frac{a^i}{\left(\frac{\nu}{2}+a\right)^{i+\frac{\nu}{2}}}. \quad (6.137)$$

Thus, replacing $IG_{k+1}(\cdot)$ in (6.133) with $\widetilde{IG}_{k+1}(\cdot)$ (of course, without τ in the argument) yields the desired average over the texture variable τ, and one only has to average over the beta variable to obtain the final PD. Figure 6.5 shows the GLRT PD plotted versus the integrated SINR in decibels.

The solid curve shows the GLRT PD for the usual complex circular Gaussian data assumption (or equivalently the t-distribution for $\nu = \infty$). The remaining curves show the GLRT PD for $\nu = 50, 15, 5, 3$, and 2.1, i.e., as ν approaches 2 and the tails of the CMES distribution become much longer than the complex circular Gaussian. These ROC curves allow one to determine the detection loss experienced due to the longer tails of the t-distribution.

6.2.4 Asymptotic Distribution of the GLRT Statistic Under Model Misspecification

In this section, we present the asymptotic distribution of the GLRT statistic under model misspecification, i.e., when the assumed model for the received signal is different from the true model. The asymptotic distribution (i.e., the distribution when the number of data samples becomes very large or approaches infinity) of the GLRT statistic under the null hypothesis was first derived by Wilks [54], and studied extensively by various authors in [55–57]. Additionally, the distribution under the alternative hypothesis was derived in [55, 58], and a derivation for the same is also provided in Chap. 6 of [32]. Under model misspecification, the asymptotic distribution of the GLRT when the null hypothesis is true has been derived in [59]. In this section, we discuss the derivation of a general expression for the asymptotic distribution of the GLRT under model misspecification for the null as well as the alternative hypothesis provided in [60].

6.2.4.1 General Composite Hypothesis Testing Framework

We begin by considering a general composite hypothesis testing framework, where we assume a statistical model which says that the $m \times 1$ measured signal vector $\mathbf{x}$ follows a probability density function (pdf) given by $p(\mathbf{x}|\boldsymbol{\theta})$, which is parameterized by the value $\boldsymbol{\theta} \in \mathbb{R}^J$. Note that in this section, we focus only on the case of a real parameter vector $\boldsymbol{\theta}$, but the data $\mathbf{x}$ can be either real or complex. Since we are interested in the asymptotics, we assume that N independent samples of $\mathbf{x}$ denoted as $\mathbf{x}(1), \dots, \mathbf{x}(N)$ are available, and we organize these samples in the form of an extended vector as $\tilde{\mathbf{x}} = [\mathbf{x}(1); \mathbf{x}(2); \dots; \mathbf{x}(N)]$. The hypothesis test is then formulated in terms of the assumed statistical model on the extended vector given by $p(\tilde{\mathbf{x}}|\boldsymbol{\theta}) = \prod_{n=1}^{N} p(\mathbf{x}(n)|\boldsymbol{\theta})$. The parameter vector $\boldsymbol{\theta}$ takes the values $\boldsymbol{\theta}_0$ and $\boldsymbol{\theta}_1$ under the null and alternative hypotheses respectively, and is partitioned as $\boldsymbol{\theta} = [\boldsymbol{\theta}_r; \boldsymbol{\theta}_s]$, where $\boldsymbol{\theta}_r \in \mathbb{R}^r$ experiences a distinct change in value under each hypothesis and is perfectly known under the null hypothesis. We refer to the elements of $\boldsymbol{\theta}_r$ as *essential parameters*. The vector $\boldsymbol{\theta}_s \in \mathbb{R}^s$ although unknown, can remain the same under both hypotheses, and its components are referred to as *nuisance parameters*. Thus, $\boldsymbol{\theta}_i = [\boldsymbol{\theta}_{r_i}; \boldsymbol{\theta}_{s_i}]$ for $i = 0, 1$. The composite hypothesis test is formulated as follows:

$$
\begin{aligned}
H_0 &: \tilde{\mathbf{x}} \sim p(\tilde{\mathbf{x}}|\boldsymbol{\theta}_0) \text{ or } \boldsymbol{\theta}_r = \boldsymbol{\theta}_{r_0}, \qquad \boldsymbol{\theta}_{r_0} \text{ known},\\
H_1 &: \tilde{\mathbf{x}} \sim p(\tilde{\mathbf{x}}|\boldsymbol{\theta}_1) \text{ or } \boldsymbol{\theta}_r \neq \boldsymbol{\theta}_{r_0},
\end{aligned}
\tag{6.138}
$$

where $p(\tilde{\mathbf{x}}|\boldsymbol{\theta}_i)$ (also referred to as the likelihood function) denotes the pdf of the random data vector $\tilde{\mathbf{x}}$ under hypothesis H_i, that is when the parameter vector $\boldsymbol{\theta}$ takes the value $\boldsymbol{\theta}_i$, where $i \in \{0, 1\}$. Under the null hypothesis H_0, it is known that $\boldsymbol{\theta}_r$ assumes the specific value $\boldsymbol{\theta}_{r_0}$. Under the alternative hypothesis H_1, however, $\boldsymbol{\theta}_r$ can assume any value $\boldsymbol{\theta}_{r_1} \neq \boldsymbol{\theta}_{r_0}$. A test statistic $t(\tilde{\mathbf{x}})$ is exclusively a function of the received signal vector $\tilde{\mathbf{x}} \in \mathbb{C}^{mN}$, and based on the value of $t(\tilde{\mathbf{x}})$, we decide which of the two hypotheses is true. Let $t_{GLRT}(\tilde{\mathbf{x}})$ denote the GLRT statistic, then the generalized log-likelihood ratio test statistic $L(\tilde{\mathbf{x}}) = \ln t_{GLRT}(\tilde{\mathbf{x}})$ is given by

$$L(\tilde{\mathbf{x}}) = \ln p(\tilde{\mathbf{x}}|\hat{\boldsymbol{\theta}}_1) - \ln p(\tilde{\mathbf{x}}|\hat{\boldsymbol{\theta}}_0) \tag{6.139}$$
$$= \ln p(\tilde{\mathbf{x}}|\hat{\boldsymbol{\theta}}_{r_1}, \hat{\boldsymbol{\theta}}_{s_1}) - \ln p(\tilde{\mathbf{x}}|\boldsymbol{\theta}_{r_0}, \hat{\boldsymbol{\theta}}_{s_0}), \tag{6.140}$$

where $\hat{\boldsymbol{\theta}}_1$ denotes the maximum likelihood estimate (MLE) of $\boldsymbol{\theta}_1$. Since $\boldsymbol{\theta}_{r_0}$ is known under H_0, we need to only estimate $\boldsymbol{\theta}_s$, and thus the MLE of $\boldsymbol{\theta}_0$ is given by $\hat{\boldsymbol{\theta}}_0 = [\boldsymbol{\theta}_{r_0}; \hat{\boldsymbol{\theta}}_{s_0}]$. According to the definition of the MLE, we have

$$\ln p(\tilde{\mathbf{x}}|\hat{\boldsymbol{\theta}}_0) = \ln p(\tilde{\mathbf{x}}|\boldsymbol{\theta}_{r_0}, \hat{\boldsymbol{\theta}}_{s_0}) = \max_{\boldsymbol{\theta}_s} \ln p(\tilde{\mathbf{x}}|\boldsymbol{\theta}_{r_0}, \boldsymbol{\theta}_s). \tag{6.141}$$

The hypothesis test formulation given in (6.138) is based on the assumed statistical model $p(\tilde{\mathbf{x}}|\boldsymbol{\theta})$. Under a model misspecification scenario, the received data may follow different true distributions under each hypothesis as follows:

$$\begin{aligned} H_0 &: \tilde{\mathbf{x}} \sim q_0(\tilde{\mathbf{x}}) \\ H_1 &: \tilde{\mathbf{x}} \sim q_1(\tilde{\mathbf{x}}). \end{aligned} \tag{6.142}$$

In the following sections, we derive the asymptotic distribution of the GLRT statistic when the assumed and true hypothesis tests are given by (6.138) and (6.142), respectively, where $p(\tilde{\mathbf{x}}|\boldsymbol{\theta}_i) \neq q(\tilde{\mathbf{x}})$ for all possible $\boldsymbol{\theta}_i$, i = 0, 1 is allowed.

6.2.4.2 Asymptotic Distribution of the MLE Under Model Misspecification

Recall the framework of parameter estimation under model misspecification, where an MLE is derived based on an assumed model $p(\tilde{\mathbf{x}}|\boldsymbol{\theta})$, but the true distribution of $\tilde{\mathbf{x}}$ is may be given by a potentially different function $q(\tilde{\mathbf{x}})$, which might not be parameterized by $\boldsymbol{\theta}$. Under such a scenario, it was mentioned earlier that the MLE has an asymptotic mean given by the pseudo-true parameter $\boldsymbol{\theta}_A$ defined in (6.84). It has been shown in [16] that the asymptotic distribution of the MLE $\hat{\boldsymbol{\theta}}$ under model misspecification is given by $\hat{\boldsymbol{\theta}} \sim \mathbb{R}N_J\,(\boldsymbol{\theta}_A, \mathbf{MCRB}(\boldsymbol{\theta}_A))$, where $\mathbf{MCRB}(\boldsymbol{\theta}) = \mathbf{C}^{-1}(\boldsymbol{\theta}) \cdot \mathbf{MFIM}(\boldsymbol{\theta}) \cdot \mathbf{C}^{-1}(\boldsymbol{\theta})$ is the MCRB matrix appearing on the right side of (6.87). The MFIM and the average Hessian matrix for $\boldsymbol{\theta} = \boldsymbol{\theta}_A$ are given by

$$\mathbf{MFIM}(\boldsymbol{\theta}) = \mathrm{E}_q \left\{ \nabla \ln p(\tilde{\mathbf{x}}|\boldsymbol{\theta}) \cdot \nabla^T \ln p(\tilde{\mathbf{x}}|\boldsymbol{\theta}) \right\} \tag{6.143}$$
$$\mathbf{C}(\boldsymbol{\theta}) = \mathrm{E}_q \left\{ \nabla^2 \ln p(\tilde{\mathbf{x}}|\boldsymbol{\theta}) \right\}, \tag{6.144}$$

where $(\nabla f(\tilde{\mathbf{x}}))_i = \partial f(\tilde{\mathbf{x}})/\partial \theta_i$ and $(\nabla^2 f(\tilde{\mathbf{x}}))_{i,k} = \partial^2 f(\tilde{\mathbf{x}})/\partial \theta_i \partial \theta_k$ for a scalar-valued function $f(\tilde{\mathbf{x}})$. We are interested in deriving the asymptotic distribution of the GLRT $L(\tilde{\mathbf{x}})$ that is based on the assumed model $p(\tilde{\mathbf{x}}|\boldsymbol{\theta})$ given by (6.139), when the actual true pdf of $\tilde{\mathbf{x}}$ is given by $q_0(\tilde{\mathbf{x}})$ and $q_1(\tilde{\mathbf{x}})$ under H_0 and H_1, respectively. In order to derive the distribution of $L(\tilde{\mathbf{x}})$ under such model misspecification, we first require the distribution of the MLE of $\boldsymbol{\theta}$ under both hypotheses. Let $\boldsymbol{\theta}_{A_0}$ and $\boldsymbol{\theta}_{A_1}$ denote the pseudo-true parameters under the null and alternative hypothesis, respectively, i.e.,

$$\boldsymbol{\theta}_{A_i} = \arg\min_{\boldsymbol{\theta}} D(q_i(\tilde{\mathbf{x}})||p(\tilde{\mathbf{x}}|\boldsymbol{\theta})). \tag{6.145}$$

Note that $\hat{\boldsymbol{\theta}}_1$ represents the "unrestricted" MLE [32][7] and thus it has an asymptotic mean of $\boldsymbol{\theta}_{A_0}$ under H_0. Thus, the asymptotic distribution of $\hat{\boldsymbol{\theta}}_1$ is given by

$$\hat{\boldsymbol{\theta}}_1 = [\hat{\boldsymbol{\theta}}_{r_1}; \hat{\boldsymbol{\theta}}_{s_1}] \sim \mathbb{R}N_J\left(\boldsymbol{\theta}_{A_i}, \mathbf{MCRB}^{(i)}(\boldsymbol{\theta}_{A_i})\right) \text{ under } H_i, \text{ where} \tag{6.146}$$

$\mathbf{MCRB}^{(i)}(\boldsymbol{\theta}) \stackrel{\text{def}}{=} \left[\mathbf{C}^{(i)}(\boldsymbol{\theta})\right]^{-1} \mathbf{MFIM}^{(i)}(\boldsymbol{\theta}) \left[\mathbf{C}^{(i)}(\boldsymbol{\theta})\right]^{-1}$ where $\mathbf{MFIM}^{(i)}(\boldsymbol{\theta})$ and $\mathbf{C}^{(i)}(\boldsymbol{\theta})$ are given by (6.143) and (6.144) respectively, where the expectations are with respect to $q_i(\tilde{\mathbf{x}})$.

6.2.4.3 Derivation for the Asymptotic Distribution of the GLRT Statistic

We begin by considering the Taylor series expansion of the log-likelihood function, which is used to derive an asymptotically equivalent expression for the GLRT. The Taylor series expansion up to second-order of the log-likelihood function around the point $\boldsymbol{\theta} = \tilde{\boldsymbol{\theta}}$ can be written as

$$\begin{aligned} \ln p(\tilde{\mathbf{x}}|\boldsymbol{\theta}) \simeq & \ln p(\tilde{\mathbf{x}}|\tilde{\boldsymbol{\theta}}) + \nabla^T \ln p(\tilde{\mathbf{x}}|\boldsymbol{\theta})\Big|_{\boldsymbol{\theta}=\tilde{\boldsymbol{\theta}}} \cdot (\boldsymbol{\theta} - \tilde{\boldsymbol{\theta}}) \\ & + \frac{1}{2}(\tilde{\boldsymbol{\theta}} - \boldsymbol{\theta})^T \cdot \nabla^2 \ln p(\tilde{\mathbf{x}}|\boldsymbol{\theta})\Big|_{\boldsymbol{\theta}=\tilde{\boldsymbol{\theta}}} \cdot (\tilde{\boldsymbol{\theta}} - \boldsymbol{\theta}). \end{aligned} \tag{6.147}$$

Following the same approach taken in [32], we wish to approximate the function at $\boldsymbol{\theta} = \boldsymbol{\theta}_0$ about the point $\tilde{\boldsymbol{\theta}} = \hat{\boldsymbol{\theta}}_1$ using the above second-order Taylor series. Detailed discussions involving the higher-order terms of the Taylor series beyond the second-order term can be found in [58, 61]. Thus, the term on the left-hand side of (6.147) becomes $\ln p(\tilde{\mathbf{x}}|\boldsymbol{\theta}_0)$; and the first-order term $\nabla \ln p(\tilde{\mathbf{x}}|\boldsymbol{\theta})\Big|_{\boldsymbol{\theta}=\hat{\boldsymbol{\theta}}_1} = \mathbf{0}$, since we assume the MLE is a regular point that maximizes the log-likelihood function. Note that

$$\nabla^2 \ln p(\tilde{\mathbf{x}}|\boldsymbol{\theta}) = \nabla^2 \ln \prod_{n=1}^{N} p(\mathbf{x}(n)|\boldsymbol{\theta}) = \sum_{n=1}^{N} \nabla^2 \ln p(\mathbf{x}(n)|\boldsymbol{\theta}) \tag{6.148}$$

since differentiation is a linear operator. Each $\mathbf{x}(n)$ is independent and identically distributed with pdf $p(\mathbf{x}|\boldsymbol{\theta})$. Thus, taking the expectation on both sides with respect to

[7] The descriptor "unrestricted" simply means that the ML estimation process is free to estimate all dimensions of parameter vector $\boldsymbol{\theta}$ without any constraints. This is to be contrasted with the MLE under hypothesis H_0 that is only free to estimate nuisance parameters $\boldsymbol{\theta}_s$, because it is known that $\boldsymbol{\theta}_r = \boldsymbol{\theta}_{r0}$.

$q_i(\tilde{\mathbf{x}})$ results in $\mathbf{C}^{(i)}(\boldsymbol{\theta}) = N \cdot \mathrm{E}_{q_i}\{\nabla^2 \ln p(\mathbf{x}|\boldsymbol{\theta})\}$. By the weak law of large numbers [6] as $N \to \infty$, note that

$$\nabla^2 \ln p(\tilde{\mathbf{x}}|\boldsymbol{\theta}) = N \cdot \frac{1}{N}\sum_{n=1}^{N} \nabla^2 \ln p(\mathbf{x}(n)|\boldsymbol{\theta}) \to N \cdot \mathrm{E}_{q_i}\{\nabla^2 \ln p(\mathbf{x}|\boldsymbol{\theta})\} = \mathbf{C}^{(i)}(\boldsymbol{\theta}) \tag{6.149}$$

under H_i. Thus, approximating the function about the point $\tilde{\boldsymbol{\theta}} = \hat{\boldsymbol{\theta}}_1$, we get

$$\ln p(\tilde{\mathbf{x}}|\boldsymbol{\theta}_0) \simeq \ln p(\tilde{\mathbf{x}}|\hat{\boldsymbol{\theta}}_1) + \frac{1}{2}(\hat{\boldsymbol{\theta}}_1 - \boldsymbol{\theta}_0)^T \mathbf{C}^{(i)}(\hat{\boldsymbol{\theta}}_1)(\hat{\boldsymbol{\theta}}_1 - \boldsymbol{\theta}_0). \tag{6.150}$$

Substituting in (6.139) and using (6.141), we get

$$\begin{aligned} L(\tilde{\mathbf{x}}) &= \ln p(\tilde{\mathbf{x}}|\hat{\boldsymbol{\theta}}_{r_1}, \hat{\boldsymbol{\theta}}_{s_1}) - \ln p(\tilde{\mathbf{x}}|\boldsymbol{\theta}_{r_0}, \hat{\boldsymbol{\theta}}_{s_0}) \\ &\simeq \ln p(\tilde{\mathbf{x}}|\hat{\boldsymbol{\theta}}_{r_1}, \hat{\boldsymbol{\theta}}_{s_1}) - \max_{\boldsymbol{\theta}_s}\left[\ln p(\tilde{\mathbf{x}}|\hat{\boldsymbol{\theta}}_{r_1}, \hat{\boldsymbol{\theta}}_{s_1}) + \frac{1}{2}(\hat{\boldsymbol{\theta}}_1 - \boldsymbol{\theta}_0)^T \mathbf{C}^{(i)}(\hat{\boldsymbol{\theta}}_1)(\hat{\boldsymbol{\theta}}_1 - \boldsymbol{\theta}_0).\right] \\ &\simeq -\frac{1}{2}\max_{\boldsymbol{\theta}_s}\left[(\hat{\boldsymbol{\theta}}_1 - \boldsymbol{\theta}_0)^T \mathbf{C}^{(i)}(\hat{\boldsymbol{\theta}}_1)(\hat{\boldsymbol{\theta}}_1 - \boldsymbol{\theta}_0)\right]. \end{aligned} \tag{6.151}$$

Since we assume that the MLE $\hat{\boldsymbol{\theta}}_1$ is a regular point that maximizes $\ln p(\tilde{\mathbf{x}}|\boldsymbol{\theta})$, the Hessian $\nabla^2 \ln p(\tilde{\mathbf{x}}|\boldsymbol{\theta})$ when evaluated at $\boldsymbol{\theta} = \hat{\boldsymbol{\theta}}_1$ for any $\tilde{\mathbf{x}}$ must be negative semi-definite. As a result, the average Hessian $\mathbf{C}^{(i)}(\hat{\boldsymbol{\theta}}_1)$ is also negative semi-definite, and the function to be maximized in the above equation is a concave function. We can partition the parameter vector $\boldsymbol{\theta}_0$ (and $\hat{\boldsymbol{\theta}}_1$) as $\boldsymbol{\theta}_0 = [\boldsymbol{\theta}_{r_0}; \boldsymbol{\theta}_{s_0}]$, and the average Hessian matrix as follows:

$$\mathbf{C}^{(i)}(\hat{\boldsymbol{\theta}}_1) = \begin{bmatrix} \mathbf{C}^{(i)}_{rr}(\hat{\boldsymbol{\theta}}_1) & \mathbf{C}^{(i)}_{rs}(\hat{\boldsymbol{\theta}}_1) \\ \mathbf{C}^{(i)}_{sr}(\hat{\boldsymbol{\theta}}_1) & \mathbf{C}^{(i)}_{ss}(\hat{\boldsymbol{\theta}}_1) \end{bmatrix} \equiv \begin{bmatrix} r \times r & r \times s \\ s \times r & s \times s \end{bmatrix}. \tag{6.152}$$

Expanding the expression for $L(\tilde{\mathbf{x}})$, the argument of the "max" function turns out to be a quadratic in $\boldsymbol{\theta}_{s_0}$ as follows:

$$\begin{aligned} L(\tilde{\mathbf{x}}) \simeq &-\frac{1}{2}(\hat{\boldsymbol{\theta}}_{r_1} - \boldsymbol{\theta}_{r_0})^T \mathbf{C}^{(i)}_{rr}(\hat{\boldsymbol{\theta}}_1)(\hat{\boldsymbol{\theta}}_{r_1} - \boldsymbol{\theta}_{r_0}) \\ &-\frac{1}{2}\max_{\boldsymbol{\theta}_s}\Big(2(\hat{\boldsymbol{\theta}}_{r_1} - \boldsymbol{\theta}_{r_0})^T \mathbf{C}^{(i)}_{rs}(\hat{\boldsymbol{\theta}}_1)(\hat{\boldsymbol{\theta}}_{s_1} - \boldsymbol{\theta}_s) \\ &+ (\hat{\boldsymbol{\theta}}_{s_1} - \boldsymbol{\theta}_s)^T \mathbf{C}^{(i)}_{ss}(\hat{\boldsymbol{\theta}}_1)(\hat{\boldsymbol{\theta}}_{s_1} - \boldsymbol{\theta}_s)\Big) \end{aligned} \tag{6.153}$$

$$
\begin{aligned}
&\simeq -\frac{1}{2}(\hat{\boldsymbol{\theta}}_{r_1}-\boldsymbol{\theta}_{r_0})^T\mathbf{C}_{rr}^{(i)}(\hat{\boldsymbol{\theta}}_1)(\hat{\boldsymbol{\theta}}_{r_1}-\boldsymbol{\theta}_{r_0})\\
&+\frac{1}{2}(\hat{\boldsymbol{\theta}}_{r_1}-\boldsymbol{\theta}_{r_0})^T\mathbf{C}_{rs}^{(i)}(\hat{\boldsymbol{\theta}}_1)\left[\mathbf{C}_{ss}^{(i)}(\hat{\boldsymbol{\theta}}_1)\right]^{-1}\mathbf{C}_{sr}^{(i)}(\hat{\boldsymbol{\theta}}_1)(\hat{\boldsymbol{\theta}}_{r_1}-\boldsymbol{\theta}_{r_0})\\
&-\frac{1}{2}\max_{\boldsymbol{\theta}_s}\Big(\left[\mathbf{C}_{ss}^{(i)}(\hat{\boldsymbol{\theta}}_1)\right]^{-1}\mathbf{C}_{sr}^{(i)}(\hat{\boldsymbol{\theta}}_1)(\hat{\boldsymbol{\theta}}_{r_1}-\boldsymbol{\theta}_{r_0})\\
&+\hat{\boldsymbol{\theta}}_{s_1}-\boldsymbol{\theta}_s\Big)^T\mathbf{C}_{ss}^{(i)}(\hat{\boldsymbol{\theta}}_1)\Big(\left[\mathbf{C}_{ss}^{(i)}(\hat{\boldsymbol{\theta}}_1)\right]^{-1}\mathbf{C}_{sr}^{(i)}(\hat{\boldsymbol{\theta}}_1)(\hat{\boldsymbol{\theta}}_{r_1}-\boldsymbol{\theta}_{r_0})+\hat{\boldsymbol{\theta}}_{s_1}-\boldsymbol{\theta}_s\Big),
\end{aligned}
\tag{6.154}
$$

where from (6.152) it is clear that submatrix $\mathbf{C}_{ss}^{(i)}(\hat{\boldsymbol{\theta}}_1)$ is also negative semi-definite. Having completed the square in $\boldsymbol{\theta}_s$, the maximum is clearly attained at $\boldsymbol{\theta}_s=\hat{\boldsymbol{\theta}}_{s_1}+\left[\mathbf{C}_{ss}^{(i)}(\hat{\boldsymbol{\theta}}_1)\right]^{-1}\mathbf{C}_{sr}^{(i)}(\hat{\boldsymbol{\theta}}_1)(\hat{\boldsymbol{\theta}}_{r_1}-\boldsymbol{\theta}_{r_0})$, and we get the final expression for the GLRT as

$$
2L(\tilde{\mathbf{x}})\simeq-(\hat{\boldsymbol{\theta}}_{r_1}-\boldsymbol{\theta}_{r_0})^T\mathbf{C}_{rr.s}^{(i)}(\hat{\boldsymbol{\theta}}_1)(\hat{\boldsymbol{\theta}}_{r_1}-\boldsymbol{\theta}_{r_0}),
\tag{6.155}
$$

where $\mathbf{C}_{rr.s}^{(i)}(\boldsymbol{\theta})\stackrel{\text{def}}{=}\mathbf{C}_{rr}^{(i)}(\boldsymbol{\theta})-\mathbf{C}_{rs}^{(i)}(\boldsymbol{\theta})\left[\mathbf{C}_{ss}^{(i)}(\boldsymbol{\theta})\right]^{-1}\mathbf{C}_{sr}^{(i)}(\boldsymbol{\theta})$.

Recall the distribution of the MLE under model misspecification given by (6.146). Partitioning the pseudo-true parameter as $\boldsymbol{\theta}_{A_i}=[\boldsymbol{\theta}_{A_{i,r}};\boldsymbol{\theta}_{A_{i,s}}]$ and using the linear regeneration property of Gaussian random vectors, we can write

$$
\hat{\boldsymbol{\theta}}_{r_1}=_d\boldsymbol{\theta}_{A_{i,r}}+\left[\mathbf{M}^{(i)}(\boldsymbol{\theta}_{A_i})\right]^{1/2}\mathbf{z}\qquad\text{under } H_i,
\tag{6.156}
$$

where $\mathbf{M}^{(i)}(\boldsymbol{\theta})$ is the upper $r\times r$ block of $\mathbf{MCRB}^{(i)}(\boldsymbol{\theta})$ and $\mathbf{z}\sim\mathbb{R}N_r(\mathbf{0},\mathbf{I}_r)$. Thus, we have

$$
\begin{aligned}
\hat{\boldsymbol{\theta}}_{r_1}-\boldsymbol{\theta}_{r_0}&=_d\boldsymbol{\theta}_{A_{i,r}}-\boldsymbol{\theta}_{r_0}+\left[\mathbf{M}^{(i)}(\boldsymbol{\theta}_{A_i})\right]^{1/2}\mathbf{z} &(6.157)\\
&=_d\left[\mathbf{M}^{(i)}(\boldsymbol{\theta}_{A_i})\right]^{1/2}\left\{\left[\mathbf{M}^{(i)}(\boldsymbol{\theta}_{A_i})\right]^{-1/2}\Delta\boldsymbol{\theta}_{r_i}+\mathbf{z}\right\}, &(6.158)
\end{aligned}
$$

where $\Delta\boldsymbol{\theta}_{r_i}=\boldsymbol{\theta}_{A_{i,r}}-\boldsymbol{\theta}_{r_0}$. Substituting in (6.155) and noting that under asymptotic conditions $\hat{\boldsymbol{\theta}}_1\rightarrow\boldsymbol{\theta}_{A_i}$, we get

$$
\begin{aligned}
2L(\tilde{\mathbf{x}})&=_d-\left(\left[\mathbf{M}^{(i)}(\boldsymbol{\theta}_{A_i})\right]^{1/2}\left\{\left[\mathbf{M}^{(i)}(\boldsymbol{\theta}_{A_i})\right]^{-1/2}\Delta\boldsymbol{\theta}_{r_i}+\mathbf{z}\right\}\right)^T\\
&\cdot\mathbf{C}_{rr.s}^{(i)}(\boldsymbol{\theta}_{A_i})\left(\left[\mathbf{M}^{(i)}(\boldsymbol{\theta}_{A_i})\right]^{1/2}\left\{\left[\mathbf{M}^{(i)}(\boldsymbol{\theta}_{A_i})\right]^{-1/2}\Delta\boldsymbol{\theta}_{r_i}+\mathbf{z}\right\}\right)\\
&=_d-\left\{\left[\mathbf{M}^{(i)}(\boldsymbol{\theta}_{A_i})\right]^{-1/2}\Delta\boldsymbol{\theta}_{r_i}+\mathbf{z}\right\}^T\left[\mathbf{M}^{(i)}(\boldsymbol{\theta}_{A_i})\right]^{1/2}\mathbf{C}_{rr.s}^{(i)}(\boldsymbol{\theta}_{A_i})\\
&\cdot\left[\mathbf{M}^{(i)}(\boldsymbol{\theta}_{A_i})\right]^{1/2}\left[\left[\mathbf{M}^{(i)}(\boldsymbol{\theta}_{A_i})\right]^{-1/2}\Delta\boldsymbol{\theta}_{r_i}+\mathbf{z}\right\}.
\end{aligned}
\tag{6.159}
$$

Let $\mathbf{K}(\boldsymbol{\theta}_{A_i})\stackrel{\text{def}}{=}-\left[\mathbf{M}^{(i)}(\boldsymbol{\theta}_{A_i})\right]^{1/2}\mathbf{C}_{rr.s}^{(i)}(\boldsymbol{\theta}_{A_i})\left[\mathbf{M}^{(i)}(\boldsymbol{\theta}_{A_i})\right]^{1/2}$, which is a Hermitian positive definite matrix. Using eigenvalue decomposition, we have $\mathbf{K}(\boldsymbol{\theta}_{A_i})=\mathbf{Q}_K\mathbf{D}\mathbf{Q}_K^T$, where $\mathbf{Q}_K$ is the orthonormal eigenvector matrix, and $\mathbf{D}$ is a diagonal matrix containing the eigenvalues of $\mathbf{K}(\boldsymbol{\theta}_{A_i})$. Also, let

$$\tilde{\mathbf{m}} = [\tilde{m}_1, \ldots, \tilde{m}_r]^T \stackrel{\text{def}}{=} \mathbf{Q}_K^T \left[\mathbf{M}^{(i)}(\boldsymbol{\theta}_{A_i})\right]^{-1/2} \Delta\boldsymbol{\theta}_{r_i}. \tag{6.160}$$

Thus, we get

$$\begin{aligned} 2L(\tilde{\mathbf{x}}) &=_d \left[\tilde{\mathbf{m}} + \mathbf{z}\right]^T \mathbf{D} \left[\tilde{\mathbf{m}} + \mathbf{z}\right] \\ &= \lambda_1 |\tilde{m}_1 + z_1|^2 + \cdots + \lambda_r |\tilde{m}_r + z_r|^2, \end{aligned} \tag{6.161}$$

where we have noted $\mathbf{Q}_K^T \mathbf{z} =_d \mathbf{z}$ since an orthonormal transformation does not change the distribution of a standardized real Gaussian random vector. $\lambda_1, \ldots, \lambda_r$ are the eigenvalues of $\mathbf{K}(\boldsymbol{\theta}_{A_i})$ (assuming each eigenvalue has an algebraic multiplicity of unity). Generally, one or more eigenvalues can have algebraic multiplicity greater than one. Even in such a case, $2L(\tilde{\mathbf{x}})$ will be a linear combination of real non-central chi-squared random variables as given in (6.162), but the number of terms in the linear combination will be less than r. More details involving this aspect can be found in Appendix B of [62]. Since the magnitude squared of a non-zero mean unit-variance real Gaussian random variable is a non-central chi-squared random variable, the GLRT statistic is distributed as a linear combination of real non-central chi-squared random variables each having one degree of freedom

$$2L(\tilde{\mathbf{x}}) =_d \lambda_1 \chi_1^2(\delta_1) + \cdots + \lambda_r \chi_1^2(\delta_r), \tag{6.162}$$

where the non-centrality parameters are such that $\delta_i^2 = |\tilde{m}_i|^2$, $i = 1, \ldots, r$. This distribution reduces to the one derived in [59] for the null hypothesis when $\boldsymbol{\theta}_{A_0,r} = \boldsymbol{\theta}_{r_0}$, i.e., $\Delta\boldsymbol{\theta}_{r_0} = \mathbf{0}$. For a given probability of false alarm, we can use the above distribution under H_0 to compute the threshold for the GLRT. Based on the threshold, the probability of detection can be computed using the above distribution under H_1. Such a process can be used to predict the ROC performance of the GLRT statistic.

6.2.4.4 Numerical Results

In this section, we consider the problem of radar target detection under model misspecification. Based on the asymptotic distribution of the GLRT obtained in (6.162), we can predict the ROC performance of the GLRT. We compare the predicted asymptotic ROC performance with the simulated ROC performance of the GLRT via Monte Carlo (MC) simulations. Similar to Sect. 6.1.7, we consider two scenarios for the array position error example, i.e., the case when the true distribution is a circular complex Gaussian and the case when the true distribution is a C-CCG distribution.

We extend the example considered in Sect. 6.1.7 to the case of multiple snapshots, i.e., we assume the availability of N independent snapshots denoted as $\mathbf{x}(1), \ldots, \mathbf{x}(N) \in \mathbb{C}^m$. Considering all the snapshots available, the total received signal vector is given by $\tilde{\mathbf{x}} = [\mathbf{x}(1); \mathbf{x}(2); \ldots; \mathbf{x}(N)] \in \mathbb{C}^g$ where $g = Nm$. The target detection problem for the assumed model can be formulated as a composite hypothesis test given by

$$\begin{aligned} &H_0 : \tilde{\mathbf{x}} = \tilde{\mathbf{n}} \sim \mathbb{C}N_g(\mathbf{0}, \tilde{\boldsymbol{\Sigma}}) && \text{(Target absent)} \\ &H_1 : \tilde{\mathbf{x}} = S \cdot \tilde{\mathbf{v}}(\phi) + \tilde{\mathbf{n}} \sim \mathbb{C}N_g(S \cdot \tilde{\mathbf{v}}(\phi), \tilde{\boldsymbol{\Sigma}}) && \text{(Target present)}, \end{aligned} \tag{6.163}$$

where $\tilde{\mathbf{v}}(\phi) = \mathbf{1}_N \otimes \mathbf{v}(\phi)$ and $\tilde{\boldsymbol{\Sigma}} = \mathbf{I}_N \otimes \boldsymbol{\Sigma}$, where $\mathbf{1}_N$ denotes a column vector of size N containing all ones, and $\mathbf{I}_N$ denotes an $N \times N$ identity matrix. The unknown parameter vector $\boldsymbol{\theta} = [\text{Re}(S), \text{Im}(S), \phi]$ is partitioned as $\boldsymbol{\theta} = [\boldsymbol{\theta}_r; \boldsymbol{\theta}_s]$, where $\boldsymbol{\theta}_r = [\text{Re}(S), \text{Im}(S)]$ and $\boldsymbol{\theta}_s = \phi$ are the essential and nuisance parameters, respectively, with $r = 2$. We have considered the real and imaginary parts of S instead of S and S^* as unknown parameters since the asymptotic result in (6.162) has been derived for the case when the parameters are real.

We first consider the asymptotic distribution of the GLRT under H_1. For the complex circular Gaussian case, the true distribution of the received data can be written as $\tilde{\mathbf{x}} \sim q_1(\tilde{\mathbf{x}}) = \mathbb{C}N_g(\tilde{\mathbf{d}}(\phi_{\mathrm{T}}), \tilde{\mathbf{B}})$, where $\tilde{\mathbf{d}}(\phi_{\mathrm{T}}) = \mathbf{1}_N \otimes \mathbf{d}(\phi_{\mathrm{T}})$ and $\tilde{\mathbf{B}} = \mathbf{I}_N \otimes \mathbf{B}$ and $\mathbf{d}(\phi_{\mathrm{T}})$ was defined in Sect. 6.1.7.1. Since the assumed and true distributions of $\tilde{\mathbf{x}}$ are complex circular Gaussian, we can obtain expressions for $\mathbf{MFIM}_D^{(1)}(\boldsymbol{\theta})$ and $\mathbf{C}_D^{(1)}(\boldsymbol{\theta})$ using the generalized Slepian formula in Sect. 6.1.6.6. Based on these matrices, we can obtain the MCRB matrix $\mathbf{MCRB}^{(1)}(\boldsymbol{\theta})$ as shown in Sect. 6.2.4.2. The pseudo-true parameters can also obtained by (6.100), by replacing $\mathbf{v}(\phi)$ and $\mathbf{d}(\phi_{\mathrm{T}})$ with $\tilde{\mathbf{v}}(\phi)$ and $\tilde{\mathbf{d}}(\phi_{\mathrm{T}})$ respectively. Thus, we can obtain all the required quantities involved in the distribution given by (6.162). The asymptotic distribution of the GLRT under H_1 as per (6.162) will be a linear combination of two real non-central chi-squared random variables of degree one. Considering the case when the true distribution is a C-CCG distribution with mean $\tilde{\mathbf{d}}(\phi_{\mathrm{T}})$, texture variable τ and scatter matrix $\tilde{\mathbf{B}}$, we can see based on Sect. 6.1.6.7 that the MCRB matrix will be given by $\mathrm{E}\{\tau\}$ times the MCRB matrix obtained for the case when the true distribution is $\mathbb{C}N_g(\tilde{\mathbf{d}}(\phi_{\mathrm{T}}), \tilde{\mathbf{B}})$. Also, we have seen that the average Hessian and the pseudo-true parameters are the same as that in the complex circular Gaussian case. Thus, using these quantities, we can obtain the asymptotic distribution of the GLRT under H_1 when the true data is C-CCG distributed, and this distribution will also be a linear combination of two real non-central chi-squared random variables of degree one. Note that the pdf of a random variable of the form of (6.162) is difficult to express in a closed form, but can be computed via different techniques such as saddlepoint methods [63]. A set of MATLAB codes to compute the pdfs of such distributions has been provided in [64].

We now consider the distribution of the GLRT under H_0. At first, it might seem that H_0 is just a special case of H_1 with $S = 0$, and the discussion for the H_1 case above can be applied as it is. However, since $\tilde{\mathbf{d}}(\phi_{\mathrm{T}}) = \mathbf{0}$ in this case, a unique solution for ϕ_A does not exist as per (6.100). Indeed, we get $S_A = 0$, and for any choice of ϕ_A the Hessian matrix $\mathbf{C}_D^{(0)}(\boldsymbol{\theta}_A)$ is singular. This violates the underlying assumption that $\boldsymbol{\theta}_A$ is a regular point of the divergence, which is implicitly assumed in (6.84). In

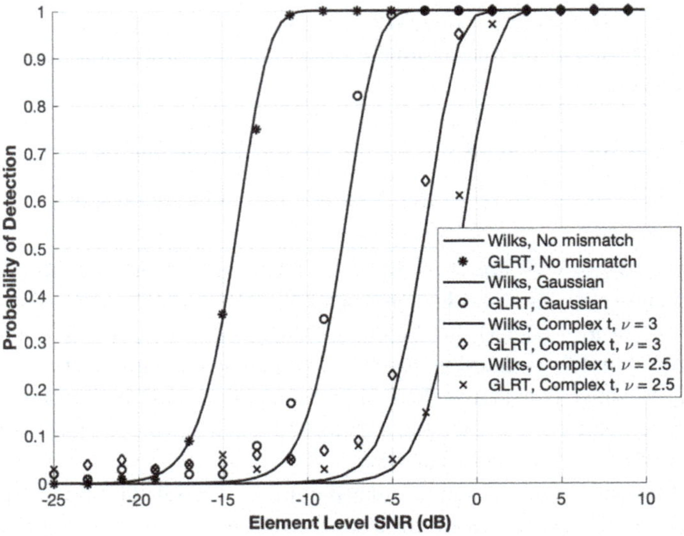

Fig. 6.6 Comparison of the ROC performance of the GLRT obtained via MC simulations and the asymptotic Wilks' prediction obtained via the result in (6.162)

order to avoid this situation, we consider the set of unknown parameters under H_0 as $\boldsymbol{\theta} = \boldsymbol{\theta}_r = [\text{Re}(S), \text{Im}(S)]$ with ϕ and $\tilde{\mathbf{v}}(\phi)$ assumed known, resulting in the absence of any nuisance parameters. With the updated set of parameters, we can compute the necessary quantities appearing in (6.162) in a similar way as discussed above in the H_1 case. Applying (6.162) to the case when the true distribution is a complex circular Gaussian, it can be shown that under H_0, asymptotically

$$2L(\mathbf{x}) =_d \frac{\tilde{\mathbf{v}}^H(\phi)\tilde{\boldsymbol{\Sigma}}^{-1}\tilde{\mathbf{B}}\tilde{\boldsymbol{\Sigma}}^{-1}\tilde{\mathbf{v}}(\phi)}{\tilde{\mathbf{v}}^H(\phi)\tilde{\boldsymbol{\Sigma}}^{-1}\tilde{\mathbf{v}}(\phi)} \cdot \chi_2^2 \stackrel{\text{def}}{=} \tilde{c}(\tilde{\mathbf{B}}) \cdot \chi_2^2, \tag{6.164}$$

where χ_2^2 denotes a real central chi-squared random variable with two degrees of freedom. Note that the distribution is a single χ_2^2 instead of a linear combination of two real chi-squared random variables with one degree of freedom as given in (6.162). This is because the matrix $\mathbf{K}(\boldsymbol{\theta}_{A_0})$ turns out to be a 2×2 identity matrix scaled by constant $\tilde{c}(\tilde{\mathbf{B}})$. Thus, instead of two distinct eigenvalues, we have a single eigenvalue with multiplicity two which leads to a single-scaled chi-squared random variable with two degrees of freedom. Note that when $\tilde{\boldsymbol{\Sigma}} = \tilde{\mathbf{B}}$, $2L(\mathbf{x}) =_d \chi_2^2$, which is the well-known Wilks' theorem under perfect model specification. For the case when the true distribution is a C-CCG with scatter matrix $\tilde{\mathbf{B}}$ and texture variable τ, we get $2L(\mathbf{x}) \sim E\{\tau\} \cdot \tilde{c}(\tilde{\mathbf{B}}) \cdot \chi_2^2$ under H_0.

Figure 6.6 shows the ROC performance of the GLRT in terms of the probability of detection versus the element level SNR for a fixed probability of false alarm 10^{-6}. We consider the same simulation scenario as considered in Sect. 6.1.7, with

$\phi_{\mathbb{T}} = 90$ degs, but with $N = 20$ snapshots and the jammer situated at $\phi_J = 83$ degs. We compare the ROC performance of the GLRT obtained via MC simulations and that predicted by the result in (6.162) for the case when the true distribution is a complex circular Gaussian, and the case when the true distribution is complex t-distributed. We also show the ROC curves for the no model mismatch case. We can see that the predicted performance provides good agreement with the MC simulations for the complex circular Gaussian mismatch case as well as for different values of ν for the complex t-distributed mismatch case.

Appendix: GLRT Derived Distribution

Under the C-CCG data assumption for H_1, note that when conditioned on τ, $\tilde{\boldsymbol{\mu}}_s = \mathbf{X}_s\mathbf{w}^*/\|\mathbf{w}\|^2 \sim \mathbb{C}N_m(\mathbf{Vs}, \tau\mathbf{R}/\|\mathbf{w}\|^2)$ since Gaussians regenerate under linear transformations [7]. Thus, it can be represented stochastically as

$$\tilde{\boldsymbol{\mu}}_s =_d \mathbf{Vs} + \frac{\sqrt{\tau}}{\|\mathbf{w}\|}\mathbf{R}^{1/2}\mathbf{z}_0, \quad \text{where } \mathbf{z}_0 \sim \mathbb{C}N_m(\mathbf{0}, \mathbf{I}_m). \tag{6.165}$$

Also, note that by Theorem 3.1.2 of [65] that matrix $\|\mathbf{w}\|^2\mathbf{D}_{s1} = \mathbf{X}_s\mathfrak{P}(\mathbf{w}^*|\mathbf{I}_{L_s})\mathbf{X}_s^H$ is complex central Wishart distributed with $L_s - 1$ degrees of freedom and with covariance parameter $\tau\mathbf{R}$, and that both $\tilde{\boldsymbol{\mu}}_s$ and $\|\mathbf{w}\|^2\mathbf{D}_{s1}$ are statistically independent.[8] Consequently, $\widehat{\mathbf{R}}$ is also independent of $\tilde{\boldsymbol{\mu}}_s$, and $\widehat{\mathbf{R}}$ must also be complex central Wishart distributed with $L_s + L_n - 1$ degrees of freedom with covariance parameter $\tau\mathbf{R}$ by virtue of Theorem 3.2.4 of [65]. Let $\text{vec}(\mathbf{Z}_{m\times(L_s+L_n-1)}) \sim \mathbb{C}N_{m(L_s+L_n-1)}(\mathbf{0}, \mathbf{I}_{L_s+L_n-1} \otimes \mathbf{I}_m)$, independent of $\tilde{\boldsymbol{\mu}}_s$. Clearly, $\mathbf{Z} =_d \mathbf{Q}^H\mathbf{Z}$ for any $m \times m$ unitary matrix $\mathbf{Q}$ (i.e., such that $\mathbf{Q}^H\mathbf{Q} = \mathbf{Q}\mathbf{Q}^H = \mathbf{I}_m$). If we define data matrix $\mathbf{X}_R \stackrel{\text{def}}{=} \sqrt{\tau}\mathbf{R}^{1/2}\mathbf{Z} =_d \sqrt{\tau}\mathbf{R}^{1/2}\mathbf{Q}^H\mathbf{Z}$, then we can represent the data covariance parameter estimate as $\widehat{\mathbf{R}} =_d \mathbf{X}_R\mathbf{X}_R^H$. Therefore,

$$\begin{aligned}\widehat{\mathbf{R}}^{-1} &=_d (\sqrt{\tau}\mathbf{R}^{1/2}\mathbf{Q}^H\mathbf{Z}\mathbf{Z}^H\mathbf{Q}\mathbf{R}^{1/2}\sqrt{\tau})^{-1} \\ &=_d \frac{1}{\sqrt{\tau}}\mathbf{R}^{-1/2}\mathbf{Q}^H(\mathbf{Z}\mathbf{Z}^H)^{-1}\mathbf{Q}\mathbf{R}^{-1/2}\frac{1}{\sqrt{\tau}}.\end{aligned} \tag{6.166}$$

We will choose $\mathbf{Q}$ such that

$$\begin{aligned}\mathbf{Q}\mathbf{R}^{-1/2}\mathbf{V} &= \begin{bmatrix}\Delta_{G\times G} \\ \mathbf{0}_{(m-G)\times G}\end{bmatrix} \Longrightarrow \\ \mathbf{V}^H\mathbf{R}^{-1}\mathbf{V} = \mathbf{V}^H\mathbf{R}^{-1/2}\mathbf{Q}^H\mathbf{Q}\mathbf{R}^{-1/2}\mathbf{V} &= [\Delta^H, \mathbf{0}^H]\begin{bmatrix}\Delta \\ \mathbf{0}\end{bmatrix} = \Delta^H\Delta.\end{aligned} \tag{6.167}$$

[8] Although the proof of [65] Theorem 3.1.2 considers the real Gaussian and the special case of $\mathbf{w} = [1, 1, \ldots, 1]^T$, the proof actually holds for the complex circular Gaussian and all $\mathbf{w}$ such that $\|\mathbf{w}\|^2 \neq 0$.

Such a unitary matrix can be found via a QR-decomposition using a series of Householder transformations [53]. Now define the transformed sample mean as

$$\begin{aligned}\boldsymbol{\mu}_0 &\stackrel{\text{def}}{=} \frac{\|\mathbf{w}\|}{\sqrt{\tau}}\mathbf{Q}\mathbf{R}^{-1/2}\tilde{\boldsymbol{\mu}}_s =_d \frac{\|\mathbf{w}\|}{\sqrt{\tau}}\mathbf{Q}\mathbf{R}^{-1/2}\mathbf{V}\mathbf{s} + \frac{\|\mathbf{w}\|}{\sqrt{\tau}}\mathbf{Q}\mathbf{R}^{-1/2}\frac{\sqrt{\tau}}{\|\mathbf{w}\|}\mathbf{R}^{1/2}\mathbf{z}_0 \\ &=_d \frac{\|\mathbf{w}\|}{\sqrt{\tau}}\begin{bmatrix}\Delta\\ \mathbf{0}\end{bmatrix}\mathbf{s} + \mathbf{z}_0 \stackrel{\text{def}}{=} \begin{bmatrix}(\boldsymbol{\mu}_1)_{G\times 1}\\ (\boldsymbol{\mu}_2)_{(m-G)\times 1}\end{bmatrix},\end{aligned} \tag{6.168}$$

where second equality follows from (6.165); and for third equality, we note (6.167) and that $\mathbf{Q}\mathbf{z}_0 =_d \mathbf{z}_0$; lastly, the partition of $\boldsymbol{\mu}_0$ is defined.

Let $L' \stackrel{\text{def}}{=} L_s + L_n - 1$ and partition the $m \times L'$ white data matrix $\mathbf{Z}$ such that

$$\mathbf{Z} = \begin{bmatrix}[\mathbf{Z}_1^H]_{G\times L'}\\ [\mathbf{Z}_2^H]_{(m-G)\times L'}\end{bmatrix} \Longrightarrow \mathbf{Z}\mathbf{Z}^H = \begin{bmatrix}\mathbf{Z}_1^H\mathbf{Z}_1 & \mathbf{Z}_1^H\mathbf{Z}_2\\ \mathbf{Z}_2^H\mathbf{Z}_1 & \mathbf{Z}_2^H\mathbf{Z}_2\end{bmatrix}. \tag{6.169}$$

Applying the matrix inversion lemma [37, 53] to (6.169), it is shown that

$$\left(\mathbf{Z}\mathbf{Z}^H\right)^{-1} = \begin{bmatrix}\boldsymbol{\Theta}^{-1}_{G\times G} & -\boldsymbol{\Theta}^{-1}\boldsymbol{\mathcal{T}}^H\\ -\boldsymbol{\mathcal{T}}\boldsymbol{\Theta}^{-1} & \boldsymbol{\Psi}_{(m-G)\times(m-G)}\end{bmatrix}, \tag{6.170}$$

where

$$\begin{aligned}&\boldsymbol{\mathcal{T}}_{(m-G)\times G} \stackrel{\text{def}}{=} (\mathbf{Z}_2^H\mathbf{Z}_2)^{-1}\mathbf{Z}_2^H\mathbf{Z}_1, \qquad \boldsymbol{\Psi} \stackrel{\text{def}}{=} \boldsymbol{\mathcal{T}}\boldsymbol{\Theta}^{-1}\boldsymbol{\mathcal{T}}^H + (\mathbf{Z}_2^H\mathbf{Z}_2)^{-1},\\ &\boldsymbol{\Theta} \stackrel{\text{def}}{=} \mathbf{Z}_1^H\mathfrak{P}(\mathbf{Z}_2|\mathbf{I}_{L'})\mathbf{Z}_1.\end{aligned} \tag{6.171}$$

Substituting the stochastic representation for $\widehat{\mathbf{R}}$ given in (6.166) into the AMF statistic form (6.128) yields

$$\begin{aligned}\|\mathbf{w}\|^2 t_{AMF}(\mathbf{X}) =_d\ & \|\mathbf{w}\|\tilde{\boldsymbol{\mu}}_s^H\frac{1}{\sqrt{\tau}}\mathbf{R}^{-1/2}\mathbf{Q}^H(\mathbf{Z}\mathbf{Z}^H)^{-1}\mathbf{Q}\mathbf{R}^{-1/2}\frac{1}{\sqrt{\tau}}\mathbf{V}\\ &\times(\mathbf{V}^H\frac{1}{\sqrt{\tau}}\mathbf{R}^{-1/2}\mathbf{Q}^H(\mathbf{Z}\mathbf{Z}^H)^{-1}\mathbf{Q}\mathbf{R}^{-1/2}\frac{1}{\sqrt{\tau}}\mathbf{V})^{-1}\\ &\times\mathbf{V}^H\frac{1}{\sqrt{\tau}}\mathbf{R}^{-1/2}\mathbf{Q}^H(\mathbf{Z}\mathbf{Z}^H)^{-1}\mathbf{Q}\mathbf{R}^{-1/2}\frac{1}{\sqrt{\tau}}\tilde{\boldsymbol{\mu}}_s\|\mathbf{w}\|.\end{aligned} \tag{6.172}$$

Recalling definition of $\boldsymbol{\mu}_0$ in (6.168), and how unitary matrix $\mathbf{Q}$ is chosen in (6.167), we can express the AMF statistic as

$$\begin{aligned}\|\mathbf{w}\|^2 t_{AMF}(\mathbf{X}) =_d\ & \boldsymbol{\mu}_0^H(\mathbf{Z}\mathbf{Z}^H)^{-1}\begin{bmatrix}\Delta\\ \mathbf{0}\end{bmatrix}\\ &\times\left([\Delta^H, \mathbf{0}^H](\mathbf{Z}\mathbf{Z}^H)^{-1}\begin{bmatrix}\Delta\\ \mathbf{0}\end{bmatrix}\right)^{-1}\\ &\times[\Delta^H, \mathbf{0}^H](\mathbf{Z}\mathbf{Z}^H)^{-1}\boldsymbol{\mu}_0.\end{aligned} \tag{6.173}$$

Using the defined partition of the matrix inverse defined in (6.170), the AMF statistic can be written as

$$\begin{aligned}\|\mathbf{w}\|^2 t_{AMF}(\mathbf{X}) &=_d \boldsymbol{\mu}_0^H \begin{bmatrix} \mathbf{I}_G \\ -\mathcal{T} \end{bmatrix} \boldsymbol{\Theta}^{-1} \Delta(\Delta^H \boldsymbol{\Theta}^{-1} \Delta)^{-1} \Delta^H \boldsymbol{\Theta}^{-1} [\mathbf{I}_G, -\mathcal{T}^H] \boldsymbol{\mu}_0 \\ &=_d \boldsymbol{\mu}_0^H \begin{bmatrix} \mathbf{I}_G \\ -\mathcal{T} \end{bmatrix} \boldsymbol{\Theta}^{-1} [\mathbf{I}_G, -\mathcal{T}^H] \boldsymbol{\mu}_0 \\ &=_d \left\| \boldsymbol{\Theta}^{-1/2} [\boldsymbol{\mu}_1 - \mathbf{Z}_1^H \mathbf{Z}_2 (\mathbf{Z}_2^H \mathbf{Z}_2)^{-1} \boldsymbol{\mu}_2] \right\|^2,\end{aligned} \tag{6.174}$$

where the last equality uses the partition of $\boldsymbol{\mu}_0$ defined in (6.168) and definitions given in (6.171).

The term in the GLRT that is quadratic in sample mean $\tilde{\boldsymbol{\mu}}_s$ can be represented as

$$\begin{aligned}\|\mathbf{w}\|^2 \tilde{\boldsymbol{\mu}}_s^H \widehat{\mathbf{R}}^{-1} \tilde{\boldsymbol{\mu}}_s &=_d \tilde{\boldsymbol{\mu}}_s^H \frac{\|\mathbf{w}\|}{\sqrt{\tau}} \mathbf{R}^{-1/2} \mathbf{Q}^H (\mathbf{Z}\mathbf{Z}^H)^{-1} \mathbf{Q} \mathbf{R}^{-1/2} \frac{\|\mathbf{w}\|}{\sqrt{\tau}} \tilde{\boldsymbol{\mu}}_s \\ &=_d \boldsymbol{\mu}_0^H (\mathbf{Z}\mathbf{Z}^H)^{-1} \boldsymbol{\mu}_0,\end{aligned} \tag{6.175}$$

where (6.166) and (6.168) have been recalled. Note that the matrix inverse in (6.170) can be expressed in the equivalent form

$$(\mathbf{Z}\mathbf{Z}^H)^{-1} = \begin{bmatrix} \mathbf{I}_G \\ -\mathcal{T} \end{bmatrix} \boldsymbol{\Theta}^{-1} [\mathbf{I}_G, -\mathcal{T}^H] + \begin{bmatrix} \mathbf{0} & \mathbf{0} \\ \mathbf{0} & (\mathbf{Z}_2^H \mathbf{Z}_2)^{-1} \end{bmatrix}. \tag{6.176}$$

Thus, the quadratic in $\tilde{\boldsymbol{\mu}}_s$ can be expressed as

$$\begin{aligned}\|\mathbf{w}\|^2 \tilde{\boldsymbol{\mu}}_s^H \widehat{\mathbf{R}}^{-1} \tilde{\boldsymbol{\mu}}_s &=_d \boldsymbol{\mu}_0^H (\mathbf{Z}\mathbf{Z}^H)^{-1} \boldsymbol{\mu}_0 \\ &=_d \boldsymbol{\mu}_0^H \begin{bmatrix} \mathbf{I}_G \\ -\mathcal{T} \end{bmatrix} \boldsymbol{\Theta}^{-1} [\mathbf{I}_G, -\mathcal{T}^H] \boldsymbol{\mu}_0 + \boldsymbol{\mu}_2^H (\mathbf{Z}_2^H \mathbf{Z}_2)^{-1} \boldsymbol{\mu}_2 \\ &=_d \|\mathbf{w}\|^2 t_{AMF}(\mathbf{X}) + \boldsymbol{\mu}_2^H (\mathbf{Z}_2^H \mathbf{Z}_2)^{-1} \boldsymbol{\mu}_2,\end{aligned} \tag{6.177}$$

where the second equality recalls (6.176) and the partition in (6.168); and the last equality follows from (6.174). Note from (6.129) that

$$\begin{aligned}\tilde{\tilde{t}}_{GLRT}(\mathbf{X}) &\stackrel{\text{def}}{=} \tilde{t}_{GLRT}(\mathbf{X}) - 1 \\ &= \frac{\|\mathbf{w}\|^2 t_{AMF}(\mathbf{X})}{1 + \|\mathbf{w}\|^2 \tilde{\boldsymbol{\mu}}_s^H \widehat{\mathbf{R}}^{-1} \tilde{\boldsymbol{\mu}}_s - \|\mathbf{w}\|^2 t_{AMF}(\mathbf{X})}.\end{aligned} \tag{6.178}$$

Using (6.177), $\tilde{\tilde{t}}_{GLRT}(\mathbf{X})$ can be expressed as

$$\begin{aligned}\tilde{\tilde{t}}_{GLRT}(\mathbf{X}) &=_d \frac{\|\mathbf{w}\|^2 t_{AMF}(\mathbf{X})}{1 + \|\mathbf{w}\|^2 t_{AMF}(\mathbf{X}) + \boldsymbol{\mu}_2^H (\mathbf{Z}_2^H \mathbf{Z}_2)^{-1} \boldsymbol{\mu}_2 - \|\mathbf{w}\|^2 t_{AMF}(\mathbf{X})} \\ &=_d \frac{\|\mathbf{w}\|^2 t_{AMF}(\mathbf{X})}{1 + \boldsymbol{\mu}_2^H (\mathbf{Z}_2^H \mathbf{Z}_2)^{-1} \boldsymbol{\mu}_2}.\end{aligned} \tag{6.179}$$

Thus, we've established the following relationship between the GLRT and AMF statistics:

$$\tilde{\tilde{t}}_{GLRT}(\mathbf{X}) =_d \|\mathbf{w}\|^2 t_{AMF}(\mathbf{X}) \cdot \beta, \quad \text{where } \beta \stackrel{\text{def}}{=} \frac{1}{1 + \boldsymbol{\mu}_2^H (\mathbf{Z}_2^H \mathbf{Z}_2)^{-1} \boldsymbol{\mu}_2}. \tag{6.180}$$

Combining this with (6.174), we can express the GLRT statistic as

$$\tilde{\tilde{t}}_{GLRT}(\mathbf{X}) =_d \left\| \boldsymbol{\Theta}^{-1/2} [\boldsymbol{\mu}_1 - \mathbf{Z}_1^H \mathbf{Z}_2 (\mathbf{Z}_2^H \mathbf{Z}_2)^{-1} \boldsymbol{\mu}_2] \sqrt{\beta} \right\|^2. \tag{6.181}$$

Note that if we condition on $\boldsymbol{\mu}_2$ and $\mathbf{Z}_2$, then note that

$$\begin{gathered} \boldsymbol{\mu}_1 - \mathbf{Z}_1^H \mathbf{Z}_2 (\mathbf{Z}_2^H \mathbf{Z}_2)^{-1} \boldsymbol{\mu}_2 \sim \mathbb{C}N_G \left[\frac{\|\mathbf{w}\|}{\sqrt{\tau}} \Delta \mathbf{s}, \boldsymbol{\Upsilon} \right], \\ \text{where} \quad \boldsymbol{\Upsilon} = \mathbf{I}_G + E\left\{ \mathbf{Z}_1^H \mathbf{Z}_2 (\mathbf{Z}_2^H \mathbf{Z}_2)^{-1} \boldsymbol{\mu}_2 \boldsymbol{\mu}_2^H (\mathbf{Z}_2^H \mathbf{Z}_2)^{-1} \mathbf{Z}_2^H \mathbf{Z}_1 \middle| \boldsymbol{\mu}_2, \mathbf{Z}_2 \right\}. \end{gathered} \tag{6.182}$$

We can evaluate $\boldsymbol{\Upsilon}$ by labeling the rows of $\mathbf{Z}_1^H$ as

$$\mathbf{Z}_1^H \stackrel{\text{def}}{=} \begin{bmatrix} [\tilde{\mathbf{z}}_1^H]_{1 \times L'} \\ [\tilde{\mathbf{z}}_2^H]_{1 \times L'} \\ \vdots \\ [\tilde{\mathbf{z}}_G^H]_{1 \times L'} \end{bmatrix}, \quad \text{where } E\{\tilde{\mathbf{z}}_k \tilde{\mathbf{z}}_i^H\} = \delta_{k,i} \cdot \mathbf{I}_{L'} \tag{6.183}$$

and for $k, i = 1, 2, \ldots, G$, where

$$\delta_{k,i} \stackrel{\text{def}}{=} \begin{cases} 1, & i = k \\ 0, & \text{otherwise.} \end{cases} \tag{6.184}$$

Note that, when conditioned on $\boldsymbol{\mu}_2$ and $\mathbf{Z}_2$, it follows that

$$\begin{aligned} &\left(E\left\{ \mathbf{Z}_1^H \mathbf{Z}_2 (\mathbf{Z}_2^H \mathbf{Z}_2)^{-1} \boldsymbol{\mu}_2 \boldsymbol{\mu}_2^H (\mathbf{Z}_2^H \mathbf{Z}_2)^{-1} \mathbf{Z}_2^H \mathbf{Z}_1 \middle| \boldsymbol{\mu}_2, \mathbf{Z}_2 \right\}\right)_{i,k} \\ &= E\left\{ \mathbf{e}_i^T \mathbf{Z}_1^H \mathbf{Z}_2 (\mathbf{Z}_2^H \mathbf{Z}_2)^{-1} \boldsymbol{\mu}_2 \boldsymbol{\mu}_2^H (\mathbf{Z}_2^H \mathbf{Z}_2)^{-1} \mathbf{Z}_2^H \mathbf{Z}_1 \mathbf{e}_k \middle| \boldsymbol{\mu}_2, \mathbf{Z}_2 \right\} \\ &= E\left\{ \tilde{\mathbf{z}}_i^H \mathbf{Z}_2 (\mathbf{Z}_2^H \mathbf{Z}_2)^{-1} \boldsymbol{\mu}_2 \boldsymbol{\mu}_2^H (\mathbf{Z}_2^H \mathbf{Z}_2)^{-1} \mathbf{Z}_2^H \tilde{\mathbf{z}}_k \middle| \boldsymbol{\mu}_2, \mathbf{Z}_2 \right\} \\ &= E\left\{ \boldsymbol{\mu}_2^H (\mathbf{Z}_2^H \mathbf{Z}_2)^{-1} \mathbf{Z}_2^H \tilde{\mathbf{z}}_k \tilde{\mathbf{z}}_i^H \mathbf{Z}_2 (\mathbf{Z}_2^H \mathbf{Z}_2)^{-1} \boldsymbol{\mu}_2 \middle| \boldsymbol{\mu}_2, \mathbf{Z}_2 \right\} \\ &= \boldsymbol{\mu}_2^H (\mathbf{Z}_2^H \mathbf{Z}_2)^{-1} \mathbf{Z}_2^H E\left\{ \tilde{\mathbf{z}}_k \tilde{\mathbf{z}}_i^H \right\} \mathbf{Z}_2 (\mathbf{Z}_2^H \mathbf{Z}_2)^{-1} \boldsymbol{\mu}_2 \\ &= \delta_{k,i} \cdot \boldsymbol{\mu}_2^H (\mathbf{Z}_2^H \mathbf{Z}_2)^{-1} \boldsymbol{\mu}_2. \end{aligned} \tag{6.185}$$

Thus, conditioned on $\boldsymbol{\mu}_2$ and $\mathbf{Z}_2$, it follows that

$$\boldsymbol{\Upsilon} = \mathbf{I}_G \cdot \left(1 + \boldsymbol{\mu}_2^H (\mathbf{Z}_2^H \mathbf{Z}_2)^{-1} \boldsymbol{\mu}_2\right) = \mathbf{I}_G \cdot \frac{1}{\beta} \tag{6.186}$$

and that

$$\boldsymbol{\mu}_1 - \mathbf{Z}_1^H \mathbf{Z}_2 (\mathbf{Z}_2^H \mathbf{Z}_2)^{-1} \boldsymbol{\mu}_2 \sim \mathbb{C}N_G \left[\frac{\|\mathbf{w}\|}{\sqrt{\tau}} \Delta \mathbf{s}, \frac{1}{\beta} \mathbf{I}_G \right]. \tag{6.187}$$

The conditioning on $\boldsymbol{\mu}_2$ and $\mathbf{Z}_2$ is completely summarized by the random variable β. Thus, we can define $G \times 1$ vector $\boldsymbol{\vartheta}$ as

$$\begin{aligned} \boldsymbol{\vartheta} &\stackrel{\text{def}}{=} (\boldsymbol{\mu}_1 - \mathbf{Z}_1^H \mathbf{Z}_2 (\mathbf{Z}_2^H \mathbf{Z}_2)^{-1} \boldsymbol{\mu}_2) \sqrt{\beta} \\ &\sim \mathbb{C}N_G \left[(\sqrt{\beta} \|\mathbf{w}\| / \sqrt{\tau}) \Delta \mathbf{s}, \mathbf{I}_G \right] = p(\boldsymbol{\vartheta} | \beta, \tau), \end{aligned} \tag{6.188}$$

and by (6.181) we can write the GLRT statistic as

$$\tilde{\tilde{t}}_{GLRT}(\mathbf{X}) =_d \left\| \boldsymbol{\Theta}^{-1/2} \boldsymbol{\vartheta} \right\|^2 = \boldsymbol{\vartheta}^H \boldsymbol{\Theta}^{-1} \boldsymbol{\vartheta}. \tag{6.189}$$

The Wishart matrix partition Theorem 3.2.10 in [65] informs that $\boldsymbol{\Theta}$ and $\boldsymbol{\mathcal{T}}$ are statistically independent. Thus, $\boldsymbol{\Theta}$ must be independent of β (we already know both $\boldsymbol{\Theta}$ and β are independent of τ).

Recall that $\boldsymbol{\Theta} = \mathbf{Z}_1^H \mathfrak{P}(\mathbf{Z}_2 | \mathbf{I}_{L'}) \mathbf{Z}_1$. Since $\mathfrak{P}(\mathbf{Z}_2 | \mathbf{I}_{L'})$ is a projection matrix onto the space orthogonal to the columns of $\mathbf{Z}_2$, we know that it has an eigendecomposition of form

$$\mathfrak{P}(\mathbf{Z}_2 | \mathbf{I}_{L'}) = \mathbf{Q}_z^H \operatorname{diag}(1, 0, 1, 1, \ldots, 1, 0, 0) \mathbf{Q}_z, \tag{6.190}$$

where $\mathbf{Q}_z$ is an $L' \times L'$ unitary matrix, and the eigenvalues are zero or unity [53]. The number of unity valued eigenvalues must be $L' - (m - G)$ due to known rank of $\mathbf{Z}_2$. By invariance of white complex circular Gaussian data, it follows that $\mathbf{Q}_z \mathbf{Z}_1 =_d \mathbf{Z}_1$. Thus,

$$\begin{aligned} \boldsymbol{\Theta} &=_d \mathbf{Z}_1^H \mathfrak{P}(\mathbf{Z}_2 | \mathbf{I}_{L'}) \mathbf{Z}_1 \\ &=_d \mathbf{Z}_1^H \mathbf{Q}_z^H \operatorname{diag}(1, 0, 1, 1, \ldots, 1, 0, 0) \mathbf{Q}_z \mathbf{Z}_1 \\ &=_d \mathbf{Z}_1^H \operatorname{diag}(1, 0, 1, 1, \ldots, 1, 0, 0) \mathbf{Z}_1 =_d \mathbf{Z}_3 \mathbf{Z}_3^H, \end{aligned} \tag{6.191}$$

where $\mathbf{Z}_3$ is a $G \times (L' - m + G)$ white complex circular Gaussian data matrix, and $\boldsymbol{\Theta}$ must be complex central Wishart with $L' - m + G$ degrees of freedom and covariance parameter $\mathbf{I}_G$. Clearly, $\boldsymbol{\Theta}$ is independent of any specific value of $\mathbf{Z}_2$. Thus, conditioning on $\boldsymbol{\vartheta}$ it follows by Theorems 3–4 (Wishart Inversion Theorem) of [37], that

$$\tilde{\tilde{t}}_{GLRT}(\mathbf{X}) =_d \boldsymbol{\vartheta}^H \boldsymbol{\Theta}^{-1} \boldsymbol{\vartheta} =_d \frac{\|\boldsymbol{\vartheta}\|^2}{\chi^2_{L'-m+1}}, \tag{6.192}$$

where $\chi^2_{L'-m+1}$ is a complex central chi-squared random variable with $L' - m + 1$ degrees of freedom. Conditioned on β and τ it follows from basic complex circular Gaussian related statistics (see Appendix A in [66]) that

$$\|\boldsymbol{\vartheta}\|^2 =_d \chi^2_G \left(\delta \sqrt{\beta / \tau} \right), \tag{6.193}$$

i.e., it is distributed as a complex noncentral chi-squared random variable of G degrees of freedom with a noncentrality parameter proportional to the integrated signal-to-interference plus noise ratio (SINR) defined as

$$\delta^2 \stackrel{\text{def}}{=} \|\mathbf{w}\|^2 \mathbf{s}^H \Delta^H \Delta \mathbf{s} = \|\mathbf{w}\|^2 \mathbf{s}^H \mathbf{V}^H \mathbf{R}^{-1} \mathbf{V} \mathbf{s}, \tag{6.194}$$

where second equality follows from (6.167). The GLRT statistic is, therefore, seen to be distributed as

$$\tilde{\tilde{t}}_{GLRT}(\mathbf{X}) =_d \frac{\chi_G^2\left(\delta\sqrt{\frac{\beta}{\tau}}\right)}{\chi^2_{L'-m+1}} =_d F_{G,L_s+L_n-m}\left(\delta\sqrt{\frac{\beta}{\tau}}\right) \Longrightarrow p(\tilde{\tilde{t}}_{GLRT}|\beta,\tau), \tag{6.195}$$

where $F_{G,L_s+L_n-m}(\delta\sqrt{\beta/\tau})$ is a complex noncentral F-distributed random variable with G, $L_s + L_n - m$ degrees of freedom and a noncentrality parameter given by $\delta^2\beta/\tau$ (see Appendix A of [66]). Regarding the distribution of random variable β, recall that $\boldsymbol{\mu}_2$ and $\mathbf{Z}_2$ are independent, and note that $\mathbf{Z}_2^H\mathbf{Z}_2$ is a $(m - G) \times (m - G)$ complex central Wishart matrix with L' degrees of freedom and covariance parameter $\mathbf{I}_{m-G}$. Thus, we have

$$\begin{aligned} \boldsymbol{\mu}_2^H(\mathbf{Z}_2^H\mathbf{Z}_2)^{-1}\boldsymbol{\mu}_2 &=_d \frac{\|\boldsymbol{\mu}_2\|^2}{\chi^2_{L'-(m-G)+1}} \\ &=_d \frac{\chi^2_{m-G}}{\chi^2_{L'-(m-G)+1}} =_d F_{m-G,L'-m+G+1}, \end{aligned} \tag{6.196}$$

where the first equality follows by Theorems 3–4 of [37]; and the second equality follows from (6.168) that indicates $\boldsymbol{\mu}_2 \sim \mathbb{C}N_{m-G}(\mathbf{0}, \mathbf{I}_{m-G})$; recalling complex circular Gaussian related statistics (see Appendix A [66]), $\boldsymbol{\mu}_2^H(\mathbf{Z}_2^H\mathbf{Z}_2)^{-1}\boldsymbol{\mu}_2$ is seen to be complex central F distributed. Consequently,

$$\begin{aligned} \beta &= \frac{1}{1 + \boldsymbol{\mu}_2^H(\mathbf{Z}_2^H\mathbf{Z}_2)^{-1}\boldsymbol{\mu}_2} =_d \frac{1}{1 + F_{m-G,L'-m+G+1}} \\ &=_d \beta_{L'-m+G+1,m-G} = \beta_{L_s+L_n-(m-G),m-G}, \end{aligned} \tag{6.197}$$

where the probability density function is given by the complex central beta distribution in (6.132). The PD for the GLRT conditioned on β and τ is given by

$$\begin{aligned} \text{PD}_{GLRT|\beta,\tau} &= \Pr(\tilde{\tilde{t}}_{GLRT}(\mathbf{X}) > \tilde{\tilde{\eta}} \mid \beta, \tau) \\ &= \Pr(F_{G,L_s+L_n-m}(\delta\sqrt{\beta/\tau}) > \tilde{\tilde{\eta}} \mid \beta, \tau) \end{aligned} \tag{6.198}$$

which leads to (6.133). The unconditional PD for the GLRT is given by (6.135).

The PFA follows from the PD by setting $\delta = 0$. Specifically,

$$\text{PFA}_{GLRT|\beta,\tau} = \Pr(F_{G,L_s+L_n-m}(0) > \tilde{\tilde{\eta}} \mid \beta, \tau) =$$

$$1 - \frac{(\tilde{\tilde{\eta}})^G}{(1+\tilde{\tilde{\eta}})^{L_s+L_n-m+G-1}} \sum_{k=0}^{L_s+L_n-m-1} \binom{L_s+L_n-m+G-1}{k+G} \cdot (\tilde{\tilde{\eta}})^k. \quad (6.199)$$

An alternative form for the GLRT PFA is possible by noting the following:

$$\begin{aligned} \text{PFA}_{GLRT|\beta,\tau} &= \Pr(F_{G,L_s+L_n-m} > \tilde{\tilde{\eta}} \mid \beta, \tau) \\ &= \Pr(-1 + 1/\beta_{L_s+L_n-m,G} > \tilde{\tilde{\eta}} \mid \beta, \tau) \\ &= \Pr(\beta_{L_s+L_n-m,G} < 1/(\tilde{\tilde{\eta}}+1) \mid \beta, \tau) \end{aligned} \quad (6.200)$$

which leads to (6.136). It's independent of β and τ. Thus, $\text{PFA}_{GLRT} = \text{PFA}_{GLRT|\beta,\tau}$.

References

1. Van Trees, H.L., Bell, K.L., Tian, Z.: Detection, Estimation and Modulation Theory, 2nd edn. John Wiley & Sons Inc., Hoboken, New Jersey (2013)
2. S.M. Kay, Fundamentals of Statistical Signal Processing, Volume I: Estimation Theory. Prentice-Hall, Upper Saddle River, NJ (1993)
3. Scharf, L.L.: Statistical Signal Processing: Detection, Estimation, and Time Series Analysis. Addison Wesley, Reading, MA (1991)
4. Lehmann, E.L., Casella, G.: Theory of Point Estimation, 2nd edn. Springer, New York (1998)
5. Richmond, C.D.: Mean Squared Error and Threshold SNR Prediction of Maximum-Likelihood Signal Parameter Estimation with Estimated Colored Noise Covariances, vol. 52, no. 5, pp. 2146–2164 (May 2006)
6. Yates, R.D., Goodman, D.J.: Probability and Stochastic Processes, 3rd edn. John Wiley & Sons Inc., Hoboken, NJ (2014)
7. Papoulis, A., Pillai, S.U.: Probability, Random Variables, and Stochastic Processes. McGraw-Hill, Boston, MA (2014)
8. Blyth, C.R., Roberts, D.M.: On inequalities of Cramér-Rao type and admissibility proofs. In: Proceedings of the Sixth Berkeley Symposium on Mathematical Statistics and Probability, Berkeley, CA, pp. 17–30 (1972)
9. Blyth, C.R.: Necessary and sufficient conditions for inequalities of the Cramér-Rao type. Ann. Stat. **2**(3), 464–473 (May1974)
10. Abel, J.S.: A Bound on Mean-Squared-Estimate Error, vol. 39, no. 5, pp. 1675–1680 (Sept 1993)
11. McWhorter, L.T., Scharf, L.L.: Properties of quadratic covariance bounds. In: Proceedings of the Asilomar Conference on Signals, Systems and Computers, vol. 2, pp. 1176–1180 (Nov 1993)
12. Rao, C.R.: Linear Statistical Inference and Its Applications, 2nd edn. John Wiley & Sons Inc., New York, NY (1973)
13. Slepian, D.: Estimation of signal parameters in the presence of noise. Trans. IRE Prof. Group Inform. Theory, **IT-3**, 68–89 (Mar 1954)
14. Bangs, W.J.: Array Processing with Generalized Beamformers. Ph.D. Thesis, Yale University, New Haven, CT (1971)
15. Stoica, P., Moses, R.: Spectral Analysis of Signals. Pearson Prentice-Hall, Upper Saddle River, New Jersey (2005)

16. Huber, P.J.: The behavior of maximum likelihood estimates under nonstandard conditions. In: Proceedings of the Fifth Berkley Symposium on Mathematical Statistics and Probability, pp. 221–233 (1967)
17. White, H.: Maximum-likelihood estimation of misspecified models. Econometrica **50**(1), 1–26 (Jan 1982)
18. Vuong, Q.H.: Cramér-Rao bounds for misspecified models (Working Paper 652). In: Division of the Humanities and Social Sciences, California Institute of Technology (1986). https://www.hss.caltech.edu/content/cramer-rao-bounds-misspecified-models
19. Richmond, C.D., Horowitz, L.L.: Parameter bounds on estimation accuracy under model misspecification, vol. 63, no. 9, pp. 2263–2278 (May 2015)
20. Fortunati, S., Gini, F., Greco, M.S.: The constrained misspecified Cramér-Rao bound, vol. 23, no. 5, pp. 718–721 (May 2016)
21. S. Fortunati, F. Gini, M.S. Greco, The misspecified Cramér-Rao bound and its application to the scatter matrix estimation in complex elliptically symmetric distributions, vol. 64, no. 9, pp. 2387–2399 (Nov 2016)
22. S. Fortunati, F. Gini, M.S. Greco, C.D. Richmond, Performance bounds for parameter estimation under misspecified models, vol. 34, no. 6, pp. 142–157 (Nov 2017)
23. Remmert, R.: Theory of Complex Functions. Springer-Verlag, Inc., New York (1991)
24. Brandwood, D.H.: A complex gradient operator and its application in adaptive array theory. In: IEE Proceedings H (Microwaves, Optics, and Antennas) (British), vol. 130, no. 1, pp. 11–16 (Feb 1983)
25. van den Bos, A.: Complex gradient and hessian. In: IEE Proceeding-Vision Image Signal Processing (British), vol. 141, no. 6, pp. 380–382 (Dec 1994)
26. Forster, P., Larzabal, P.: On lower bounds for deterministic parameter estimation. In: Proceedings of the IEEE International Conference on Acoustics, Speech, and Signal Processing (ICASSP), vol. 2, pp. 1137–1140 (May 2002)
27. Renaux, A., Forster, P., Larzabal, P., Richmond, C.D., Nehorai, A.: A fresh look at the Bayesian bounds of the Weiss-Weinstein family, vol. 56, no. 11, pp. 5334–5352 (Nov 2008)
28. Richmond, C.D.: On constraints in parameter estimation and model misspecification. In: 2018 21st International Conference on Information Fusion (FUSION), pp. 1080–1085 (2018)
29. Rudin, W.: Principles of Mathematical Analysis, 3rd edn. McGraw-Hill Inc., New York (1976)
30. Cover, T.M., Thomas, J.A.: Elements of Information Theory, 2nd edn. John Wiley & Sons Inc., New Jersey (2006)
31. Parker, P.A., Richmond, C.D.: Methods and bounds for waveform parameter estimation with a misspecified model. In: 2015 49th Asilomar Conference on Signals, Systems and Computers, pp. 1702–1706 (2015)
32. Kay, S.M.: Fundamentals of Statistical Signal Processing, Volume II: Detection Theory. Prentice-Hall, Upper Saddle River, NJ (1998)
33. Reed, I., Mallett, J., Brennan, L.: Rapid convergence rate in adaptive arrays. IEEE Trans. Aerosp. Electron. Syst. **AES-10**(6), 853–863 (1974)
34. Kelly, E.: An adaptive detection algorithm. IEEE Trans. Aerosp. Electron. Syst. **AES-22**(2), 115–127 (1986)
35. Chen, W.S., Reed, I.S.: A new CFAR detection test for radar, vol. 1, no. 4, pp. 198–214, (Oct 1991)
36. Robey, F.C., Fuhrmann, D.R., Kelly, E.J., Nitzberg, R.A.: A CFAR adaptive matched filter detector, vol. 28, no. 1, pp. 208–216 (Jan 1992)
37. Steinhardt, A.O.: Adaptive radar detection and estimation. In: Haykin, S., Steinhardt, A.O. (eds.) Adaptive Multisensor Detection and Estimation, Chapter 3. John Wiley & Sons, Inc., New York, NY (1992)
38. Ward, J.: Space-Time Adaptive Processing for Airborne Radar. Technical Report 1015, MIT Lincoln Laboratory, Lexington, MA (1994)
39. Conte, E., Lops, M., Ricci, G.: Asymptotically optimum radar detection in compound-gaussian clutter. IEEE Trans. Aerosp. Electron. Syst. **31**(2), 617–625 (1995)

40. Conte, E., Lops, M., Ricci, G.: Adaptive matched filter detection in spherically invariant noise. IEEE Signal Process. Lett. **3**(8), 248–250 (1996)
41. Conte, E., Lops, M., Ricci, G.: Adaptive detection schemes in compound-gaussian clutter. IEEE Trans. Aerosp. Electron. Syst. **34**(4), 1058–1069 (1998)
42. Conte, E., De Maio, A., Ricci, G.: Covariance matrix estimation for adaptive cfar detection in compound-gaussian clutter. IEEE Trans. Aerosp. Electron. Syst. **38**(2), 415–426 (2002)
43. Pascal, F., Chitour, Y., Ovarlez, J.-P., Forster, P., Larzabal, P.: Covariance structure maximum-likelihood estimates in compound gaussian noise: Existence and algorithm analysis. IEEE Trans. Signal Process. **56**(1), 34–48 (2008)
44. Chitour, Y., Pascal, F.: Exact maximum likelihood estimates for SIRV covariance matrix: existence and algorithm analysis. IEEE Trans. Signal Process. **56**(10), 4563–4573 (2008)
45. Khatri, C.G., Rao, C.R.: Test for a Specified Signal when the Noise Covariance Matrix is Unknown. Technical Report 85-47, Center for Multivariate Analysis, University of Pittsburgh, Pittsburgh, PA (Nov 1985)
46. Khatri, C.G., Rao, C.R.: Effects of estimated noise covariance matrix in optimal signal detection. IEEE Trans. Acoust. Speech Signal Process. **35**(5), 671–679 (1987)
47. Richmond, C.D.: Adaptive array signal processing and performance analysis in non-Gaussian environments. Ph.D. Dissertation, Massachusetts Institute of Technology, Cambridge, MA (1996)
48. Richmond, C.D.: A note on non-Gaussian adaptive array detection and signal parameter estimation. IEEE Signal Process. Lett. **3**(8), 251–252 (1996)
49. Pulsone, N., Raghavan, R.: Analysis of an adaptive CFAR detector in non-gaussian interference. IEEE Trans. Aerosp. Electron. Syst. **35**(3), 903–916 (1999)
50. Anderson, T.W., Fang, K.-T., Hsu, H.: Maximum-likelihood estimates and likelihood-ratio criteria for multivariate elliptically contoured distributions. Canad. J. Statist. /La Revue Canadienne de Statistique **14**(1), 55–59 (1986)
51. Fang, K.-T., Anderson, T.W. (eds.): Statistical Inference in Elliptically Contoured and Related Distributions, pp. 201–206. Allerton, New York (1990)
52. Gupta, A.K., Varga, T.: Elliptically Contoured Models in Statistics, pp. 231–233. Kluwer, Boston, MA (1993)
53. Noble, B., Daniel, J.W.: Applied Linear Algebra, 3rd edn. Prentice-Hall, Englewood Cliffs, NJ (1988)
54. Wilks, S.S.: The large-sample distribution of the likelihood ratio for testing composite hypotheses. Ann. Math. Stat. **9**(1), 60–62 (1938)
55. Wald, A.: Tests of statistical hypotheses concerning several parameters when the number of observations is large. Trans. Am. Math. Soc. **54**(3), 426–482 (1943)
56. Chernoff, H.: On the distribution of the likelihood ratio. In: The Annals of Mathematical Statistics, pp. 573–578 (1954)
57. Roy, K.: A note on the asymptotic distribution of likelihood ratio. Calcutta Statist. Assoc. Bull. **7**(2), 73–77 (1957)
58. Davidson, R.R., Lever, W.E.: The limiting distribution of the likelihood ratio statistic under a class of local alternatives. Sankhyā: Indian J. Statist. Ser. A 209–224 (1970)
59. Foutz, R.V., Srivastava, R.: The performance of the likelihood ratio test when the model is incorrect. Ann. Statist. 1183–1194 (1977)
60. Bondre, A.S., Richmond, C.D.: Asymptotic distribution of generalized likelihood ratio test under model misspecification with application to cooperative radar-communications. In: Proceedings of the IEEE International Conference on Acoustics, Speech, and Signal Processing (ICASSP), pp. 8463–8467 (June 2021)
61. Noam, Y., Tabrikian, J.: Marginal likelihood for estimation and detection theory. IEEE Trans. Signal Process. **55**(8), 3963–3974 (2007)
62. Richmond, C.D.: Performance of a class of adaptive detection algorithms in nonhomogeneous environments. IEEE Trans. Signal Process. **48**(5), 1248–1262 (2000)
63. Daniels, H.E.: Saddlepoint approximations in statistics. Ann. Math. Statist. 631–650 (1954)

64. Das, A., Geisler, W.S.: A method to integrate and classify normal distributions. J. Vis. **21**(10) (2021)
65. Muirhead, R.J.: Aspects of Multivariate Statistical Theory. John Wiley & Sons Inc., New York, NY (1982)
66. Richmond, C.D.: Performance of the adaptive sidelobe blanker detection algorithm in homogeneous environments. IEEE Trans. Signal Process. **48**(5), 1235–1247 (2000)

Chapter 7
Performance Analysis of Subspace-Based Algorithms in CES Data Models

Jean-Pierre Delmas and Habti Abeida

Abstract Subspace-based algorithms that exploit the orthogonality between a sample subspace and a parameter-dependent subspace have proved very useful in many applications in signal processing. The statistical performance of these subspace-based algorithms depends on the deterministic and stochastic statistical model of the noisy linear mixture of the data, the estimate of the projector associated with different estimates of the scatter/covariance of the data, and the algorithm that estimates the parameters from the projector. This chapter presents various different complex circular (C-CES) and non-circular (NC-CES) elliptically symmetric models of the data, as well as different associated non-robust and robust covariance estimators. These estimators include the sample covariance matrix (SCM), the maximum likelihood (ML) estimate, robust M-estimates, Tyler's M-estimate and the sample sign covariance matrix (SSCM). The asymptotic distributions of these estimators are also derived. This enables us to unify the asymptotic distribution of subspace projectors adapted to the different models of the data and demonstrate various invariance properties that have impacts on the parameters to be estimated. Particular attention is paid to the comparison between the projectors derived from Tyler's M-estimate and SSCM. Finally, this study investigates the asymptotic distributions of paratmeter estimates of parameters characterized by the principal subspace derived from the distributions of subspace-based parameter estimates. In particular, the efficiency with respect to the stochastic and semiparametric Cramér–Rao bound is considered.

7.1 Introduction

Noisy linear mixtures of signals in which the parameter of interest is characterized by the range space of the mixing matrix are very common in many applications,

J.-P. Delmas (✉)
laboratoire Samovar, Institut Polytechnique de Paris, Telecom SudParis, 91120 Palaiseau, France
e-mail: jean-pierre.delmas@it-sudparis.eu

H. Abeida
Electrical Engineering, University of Taif, Al-Haweiah 21974, Saudi Arabia
e-mail: h.abeida@tu.edu.sa

J.-P. Delmas et al. (eds.), *Elliptically Symmetric Distributions in Signal Processing and Machine Learning*, https://doi.org/10.1007/978-3-031-52116-4_7

including array processing and linear system identification (see e.g., [1, 29, 38]). To get rid of the nuisance parameters, subspace-based estimates obtained by exploiting the orthogonality between a sample subspace and a parameter-dependent subspace have been exploited since the seminal paper [46] that introduces the multiple signal classification (MUSIC) algorithm for direction of arrival (DOA) estimation. These methods are always the object of active research in many applications (see e.g., [14, 30, 56]), with generally many possible algorithms (see e.g., [24] for special structures of the mixing matrix). In these noisy linear mixtures, two statistical models have been commonly used [49]. If the signals in the mixture are nonrandom, but rather unknown deterministic parameters, the model is called deterministic or conditional. Otherwise, they are random and the model is a stochastic or unconditional model. Subspace-based algorithms associated with these two models have been intensely studied in the circular complex Gaussian framework (see e.g., [1, 14, 24, 29, 30, 38, 46, 49, 56] and references therein). But this framework is often insufficient for non-Gaussian heavy-tailed distributed data that are well modeled by circular (C-CES) or non-circular complex (NC-CES) elliptically symmetric distributions.

The aim of this chapter is to unify and complement various deterministic and stochastic CES models of the data and asymptotic distributions of the associated projectors derived from different estimates of the covariance matrix of the parametric noisy linear mixture data presented in the literature. The asymptotic distribution (w.r.t. the number of measurements) of the sample covariance matrix (SCM), maximum likelihood (ML), robust M, Tyler's M, and sample sign covariance matrix (SSCM) estimate of the covariance are considered. This allows us to derive the asymptotic distributions of the associated projectors, whose covariances have a unified structure, and consequently to prove several invariance properties. Particular attention is paid to the comparison between the projectors derived from Tyler's M-estimate and SSCM, yielding that the performance of Tyler's M-estimate is much better than that of SSCM, except for high-dimensional data and not too small dimension of the principal subspace w.r.t. this data dimension. In that case, they are close and where the SSCM can be advantageously used instead of Tyler's M-estimate for its lower computational complexity. Finally, asymptotic distributions of estimates of parameters characterized by the principal subspace derived from the distributions of projectors are studied, in particular, the efficiency with respect to the stochastic and semiparametric Cramér–Rao bound.

The rest of this chapter is organized as follows. Section 7.2 specifies the deterministic and stochastic C-CES and NC-CES distributed noisy linear mixture models, and the different associated parameterized mixing matrices. Subspace-based algorithms are interpreted in Sect. 7.3 as compositions of mappings. The asymptotic distributions of different non-robust and robust covariance estimates adapted to C-CES and NC-CES distributed data models are presented in Sect. 7.4. This allows us to derive the asymptotic distribution of the associated projectors and to prove different properties in Sect. 7.5. Asymptotic distributions of subspace-based parameter estimates are studied in Sect. 7.6. Finally, the chapter is concluded in Sect. 7.7.

The notation used in this chapter are those presented in Chap. 1 to which some specific notation are used in the following. In particular, $\mathrm{RES}_m(\boldsymbol{\mu}, \mathbf{R}, g)$, C-

$\mathrm{CES}_m(\boldsymbol{\mu}, \mathbf{R}, g)$, NC-$\mathrm{CES}_m(\boldsymbol{\mu}, \mathbf{R}, \mathbf{C}, g)$, $\mathbb{C}N_m(\boldsymbol{\mu}, \mathbf{R})$, and $\mathbb{C}N_m(\boldsymbol{\mu}, \mathbf{R}, \mathbf{C})$ denote the real elliptically symmetric (RES) distribution, C-CES, NC-CES, circular, and non-circular Gaussian distributions of dimension m with finite second-order moments, respectively, where $\boldsymbol{\mu}$, $\mathbf{R}$ and $\mathbf{C}$ are the mean, the covariance and complementary covariance matrices, respectively, and g the density generator. The matrix $\mathbf{J}$ denotes the $2m \times 2m$ exchange matrix $\begin{pmatrix} \mathbf{0} & \mathbf{I} \\ \mathbf{I} & \mathbf{0} \end{pmatrix}$. ${}_2F_1(a, b, c, x)$ is the Gauss hypergeometric functions with ${}_2F_1(a, b, c, x) = \frac{1}{B(b,c-b)} \int_0^1 t^{b-1}(1-t)^{c-b-1}(1-tx)^{-a} dt$ for $c > b > 0$ and $|x| < 1$ where $B(x, y)$ is defined in Chap. 1. Finally, throughout the chapter, circularity is defined only up to the second-order.

7.2 Noisy Linear Mixture Model

Consider the following general noisy linear mixture model[1]

$$\mathbf{x}_i = \mathbf{A}\mathbf{s}_i + \mathbf{n}_i \in \mathbb{C}^m, \quad i = 1, ..., n, \tag{7.1}$$

where $(\mathbf{x}_i)_{i=1}^n$ are independent observations, $\mathbf{s}_i$ and $\mathbf{n}_i$ represent a signal of interest and an additive measurement noise, respectively, which are assumed to be zero-mean mutually and uncorrelated. $\mathbf{n}_i$ is assumed to be complex circular spatially uncorrelated with $\mathrm{E}(\mathbf{n}_i\mathbf{n}_i^H) = \sigma_n^2\mathbf{I}$ and $\mathrm{E}(\mathbf{n}_i\mathbf{n}_i^T) = \mathbf{0}$.

Deterministic and stochastic parametric data models have been commonly used to model the distribution of $(\mathbf{s}_i, \mathbf{n}_i)$, where $\mathbf{n}_i$ is complex circular Gaussian distributed [49]. These two statistical data models are extended here within the framework of CES distributions.

7.2.1 *Deterministic CES Data Model*

In the conditional or deterministic model, the signal sequence $(\mathbf{s}_i)_{i=1,...,n}$ is conditioned from an independent zero-mean Gaussian process. As explained in [49], the sequence $(\mathbf{s}_i)_{i=1,...,n}$ is the same here in all the realizations of the random data $(\mathbf{x}_i)_{i=1,...,n}$. For complex-valued $\mathbf{s}_i$ with arbitrary circularity, we assume that $\lim_{n\to\infty} \frac{1}{n} \sum_{i=1}^n \mathbf{s}_i\mathbf{s}_i^H = \mathbf{R}_{s,\infty}$ exists and is also positive definite. The law of large number implies that:

$$\mathbf{R}_{x,\infty} \stackrel{\text{def}}{=} \lim_{n\to\infty} \frac{1}{n} \sum_{i=1}^{n} \mathbf{x}_i\mathbf{x}_i^H = \mathbf{A}\mathbf{R}_{s,\infty}\mathbf{A}^H + \sigma_n^2\mathbf{I}. \tag{7.2}$$

---

[1] $\mathbf{x}_i$ can represent the sampled complex baseband amplitudes of the data at the output of m passband antennas.

For strictly non-circular complex (also called rectilinear)-valued $\mathbf{s}_i$, i.e., satisfying the condition

$$s_{i,k} = r_{i,k} e^{i\phi_k}, \; k = 1, .., p, \text{ where } r_{i,k} \text{ are real-valued,} \tag{7.3}$$

and let $\mathbf{r}_i \stackrel{\text{def}}{=} (r_{i,1}, ..., r_{i,p})^T$ where $\lim_{n\to\infty} \frac{1}{n} \sum_{i=1}^n \mathbf{r}_i \mathbf{r}_i^T = \mathbf{R}_{r,\infty}$ exists and is also positive definite. The phases ϕ_k associated with each k-th source are assumed fixed, but unknown during the observation. To take into account this property (7.3) of the signals $s_{i,k}$, we consider the extended observation $\widetilde{\mathbf{x}}_i \stackrel{\text{def}}{=} [\mathbf{x}_i^T, \mathbf{x}_i^H]^T = \widetilde{\mathbf{A}}_r \mathbf{r}_i + [\mathbf{n}_i^T, \mathbf{n}_i^H]^T$ which leads as in (7.2) to

$$\mathbf{R}_{\tilde{x},\infty} \stackrel{\text{def}}{=} \lim_{n\to\infty} \frac{1}{n} \sum_{i=1}^{n} \widetilde{\mathbf{x}}_i \widetilde{\mathbf{x}}_i^H = \widetilde{\mathbf{A}}_r \mathbf{R}_{r,\infty} \widetilde{\mathbf{A}}_r^H + \sigma_n^2 \mathbf{I}, \tag{7.4}$$

where $\widetilde{\mathbf{A}}_r \stackrel{\text{def}}{=} \begin{bmatrix} \mathbf{A}\boldsymbol{\Delta} \\ \mathbf{A}^* \boldsymbol{\Delta}^* \end{bmatrix}$ with $\boldsymbol{\Delta} \stackrel{\text{def}}{=} \text{Diag}(e^{i\phi_1}, ..., e^{i\phi_p})$. In this deterministic model $(\mathbf{s}_i)_{i=1,...,n}$ or $(\mathbf{r}_i)_{i=1,...,n}$ and $(\phi_1, ..., \phi_p)$ are unknown deterministic parameters. However, the noise $\mathbf{n}_i$ is assumed C-CES distributed with density generator g_n. Consequently, the distribution of the observed data $\mathbf{x}_i$ is either C-CES$_m(\mathbf{A}\mathbf{s}_i, \sigma_n^2 \mathbf{I}, g_n)$ or C-CES$_m(\mathbf{A}\Delta\mathbf{r}_i, \sigma_n^2 \mathbf{I}, g_n)$ distributed, for complex-valued $\mathbf{s}_i$ with arbitrary circularity or rectilinear, respectively.

7.2.2 Stochastic Gaussian Data Model

In the unconditional or stochastic model, both $\mathbf{s}_i$ and $\mathbf{n}_i$ are usually assumed Gaussian distributed and independent of each other. $\mathbf{s}_i$ is here either circular, rectilinear or non-circular and non-rectilinear complex-valued. In the complex circular case, $\mathbf{x}_i$ are $\mathbb{C}N_m(\mathbf{0}, \mathbf{R}_x)$ distributed, where the covariance $\mathbf{R}_x$ is given by

$$\mathbf{R}_x \stackrel{\text{def}}{=} \text{E}(\mathbf{x}_i \mathbf{x}_i^H) = \mathbf{A}\mathbf{R}_s \mathbf{A}^H + \sigma_n^2 \mathbf{I}, \tag{7.5}$$

where $\mathbf{R}_s \stackrel{\text{def}}{=} \text{E}(\mathbf{s}_i \mathbf{s}_i^H)$ is positive definite.

In the complex rectilinear case, the signals $s_{i,k}$, $k = 1, .., p$, satisfy constraint (7.3). In this case, the distribution of $\mathbf{x}_i$ is characterized by the covariance of the extended observation $\widetilde{\mathbf{x}}_i$ given by

$$\mathbf{R}_{\tilde{x}} \stackrel{\text{def}}{=} \text{E}(\widetilde{\mathbf{x}}_i \widetilde{\mathbf{x}}_i^H) = \widetilde{\mathbf{A}}_r \mathbf{R}_r \widetilde{\mathbf{A}}_r^H + \sigma_n^2 \mathbf{I}, \tag{7.6}$$

with $\mathbf{R}_r \stackrel{\text{def}}{=} \text{E}(\mathbf{r}_i \mathbf{r}_i^T)$, and thus, $\mathbf{x}_i$ are $\mathbb{C}N_m(\mathbf{0}, \mathbf{R}_x, \mathbf{C}_x)$ distributed with

$$\mathbf{R}_x = \mathbf{A}\boldsymbol{\Delta}\mathbf{R}_r\boldsymbol{\Delta}^*\mathbf{A}^H + \sigma_n^2\mathbf{I} \text{ and } \mathbf{C}_x = \mathbf{A}\boldsymbol{\Delta}\mathbf{R}_r\boldsymbol{\Delta}\mathbf{A}^T. \tag{7.7}$$

For the arbitrary non-circular case, the distribution of $\mathbf{x}_i$ is also characterized by the extended covariance matrix $\mathbf{R}_{\tilde{x}}$ given by

$$\mathbf{R}_{\tilde{x}} = \widetilde{\mathbf{A}}_c\mathbf{R}_{\tilde{s}}\widetilde{\mathbf{A}}_c^H + \sigma_n^2\mathbf{I}, \tag{7.8}$$

with $\mathbf{R}_{\tilde{s}} \stackrel{\text{def}}{=} \mathrm{E}(\widetilde{\mathbf{s}}_i\widetilde{\mathbf{s}}_i^H) = \begin{pmatrix} \mathbf{R}_s & \mathbf{C}_s \\ \mathbf{C}_s^* & \mathbf{R}_s^* \end{pmatrix}$, where $\widetilde{\mathbf{s}}_i \stackrel{\text{def}}{=} [\mathbf{s}_i^T, \mathbf{s}_i^H]^T$ and $\widetilde{\mathbf{A}}_c \stackrel{\text{def}}{=} \begin{pmatrix} \mathbf{A} & \mathbf{0} \\ \mathbf{0} & \mathbf{A}^* \end{pmatrix}$. Thus, $\mathbf{x}_i$ are $\mathbb{C}N_m(\mathbf{0}, \mathbf{R}_x, \mathbf{C}_x)$ distributed with

$$\mathbf{R}_x = \mathbf{A}\mathbf{R}_s\mathbf{A}^H + \sigma_n^2\mathbf{I} \text{ and } \mathbf{C}_x = \mathbf{A}\mathbf{C}_s\mathbf{A}^T, \tag{7.9}$$

where $\mathbf{C}_s \stackrel{\text{def}}{=} \mathrm{E}(\mathbf{s}_i\mathbf{s}_i^T)$ is a symmetric complex matrix.

7.2.3 Stochastic CES Data Model

A first extension of the stochastic Gaussian data model consists in modeling the independent signals $\mathbf{s}_i$ and $\mathbf{n}_i$ by CES distributions to take into account possible heavy-tailed (with respect to the Gaussian one) signals. For circular and non-circular (both rectilinear and non-rectilinear signal $\mathbf{s}_i$) cases, $\mathbf{s}_i$ is, respectively, $\text{C-CES}_p(\mathbf{0}, \mathbf{R}_s, g_s)$ and $\text{NC-CES}_p(\mathbf{0}, \mathbf{R}_s, \mathbf{C}_s, g_s)$ distributed,[2] and $\mathbf{n}_i$ is $\text{C-CES}_m(\mathbf{0}, \sigma_n^2\mathbf{I}, g_n)$ distributed. It is worth noting here that in this stochastic data model $\mathbf{x}_i$ is not CES distributed (except for Gaussian distributions) because this family of distributions is not closed under summations.

To take advantage of robust covariance matrix estimates available in the context of CES distributions, the CES distribution has been preferred over the Gaussian distribution to model the data $\mathbf{x}_i$ in many DOA finding and beamforming processing (see e.g., [17, 39, 40, 42]). In this case, the distributions of $\mathbf{s}_i$ and $\mathbf{n}_i$ are generally not specified, but only their second-order statistics are imposed by setting the structured covariance in (7.5) or extended covariance matrices in (7.6) and (7.8). More specifically, in the case of circular complex and non-circular complex signals $\mathbf{s}_i$, the observations $\mathbf{x}_i$ are $\text{C-CES}_m(\mathbf{0}, \mathbf{R}_x, g_x)$ and $\text{NC-CES}_m(\mathbf{0}, \mathbf{R}_x, \mathbf{C}_x, g_x)$ distributed, respectively. Moreover, as a particular case of this modeling, the complex compound Gaussian distribution which is a subclass of CES distributions that was used to model the clutter in radar [22] has also been used in robust and sparse M-estimation of DOA [37, Eq. 1] in the form of the model:

$$\mathbf{x}_i = \mathbf{A}\mathbf{s}_i + \mathbf{n}_i \stackrel{\text{def}}{=} \sqrt{\tau_i}(\mathbf{A}\mathbf{s}_i' + \mathbf{n}_i'), \tag{7.10}$$

[2] Note that for rectilinear $\mathbf{s}_i$, this modeling is equivalent to $\mathbf{r}_i$ being $\text{RES}_p(\mathbf{0}, \mathbf{R}_r, g_s)$ distributed.

where the so-called texture $\tau_i > 0$ (with $\mathrm{E}(\tau_i) = 1$) is independent of $(\mathbf{s}_i', \mathbf{n}_i')$ which are complex Gaussian distributed. In this case $\mathbf{s}_i$ and $\mathbf{n}_i$ are uncorrelated but not independent, unlike $\mathbf{s}_i'$ and $\mathbf{n}_i'$ which are independent.

7.2.4 Parameterized Mixing Matrix

Since the complex-valued signals $\mathbf{s}_i$, can be either circular, non-circular and non-rectilinear or rectilinear signals, together with the dependence of (7.1) on $m \times p$ mixing matrix $\mathbf{A}$ and on the parameter of interest $\boldsymbol{\theta} \in \mathbb{R}^k$, leads us to distinguish the following two considered parameterized cases:

- For circular, and non-circular and non-rectilinear complex-valued signals $\mathbf{s}_i$, $\boldsymbol{\theta}$ is characterized by the subspace generated by the columns of the full column rank matrix $\mathbf{A}$ with $p < m$. We will use the parameterizations

$$\mathbf{B}(\boldsymbol{\theta}) \stackrel{\text{def}}{=} \mathbf{A} \tag{7.11}$$

  in the circular case and

$$\mathbf{B}(\boldsymbol{\theta}) \stackrel{\text{def}}{=} \widetilde{\mathbf{A}}_c = \begin{pmatrix} \mathbf{A} & \mathbf{0} \\ \mathbf{0} & \mathbf{A}^* \end{pmatrix} \tag{7.12}$$

  in the non-circular and non-rectilinear case.
- For rectilinear complex-valued signals $\mathbf{s}_i$, $\boldsymbol{\theta}$ is characterized by the subspace generated by the columns of the full column rank $2m \times p$ extended mixing matrix

$$\mathbf{B}(\boldsymbol{\theta}) \stackrel{\text{def}}{=} \widetilde{\mathbf{A}}_r = \begin{bmatrix} \mathbf{A}\boldsymbol{\Delta} \\ \mathbf{A}^*\boldsymbol{\Delta}^* \end{bmatrix} \quad \text{with } \ p < 2m. \tag{7.13}$$

This low-rank signal in full-rank noise data model (7.1) encompasses many far or near-field, narrow, or wide-band DOA models with scalar or vector sensors for an arbitrary number of parameters per source $s_{i,k}$ (with $\mathbf{s}_i \stackrel{\text{def}}{=} (s_{i,1}, ..s_{i,k}, .., s_{i,p})^T$) and many other models as the bandlimited SISO, SIMO [38], and MIMO [1] channel models. For example, parametrization (7.13) can be applied for DOA estimation modeling with rectilinear or strictly second-order sources and for SIMO channels estimation modeling with BPSK or MSK symbols [13] where $\boldsymbol{\theta}$ represents both the localization parameters (azimuth, elevation, range) and the phase of the sources, and the real and imaginary parts of channel impulse response coefficients, respectively. Whereas parametrization (7.12) is used for DOA modeling with generally non-circular and non-rectilinear complex sources.

7.3 Subspace-Based Estimation Approaches

Since the parameter of interest $\boldsymbol{\theta}$ is characterized by the subspace generated by the columns of the full column rank matrices $\mathbf{A}$, $\widetilde{\mathbf{A}}_c$ or $\widetilde{\mathbf{A}}_r$, a simple way to get rid of the nuisance parameters, is to consider subspace-based algorithms as the following mapping:

$$(\mathbf{x}_1, .., \mathbf{x}_i, ..., \mathbf{x}_n) \longmapsto \widehat{\mathbf{R}} \overset{\text{EVD}}{\longmapsto} \widehat{\boldsymbol{\Pi}} = \boldsymbol{\Pi}(\widehat{\mathbf{R}}) \overset{\text{alg}}{\longmapsto} \widehat{\boldsymbol{\theta}} = \text{alg}(\widehat{\boldsymbol{\Pi}}), \tag{7.14}$$

where $\widehat{\mathbf{R}}$ can be either any estimate $\widehat{\mathbf{R}}_x$ of $\mathbf{R}_x \overset{\text{def}}{=} \mathrm{E}(\mathbf{x}_i\mathbf{x}_i^H)$ or any estimate $\widehat{\mathbf{R}}_{\tilde{x}}$ of $\mathbf{R}_{\tilde{x}} \overset{\text{def}}{=} \mathrm{E}(\tilde{\mathbf{x}}_i\tilde{\mathbf{x}}_i^H)$, and $\widehat{\boldsymbol{\Pi}}$ denotes the orthogonal projection matrix $\widehat{\boldsymbol{\Pi}}_x$ [resp., $\widehat{\boldsymbol{\Pi}}_{\tilde{x}}$] associated with the so-called noise subspace of $\widehat{\mathbf{R}}_x$ [resp., $\widehat{\mathbf{R}}_{\tilde{x}}$]. That is, if $\widehat{\mathbf{R}}_x = [\widehat{\mathbf{V}}_s, \widehat{\mathbf{V}}_n]\widehat{\boldsymbol{\Lambda}}[\widehat{\mathbf{V}}_s, \widehat{\mathbf{V}}_n]^H$ denotes the EVD of $\widehat{\mathbf{R}}_x$ where $\widehat{\boldsymbol{\Lambda}}$ gathers the eigenvalues of $\widehat{\mathbf{R}}_x$ in decreasing order, $\widehat{\mathbf{V}}_s$ and $\widehat{\mathbf{V}}_n$ are $m \times p$ and $m \times (m-p)$ unitary matrices, respectively, $\widehat{\boldsymbol{\Pi}}_x$ is given by $\widehat{\mathbf{V}}_n\widehat{\mathbf{V}}_n^H$. The functional dependence $\widehat{\boldsymbol{\theta}} = \text{alg}(\widehat{\boldsymbol{\Pi}})$ constitutes an extension of the mapping

$$\boldsymbol{\Pi}(\boldsymbol{\theta}) \overset{\text{def}}{=} \mathbf{I} - \mathbf{B}(\boldsymbol{\theta})[\mathbf{B}^H(\boldsymbol{\theta})\mathbf{B}(\boldsymbol{\theta})]^{-1}\mathbf{B}^H(\boldsymbol{\theta}) \overset{\text{alg}}{\longmapsto} \boldsymbol{\theta}, \tag{7.15}$$

in the neighborhood of $\boldsymbol{\Pi}(\boldsymbol{\theta})$ with $\mathbf{B}(\boldsymbol{\theta})$ can either be $\mathbf{A}$, $\tilde{\mathbf{A}}_r$, or $\tilde{\mathbf{A}}_c$. Each extension alg(.) specifies a particular subspace-based algorithm. Conventional MUSIC algorithm [46] based on $\widehat{\boldsymbol{\Pi}}_x$ and non-circular MUSIC algorithms [2] based on $\widehat{\boldsymbol{\Pi}}_{\tilde{x}}$ for parametrization (7.6) can be seen as examples in DOA estimation. Among these algorithms are the asymptotically minimum variance (AMV) algorithm or the asymptotically best consistent estimators (ABC) introduced by Porat and Friedlander [45] and Stoica et al. [48], respectively. This algorithm minimizes the covariance matrix of the asymptotic distribution of the estimate $\widehat{\boldsymbol{\theta}}$ among all the estimates alg($\widehat{\boldsymbol{\Pi}}$) (7.14) satisfying $\text{alg}(\boldsymbol{\Pi}(\boldsymbol{\theta})) = \boldsymbol{\theta}$ and a regularity condition (see (7.74)). The associated covariance is a lower bound that plays *the* role of benchmark against competing algorithms "alg". According to mapping (7.14), the statistical properties of the estimate $\widehat{\boldsymbol{\theta}}$ depends on both the choice of the covariance estimate $\widehat{\mathbf{R}}$ and that of the subspace-based algorithm "alg".

Note that direct ML approaches have been extensively studied to estimate DOA parameters from the data. But they generally require non-convex multidimensional minimizations except for a single source for which the deterministic and stochastic ML algorithms coincide with the conventional Capon algorithm (see e.g., [12, Chap. 16]).

7.4 Asymptotic Distributions of Covariance Estimates

Analyzing the asymptotic performance of the subspace-based estimation approaches in (7.14) first requires determining the asymptotic distribution of different covariance estimators $\widehat{\mathbf{R}}$ adapted to the different data models presented in Sect. 7.2.

7.4.1 Deterministic Data Model

We only consider in this model the sample covariance matrix (SCM) estimates $\widehat{\mathbf{R}} = \frac{1}{n}\sum_{i=1}^{n} \mathbf{x}_i\mathbf{x}_i^H$ for complex-valued of arbitrary circularity signals $\mathbf{s}_i$ and extended SCM estimates $\widehat{\mathbf{R}} = \frac{1}{n}\sum_{i=1}^{n} \widetilde{\mathbf{x}}_i\widetilde{\mathbf{x}}_i^H$ for complex rectilinear-valued signals $\mathbf{s}_i$. Under finite fourth-order moments of $\mathbf{n}_i$, we get for complex-valued of arbitrary circularity signals $\mathbf{s}_i$:

$$\mathrm{E}(\mathrm{vec}(\widehat{\mathbf{R}})) = (\mathbf{A}^* \otimes \mathbf{A})[\frac{1}{n}\sum_{i}^{n} \mathbf{s}_i^* \otimes \mathbf{s}_i] + \sigma_n^2 \mathrm{vec}(\mathbf{I}), \tag{7.16}$$

$$\begin{aligned}\mathrm{Cov}(\mathrm{vec}(\widehat{\mathbf{R}})) &= \frac{1}{n}\left([\mathbf{A}^*(\frac{1}{n}\sum_{i}^{n} \mathbf{s}_i\mathbf{s}_i^H)\mathbf{A}^T] \otimes \sigma_n^2\mathbf{I} + \sigma_n^2\mathbf{I} \otimes [\mathbf{A}(\frac{1}{n}\sum_{i}^{n} \mathbf{s}_i\mathbf{s}_i^H)\mathbf{A}^H]\right)\\ &+ \frac{1}{n}\mathrm{Cov}(\mathbf{n}_i^* \otimes \mathbf{n}_i),\end{aligned} \tag{7.17}$$

where $\mathrm{Cov}(\mathbf{n}_i^* \otimes \mathbf{n}_i) = \sigma_n^4[(1+\kappa_n)\mathbf{I} + \kappa_n \mathrm{vec}(\mathbf{I})\mathrm{vec}^T(\mathbf{I})]$ with

$$\kappa_n = \frac{\mathrm{E}(\mathcal{Q}_n^2)}{m(m+1)} - 1 \tag{7.18}$$

is the kurtosis parameter[3] of $\mathbf{n}_i$ and $\mathcal{Q}_n$ is its second-order modular variate. Using the Liapounov central limit theorem (CLT) for independent non-identically distributed r.v. $\mathbf{x}_i^* \otimes \mathbf{x}_i$ (see e.g., [31, Theorem 2.7.1]) and the Slutsky theorem (see e.g., [31, Theorem 5.1.6]), we get the following convergence in distribution: [3, 5].

$$\sqrt{n}(\mathrm{vec}(\widehat{\mathbf{R}}) - \mathrm{vec}(\mathbf{R})) \to_d \mathbb{C}N_{m^2}(\mathbf{0}, \mathbf{R}_{r_x}, \mathbf{R}_{r_x}\mathbf{K}), \tag{7.19}$$

with $\mathbf{R} \stackrel{\mathrm{def}}{=} \mathbf{R}_{x,\infty}$ defined in (7.2), $\mathbf{K}$ defined in Chap. 1 and

[3] Note that for C and NC-CES$_m$ distributions, the kurtosis parameter $\kappa_{c,m}$ is defined by the kurtosis parameter of the associated RES$_{2m}$ distribution $\kappa_{r,2m} = \frac{\mathrm{E}(\mathcal{Q}_{r,2m}^2)}{2m(2m+2)} - 1 = \frac{\mathrm{E}(\mathcal{Q}_{c,m}^2)}{m(m+1)} - 1$ with $\mathcal{Q}_{c,m} = \frac{1}{2}\mathcal{Q}_{r,2m}$, where $\mathcal{Q}_{c,m}$ and $\mathcal{Q}_{r,2m}$ are the second-order modular variates of the associated C and NC-CES$_m$, and RES$_{2m}$ distributions (see Chap. 1).

$$\mathbf{R}_{r_x} = \mathbf{A}^*\mathbf{R}_{s,\infty}\mathbf{A}^T \otimes \sigma_n^2\mathbf{I} + \sigma_n^2\mathbf{I} \otimes \mathbf{A}\mathbf{R}_{s,\infty}\mathbf{A}^H + \sigma_n^4[(1+\kappa_n)\mathbf{I} + \kappa_n \text{vec}(\mathbf{I})\text{vec}^T(\mathbf{I})], \tag{7.20}$$

where $\mathbf{R}_{s,\infty}$ is defined in Sect. 7.2.1. Similarly, we obtain for complex rectilinear-valued $\mathbf{s}_i$ that

$$\sqrt{n}(\text{vec}(\widehat{\mathbf{R}}) - \text{vec}(\mathbf{R})) \to_d \mathbb{C}N_{4m^2}(\mathbf{0}, \mathbf{R}_{r_{\tilde{x}}}, \mathbf{R}_{r_{\tilde{x}}}\mathbf{K}), \tag{7.21}$$

with $\mathbf{R} \stackrel{\text{def}}{=} \mathbf{R}_{\tilde{x},\infty}$ defined in (7.4) and

$$\mathbf{R}_{r_{\tilde{x}}} = [\mathbf{I} + \mathbf{K}(\mathbf{J} \otimes \mathbf{J})][(\widetilde{\mathbf{A}}_r^*\mathbf{R}_{r,\infty}\widetilde{\mathbf{A}}_r^T \otimes \sigma_n^2\mathbf{I}) + (\sigma_n^2\mathbf{I} \otimes \widetilde{\mathbf{A}}_r\mathbf{R}_{r,\infty}\widetilde{\mathbf{A}}_r^H) \\ + \sigma_n^4(1+\kappa_n)\mathbf{I}] + \sigma_n^4\kappa_n \text{vec}(\mathbf{I})\text{vec}^T(\mathbf{I}). \tag{7.22}$$

7.4.2 *Stochastic Data Model*

We consider here two cases. In the first one, $\mathbf{s}_i$ and $\mathbf{n}_i$ are both CES distributed for which only the SCM estimate is considered. In this case, $\mathbf{x}_i$ is not CES distributed. In the second one, $\mathbf{x}_i$ is CES distributed and many robust covariance estimates are considered.

7.4.2.1 SCM Estimators for Both CES Distributed $\mathbf{s}_i$ and $\mathbf{n}_i$

Let's start first with the SCM estimates $\widehat{\mathbf{R}} = \frac{1}{n}\sum_{i=1}^n \mathbf{x}_i\mathbf{x}_i^H$ and $\widehat{\mathbf{R}} = \frac{1}{n}\sum_{i=1}^n \widetilde{\mathbf{x}}_i\widetilde{\mathbf{x}}_i^H$ for, respectively, circular and complex of arbitrary circularity signals $\mathbf{s}_i$, where $\mathbf{s}_i$ and $\mathbf{n}_i$ are both CES distributed and have finite fourth-order moments. By applying the classic CLT to the r.v. $\mathbf{x}_i^* \otimes \mathbf{x}_i$ and $\widetilde{\mathbf{x}}_i^* \otimes \widetilde{\mathbf{x}}_i$, we get, respectively, the following asymptotic distribution of $\widehat{\mathbf{R}}$ which did not appear in the literature:

$$\sqrt{n}(\text{vec}(\widehat{\mathbf{R}}) - \text{vec}(\mathbf{R})) \to_d \mathbb{C}N_{m^2}(\mathbf{0}, \mathbf{R}_{r_x}, \mathbf{R}_{r_x}\mathbf{K}), \tag{7.23}$$

$$\sqrt{n}(\text{vec}(\widehat{\mathbf{R}}) - \text{vec}(\mathbf{R})) \to_d \mathbb{C}N_{4m^2}(\mathbf{0}, \mathbf{R}_{r_{\tilde{x}}}, \mathbf{R}_{r_{\tilde{x}}}\mathbf{K}), \tag{7.24}$$

with $\mathbf{R} \stackrel{\text{def}}{=} \mathbf{R}_x$ defined in (7.5) and $\mathbf{R} \stackrel{\text{def}}{=} \mathbf{R}_{\tilde{x}}$ defined in (7.6) and (7.8), respectively, and

$$\mathbf{R}_{r_{\tilde{z}}} = (\mathbf{R}_x^* \otimes \mathbf{R}_x) + \kappa_s[(\mathbf{A}^*\mathbf{R}_s^*\mathbf{A}^T) \otimes (\mathbf{A}\mathbf{R}_s\mathbf{A}^H) \\ + \text{vec}(\mathbf{A}\mathbf{R}_s\mathbf{A}^H)\text{vec}^H(\mathbf{A}\mathbf{R}_s\mathbf{A}^H)] + \sigma_n^4\kappa_n[\mathbf{I} + \text{vec}(\mathbf{I})\text{vec}^T(\mathbf{I})], \tag{7.25}$$

$$\mathbf{R}_{r_{\tilde{x}}} = [\mathbf{I} + \mathbf{K}(\mathbf{J} \otimes \mathbf{J})][(\mathbf{R}_{\tilde{x}}^* \otimes \mathbf{R}_{\tilde{x}}) + \kappa_s(\widetilde{\mathbf{A}}_c^*\mathbf{R}_{\tilde{s}}^*\widetilde{\mathbf{A}}_c^T) \otimes (\widetilde{\mathbf{A}}_c\mathbf{R}_{\tilde{s}}\widetilde{\mathbf{A}}_c^H) \\ + \kappa_s \text{vec}(\widetilde{\mathbf{A}}_c\mathbf{R}_{\tilde{s}}\widetilde{\mathbf{A}}_c^H)\text{vec}^H(\widetilde{\mathbf{A}}_c\mathbf{R}_{\tilde{s}}\widetilde{\mathbf{A}}_c^H) + \sigma_n^4\kappa_n\mathbf{I}] \\ + \sigma_n^4\kappa_n \text{vec}(\mathbf{I})\text{vec}^T(\mathbf{I}), \tag{7.26}$$

where κ_s is the kurtosis parameter of $\mathbf{s}_i$ given by $\kappa_s = \frac{\mathrm{E}(\mathcal{Q}_s^2)}{p(p+1)} - 1$ and where $\mathcal{Q}_s$ is the second-order modular variate of $\mathbf{s}_i$. Note that (7.24) and (7.26) remain valid for complex rectilinear signals $\mathbf{s}_i$ if κ_s is replaced by $\kappa_r = \frac{\mathrm{E}(\mathcal{Q}_r^2)}{p(p+2)} - 1$ where $\mathcal{Q}_r$ is the second-order modular variate of $\mathbf{r}_i$ which is $\mathrm{RES}_p(\mathbf{0}, \mathbf{R}_r, g_s)$ distributed. Furthermore, in this case $\widetilde{\mathbf{A}}_c \mathbf{R}_{\tilde{s}} \widetilde{\mathbf{A}}_c^H$ reduces to $\widetilde{\mathbf{A}}_r \mathbf{R}_r \widetilde{\mathbf{A}}_r^H$.

7.4.2.2 Covariance Estimates for CES Distributed $\mathbf{x}_i$

Now consider the cases where the observations $\mathbf{x}_i$ are $\mathrm{C\text{-}CES}_m(\mathbf{0}, \mathbf{R}_x, g_x)$ and $\mathrm{NC\text{-}CES}_m(\mathbf{0}, \mathbf{R}_x, \mathbf{C}_x, g_x)$ distributed. For these distributions, many covariance estimators have been proposed in the literature. This study examines the asymptotic distributions of various estimators, including the SCM, ML and M, Tyler's and SSCM estimators.

SCM Estimators

Under finite fourth-order moments of $\mathbf{x}_i$, applying again the classic CLT to the r.v. $\mathbf{x}_i^* \otimes \mathbf{x}_i$ and $\widetilde{\mathbf{x}}_i^* \otimes \widetilde{\mathbf{x}}_i$, we get the convergences in distribution (7.23) and (7.24) for the SCM estimate $\widehat{\mathbf{R}}$ with also $\mathbf{R} \stackrel{\text{def}}{=} \mathbf{R}_x$ defined in (7.5) and $\mathbf{R} \stackrel{\text{def}}{=} \mathbf{R}_{\tilde{x}}$ defined in (7.6) and (7.8), respectively, and with now $\mathbf{R}_{r_x}$ and $\mathbf{R}_{r_{\tilde{x}}}$ are given by [3, 5]

$$\mathbf{R}_{r_x} = \sigma_1(\mathbf{R}_x^* \otimes \mathbf{R}_x) + \sigma_2 \mathrm{vec}(\mathbf{R}_x)\mathrm{vec}^H(\mathbf{R}_x), \tag{7.27}$$

$$\mathbf{R}_{r_{\tilde{x}}} = \sigma_1[\mathbf{I} + \mathbf{K}(\mathbf{J} \otimes \mathbf{J})](\mathbf{R}_{\tilde{x}}^* \otimes \mathbf{R}_{\tilde{x}}) + \sigma_2 \mathrm{vec}(\mathbf{R}_{\tilde{x}})\mathrm{vec}^H(\mathbf{R}_{\tilde{x}}), \tag{7.28}$$

with $\sigma_1 = 1 + \kappa_x$ and $\sigma_2 = \kappa_x$, where κ_x is the kurtosis parameter of $\mathbf{x}_i$ given by $\kappa_x = \frac{\mathrm{E}(\mathcal{Q}_x^2)}{m(m+1)} - 1$, where $\mathcal{Q}_x$ is the second-order modular variate of $\mathbf{x}_i$. Note that thanks to the linear one-to-one mapping $\bar{\mathbf{x}}_i \longleftrightarrow \widetilde{\mathbf{x}}_i$ with $\bar{\mathbf{x}}_i \stackrel{\text{def}}{=} (\mathrm{Re}(\mathbf{x}_i)^T, \mathrm{Im}(\mathbf{x}_i)^T)^T \in \mathbb{R}^{2m}$, (7.27) and (7.28) can be directly deduced from the convergence in distribution of the SCM estimate $\frac{1}{n}\sum_{i=1}^n \bar{\mathbf{x}}_i \bar{\mathbf{x}}_i^T$ of $\mathbf{R}_{\bar{x}} \stackrel{\text{def}}{=} \mathrm{E}(\bar{\mathbf{x}}_i \bar{\mathbf{x}}_i^T)$, given by [43, p. 5]

$$\sqrt{n}(\mathrm{vec}(\widehat{\mathbf{R}}) - \mathrm{vec}(\mathbf{R}_{\bar{x}})) \rightarrow_d \mathbb{R}N_{4m^2}(\mathbf{0}, \mathbf{R}_{r_{\bar{x}}}), \tag{7.29}$$

with

$$\mathbf{R}_{r_{\bar{x}}} = (1 + \kappa_{\bar{x}})(\mathbf{I} + \mathbf{K})(\mathbf{R}_{\bar{x}} \otimes \mathbf{R}_{\bar{x}}) + \kappa_{\bar{x}} \mathrm{vec}(\mathbf{R}_{\bar{x}})\mathrm{vec}^T(\mathbf{R}_{\bar{x}}), \tag{7.30}$$

where $\kappa_{\bar{x}} = \kappa_x$ by definition of the kurtosis κ_x of the complex-valued r.v. $\mathbf{x}_i$ given in Footnote 4. For heavier tails than Gaussian distributions, we have $\kappa_x > 0$ and can be very large without any upper bound, and consequently, the SCM estimator can be a very poor estimator of the covariance matrix.

ML Estimators

To take into account the particular CES distribution of $\mathbf{x}_i$, the ML estimator is often considered as the benchmark estimator. From the p.d.f. of the complex-valued data $\mathbf{x}_i$: $p(\mathbf{x}) = |\mathbf{R}_{\tilde{x}}|^{-1/2} g_x(\frac{1}{2}\widetilde{\mathbf{x}}^H \mathbf{R}_{\tilde{x}}^{-1} \widetilde{\mathbf{x}})$ which reduces to $p(\mathbf{x}) = |\mathbf{R}_x|^{-1} g_x(\mathbf{x}^H \mathbf{R}_x^{-1} \mathbf{x})$ in the circular case (see Chap. 1), the ML estimate $\widehat{\mathbf{R}}$ of $\mathbf{R}_x$ and $\mathbf{R}_{\bar{x}}$ are, respectively, solutions of the implicit equations:

$$\widehat{\mathbf{R}} = \frac{1}{n}\sum_{i=1}^{n} \varphi_x(\mathbf{x}_i^H \widehat{\mathbf{R}}^{-1}\mathbf{x}_i)\mathbf{x}_i\mathbf{x}_i^H \text{ and } \widehat{\mathbf{R}} = \frac{1}{n}\sum_{i=1}^{n} \varphi_x\left(\frac{1}{2}\widetilde{\mathbf{x}}_i^H \widehat{\mathbf{R}}^{-1}\widetilde{\mathbf{x}}_i\right)\widetilde{\mathbf{x}}_i\widetilde{\mathbf{x}}_i^H, \quad (7.31)$$

with $\varphi_x(t) \stackrel{\text{def}}{=} -\frac{1}{g_x(t)}\frac{dg_x(t)}{dt}$. Under existence, uniqueness, and usual regularity conditions, the ML estimate of $\mathbf{R}_x$ and $\mathbf{R}_{\bar{x}}$ is generally asymptotically efficient with a speed of convergence in $\sqrt{n}$. Thus, we get the convergences in distribution (7.23) and (7.24), with also $\mathbf{R} \stackrel{\text{def}}{=} \mathbf{R}_x$ defined in (7.5) and $\mathbf{R} \stackrel{\text{def}}{=} \mathbf{R}_{\bar{x}}$ defined in (7.6) and (7.8), respectively, where $\mathbf{R}_{r_x}$ and $\mathbf{R}_{r_{\bar{x}}}$ can be deduced from the matrix Slepian–Bangs formula [4, 7, 9, 23] associated with the single parameters $\mathbf{R}_x$ and $\mathbf{R}_{\bar{x}}$ and are still given by (7.27) and (7.28) with

$$\sigma_1 = \frac{m(m+1)}{\mathrm{E}[\mathcal{Q}_x^2\varphi_x^2(\mathcal{Q}_x)]} \text{ and } \sigma_2 = -\frac{2\sigma_1(1-\sigma_1)}{1+2m(1-\sigma_1)}. \quad (7.32)$$

M-Estimator

Since the ML estimator may be drastically affected by the presence of outliers or when the data distribution deviates slightly from the CES distribution of the model, robust estimators of the covariance of the data have been proposed. Among the different families of robust estimators, we focus our attention on the class of M-estimators introduced by Maronna [36] for RES distributions. Applied to $\mathrm{RES}_{2m}(\mathbf{0}, \mathbf{R}_{\bar{x}}, g_{\bar{x}})$ distributed data $\bar{\mathbf{x}}_i$, with $g_{\bar{x}}(t) = 2^{-m}g_x(\frac{t}{2})$ (see Chap. 1), these M-estimators were defined as solutions $\widehat{\mathbf{R}}$ of the implicit equation:

$$\widehat{\mathbf{R}} = \frac{1}{n}\sum_{i=1}^{n} u_{\bar{x}}(\bar{\mathbf{x}}_i^T \widehat{\mathbf{R}}^{-1}\bar{\mathbf{x}}_i)\bar{\mathbf{x}}_i\bar{\mathbf{x}}_i^T, \quad (7.33)$$

where $u_{\bar{x}}(.)$ is any real-valued weight function on $[0, \infty)$ not related to a particular RES distribution. Under sufficient conditions (called Maronna conditions), it was proved in [36, Theorem 4], the existence and uniqueness of solution $\widehat{\mathbf{R}}$ of (7.33). It has also been proved in [36, Theorem 2] the existence and uniqueness of the solution $\mathbf{V}$ of

$$\mathbf{V} = \mathrm{E}[u_{\bar{x}}(\bar{\mathbf{x}}_i^T \mathbf{V}^{-1}\bar{\mathbf{x}}_i)\bar{\mathbf{x}}_i\bar{\mathbf{x}}_i^T]. \quad (7.34)$$

Sufficient conditions are also given in [36, Theorem 5] to ensure the strong consistency of the estimate $\widehat{\mathbf{R}}$ solution of (7.33) to the solution $\mathbf{V}$ of (7.34). $\mathbf{V}$ is related to $\mathbf{R}_{\bar{x}}$ by $\mathbf{V} = c^{-1}\mathbf{R}_{\bar{x}}$ where c is the unique solution of $\mathrm{E}[c\mathcal{Q}_{\bar{x}}u_{\bar{x}}(c\mathcal{Q}_{\bar{x}})] = 2m$. [36, Appendix 3]. This is equivalent to the solution of

$$\mathrm{E}[c\mathcal{Q}_x u_x(c\mathcal{Q}_x)] = m, \quad (7.35)$$

for CES distributions [42, (rel (46)]. Using a general result on M-estimators given in [25, Sect. 4], Maronna proved in [36, Theorem 6] the asymptotic gaussianity of $\widehat{\mathbf{R}}$. Then, using the affine invariance property of any M-estimators and the general structure of the covariance of radial random matrices, the covariance of the asymptotic distribution (called also asymptotic error covariance) of $\widehat{\mathbf{R}}$ was specified in [50, Appendix 2].

Extensions of M-estimators to C-CES distributed data were introduced in [39] and later studied and used in various signal processing applications (see [42] and references therein). Then, it was extended to NC-CES distributed data [5]. Since $\mathbf{x}_i$ are NC-CES$_m(\mathbf{0}, \mathbf{R}_x, \mathbf{C}_x, g_x)$ distributed by definition if $\bar{\mathbf{x}} = \frac{1}{\sqrt{2}}\mathbf{M}^H\widetilde{\mathbf{x}}_i$ (where $\mathbf{M} \stackrel{\text{def}}{=} \frac{1}{\sqrt{2}}\begin{pmatrix} \mathbf{I} & i\mathbf{I} \\ \mathbf{I} & -i\mathbf{I} \end{pmatrix}$ is unitary) are RES$_{2m}(\mathbf{0}, \mathbf{R}_{\bar{x}}, g_{\bar{x}})$ distributed, (7.33) is equivalent to (7.36) where $u_x(t) \stackrel{\text{def}}{=} u_{\bar{x}}(\frac{t}{2})$ (see Chap. 1) and where $\widehat{\mathbf{R}}$ now denotes the M-estimate of $\mathbf{R}_{\tilde{x}} = 2\mathbf{M}\mathbf{R}_{\bar{x}}\mathbf{M}^H$:

$$\widehat{\mathbf{R}} = \frac{1}{n}\sum_{i=1}^{n} u_x\left(\frac{1}{2}\widetilde{\mathbf{x}}_i^H\widehat{\mathbf{R}}^{-1}\widetilde{\mathbf{x}}_i\right)\widetilde{\mathbf{x}}_i\widetilde{\mathbf{x}}_i^H. \tag{7.36}$$

Moreover, in the particular complex circular case where $\mathbf{R}_{\tilde{x}} = \begin{pmatrix} \mathbf{R}_x & \mathbf{0} \\ \mathbf{0} & \mathbf{R}_x^* \end{pmatrix}$, imposing $\widehat{\mathbf{R}}$ in (7.36) to be block diagonal structured as $\mathbf{R}_{\tilde{x}}$, the M-estimate $\widehat{\mathbf{R}}$ of $\mathbf{R}_x$ introduced in [39] is derived as solution of

$$\widehat{\mathbf{R}} = \frac{1}{n}\sum_{i=1}^{n} u_x(\mathbf{x}_i^H\widehat{\mathbf{R}}^{-1}\mathbf{x}_i)\mathbf{x}_i\mathbf{x}_i^H. \tag{7.37}$$

Thus, all the properties of the M-estimate $\widehat{\mathbf{R}}$ of $\mathbf{R}_{\bar{x}}$ are transferred to the M-estimates $\widehat{\mathbf{R}}$ of $\mathbf{R}_x$ and $\mathbf{R}_{\tilde{x}}$ for C-CES and NC-CES [5] distributions, respectively. In particular, we get the convergences in distribution (7.23) and (7.24) where $\mathbf{R} \stackrel{\text{def}}{=} c^{-1}\mathbf{R}_x$ in the circular case and $\mathbf{R} \stackrel{\text{def}}{=} c^{-1}\mathbf{R}_{\tilde{x}}$ in the complex of arbitrary circularity case, and where $\mathbf{R}_{r_x}$ and $\mathbf{R}_{r_{\tilde{x}}}$ are given by (7.27) and (7.28), respectively, with [35, 42]

$$\sigma_1 = \frac{\mathrm{E}[\mathcal{Q}_x^2 u_x^2(c\mathcal{Q}_x)]}{m(m+1)(1+[m(m+1]^{-1}c_u)^2} \tag{7.38}$$

and

$$\sigma_2 = \frac{\mathrm{E}[(\mathcal{Q}_x u_x(c\mathcal{Q}_x) - mc_u)^2]}{(m+c_u)^2} - \frac{\sigma_1}{m}, \tag{7.39}$$

with $c_u \stackrel{\text{def}}{=} \mathrm{E}[c^2\mathcal{Q}_x^2 u_x^{'}(c\mathcal{Q}_x)]$ with $u_x^{'}(t) \stackrel{\text{def}}{=} du_x(t)/dt$. Note that for $u_x(t) = \varphi_x(t)$, the M-estimate gives the ML estimate for which $c = 1$. Using the identity

$E[\mathcal{Q}_x^2\varphi_x'(\mathcal{Q}_x)] = E[\mathcal{Q}_x^2\varphi_x^2(\mathcal{Q}_x)] - m(m+1)$ proved in [5, Appendix] with $\varphi_x'(t) \stackrel{\text{def}}{=} d\varphi_x(t)/dt$, (7.38)and (7.39) reduce to (7.32).

Tyler's M-Estimator

Tyler's M-estimator was introduced in [51] for RES distributions as a solution of (7.33) with the weight function $u_{\bar{x}}(t) = \frac{2m}{t}$. It became very popular in signal processing applications because it enjoys important properties among which to have a distribution that does not depend on the RES distribution of the data. Extensions of this estimator to C-CES [42, 44] and to NC-CES distributions [5] are straightforward using the real $\bar{\mathbf{x}}_i \in \mathbb{R}^{2m}$ representation of CES distributed $\mathbf{x}_i \in \mathbb{C}^m$. They are given by the solution $\widehat{\mathbf{R}}$ of the implicit equation

$$\widehat{\mathbf{R}} = \frac{2m}{n}\sum_{i=1}^{n}\frac{\widetilde{\mathbf{x}}_i\widetilde{\mathbf{x}}_i^H}{\widetilde{\mathbf{x}}_i^H\widehat{\mathbf{R}}^{-1}\widetilde{\mathbf{x}}_i} \tag{7.40}$$

constrained to $\text{Tr}(\widehat{\mathbf{R}}) = 2m$ for NC-CES distributed $\mathbf{x}_i$, which reduces to the solution $\widehat{\mathbf{R}}$ of

$$\widehat{\mathbf{R}} = \frac{m}{n}\sum_{i=1}^{n}\frac{\mathbf{x}_i\mathbf{x}_i^H}{\mathbf{x}_i^H\widehat{\mathbf{R}}^{-1}\mathbf{x}_i} \tag{7.41}$$

constrained to $\text{Tr}(\widehat{\mathbf{R}}) = m$ for C-CES distributed $\mathbf{x}_i$. Here too, all the properties of the Tyler's M-estimate of $\mathbf{R}_{\bar{x}}$ are transferred to the Tyler's M-estimates of $\mathbf{R}_x$ and $\mathbf{R}_{\tilde{x}}$ for C-CES and NC-CES distributions, respectively. In particular, we get the convergences in distribution (7.23) and (7.24) where $\mathbf{R} \stackrel{\text{def}}{=} \frac{m}{\text{Tr}(\mathbf{R}_x)}\mathbf{R}_x$ in the complex circular case and $\mathbf{R} \stackrel{\text{def}}{=} \frac{m}{\text{Tr}(\mathbf{R}_x)}\mathbf{R}_{\tilde{x}}$ in the complex of arbitrary circularity case, and where $\mathbf{R}_{r_x}$ and $\mathbf{R}_{r_{\tilde{x}}}$ are given by (7.27) and (7.28), respectively, with [51]

$$\sigma_1 = \left(\frac{m}{\text{Tr}(\mathbf{R}_x)}\right)^2\left(1+\frac{1}{m}\right) \text{ and } \sigma_2 = -\left(\frac{m}{\text{Tr}(\mathbf{R}_x)}\right)^2\frac{1}{m}\left(1+\frac{1}{m}\right). \tag{7.42}$$

SSCM Estimator

The SSCM estimator is another distribution-free estimator of $\mathbf{R}_x$ and $\mathbf{R}_{\tilde{x}}$ that is easier to compute than Tyler's M-estimator. It was first introduced [10, 32] under various names and then studied [16, 33, 34, 41, 54, 55] in the context of RES distributions. It was proved in particular that the expectations of the SSCM and SCM share the same eigenvectors with different eigenvalues with the same multiplicity and with a monotone one-to-one but rather complicated correspondence [16]. As for the covariance matrices of the SSCM and SCM estimates, it was proved that they are similarly structured. They also share the same eigenvectors with different eigenvalues [33, 34]. C-CES [8] and NC-CES [6] extensions of the definition and properties of the SSCM given in the RES context are also straightforward. The SSCM estimate $\widehat{\mathbf{R}}$ of $\mathbf{R}_x$ and $\mathbf{R}_{\tilde{x}}$ are given by

$$\widehat{\mathbf{R}} = \frac{1}{n}\sum_{i=1}^{n}\frac{\mathbf{x}_i\mathbf{x}_i^H}{\|\mathbf{x}_i\|^2} \text{ and } \widehat{\mathbf{R}} = \frac{1}{n}\sum_{i=1}^{n}\frac{\widetilde{\mathbf{x}}_i\widetilde{\mathbf{x}}_i^H}{\|\widetilde{\mathbf{x}}_i\|^2}, \tag{7.43}$$

for C-CES and NC-CES distributed $\mathbf{x}_i$, respectively. Under only finite second-order moments of $\mathbf{x}_i$, applying the classic CLT to r.v. $\frac{\mathbf{x}_i^*\otimes\mathbf{x}_i}{\|\mathbf{x}_i\|^2}$ and $\frac{\widetilde{\mathbf{x}}_i^*\otimes\widetilde{\mathbf{x}}_i}{\|\widetilde{\mathbf{x}}_i\|^2}$, gives the asymptotic distributions of $\widehat{\mathbf{R}}$ which is also given by (7.23) and (7.24) where now

$$\mathbf{R} \stackrel{\text{def}}{=} \sum_{k=1}^{m}\chi_k\mathbf{v}_k\mathbf{v}_k^H \text{ and } \mathbf{R} \stackrel{\text{def}}{=} \sum_{k=1}^{2m}\widetilde{\chi}_k\widetilde{\mathbf{v}}_k\widetilde{\mathbf{v}}_k^H, \tag{7.44}$$

respectively, where $\sum_{k=1}^{m}\lambda_k\mathbf{v}_k\mathbf{v}_k^H$ and $\sum_{k=1}^{2m}\widetilde{\lambda}_k\widetilde{\mathbf{v}}_k\widetilde{\mathbf{v}}_k^H$ denote, respectively, the EVD of $\mathbf{R}_x$ and $\mathbf{R}_{\tilde{x}}$, and where closed-form expressions of the eigenvalues χ_k and $\widetilde{\chi}_k$ are given by [6, Rels. (11), (12)]. The covariances $\mathbf{R}_{r_x}$ and $\mathbf{R}_{r_{\tilde{x}}}$ of the asymptotic distributions (7.23) and (7.24) share, respectively, the same eigenvectors as (7.27) and (7.28), whose expressions of the different eigenvalues $\gamma_{k,\ell}$, $\gamma_{k,\ell} - \chi_k\chi_\ell$ and $\widetilde{\gamma}_{k\ell}$, $\widetilde{\gamma}_{k,\ell} - \widetilde{\chi}_k\widetilde{\chi}_\ell$ in the EVD [6, Rels. (15–16)] of $\mathbf{R}_{r_x}$ and $\mathbf{R}_{r_{\tilde{x}}}$ are given by [6, Rels. (17–20)].

7.5 Asymptotic Distributions of Subspace Projector Estimates

The asymptotic distributions of covariance estimates enable us to infer certain properties of the corresponding subspace projector estimates.

7.5.1 Asymptotic Inadmissibility of Subspace Projector Estimates

Tyler's M-estimate is the ML of $\mathbf{R}_x$ when the data are real-valued angular central Gaussian distribution distributed [52]. It is similarly straightforward to prove that Tyler's M-estimates $\widehat{\mathbf{R}}_x^{\text{Ty}}$ and $\widehat{\mathbf{R}}_{\tilde{x}}^{\text{Ty}}$ are also the ML of $\mathbf{R}_x$ and $\mathbf{R}_{\tilde{x}}$, for complex circular [27] and non-circular [6] angular central Gaussian distributed data whose p.d.f. with respect to the Lebesgue measure on complex unit sphere $\mathbb{C}S^m$ are respectively given by

$$p(\mathbf{x}) = \frac{\Gamma(m)}{2\pi^m}|\mathbf{R}_x|^{-1}(\mathbf{x}^H\mathbf{R}_x^{-1}\mathbf{x})^{-m} \tag{7.45}$$

and

$$p(\mathbf{x}) = \frac{2^m\Gamma(m)}{2\pi^m}|\mathbf{R}_{\tilde{x}}|^{-1/2}(\widetilde{\mathbf{x}}^H\mathbf{R}_{\tilde{x}}^{-1}\widetilde{\mathbf{x}})^{-m}, \tag{7.46}$$

using the linear one to one mapping $\bar{\mathbf{x}}_i \longleftrightarrow \widetilde{\mathbf{x}}_i$. Thus, by the invariance property of the ML estimator, its associated orthogonal projector $\widehat{\boldsymbol{\Pi}}_x^{\mathrm{Ty}} = \boldsymbol{\Pi}(\widehat{\mathbf{R}}_x^{\mathrm{Ty}})$ (resp., $\widehat{\boldsymbol{\Pi}}_{\tilde{x}}^{\mathrm{Ty}} = \boldsymbol{\Pi}(\widehat{\mathbf{R}}_{\tilde{x}}^{\mathrm{Ty}})$) (7.14) is the ML of the orthogonal projector $\boldsymbol{\Pi}_x$ (resp., $\boldsymbol{\Pi}_{\tilde{x}}$). This implies that for complex angular central Gaussian distributed data, the following relationship between the covariance of the asymptotic distribution of the projector based on ML, Tyler's M, and SSCM estimates

$$\mathbf{R}_{\pi_x}^{\mathrm{ML}} = \mathbf{R}_{\pi_x}^{\mathrm{Ty}} \leq \mathbf{R}_{\pi_x}^{\mathrm{SSCM}} \text{ and } \mathbf{R}_{\pi_{\tilde{x}}}^{\mathrm{ML}} = \mathbf{R}_{\pi_{\tilde{x}}}^{\mathrm{Ty}} \leq \mathbf{R}_{\pi_{\tilde{x}}}^{\mathrm{SSCM}}. \tag{7.47}$$

Consequently, thanks to the free distribution property of Tyler's M-estimate and SSCM estimate in the C-CES and NC-CES family, where is added the circular and non-circular complex angular central Gaussian distribution [51], we can extend to arbitrary C-CES and NC-CES with finite second-order moments distributed data, the matrix inequalities between covariance of the asymptotic distribution of the projector based on Tyler's M and SSCM estimates (7.47). This gives:

$$\mathbf{R}_{\pi_x}^{\mathrm{Ty}} \leq \mathbf{R}_{\pi_x}^{\mathrm{SSCM}} \text{ and } \mathbf{R}_{\pi_{\tilde{x}}}^{\mathrm{Ty}} \leq \mathbf{R}_{\pi_{\tilde{x}}}^{\mathrm{SSCM}}. \tag{7.48}$$

Furthermore it was proved in [6] that when $\mathbf{R}_x \to \lambda\mathbf{I}$ and $\mathbf{R}_{\tilde{x}} \to \lambda\mathbf{I}$, inequalities (7.48) approach equalities. These inequalities (7.48) show that the estimator $\widehat{\Pi}_x^{\mathrm{Ty}}$ (resp., $\widehat{\boldsymbol{\Pi}}_{\tilde{x}}^{\mathrm{Ty}}$) asymptotically dominates the estimator $\widehat{\Pi}_x^{\mathrm{SSCM}}$ (resp., $\widehat{\Pi}_{\tilde{x}}^{\mathrm{SSCM}}$) in the sense of the mean squared error. This property of asymptotic inadmissibility of the projector associated with the SSCM proved firstly for RES distributed data in [33, 34], extends to arbitrary C-CES and NC-CES distributed data. However, since the SCM is very sensitive to heavy-tailed CES distributions, $\mathbf{R}_{\pi_x}^{\mathrm{SSCM}}$ and $\mathbf{R}_{\pi_{\tilde{x}}}^{\mathrm{SSCM}}$ is often bounded above by $\mathbf{R}_{\pi_x}^{\mathrm{SCM}}$ and $\mathbf{R}_{\pi_{\tilde{x}}}^{\mathrm{SCM}}$, respectively, for such distributions. An example of such behavior is given in Fig. 7.2 and conditions for which the performance of Tyler's M-estimate and SSCM are close are specified in Sect. 7.5.3.

7.5.2 *Asymptotic Distributions of Projector Estimates*

From the asymptotic distributions (7.19), (7.21), (7.23) and (7.24) of the different estimates $\widehat{\mathbf{R}}$ adapted to the different models presented in Sect. 7.2, we note that $\widehat{\mathbf{R}}$ converges in probability to the matrices $\mathbf{R}$. All these matrices are structured as

$$\mathbf{R} = \mathbf{S} + \sigma^2\mathbf{I}, \tag{7.49}$$

where $\mathrm{Span}(\mathbf{S}) = \mathrm{Span}(\mathbf{B}(\boldsymbol{\theta}))$, where $\mathbf{B}(\boldsymbol{\theta})$ denotes the mixing matrices $\mathbf{A}$, $\widetilde{\mathbf{A}}_r$ and $\widetilde{\mathbf{A}}_c$ for circular, rectilinear, and non-rectilinear and non-circular complex-valued $\mathbf{s}_i$, respectively. Then, using the standard perturbation result associated with mapping (7.14)

$$\widehat{\mathbf{R}} = \mathbf{R} + \delta(\mathbf{R}) \overset{\mathrm{EVD}}{\mapsto} \widehat{\boldsymbol{\Pi}} = \boldsymbol{\Pi}(\boldsymbol{\theta}) + \delta(\boldsymbol{\Pi}), \tag{7.50}$$

for orthogonal projectors [26] (see also the operator approach in [28]) applied to $\mathbf{\Pi}(\boldsymbol{\theta})$ associated with the noise subspace of $\mathbf{R}$,

$$\delta(\mathbf{\Pi}) = -\mathbf{\Pi}(\boldsymbol{\theta})\delta(\mathbf{R})\mathbf{S}^{\#} - \mathbf{S}^{\#}\delta(\mathbf{R})\mathbf{\Pi}(\boldsymbol{\theta}) + o(\delta(\mathbf{R})), \tag{7.51}$$

the asymptotic behaviors of $\widehat{\mathbf{\Pi}}$ and $\widehat{\mathbf{R}}$ are directly related. A standard theorem of continuity (see e.g., [47, p. 122]) (called also Delta-method) on regular functions of asymptotically Gaussian statistics applies and we get

$$\sqrt{n}(\mathrm{vec}(\widehat{\mathbf{\Pi}}) - \mathrm{vec}(\mathbf{\Pi}(\boldsymbol{\theta}))) \to_d \mathbb{C}N_{m^2}(\mathbf{0}, \mathbf{R}_{\pi_x}, \mathbf{R}_{\pi_x}\mathbf{K}) \tag{7.52}$$
$$\sqrt{n}(\mathrm{vec}(\widehat{\mathbf{\Pi}}) - \mathrm{vec}(\mathbf{\Pi}(\boldsymbol{\theta}))) \to_d \mathbb{C}N_{4m^2}(\mathbf{0}, \mathbf{R}_{\pi_{\tilde{x}}}, \mathbf{R}_{\pi_{\tilde{x}}}\mathbf{K}), \tag{7.53}$$

for circular and non-circular complex-valued $\mathbf{s}_i$, respectively, where $\boldsymbol{\Pi}(\boldsymbol{\theta})$ is given by (7.15) with its associated $\mathbf{B}(\boldsymbol{\theta})$ and where

$$\mathbf{R}_{\pi_x} = [(\mathbf{S}_x^{T\#} \otimes \mathbf{\Pi}_x) + (\mathbf{\Pi}_x^T \otimes \mathbf{S}_x^{\#})]\mathbf{R}_{r_x}[(\mathbf{S}_x^{T\#} \otimes \mathbf{\Pi}_x) + (\mathbf{\Pi}_x^T \otimes \mathbf{S}_x^{\#})], \tag{7.54}$$
$$\mathbf{R}_{\pi_{\tilde{x}}} = [(\mathbf{S}_{\tilde{x}}^{T\#} \otimes \mathbf{\Pi}_{\tilde{x}}) + (\mathbf{\Pi}_{\tilde{x}}^T \otimes \mathbf{S}_{\tilde{x}}^{\#})]\mathbf{R}_{r_{\tilde{x}}}[(\mathbf{S}_{\tilde{x}}^{T\#} \otimes \mathbf{\Pi}_{\tilde{x}}) + (\mathbf{\Pi}_{\tilde{x}}^T \otimes \mathbf{S}_{\tilde{x}}^{\#})], \tag{7.55}$$

where $\mathbf{R}_{r_x}$ and $\mathbf{R}_{r_{\tilde{x}}}$ are given by (7.20), (7.25), (7.27) and by (7.22), (7.26), (7.28), respectively, and where each of the two matrices $(\mathbf{S}_x, \mathbf{\Pi}_x)$ and $(\mathbf{S}_{\tilde{x}}, \mathbf{\Pi}_{\tilde{x}})$ are the matrices $\mathbf{S}$ and $\mathbf{\Pi}(\boldsymbol{\theta})$ associated with circular and non-circular complex-valued $\mathbf{s}_i$, respectively.

Then plugging expressions (7.20), (7.25), (7.27) of $\mathbf{R}_{r_x}$ and (7.22), (7.26), (7.28) of $\mathbf{R}_{r_{\tilde{x}}}$ in (7.54) and (7.55), and using $\mathbf{\Pi}_x\mathbf{S}_x = \mathbf{0}$ and $\mathbf{\Pi}_{\tilde{x}}\mathbf{S}_{\tilde{x}} = \mathbf{0}$, the following theorem extending [15, Theorem IV.1] is proved:

Theorem 7.1 *The covariance matrices $\mathbf{R}_{\pi_x}$ and $\mathbf{R}_{\pi_{\tilde{x}}}$ of the asymptotic distributions (7.52) and (7.53) of the different projector estimates $\widehat{\mathbf{\Pi}}$ associated with the different data models presented in Sect. 7.2 have a unified structure given by*

$$\mathbf{R}_{\pi_x} = (\mathbf{U}^T \otimes \boldsymbol{\Pi}(\boldsymbol{\theta})) + (\boldsymbol{\Pi}^T(\boldsymbol{\theta}) \otimes \mathbf{U}), \tag{7.56}$$
$$\mathbf{R}_{\pi_{\tilde{x}}} = [\mathbf{I} + \mathbf{K}(\mathbf{J} \otimes \mathbf{J})][(\mathbf{U}^T \otimes \boldsymbol{\Pi}(\boldsymbol{\theta})) + (\boldsymbol{\Pi}^T(\boldsymbol{\theta}) \otimes \mathbf{U})]. \tag{7.57}$$

Here $\boldsymbol{\Pi}(\boldsymbol{\theta})$ are the projection matrices $\sum_{k=p+1}^{m} \mathbf{v}_k\mathbf{v}_k^H$, $\sum_{k=p+1}^{2m} \widetilde{\mathbf{v}}_k\widetilde{\mathbf{v}}_k^H$ and $\sum_{k=2p+1}^{2m} \widetilde{\mathbf{v}}_k\widetilde{\mathbf{v}}_k^H$ on the noise subspace (i.e., on the orthogonal complement of the range of $\mathbf{A}$, $\widetilde{\mathbf{A}}_r$ and $\widetilde{\mathbf{A}}_c$), associated with circular, rectilinear, and non-rectilinear and non-circular complex-valued $\mathbf{s}_i$, respectively. On the other hand, the matrices $\mathbf{U}$ depend on the covariance estimates $\widehat{\mathbf{R}}$ studied in Sect. 7.4.

For the deterministic model and the stochastic model where both $\mathbf{s}_i$ and $\mathbf{n}_i$ are CES distributed, $\mathbf{U}$ takes the common form

$$\mathbf{U} = \sigma_n^2\mathbf{S}^{\#}\mathbf{R}\mathbf{S}^{\#} + \kappa_n\sigma_n^4(\mathbf{S}^{\#})^2 = \sum_{k=1}^{p} \frac{\sigma_n^2(\lambda_k + \kappa_n\sigma_n^2)}{(\lambda_k - \sigma_n^2)^2}\mathbf{v}_k\mathbf{v}_k^H, \tag{7.58}$$

with $\mathbf{R} \stackrel{\text{def}}{=} \mathbf{R}_{x,\infty} = \mathbf{R}_x = \sum_{k=1}^{m} \lambda_k \mathbf{v}_k \mathbf{v}_k^H$ and $\mathbf{S} \stackrel{\text{def}}{=} \mathbf{A}\mathbf{R}_{s,\infty}\mathbf{A}^H = \mathbf{A}\mathbf{R}_s\mathbf{A}^H$ for $\mathbf{s}_i$ deterministic of arbitrary circularity and circular stochastic. Similarly for rectilinear $\mathbf{s}_i$, (7.58) also applies where $\mathbf{R}_{x,\infty}$, $\mathbf{R}_x$, $\mathbf{A}\mathbf{R}_{s,\infty}\mathbf{A}^H$, $\mathbf{A}\mathbf{R}_s\mathbf{A}^H$, λ_k and $\mathbf{v}_k$ are replaced by $\mathbf{R}_{\tilde{x},\infty}$, $\mathbf{R}_{\tilde{x}}$, $\widetilde{\mathbf{A}}_r\mathbf{R}_{r,\infty}\widetilde{\mathbf{A}}_r^H$, $\widetilde{\mathbf{A}}_r\mathbf{R}_r\widetilde{\mathbf{A}}_r^H$, $\widetilde{\lambda}_k$ and $\widetilde{\mathbf{v}}_k$, respectively. Furthermore, for non-circular and non-rectilinear stochastic $\mathbf{s}_i$ (7.58) still applies where $\mathbf{R} \stackrel{\text{def}}{=} \mathbf{R}_{\tilde{x}}$, $\mathbf{S} \stackrel{\text{def}}{=} \widetilde{\mathbf{A}}_c\mathbf{R}_{\tilde{s}}\widetilde{\mathbf{A}}_c^H$ and p, λ_k and $\mathbf{v}_k$ are replaced, respectively, by $2p$, $\widetilde{\lambda}_k$ and $\widetilde{\mathbf{v}}_k$.

For the stochastic model, where $\mathbf{x}_i$ is CES distributed, we get for SCM, ML, M and Tyler's estimator:

$$\mathbf{U} = \vartheta\left(\sum_{k=1}^{p}\frac{\lambda_k\sigma_n^2}{(\lambda_k-\sigma_n^2)^2}\mathbf{v}_k\mathbf{v}_k^H\right) \text{ and } \mathbf{U} = \vartheta\left(\sum_{k=1}^{p}\frac{\widetilde{\lambda}_k\sigma_n^2}{(\widetilde{\lambda}_k-\sigma_n^2)^2}\widetilde{\mathbf{v}}_k\widetilde{\mathbf{v}}_k^H\right), \tag{7.59}$$

for circular and rectilinear $\mathbf{s}_i$, respectively, with

$$\vartheta \stackrel{\text{def}}{=} \vartheta_{\text{SCM}} = 1 + \kappa_x \quad \text{for the SCM estimate,} \tag{7.60}$$

$$\vartheta \stackrel{\text{def}}{=} \vartheta_{\text{ML}} = \sigma_1 \quad \text{for the ML estimate given by (32),} \tag{7.61}$$

$$\vartheta \stackrel{\text{def}}{=} \vartheta_{\text{M}} = c^2\sigma_1 \quad \text{for the } M\text{-estimate where } \sigma_1 \text{ and } c \text{ are given by (38) and (35), respectively,} \tag{7.62}$$

$$\vartheta \stackrel{\text{def}}{=} \vartheta_{\text{Ty}} = 1 + \frac{1}{m} \quad \text{for the Tyler's } M\text{-estimate.} \tag{7.63}$$

For non-circular and non-rectilinear $\mathbf{s}_i$, the second relation of (7.59) also applies by replacing p by $2p$.

For the SSCM estimator, $\mathbf{U}$ is given by

$$\mathbf{U} = \sum_{k=1}^{p}\frac{\gamma_k}{(\chi_k-\chi)^2}\mathbf{v}_k\mathbf{v}_k^H \text{ and } \mathbf{U} = \sum_{k=1}^{p}\frac{\widetilde{\gamma}_k}{(\widetilde{\chi}_k-\widetilde{\chi})^2}\widetilde{\mathbf{v}}_k\widetilde{\mathbf{v}}_k^H, \tag{7.64}$$

for circular $\mathbf{s}_i$ where $\chi \stackrel{\text{def}}{=} \chi_{p+1} = \chi_{p+2} = \ldots = \chi_m$ (see (7.44)) and $\gamma_k \stackrel{\text{def}}{=} \gamma_{k,p+1} = \gamma_{k,p+2} = \ldots = \gamma_{k,m}$, and rectilinear $\mathbf{s}_i$ where $\widetilde{\chi} \stackrel{\text{def}}{=} \widetilde{\chi}_{p+1} = \widetilde{\chi}_{p+2} = \ldots = \widetilde{\chi}_{2m}$ (see (7.44)) and $\widetilde{\gamma}_k \stackrel{\text{def}}{=} \widetilde{\gamma}_{k,p+1} = \widetilde{\gamma}_{k,p+2} = \ldots = \widetilde{\gamma}_{k,2m}$ are given in [6, Rels. (17)–(20)]. For non-circular and non-rectilinear $\mathbf{s}_i$, the second relation of (7.64) also applies by replacing p by $2p$.

Two remarks are in order from Theorem 7.1 for the deterministic model and the stochastic model where both $\mathbf{s}_i$ and $\mathbf{n}_i$ are CES distributed:

- The projector estimators have the same asymptotic distribution in both deterministic and stochastic CES distributed models for $\mathbf{s}_i$, regardless of whether $\mathbf{s}_i$ is arbitrary circular or rectilinear $\mathbf{s}_i$. This contrasts with the covariances of the asymptotic distributions of $\mathbf{R}_x$ (7.20), (7.25) and $\mathbf{R}_{\tilde{x}}$ (7.22), (7.26), which are different. This generalizes classical results on many subspace-based DOA estimators [49] that were proven under the complex circular Gaussian noise framework.

- The asymptotic distributions of the projector estimators remain are invariant to the distribution of $\mathbf{s}_i$, regardless of whether $\mathbf{s}_i$ is of arbitrary circularity or is rectilinear. This property also generalizes classical results on subspace-based estimators [11] that were proven under the complex circular Gaussian noise framework.

7.5.3 Relative Efficiency

For circular or non-circular CES distributed data $\mathbf{x}_i$, the covariance matrices of the asymptotic distribution of the projector derived from the SCM, ML, M, and Tyler's estimators are proportional. The proportionality coefficient ϑ (7.60)–(7.63) serves as an efficiency index for the estimation of the projector. This proportionality relation only holds for the projector derived from the SSCM estimator when $\lambda_1 = ... = \lambda_p$ and $\widetilde{\lambda}_1 = ... = \widetilde{\lambda}_p$ which implies that $\chi_1 = ... = \chi_p$, $\gamma_1 = ... = \gamma_p$ and $\widetilde{\chi}_1 = ... = \widetilde{\chi}_p$, $\widetilde{\gamma}_1 = ... = \widetilde{\gamma}_p$ in (7.64), and hence

$$\vartheta \stackrel{\text{def}}{=} \vartheta_{\text{SSCM}} = \frac{\gamma_1}{(\chi_1 - \chi)^2} \frac{(\lambda_1 - \sigma_n^2)^2}{\lambda_1 \sigma_n^2} \quad \text{and} \quad \vartheta \stackrel{\text{def}}{=} \vartheta_{\text{SSCM}} = \frac{\widetilde{\gamma}_1}{(\widetilde{\chi}_1 - \widetilde{\chi})^2} \frac{(\widetilde{\lambda}_1 - \sigma_n^2)^2}{\widetilde{\lambda}_1 \sigma_n^2} \tag{7.65}$$

in the circular and rectilinear cases, respectively. For non-circular and non-rectilinear $\mathbf{s}_i$, the second relation of (7.65) also applies under the condition $\widetilde{\lambda}_1 = ... = \widetilde{\lambda}_{2p}$. To obtain insight into how the asymptotic inadmissibility of the SSCM is affected whenever $\mathbf{R}_x \neq \lambda\mathbf{I}$ and $\mathbf{R}_{\tilde{x}} \neq \lambda\mathbf{I}$ (see discussion after (7.48)), the following theorem is proved in [6, Theorem 5]:

Theorem 7.2 *The asymptotic efficiency of the SSCM estimate relative to Tyler's M-estimate defined by the ratio $r \stackrel{\text{def}}{=} \vartheta_{\text{Ty}}/\vartheta_{\text{SSCM}}$, is given by the closed-form expressions in the circular case under the assumption $\lambda_1 = ... = \lambda_p$ and rectilinear case under the assumption $\widetilde{\lambda}_1 = ... = \widetilde{\lambda}_p$, respectively, by*

$$r_c = \frac{[{}_2F_1(1, m-p+1, m+2, 1-\rho)]^2}{{}_2F_1(2, m-p+1, m+2, 1-\rho)} \tag{7.66}$$

and

$$r_r = \frac{[{}_2F_1(1, m-\frac{p}{2}+1, m+2, 1-\widetilde{\rho})]^2}{{}_2F_1(2, m-\frac{p}{2}+1, m+2, 1-\widetilde{\rho})}, \tag{7.67}$$

where $\rho \stackrel{\text{def}}{=} \frac{\sigma_n^2}{\lambda_1}$ and $\widetilde{\rho} \stackrel{\text{def}}{=} \frac{\sigma_n^2}{\widetilde{\lambda}_1}$, and these ratios are monotonic increasing functions of respectively ρ and $\widetilde{\rho}$ from the intervals (0,1) to (0,1). Furthermore in the neighborhood of $\rho = 1$, $\widetilde{\rho} = 1$ and $\rho = 0$, $\widetilde{\rho}_1 = 0$, we have, respectively,

$$r_c = 1 - \frac{(m-p+1)(p+1)}{(m+2)^2(m+3)}(1-\rho)^2 + o(1-\rho)^2, \tag{7.68}$$

$$r_r = 1 - \left(\frac{(2m-p+2)(p+2)}{4(m+2)^2(m+3)}\right)(1-\widetilde{\rho})^2 + o(1-\widetilde{\rho})^2, \tag{7.69}$$

and

$$r_c = \begin{cases} o_{m,1}(1) & \textit{for } p = 1 \\ (1+\frac{1}{m})(1-\frac{1}{p})(1+o_{m,p}(1)) & \textit{for } p > 1 \end{cases} \tag{7.70}$$

$$r_r = \begin{cases} \widetilde{o}_{m,p}(1) & \textit{for } p = 1,2 \\ (1+\frac{1}{m})(1-\frac{2}{p})(1+\widetilde{o}_{m,p}(1)) & \textit{for } p > 2 \end{cases}, \tag{7.71}$$

where first-order expansions of $o_{m,p}(1)$ *and* $\widetilde{o}_{m,p}(1)$ *satisfying* $\lim_{\rho\to 0} o_{m,p}(1) = \lim_{\widetilde{\rho}\to 0} \widetilde{o}_{m,p}(1) = 0$ *are given by [6, Rels. (57–58)]. For non-circular and non-rectilinear* $\mathbf{s}_i$*, relation* (7.67) *and the expansions of* r_r *also apply under the condition* $\widetilde{\lambda}_1 = ... = \widetilde{\lambda}_{2p}$ *where* p *is replaced par* $2p$*.*

It follows from (7.68) and (7.69) that the performance of the subspace estimation derived from SSCM and Tyler's M-estimate are very similar for close eigenvalues, and particularly for large values of m and p. It follows, conversely, from (7.70) and (7.71), that for well-separated eigenvalues, the performance of the SSCM-based subspace estimation is largely outperformed by those derived from Tyler's M estimate for $p = 1$ and $p = 1, 2$ for C-CES and NC-CES distributed data because r_c and r_r tend to zero as ρ and $\widetilde{\rho}$ tend to zero, respectively. Note, however, that for m and p large, the performance of the subspace estimation derived from SSCM and Tyler's M estimate are very similar because r_c and r_r are equivalent to $(1+\frac{1}{m})(1-\frac{1}{p}) < 1$ and $(1+\frac{1}{m})(1-\frac{2}{p}) < 1$ as ρ and $\widetilde{\rho}$ tend to zero, respectively. Consequently, despite the asymptotic inadmissibility of subspace projectors built from the SSCM estimate, the performance of this estimator and those derived from Tyler's M-estimator are close in particular for large values of m and not too small values of p. Therefore, we can conclude that SSCM estimate is of great interest from the point of view of its lower computational complexity for large values of m.

7.6 Asymptotic Distributions of Subspace-Based Parameter Estimates

The asymptotic distributions of projector estimates given in Theorem 7.1 allow us to derive the asymptotic distribution of subspace-based parameter estimates for any algorithms "alg" (7.14) which are assumed to be differentiable.[4]

[4] Note that since $\boldsymbol{\Pi}(\boldsymbol{\theta})$ is Hermitian, the mappings "alg" are differentiable w.r.t. $\boldsymbol{\Pi}(\boldsymbol{\theta})$ i.f.f. they are differentiable w.r.t. $(\mathrm{Re}(\boldsymbol{\Pi}(\boldsymbol{\theta})), \mathrm{Im}(\boldsymbol{\Pi}(\boldsymbol{\theta})))$. This last hypothesis being verified by most algorithms "alg".

7.6.1 Asymptotic Distribution of Parameter Estimate

Again using a standard theorem of continuity (see e.g., [47, p. 122]) (called also Delta-method), we get:

$$\sqrt{n}(\widehat{\boldsymbol{\theta}} - \boldsymbol{\theta}) \rightarrow_d \mathbb{R}N_k(\mathbf{0}, \mathbf{R}^{\text{alg}}_{\hat{\theta}}), \tag{7.72}$$

where $\mathbf{R}^{\text{alg}}_{\hat{\theta}}$ is given by

$$\mathbf{R}^{\text{alg}}_{\hat{\theta}} = \mathbf{D}_{\text{alg}} \mathbf{R}_{\pi} \mathbf{D}^{H}_{\text{alg}}, \tag{7.73}$$

with $\mathbf{R}_{\pi} = \mathbf{R}_{\pi_x}$ [resp., $\mathbf{R}_{\pi} = \mathbf{R}_{\pi_{\tilde{x}}}$] is given by (7.56) [resp., (7.57)], for circular [resp., non-circular] $\mathbf{s}_i$, and where $\mathbf{D}_{\text{alg}}$ is the differential (or Jacobian) matrix defined by the relation

$$\widehat{\boldsymbol{\theta}} = \text{alg}(\widehat{\boldsymbol{\Pi}}) = \underbrace{\text{alg}(\boldsymbol{\Pi}(\boldsymbol{\theta}))}_{\boldsymbol{\theta}} + \mathbf{D}_{\text{alg}} \text{vec}(\widehat{\boldsymbol{\Pi}} - \boldsymbol{\Pi}(\boldsymbol{\theta})) + o(\widehat{\boldsymbol{\Pi}} - \boldsymbol{\Pi}(\boldsymbol{\theta})). \tag{7.74}$$

Using the unified expressions of $\mathbf{R}_{\pi_x}$ and $\mathbf{R}_{\pi_{\tilde{x}}}$ of Theorem 7.1 where $\mathbf{U}$ is given by (7.58) and (7.59), we get the following theorem:

Theorem 7.3 *The covariance matrix* $\mathbf{R}^{\text{alg}}_{\hat{\theta}}$ *of the asymptotic distribution of the parameter estimate* $\widehat{\boldsymbol{\theta}}$ *given by any subspace-based algorithm "alg" satisfying the previously aforementioned assumptions, derived from the SCM, ML, M and Tyler's M covariance estimate, breaks down as follows*

$$\mathbf{R}^{\text{alg}}_{\hat{\theta}} = \vartheta \mathbf{R}^{\text{G,alg}}_{\hat{\theta}} + \kappa_n \mathbf{R}^{'\text{alg}}_{\hat{\theta}}, \tag{7.75}$$

where $\mathbf{R}^{\text{G,alg}}_{\hat{\theta}}$ is the covariance matrix $\mathbf{R}^{\text{alg}}_{\hat{\theta}}$ derived for Gaussian distributed data $\mathbf{x}_i$ and in the second term $\mathbf{R}^{'\text{alg}}_{\hat{\theta}}$ is a positive definite matrix obtained only in the deterministic and stochastic model where $\mathbf{n}_i$ is non-Gaussian C-CES distributed. $\vartheta = 1$ in the deterministic and stochastic model where $\mathbf{n}_i$ is C-CES distributed, ϑ is given by (7.60)–(7.63) for CES distributed data $\mathbf{x}_i$ and κ_n is the kurtosis parameter of the noise, which is zero for Gaussian distributed noise.

In particular, for CES distributed $\mathbf{x}_i$,

$$\mathbf{R}^{\text{alg}}_{\hat{\theta}} = \vartheta \mathbf{R}^{\text{G,alg}}_{\hat{\theta}} \tag{7.76}$$

and all the analytical theorems concerning the asymptotic distributions of subspace-based parameter estimates derived in the Gaussian framework extend with a simple multiplicative term ϑ, especially in DOA estimation.

For the SSCM estimate, (7.76) also applies where ϑ is given by (7.65) but only in the case $\lambda_1 = ... = \lambda_p$ and $\widetilde{\lambda}_1 = ... = \widetilde{\lambda}_p$. In contrast, for an arbitrary spectrum of eigenvalues, there is no longer a direct relationship between $\mathbf{R}^{\text{alg}}_{\hat{\theta}}$ and $\mathbf{R}^{\text{G,alg}}_{\hat{\theta}}$.

In the deterministic and stochastic model where $\mathbf{n}_i$ is non-Gaussian C-CES distributed, the structure of $\mathbf{R}_{\hat{\theta}}^{\text{alg}}$ is affected due to the additive term $\mathbf{R}_{\hat{\theta}}^{',\text{alg}}$. Furthermore, $\mathbf{R}_{\hat{\theta}}^{\text{alg}} > \mathbf{R}_{\hat{\theta}}^{\text{G,alg}}$ and $\mathbf{R}_{\hat{\theta}}^{\text{alg}} < \mathbf{R}_{\hat{\theta}}^{\text{G,alg}}$ for super-Gaussian ($\kappa_n > 0$) and sub-Gaussian ($\kappa_n < 0$) distributed noise, respectively.

Besides, the covariance of the asymptotic distribution of the estimate $\widehat{\boldsymbol{\theta}}$ given by the AMV algorithm [45, 48] takes the particular expression [5]:

$$\mathbf{R}_{\hat{\theta}}^{\text{AMV}} = (\boldsymbol{\Pi}'^{H}(\boldsymbol{\theta})\mathbf{R}_{\pi}^{\#}\boldsymbol{\Pi}'(\boldsymbol{\theta}))^{-1}, \tag{7.77}$$

where $\boldsymbol{\Pi}'(\boldsymbol{\theta}) \stackrel{\text{def}}{=} \frac{d\text{vec}(\boldsymbol{\Pi}(\boldsymbol{\theta}))}{d\boldsymbol{\theta}}$. It has been proved in [5] that the AMV estimates $\widehat{\boldsymbol{\theta}}$ derived from the projector estimates $\widehat{\boldsymbol{\Pi}}$ built on the ML estimate of $\mathbf{R}_x$ and $\mathbf{R}_{\tilde{x}}$ for, respectively, stochastic C-CES and NC-CES distributed data $\mathbf{x}_i$ are asymptotically efficient, i.e., the covariance matrices $\mathbf{R}_{\hat{\theta}}^{\text{ML,AMV}}$ of their asymptotic distributions reach the stochastic Cramér–Rao bound (CRB) of the parameter $\boldsymbol{\theta}$ when the density generator g is known. Consequently, the following theorem is deduced from (7.48), (7.73) and (7.77).

Theorem 7.4 *For CES distributed data* $\mathbf{x}_i$*, the covariance of the Gaussian asymptotic distribution of the estimated parameter* $\widehat{\boldsymbol{\theta}}$ *derived for any subspace-based algorithm "alg" built on the SSCM is bounded below by those built on Tyler's M estimate. These two covariance matrices being themselves bounded below by the CRB.*

$$\begin{aligned} n \times \text{CRB}(\boldsymbol{\theta}) &= (\boldsymbol{\Pi}'^{H}(\boldsymbol{\theta})\mathbf{R}_{\pi}^{\text{ML}\#}\boldsymbol{\Pi}'(\boldsymbol{\theta}))^{-1} \\ &= \mathbf{R}_{\hat{\theta}}^{\text{ML,AMV}} \leq \mathbf{R}_{\hat{\theta}}^{\text{ML,alg}} \leq \mathbf{R}_{\hat{\theta}}^{\text{Ty,alg}} \leq \mathbf{R}_{\hat{\theta}}^{\text{SSCM,alg}}, \end{aligned} \tag{7.78}$$

where $\mathbf{R}_{\pi}^{\text{ML}}$ *denotes the covariance of the asymptotic distribution of* $\widehat{\boldsymbol{\Pi}}$ *built on the ML estimate of* $\mathbf{R}_x$ *and* $\mathbf{R}_{\tilde{x}}$ *for, respectively, C-CES and NC-CES distributed data* $\mathbf{x}_i$*, and* $\mathbf{R}_{\hat{\theta}}^{\text{ML,alg}}$, $\mathbf{R}_{\hat{\theta}}^{\text{Ty,alg}}$ *and* $\mathbf{R}_{\hat{\theta}}^{\text{SSCM,alg}}$ *denote the covariance of the asymptotic distribution of the parameter* $\widehat{\boldsymbol{\theta}}$ *estimated by the algorithm "alg" built on the ML, Tyler's M and SSCM covariance estimate, respectively.*

Theorem 7.4 proves that the AMV subspace-based estimators built on Tyler's M-estimator of the covariance matrix are not efficient. To obtain a truly robust efficient subspace-based estimator, one has to find M-estimators with an appropriate weight function $u_x(.)$ such that ϑ_{M} be close or equal to ϑ_{ML}. Note that similarly to the projector estimate, there is no general order relation between $\mathbf{R}_{\hat{\theta}}^{\text{SSCM,alg}}$ and $\mathbf{R}_{\hat{\theta}}^{\text{SCM,alg}}$. However, since the SCM is very sensitive to heavy-tailed CES distributions, $\mathbf{R}_{\hat{\theta}}^{\text{SSCM,alg}}$ can be bounded above by $\mathbf{R}_{\hat{\theta}}^{\text{SCM,alg}}$ for such distributions.

Considering now the stochastic CRB in (7.78), the following common closed-form expression has been proved in [5]:

$$\text{CRB}(\boldsymbol{\theta}) = \sigma_1 \frac{\sigma_n^2}{2}\left[\text{Re}\left(\frac{d\mathbf{a}_{\theta}^H}{d\boldsymbol{\theta}}(\mathbf{H}^T \otimes \boldsymbol{\Pi}(\boldsymbol{\theta}))\frac{d\mathbf{a}_{\theta}}{d\boldsymbol{\theta}}\right)\right]^{-1}, \tag{7.79}$$

where σ_1 is given by (7.32). $\mathbf{a}_\theta \stackrel{\text{def}}{=} \text{vec}(\mathbf{A}), \mathbf{H} \stackrel{\text{def}}{=} \mathbf{R}_s^H \mathbf{A}^H \mathbf{R}_x^{-1} \mathbf{A} \mathbf{R}_s$ and $\boldsymbol{\Pi}(\boldsymbol{\theta}) \stackrel{\text{def}}{=} \boldsymbol{\Pi}_x(\boldsymbol{\theta})$ in circular case, $\mathbf{a}_\theta \stackrel{\text{def}}{=} \text{vec}(\widetilde{\mathbf{A}}_r)$, $\mathbf{H} \stackrel{\text{def}}{=} \mathbf{R}_r \widetilde{\mathbf{A}}_r^H \widetilde{\mathbf{R}}_{\tilde{x}}^{-1} \widetilde{\mathbf{A}}_r \mathbf{R}_r$ and $\boldsymbol{\Pi}(\boldsymbol{\theta}) \stackrel{\text{def}}{=} \boldsymbol{\Pi}_{\tilde{x}}(\boldsymbol{\theta})$ in the rectilinear case[5] associated with the structured extended covariance (7.6) and $\mathbf{a}_\theta \stackrel{\text{def}}{=} \text{vec}(\mathbf{A}), \mathbf{H} \stackrel{\text{def}}{=} \left(\mathbf{R}_s \mathbf{A}^H, \mathbf{C}_s \mathbf{A}^T\right) \widetilde{\mathbf{R}}_{\tilde{x}}^{-1} \begin{pmatrix} \mathbf{A}\mathbf{R}_s \\ \mathbf{A}^* \mathbf{C}_s^* \end{pmatrix}$ and $\boldsymbol{\Pi}(\boldsymbol{\theta}) \stackrel{\text{def}}{=} \boldsymbol{\Pi}_x(\boldsymbol{\theta})$ in the non-circular and no-rectilinear case associated with the structured extended covariance (7.8).

In general, this stochastic CRB is upper bounded by the semiparametric CRB introduced by [18] when the density generator g is considered as an infinite-dimensional unknown nuisance parameter. This semiparametric CRB has been studied for RES and C-CES distributions in [20] and [19], respectively. In particular, a closed-form expression of this bound has been derived in [17] for the DOA parameter of C-CES distributed observations. By slightly modifying and extending the proof given in the support document of [19] to general C-CES and NC-CES distributed noisy linear mixture models (7.1), it has been proved in [5], that this semiparametric CRB coincides with the stochastic CRB given by (7.79). We note that this property is very specific to the parameter of interest characterized by the column space of the mixing matrix. This property is explained by the fact that this column space does not depend on the density generator g. It is important, however, to note that if the AMV subspace-based estimator derived from the ML estimate of the covariance of the data $\mathbf{x}_i$ is asymptotically efficient w.r.t. the stochastic CRB, it is no longer asymptotically efficient w.r.t. the semiparametric CRB because this ML M-estimate requires the knowledge of the density generator g. To obtain asymptotically semiparametric efficient subspace-based estimator of $\boldsymbol{\theta}$, the AMV estimator would have to be built on an asymptotically semiparametric efficient estimator $\widehat{\mathbf{R}}$ of the covariance of the data in (7.14) like the one proposed in [21].

7.6.2 *Numerical Illustrations*

To illustrate the asymptotic distributions of subspace-based parameter estimates, we focus on the conventional and non-circular MUSIC-based DOA estimation algorithm [2]. We consider that two uncorrelated circular or rectilinear sources of equal power σ_s^2, are impinging on an uniform linear array with m sensors for which $\mathbf{A} = [\mathbf{a}_1, \mathbf{a}_2]$ with $\mathbf{a}_k = (1, e^{i\theta_k}, \ldots, e^{i(m-1)\theta_k})^T$, where $\theta_k = \pi \sin \alpha_k$, with α_k are the DOAs relative to the normal of array broadside. The SNR is defined as SNR = σ_s^2/σ_n^2. The Jacobian $\mathbf{D}_{\text{music}}$ is given for the conventional MUSIC algorithm by (see [2] for the non-circular MUSIC algorithm):

$$\mathbf{D}_{\text{music}} = \begin{pmatrix} \mathbf{d}_1^T \\ \mathbf{d}_2^T \end{pmatrix} \quad \text{with } \mathbf{d}_k^T = \frac{-1}{\alpha_{k,k}} (\mathbf{a}_k^{'T} \otimes \mathbf{a}_k^H + \mathbf{a}_k^T \otimes \mathbf{a}_k^{'H}) \tag{7.80}$$

[5] Note that this case $\frac{d\mathbf{a}_\theta^H}{d\theta} (\mathbf{H}^T \otimes \boldsymbol{\Pi}(\boldsymbol{\theta})) \frac{d\mathbf{a}_\theta}{d\theta}$ is real-valued.

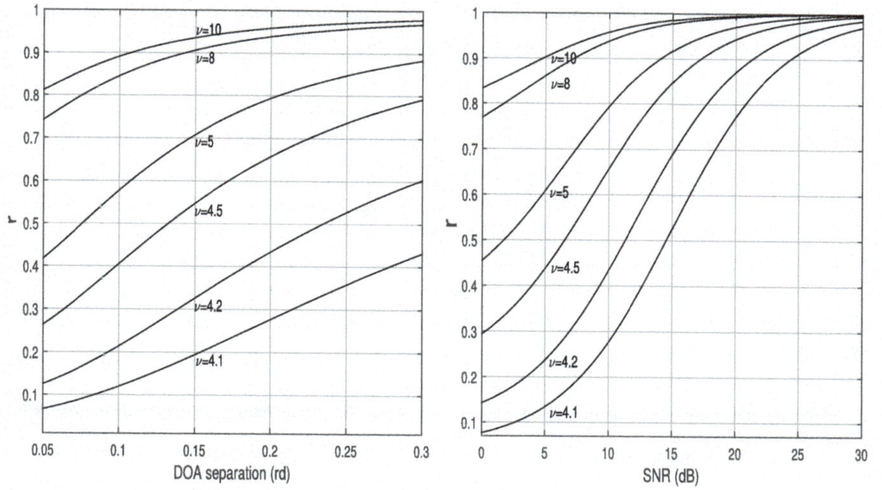

(a) r as a function of $\Delta\theta$ for SNR $=$ 10dB (b) r as a function of SNR for $\Delta\theta = 0.2$rd

Fig. 7.1 Theoretical ratio $r \stackrel{\text{def}}{=} [\mathbf{R}_{\hat{\theta}}^{\text{G,music}}]_{1,1}/[\mathbf{R}_{\hat{\theta}}^{\text{S,music}}]_{1,1}$ for several values of ν

and $\alpha_{k,k} \stackrel{\text{def}}{=} 2\mathbf{a}_k^{\prime H}\Pi_x\mathbf{a}_k^{\prime}$ where $\mathbf{a}_k^{\prime} \stackrel{\text{def}}{=} d\mathbf{a}_k/d\theta_k$. In these circular and rectilinear scenarios, the variances of the asymptotic distribution of the estimates of θ_1 and θ_2 are equal. Thus, we only consider the accuracy of source 1 in the sequel.

In this illustration, we choose the Student's t-distribution that belongs to the subclass of the compound Gaussian distributions (See Chap. 1). It is parameterized by the degree of freedom ν ($0 < \nu < \infty$) which controls the tails of the distribution that are heavier than the Gaussian ones (obtained for $\nu \to \infty$). This distribution has first-, second-, and fourth-order moments if, respectively, $\nu > 1$, $\nu > 2$ and $\nu > 4$ with $\vartheta_{\text{SCM}} = 1 + \kappa_n = \frac{\nu-2}{\nu-4}$ with $\nu > 4$ and $\vartheta_{\text{ML}} = \sigma_1 = \frac{m+1+\nu/2}{m+\nu/2}$. We also remind the reader that $\vartheta_{\text{Ty}} = 1 + 1/m$ for Tyler's M estimate (see (7.63)).

In the first setup, the sources $s_{i,k}$ are circular Gaussian distributed and the noise $\mathbf{n}_i$ is circular complex Student's t-distributed with $m = 6$ and the SCM is used. Fig. 7.1a, b show the theoretical ratio $r \stackrel{\text{def}}{=} [\mathbf{R}_{\hat{\theta}}^{\text{G,music}}]_{1,1}/[\mathbf{R}_{\hat{\theta}}^{\text{S,music}}]_{1,1}$ (where $[\mathbf{R}_{\hat{\theta}}^{\text{S,music}}]$ is given by (7.75)) of the variances of the asymptotic distribution of $\hat{\theta}_1$ for Gaussian distributed and Student's t-distributed noise for different values of the parameter ν, respectively, w.r.t. the DOA separation $\Delta\theta = |\theta_2 - \theta_1|$, and w.r.t. to the SNR. We see that the performance deteriorates strongly for heavy-tailed noise distributions (i.e., when $\nu \to 4$) w.r.t. the Gaussian distribution and that degradation increases for small DOA separation and small SNR.

In the second setup, we assume that the measurement $\mathbf{x}_i$ are circular complex Student's t-distributed. We compare now the variances of the asymptotic distribution of $\widehat{\theta_1}$ estimated from the SCM, ML and Tyler's M covariance estimate and SSCM, respectively, denoted by $[\mathbf{R}_{\hat{\theta}}^{\text{SCM,music}}]_{1,1}$, $[\mathbf{R}_{\hat{\theta}}^{\text{ML,music}}]_{1,1}$, $[\mathbf{R}_{\hat{\theta}}^{\text{Ty,music}}]_{1,1}$ and

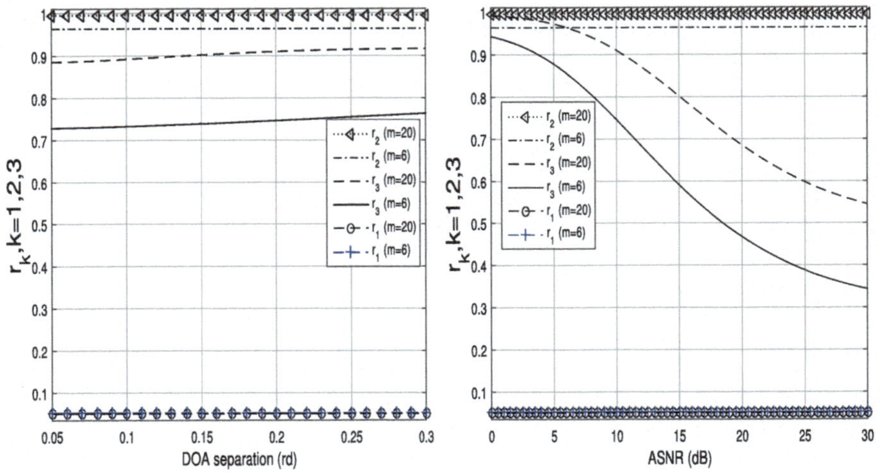

(a) r as a function of $\Delta\theta$ for ASNR = 10dB (b) r as a function of ASNR for $\Delta\theta = 0.2$rd

Fig. 7.2 Theoretical ratios $r_1 \stackrel{\text{def}}{=} [\mathbf{R}_{\hat{\theta}}^{\text{ML,music}}]_{1,1}/[\mathbf{R}_{\hat{\theta}}^{\text{SCM,music}}]_{1,1}$, $r_2 \stackrel{\text{def}}{=} [\mathbf{R}_{\hat{\theta}}^{\text{ML,music}}]_{1,1}/[\mathbf{R}_{\hat{\theta}}^{\text{Ty,music}}]_{1,1}$ and $r_3 \stackrel{\text{def}}{=} [\mathbf{R}_{\hat{\theta}}^{\text{ML,music}}]_{1,1}/[\mathbf{R}_{\hat{\theta}}^{\text{SSCM,music}}]_{1,1}$ for two values of m for $\nu = 4.1$

$[\mathbf{R}_{\hat{\theta}}^{\text{SSCM,music}}]_{1,1}$. The first three variances are proportional to $[\mathbf{R}_{\hat{\theta}}^{\text{G,music}}]_{1,1}$ given for $\mathbf{x}_i$ circular Gaussian distributed (7.76), but $[\mathbf{R}_{\hat{\theta}}^{\text{SSCM,music}}]_{1,1}$ is no longer proportional because

$$\lambda_1 = m\sigma_s^2\left(1+\frac{\sin(m\Delta\theta/2)}{m\sin(\Delta\theta/2)}\right)+\sigma_n^2 \neq \lambda_2 = m\sigma_s^2\left(1-\frac{\sin(m\Delta\theta/2)}{m\sin(\Delta\theta/2)}\right)+\sigma_n^2. \tag{7.81}$$

Consequently, the ratios $r_1 \stackrel{\text{def}}{=} [\mathbf{R}_{\hat{\theta}}^{\text{ML,music}}]_{1,1}/[\mathbf{R}_{\hat{\theta}}^{\text{SCM,music}}]_{1,1} = \vartheta_{\text{ML}}/\vartheta_{\text{SCM}}$ and $r_2 \stackrel{\text{def}}{=} [\mathbf{R}_{\hat{\theta}}^{\text{ML,music}}]_{1,1}/[\mathbf{R}_{\hat{\theta}}^{\text{Ty,music}}]_{1,1} = \vartheta_{\text{ML}}/\vartheta_{\text{Ty}}$ depend only on m and ν, in contrast to $r_3 \stackrel{\text{def}}{=} [\mathbf{R}_{\hat{\theta}}^{\text{ML,music}}]_{1,1}/[\mathbf{R}_{\hat{\theta}}^{\text{SSCM,music}}]_{1,1}$ that also depend on $\Delta\theta$ and SNR. Figure 7.2a and b show the theoretical ratio r_1, r_2 and r_3 for $\nu = 4.1$ and two values of m, respectively, w.r.t. the DOA separation $\Delta\theta = |\theta_2 - \theta_1|$ where the array SNR (ASNR) (defined by $m\sigma_s^2/\sigma_n^2$ [53, Chap. 9]) is fixed at 10 dB and w.r.t. to the ASNR where $\Delta\theta = |\theta_2 - \theta_1| = 0.2$rd. We clearly see that the performance of the SCM-based estimate is very poor unlike that of Tyler's M-based estimate whose performance is very close to that of the ML-based estimate for heavy-tailed noise distributions. As for the SSCM-based estimate, its performance is degraded compared to that of Tyler's M-based estimate, and this degradation increases when the ASNR increases and $\Delta\theta$ decreases. This behavior is then explained by an increase in the differences between eigenvalues λ_1, λ_2 and σ_n^2 consistently with the comment following expression (7.48). But the SSCM-based estimate largely outperforms the SCM-based estimate.

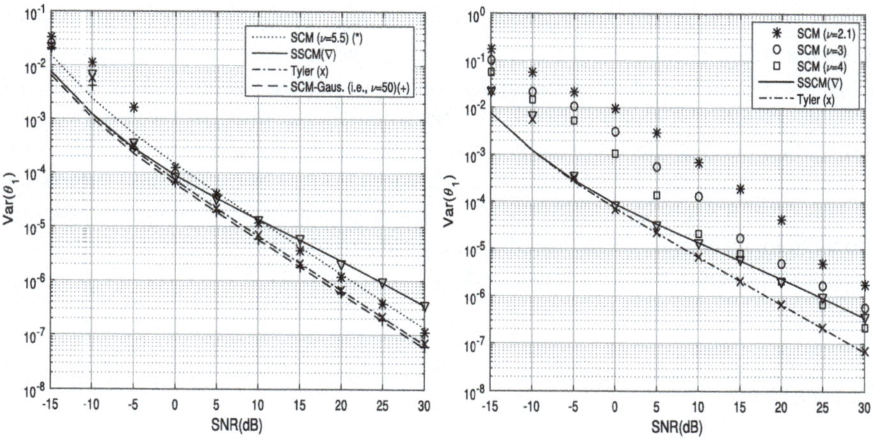

(a) case $\nu > 4$ (b) case $2 < \nu \leq 4$

Fig. 7.3 Theoretical asymptotic variances $\frac{1}{n}[\mathbf{R}_{\hat{\theta}}^{\text{SCM,music}}]_{1,1}$, $\frac{1}{n}[\mathbf{R}_{\hat{\theta}}^{\text{Ty,music}}]_{1,1}$, $\frac{1}{n}[\mathbf{R}_{\hat{\theta}}^{\text{SSCM,music}}]_{1,1}$ and MSEs (with 2000 Monte Carlo runs) of non-circular MUSIC algorithm, versus SNR where $T = 500$ and $\Delta\theta = 0.2$rd for either $\nu > 4$ or $2 < \nu \leq 4$ and $m = 6$

In the third setup, the measurement $\mathbf{x}_i$ are complex non-circular Student's t-distributed where $\mathbf{R}_{\tilde{x}}$ is given by (7.8) with $(\phi_1, \phi_2) = (\frac{\pi}{3}, \frac{2\pi}{3})$rd and $m = 6$. Figure 7.3a and b compare the theoretical asymptotic variance $\frac{1}{n}[\mathbf{R}_{\hat{\theta}}^{\text{music}}]_{1,1}$ and MSEs of non-circular MUSIC algorithms based on SCM, SSCM and Tyler's M estimate versus SNR. Note first that for $\nu \in [2, 4]$, we get $\vartheta_{\text{Ty}} = 7/6$ and $\vartheta_{\text{ML}} \in [8/7, 9/8]$. So the asymptotic variance of Tyler's M estimator and the CRB are too close to be distinguishable in Fig. 7.3. These figures also show that the theoretical asymptotic variances given by the non-circular MUSIC algorithms based on SSCM and Tyler's M estimates from (7.73), are very close to each other and to their MSE for a weak SNR and for $\nu > 4$. On the other hand, for $2 < \nu \leq 4$, for which the fourth-order moments of the data do not exist, and hence the asymptotic distribution of the non-circular MUSIC estimates based on the SCM is not available, the associated MSE increases strongly when ν approaches 2, for which the data are no longer of second-order.

7.7 Conclusion

The aim of this chapter is to unify the different performance analyses of subspace-based algorithms in C-CES and NC-CES data models within the same framework. In particular, general closed-form expressions are given for the covariances of the asymptotic distribution of different subspace projector estimates and the associated subspace-based parameter estimates have been given. This result allows us to prove various invariance properties and general inequalities between covariances of subspace-based parameter estimates, including the stochastic and semiparametric CRB. Finally, note that the presented asymptotic distributions have been obtained w.r.t. the number of measurements. But the main shortcoming of these asymptotic distributions is that they provide good approximations of the variances of the estimates only when the sample size n is sufficiently large w.r.t. the data dimension m. However, when n is comparable to m, it is necessary to use random matrix theory tools to develop new subspace-based algorithms and their associated asymptotic distributions within the CES framework.

References

1. Abed-Meraim, K., Hua, Y.: Blind identification of multi-input multi-output system using minimum noise subspace. IIEEE Trans. Signal Process. **45**(1), 254–258 (1997)
2. Abeida, H., Delmas, J.P.: MUSIC-like estimation of direction of arrival for noncircular sources. IEEE Trans. Signal Process. **54**(7), 2678–2690 (2006)
3. Abeida, H., Delmas, J.P.: Robustness of subspace-based algorithms with respect to the distribution of the noise: application to DOA estimations. Signal Process. **164**, 313–319 (2019)
4. Abeida, H., Delmas, J.P.: Slepian-Bangs formula and Cramér-Rao bound for circular and non-circular complex elliptical symmetric distributions. IEEE Signal Process. Lett. **26**, 1561–1565 (2019)
5. Abeida, H., Delmas, J.P.: Efficiency of subspace-based estimators for elliptical symmetric distributions. Signal Process. **174** (2020)
6. Abeida, H., Delmas, J.P.: Performance of subspace-based algorithms associated with the sample sign covariance matrix. Digital Signal Process. **131** (2022)
7. Abeida, H., Delmas, J.P.: Slepian-Bangs formulas for parameterized density generator of elliptically symmetric distributions. Signal Process. **205** (2023)
8. Bausson, S., Pascal, F., Forster, P., Ovarlez, J.P., Larzabal, P.: First- and second-order moments of the normalized sample covariance matrix of spherically invariant random vectors. IEEE Signal Process. Lett. **14**(6), 425–428 (2007)
9. Besson, O., Abramovich, Y.I.: On the Fisher information matrix for multivariate elliptically contoured distributions. IEEE Signal Process. Lett. **20**(11), 1130–1133 (2013)
10. Conte, E., Lops, M., Ricci, G.: Adaptive radar detection in compound-gaussian clutter. In: EUSIPCO, pp. 526–529. Edinburgh, Scotland (1994)
11. Delmas, J.P.: Asymptotic performance of second-order algorithms. IEEE Trans. Signal Process. **50**(1), 49–57 (2002)
12. Delmas, J.P.: Performance Bounds and Statistical Analysis of DOA Estimation. Academic Press Library in Signal Processing (2013)
13. Delmas, J.P., Comon, P., Meurisse, Y.: Performance limits of alphabet diversities for FIR SISO channel identification. IEEE Trans. Signal Process. **57**(1), 73–82 (2009)
14. Di Claudio, E.D., Parisi, R., Jacovitti, G.: Space time MUSIC: consistent signal subspace estimation for wideband sensor arrays. IEEE Trans. Signal Process. **66**(10), 2685–2699 (2018)
15. Draskovic, G., Breloy, A., Pascal, F.: On the asymptotics of Maronna's robust PCA. IEEE Trans. Signal Process. **67**(19), 4964–4975 (2019)
16. Durre, A., Tyler, D.E., Vogel, D.: On the eigenvalues of the spatial sign covariance matrix in more than two dimensions. Stat. Prob. Lett. **111**, 80–85 (2016)
17. Fortunati, S., Gini, F., Greco, M.S., Zoubir, A.M.: Semiparametric stochastic CRB for DOA estimation in elliptical data model. In: EUSIPCO. Coruña, Spain (2019)
18. Fortunati, S., Gini, F., Greco, M.S., Zoubir, A.M., Rangaswamy, M.: A fresh look at the semiparametric Cramér-Rao bound. In: EUSIPCO. Rome, Italy (2018)

19. Fortunati, S., Gini, F., Greco, M.S., Zoubir, A.M., Rangaswamy, M.: Semiparametric CRB and Slepian-Bangs formulas for complex elliptically symmetric distributions. IEEE Trans. Signal Process. **67**(20), 5352–5364 (2019)
20. Fortunati, S., Gini, F., Greco, M.S., Zoubir, A.M., Rangaswamy, M.: Semiparametric inference and lower bounds for real elliptically symmetric distributions. IEEE Trans. Signal Process. **67**(1), 164–177 (2019)
21. Fortunati, S., Renaux, A., Pascal, F.: Robust semiparametric efficient estimators in complex elliptically symmetric distributions. IEEE Trans. Signal Process. **68**, 5001–5015 (2020)
22. Gini, F., Greco, M.V.: Covariance matrix estimation for CFAR detection in correlated heavy tailed clutter. Signal Process. **82**, 1847–1859 (2002)
23. Greco, M., Gini, F.: Cramér-Rao lower bounds on covariance matrix estimation for complex elliptically symmetric distributions. IEEE Trans. Signal Process. **61**, 6401–6409 (2013)
24. Haardt, M., Pesavento, M., Roemer, F., El Korso, M.N.: Subspace Methods and Exploitation of Special Array Structures. pp. 651–717. Academic Press Library in Signal Processing, Elsevier (M. Viberg, ed.) (2014)
25. Huber, P.J.: The behavior of maximum likelihood estimates under nonstandard conditions. In: Proceedings of the Fifth Berkeley Symposium in Mathematical Statistics and Probability. University of California Press, Berkley (1967)
26. Kato, T.: Perturbation Theory for Linear Operators. Springer, Berlin (1995)
27. Kent, J.T.: Data analysis for shapes and images. J. Stat. Plan. Interference **57**(2), 181–197 (1997)
28. Krim, H., Forster, P., Proakis, G.: Operator approach to performance analysis of root-music and root-min-norm. EEE Trans. Signal Process. **40**(7), 1687–1696 (1992)
29. Krim, H., Viberg, M.: Two decades of array signal processing research: the parametric approach. IEEE Signal Process. Mag. **13**, 67–94 (1996)
30. Ladaycia, H., Abed-Meraim, K., Mokraoui, A., Belouchrani, A.: Efficient semi-blind subspace channel estimation for MIMO-OFDM system. In: EUSIPCO. Rome, Italy (2018)
31. Lehmann, E.L.: Elements of Large Sample Theory. Springer Texts in Statistics (2004)
32. Locantore, M., et al.: Robust principle component analysis for functional data. TEST **8**, 1–73 (1999)
33. Magyar, A.F.: The efficiencies of the spatial median and spatial sign covariance matrix for elliptically symmetric distributions. Ph.D. thesis. State University of New Jersey, New Brunswick (2012)
34. Magyar, A.F., Tyler, D.E.: The asymptotic inadmissibility of the spatial sign covariance matrix for the elliptically symmetric distributions. Biometrika **101**(3), 3673–688 (2014)
35. Mahot, M., Pascal, F., Forster, P., Ovarlez, J.P.: Asymptotic properties of robust complex covariance matrix estimates. IEEE Trans. Signal Process. **61**(13), 3348–3356 (2013)
36. Maronna, R.: Robust M-estimators of multivariate location and scatter. Ann. Stat. **4**(1), 51–67 (1976)
37. Mecklenbräuker, C.F., Gerstoft, P., Ollila, E.: DOA M-estimation using sparse Bayesian learning. In: ICASSP. Singapore (2022)
38. Moulines, E., Duhamel, P., Cardoso, J.F., Mayrargue, S.: Subspace methods for the blind identification FIR filters. IEEE Trans. Signal Process. **43**(2), 516–525 (1995)
39. Ollila, E., Koivunen, V.: Robust antenna array processing using M-estimators of pseudo covariance. In: 14th International Symposium on Personal Indoor and Mobile Radio Communication. Toronto, Canada (2003)
40. Ollila, E., Koivunen, V.: Influence function and asymptotic efficiency of scatter matrix based array processors: case MVDR beamformer. IEEE Trans. Signal Process. **57**(1), 247–259 (2009)
41. Ollila, E., Oja, H., Croux, C.: The affine equivariant sign covariance matrix: asymptotic behavior and efficiencies. J. Multivar. Anal. **87**, 328–355 (2003)
42. Ollila, E., Tyler, D.E., Koivunen, V., Poor, H.V.: Complex elliptically symmetric distributions: survey, new results and applications. IEEE Trans. Signal Process. **60**(11), 5597–5625 (2012)
43. Paindaveine, D.: Elliptical symmetry. In: Book Wiley Statistics Reference Online (2014)

44. Pascal, F., Forster, P., Ovarlez, J.P., Larzabal, P.: Performance analysis of covariance matrix estimates in impulsive noise. IEEE Trans. Signal Process. **56**(6), 2206–2216 (2008)
45. Porat, B., Friedlander, B.: Asymptotic accuracy of ARMA parameter estimation methods based on sample covariances. In: 7th IFAC/IFORS Symposium on Identification and System Parameter Estimation, York. York (1985)
46. Schmidt, R.O.: Multiple emitter location and signal parameter estimation. IEEE Trans. Antennas Propag. **34**, 276–280 (1986)
47. Serfling, R.J.: Approximation Theorems of Mathematical Statistics. Wiley (1980)
48. Stoica, P., Friedlander, B., Söderström, T.: An approximate maximum approach to ARMA spectral estimation. In: 24th IEEE Conference on Decision and Control. Fort Lauderdale (1985)
49. Stoica, P., Nehorai, A.: Performances study of conditional and unconditional direction of arrival estimation. IEEE Trans. Signal Process. **38**(10), 1783–1795 (1990)
50. Tyler, D.E.: Radial estimates and the test for sphericity. Biometrika **69**(2), 429 (1982)
51. Tyler, D.E.: A distribution-free M-estimator of multivariate scatter. Ann. Stat. **15**(1), 234–251 (1987)
52. Tyler, D.E.: Statistical analysis for the angular central gaussian distribution on the sphere. Biometrika **74**(3), 579–589 (1987)
53. Van Trees, H.L.: Optimum Array Processing Part IV of Detection, Estimation, and Modulation Theory. Wiley Interscience, Wiley, New York (2002)
54. Visuri, S., Koivunen, V., Oja, H.: Sign and rank covariance matrices. J. Stat. Plan. Inference **91**, 557–575 (2000)
55. Visuri, S., Oja, H., Koivunen, V.: Subspace-based direction of arrival estimation using nonparametric statistics. IEEE Trans. Signal Process. **49**(9), 2060–2073 (2001)
56. Zuo, W., Xin, J., Ohmori, H., Zheng, N., Sano, A.: Subspace-based algorithms for localization and tracking of multiple near-field sources. IEEE J. Select. Top. Signal Process. **13**(1) (2019)

Part III
Applications to Machine Learning

Chapter 8
Robust Bayesian Cluster Enumeration for RES Distributions

Freweyni K. Teklehaymanot, Christian Schroth, and Michael Muma

Abstract Cluster analysis loosely refers to a collection of unsupervised learning methods that group similar objects within unlabeled datasets into clusters. Clustering methods are highly useful in a variety of applications. For example, in the medical sciences, identifying clusters may allow for a comprehensive characterization of subgroups of individuals. However, in real-world data, the true cluster structure is often obscured by heavy-tailed noise, artifacts, and outliers. Discovering the true cluster structure becomes even more challenging when the number of clusters is unknown. This chapter addresses the above challenges and revisits some recent developments on robust statistical model-based cluster analysis. In particular, Bayesian robust cluster enumeration criteria based on real elliptically symmetric (RES) distributions are discussed. Such approaches formulate the problem of estimating the number of clusters as a maximization of the posterior probability of multivariate elliptically distributed candidate models. We show, starting from Bayes' theorem and using asymptotic approximations, how to come up with robust criteria that possess closed-form expressions. Then, robust clustering algorithms are discussed that implement the criteria based on arbitrary RES distributed mixture models, and even M-estimators. Real-data examples highlight the usefulness of the discussed methods. Links to open-source software packages are provided enabling the reader to explore and apply the methods.

F. K. Teklehaymanot
Former Member of Signal Processing Group, Technische Universität Darmstadt, Darmstadt, Germany

C. Schroth
Signal Processing Group, Technische Universität Darmstadt, Darmstadt, Germany
e-mail: schroth@spg.tu-darmstadt.de

M. Muma (✉)
Robust Data Science, Technische Universität Darmstadt, Darmstadt, Germany
e-mail: michael.muma@tu-darmstadt.de

J.-P. Delmas et al. (eds.), *Elliptically Symmetric Distributions in Signal Processing and Machine Learning*, https://doi.org/10.1007/978-3-031-52116-4_8

8.1 Introduction

Cluster enumeration is about determining the number of subgroups of similar points existing in a given dataset. Determining the number of clusters K, is essential in a wide range of applications, however, determining K from a given dataset correctly, is challenging. It depends on the definition of what constitutes a cluster and how to deal with outliers, noise, and other data uncertainties. Popular clustering algorithms [5, 6, 9, 19, 20, 40] rely, for example, on small distances within clusters compared to the between-cluster distances (or other measures of similarity), dense areas of the data space, or mixture models of particular statistical distributions. This chapter focuses on robust statistical model-based cluster analysis using mixtures of RES distributions. As discussed in other chapters of this book, this family of distributions is useful in statistically modeling the non-Gaussian behavior of noisy data in many practical applications [12–14, 26]. RES distributions include, for example, Gaussian, the generalized Gaussian [27], the t-distribution, the compound Gaussian [28], and Huber's distribution [30], as special cases. In contrast to purely data-driven unsupervised approaches, model-based methods allow for incorporating prior knowledge and assumptions. Statistically robust methods [18, 23, 42] are designed so as to provide reliable results, even if the assumptions are only approximately fulfilled, the cluster distribution is heavy-tailed, or if the dataset contains outliers, which may be considered as untypical data points that may not belong to any of the clusters. Figure 8.1 gives an example of an unlabeled dataset that contains three clusters and a large number of outlying data points. A real data example containing outliers and four clusters that represent persons in a radar dataset will be later discussed in Sect. 8.6. The cluster enumeration methods that are discussed in this chapter also contain finite sample versions that are applicable even for cases where the clusters may have a relatively small number of associated data samples.

A popular strategy in robust cluster enumeration is to use existing model selection criteria such as the Bayesian Information Criterion (BIC) derived by Schwarz [7, 32] in combination with robust clustering algorithms. Variations of this strategy include outlier detection and removal [16, 17, 25, 39], modeling noise or outliers using an

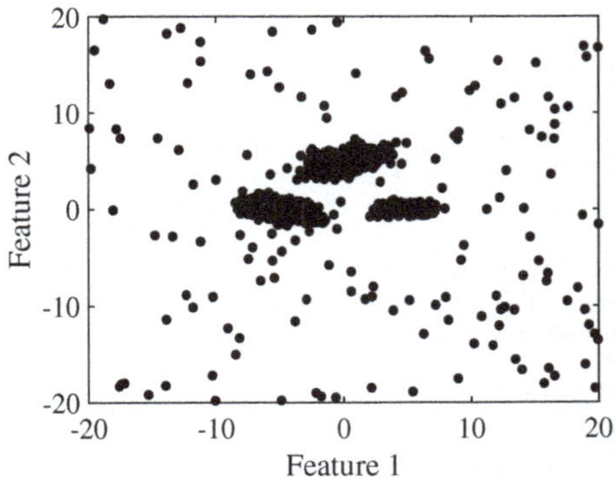

Fig. 8.1 How many clusters are there? In this example, the data has been created from three clusters and it contains many outliers that are not associated to any of the clusters

additional component in a mixture model [8, 15], or modeling the data as a mixture of heavy-tailed distributions [4, 24]. The above methods have in common that a "robustified likelihood" is complemented by a general penalty term to establish a trade-off between robust data-fit and model complexity. However, for all of these methods, the penalty term is left untouched and, in the case of Schwarz BIC, only depends on the number of model parameters and on the number of data points. Therefore, it penalizes two structurally different models the same way if they have the same number of unknown parameters [10, 34].

Recently, a BIC that is specific to cluster analysis has been derived by formulating the cluster enumeration problem as a maximization of the posterior probability of candidate models [35, 37]. By doing so, more information is incorporated into the penalty term, which depends on the number of model parameters, the assumed data distribution, the number of data points per cluster, and the estimated parameters. A first attempt at robust Bayesian cluster enumeration has been recently derived by formulating the cluster enumeration problem as a maximization of the posterior probability of multivariate t-distributed candidate models [38]. Although this heavy-tailed model provided a significant increase in robustness compared to using Gaussian candidate models, it still relied on a specific distributional model. In [30] robust Bayesian cluster enumeration was extended, so that it can be used with any arbitrary RES distributed mixture model, and even with M-estimators, which are popular in robust statistics [18, 23, 42]. They are a generalization of Maximum-Likelihood Estimators (MLE) where the negative log-likelihood function may be replaced by a robustness-inducing objective function.

Notation: Lower- and upper-case boldface letters stand for column vectors and matrices, respectively; Calligraphic letters denote sets with the exception of $\mathcal{L}$ which is used for the likelihood function; $\mathbb{R}$ represents the set of real numbers; $\mathbb{Z}^+$ denotes the set of positive integers; Probability density and mass functions are denoted by $f(\cdot)$ and $p(\cdot)$, respectively; $\boldsymbol{x} \sim \mathcal{N}(\boldsymbol{\mu}, \boldsymbol{\Sigma})$ represents a Gaussian distributed random variable $\boldsymbol{x}$ with mean $\boldsymbol{\mu}$ and covariance matrix $\boldsymbol{\Sigma}$; $\hat{\boldsymbol{\theta}}$ stands for the estimate of the parameter $\boldsymbol{\theta}$; $\ln(\cdot)$ denotes the natural logarithm; iid stands for independent and identically distributed; **(A.)** denotes an assumption, for example **(A.1)** stands for the first assumption; $\mathcal{O}(1)$ represents Landau's term which tends to a constant as the data size goes to infinity; $\boldsymbol{I}_r$ stands for an $r \times r$ identity matrix; $\boldsymbol{0}_{r \times r}$ denotes an $r \times r$ all zero matrix; $\#\mathcal{X}$ represents the cardinality of the set $\mathcal{X}$; $(\cdot)^\top$ stands for vector or matrix transpose; $|\boldsymbol{Y}|$ denotes the determinant of the matrix $\boldsymbol{Y}$; $\mathrm{Tr}(\cdot)$ represents the trace of a matrix; $\otimes$ denotes the Kronecker product; $\mathrm{vec}(\boldsymbol{Y})$ refers to the stacking of the columns of an matrix $\boldsymbol{Y}$ into one column vector; the $\mathrm{vech}(\boldsymbol{Y})$ (vector half) operator takes a symmetric $r \times r$ matrix and stacks the lower triangular half into a single vector of length $r(r+1)/2$.

8.2 Problem Formulation

Cluster analysis starts with a set of r-dimensional vectors $\mathcal{X} \triangleq \{\boldsymbol{x}_1, \ldots, \boldsymbol{x}_N\}$ which can be partitioned into K clusters $\{\mathcal{X}_1, \ldots, \mathcal{X}_K\}$, where $\mathcal{X}_k \subseteq \mathcal{X}$ for $k \in \mathcal{K} \triangleq \{1, \ldots, K\}$. The clusters $\mathcal{X}_k, k \in \mathcal{K}$, are assumed to be independent, mutually exclusive,[1] and is non-empty. Following a model-based clustering approach, let $\mathcal{M} \triangleq \{M_{L_{\min}}, \ldots, M_{L_{\max}}\}$ be a family of candidate models that represent a partitioning of $\mathcal{X}$ into $l = L_{\min}, \ldots, L_{\max}$ subsets, where $l \in \mathbb{Z}^+$, see Fig. 8.2 as an example. The parameters of each model $M_l \in \mathcal{M}$ lie in a parameter space $\Omega_l \subset \mathbb{R}^{q \times l}$ and are denoted by $\boldsymbol{\Theta}_l = [\boldsymbol{\theta}_1, \ldots, \boldsymbol{\theta}_l]$.

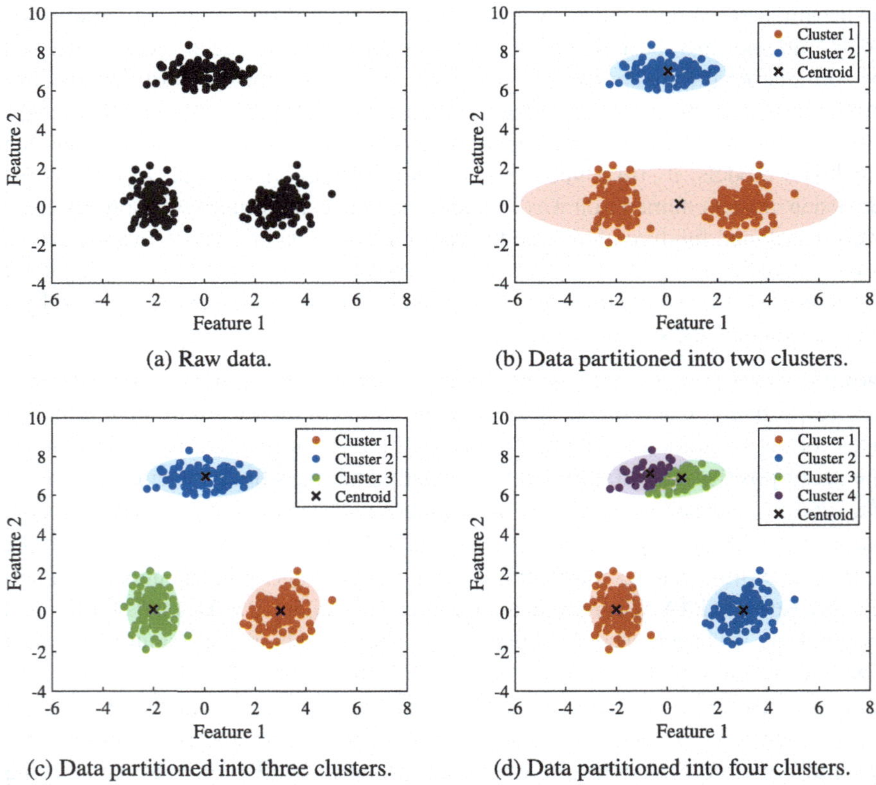

Fig. 8.2 Partitioning of a dataset based on the number of clusters specified by different candidate models. In this example, the true number of clusters is $K = 3$ and $\mathcal{X}$ is partitioned into $l = 1, \ldots, 4$ subsets

[1] Independence means that the joint probability distribution of $\{\mathcal{X}_1, \ldots, \mathcal{X}_K\}$ is a product of the marginal distributions. Two sets are mutually exclusive if their intersection is empty. This means that every data point belongs to one specific cluster.

According to Bayes' theorem, the joint posterior density of M_l and $\boldsymbol{\Theta}_l$ given the observed dataset $\mathcal{X}$ is given by

$$f(M_l, \boldsymbol{\Theta}_l|\mathcal{X}) = \frac{p(M_l) f(\boldsymbol{\Theta}_l|M_l) f(\mathcal{X}|M_l, \boldsymbol{\Theta}_l)}{f(\mathcal{X})}, \tag{8.1}$$

where $p(M_l)$ is the discrete prior of the model M_l over the set of candidate models $\mathcal{M}$, $f(\boldsymbol{\Theta}_l|M_l)$ denotes a prior on the parameter vectors in $\boldsymbol{\Theta}_l$ given $M_l \in \mathcal{M}$, $f(\mathcal{X})$ is the pdf of $\mathcal{X}$, and $f(\mathcal{X}|M_l, \boldsymbol{\Theta}_l)$ denotes the pdf of the observation set $\mathcal{X}$ given the candidate model M_l and its associated parameter matrix $\boldsymbol{\Theta}_l$. Our objective is to choose the candidate model $M_{\hat{K}} \in \mathcal{M}$, where $\hat{K} \in \{L_{\min}, \ldots, L_{\max}\}$, which is most probable a posteriori assuming that

(A.1) the true number of clusters (K) in the observed dataset $\mathcal{X}$ satisfies the constraint $L_{\min} \leq K \leq L_{\max}$.

8.3 Generic Bayesian Cluster Enumeration Criterion

In the considered Bayesian setting, estimating the number of clusters in a given dataset corresponds to choosing the model $M_{\hat{K}}$ which is a posteriori most probable [35]. Mathematically, this corresponds to solving

$$M_{\hat{K}} = \arg\max_{M_l \in \mathcal{M}} p(M_l|\mathcal{X}), \qquad l = L_{\min}, \ldots, L_{\max}, \tag{8.2}$$

where $p(M_l|\mathcal{X})$ is the posterior probability of M_l given the observations $\mathcal{X}$, i.e.,

$$\begin{aligned} p(M_l|\mathcal{X}) &= \int_{\Omega_l} f(M_l, \boldsymbol{\Theta}_l|\mathcal{X}) d\boldsymbol{\Theta}_l \\ &= f(\mathcal{X})^{-1} p(M_l) \int_{\Omega_l} f(\boldsymbol{\Theta}_l|M_l) \mathcal{L}(\boldsymbol{\Theta}_l|\mathcal{X}) d\boldsymbol{\Theta}_l. \end{aligned} \tag{8.3}$$

Here, $\mathcal{L}(\boldsymbol{\Theta}_l|\mathcal{X}) \triangleq f(\mathcal{X}|M_l, \boldsymbol{\Theta}_l)$ is the likelihood function. To simplify Eq. (8.3), $M_{\hat{K}}$ can also be determined via

$$\arg\max_{M_l \in \mathcal{M}} \ln p(M_l|\mathcal{X}), \qquad l = L_{\min}, \ldots, L_{\max} \tag{8.4}$$

since the natural logarithm is a monotonic function. This results in

$$\ln p(M_l|\mathcal{X}) = \ln p(M_l) + \ln \int_{\Omega_l} f(\boldsymbol{\Theta}_l|M_l) \mathcal{L}(\boldsymbol{\Theta}_l|\mathcal{X}) d\boldsymbol{\Theta}_l + \xi. \tag{8.5}$$

Since $-\ln f(\mathcal{X})$ is not a function of $M_l \in \mathcal{M}$ and, thus, has no effect on the maximization of $\ln p(M_l|\mathcal{X})$ over $\mathcal{M}$, it is replaced by ξ (a constant). Under the assumption that the clusters $\mathcal{X}_m \subseteq \mathcal{X}$, $m = 1, \ldots, l$, are independent, mutually exclusive, and non-empty, $f(\boldsymbol{\Theta}_l|M_l)$ and $\mathcal{L}(\boldsymbol{\Theta}_l|\mathcal{X})$ can be written as

$$f(\boldsymbol{\Theta}_l|M_l) = \prod_{m=1}^{l} f(\boldsymbol{\theta}_m|M_l), \tag{8.6}$$

$$\mathcal{L}(\boldsymbol{\Theta}_l|\mathcal{X}) = \prod_{m=1}^{l} \mathcal{L}(\boldsymbol{\theta}_m|\mathcal{X}_m). \tag{8.7}$$

Substituting Eqs. (8.6) and (8.7) into Eq. (8.5) results in

$$\ln p(M_l|\mathcal{X}) = \ln p(M_l) + \sum_{m=1}^{l} \ln \int_{\mathbb{R}^q} f(\boldsymbol{\theta}_m|M_l)\mathcal{L}(\boldsymbol{\theta}_m|\mathcal{X}_m)d\boldsymbol{\theta}_m + \xi. \tag{8.8}$$

Maximizing $\ln p(M_l|\mathcal{X})$ over all candidate models $M_l \in \mathcal{M}$ involves the computation of the logarithm of a multidimensional integral. For most practical cases the solution of the multidimensional integral does not possess a closed analytical form. This problem can be solved using either numerical integration or approximations that allow a closed-form solution. In the context of model selection, closed-form approximations are preferred over numerical integration since they are known to provide more insight into the problem [10]. Following this line of argument, in order to provide a closed-form analytic approximation to Eq. (8.8), we begin by approximating the multidimensional integral using Laplace's method of integration [3, 10, 34]. The following assumptions are made in Laplace's method of integration.

(A.2) $\ln \mathcal{L}(\boldsymbol{\theta}_m|\mathcal{X}_m)$ with $m = 1, \ldots, l$ has first- and second-order derivatives which are continuous over the parameter space Ω_l.

(A.3) $\ln \mathcal{L}(\boldsymbol{\theta}_m|\mathcal{X}_m)$ with $m = 1, \ldots, l$ has a global maximum at $\hat{\boldsymbol{\theta}}_m$, where $\hat{\boldsymbol{\theta}}_m$ is an interior point of Ω_l.

(A.4) $f(\boldsymbol{\theta}_m|M_l)$ with $m = 1, \ldots, l$ is continuously differentiable and its first-order derivatives are bounded on Ω_l.

(A.5) The negative of the Hessian matrix of $\frac{1}{N_m} \ln \mathcal{L}(\boldsymbol{\theta}_m|\mathcal{X}_m)$

$$\hat{\boldsymbol{H}}_m \triangleq -\frac{1}{N_m} \frac{d^2 \ln \mathcal{L}(\boldsymbol{\theta}_m|\mathcal{X}_m)}{d\boldsymbol{\theta}_m d\boldsymbol{\theta}_m^\top}\bigg|_{\boldsymbol{\theta}_m=\hat{\boldsymbol{\theta}}_m} \in \mathbb{R}^{q\times q} \tag{8.9}$$

is positive definite, where N_m is the cardinality of $\mathcal{X}_m$ ($N_m = \#\mathcal{X}_m$). That is, $\exists \epsilon > 0$, such that $\min_{s,m} \lambda_s \left(\hat{\boldsymbol{H}}_m\right) > \epsilon$ for $s = 1, \ldots, q$ and $m = 1, \ldots, l$, where $\lambda_s \left(\hat{\boldsymbol{H}}_m\right)$ is the sth eigenvalue of $\hat{\boldsymbol{H}}_m$.

The first step in Laplace's method of integration is to write the Taylor series expansion of $f(\boldsymbol{\theta}_m|M_l)$ and $\ln \mathcal{L}(\boldsymbol{\theta}_m|\mathcal{X}_m)$ around $\hat{\boldsymbol{\theta}}_m$, $m = 1, \ldots, l$. We begin by approximating $\ln \mathcal{L}(\boldsymbol{\theta}_m|\mathcal{X}_m)$ by its second-order Taylor series expansion around $\hat{\boldsymbol{\theta}}_m$ as follows:

$$\ln \mathcal{L}(\boldsymbol{\theta}_m|\mathcal{X}_m) \approx \ln \mathcal{L}(\hat{\boldsymbol{\theta}}_m|\mathcal{X}_m) + \tilde{\boldsymbol{\theta}}_m^\top \frac{d \ln \mathcal{L}(\boldsymbol{\theta}_m|\mathcal{X}_m)}{d\boldsymbol{\theta}_m}\bigg|_{\boldsymbol{\theta}_m=\hat{\boldsymbol{\theta}}_m}$$
$$+ \frac{1}{2}\tilde{\boldsymbol{\theta}}_m^\top \left[\frac{d^2 \ln \mathcal{L}(\boldsymbol{\theta}_m|\mathcal{X}_m)}{d\boldsymbol{\theta}_m d\boldsymbol{\theta}_m^\top}\bigg|_{\boldsymbol{\theta}_m=\hat{\boldsymbol{\theta}}_m} \right] \tilde{\boldsymbol{\theta}}_m$$
$$= \ln \mathcal{L}(\hat{\boldsymbol{\theta}}_m|\mathcal{X}_m) - \frac{N_m}{2}\tilde{\boldsymbol{\theta}}_m^\top \hat{\boldsymbol{H}}_m \tilde{\boldsymbol{\theta}}_m, \tag{8.10}$$

where $\tilde{\boldsymbol{\theta}}_m \triangleq \boldsymbol{\theta}_m - \hat{\boldsymbol{\theta}}_m$, $m = 1, \ldots, l$. The first derivative of $\ln \mathcal{L}(\boldsymbol{\theta}_m|\mathcal{X}_m)$ evaluated at $\hat{\boldsymbol{\theta}}_m$ vanishes because of assumption **(A.3)**. With

$$U \triangleq \int_{\mathbb{R}^q} f(\boldsymbol{\theta}_m|M_l) \exp\left(\ln \mathcal{L}(\boldsymbol{\theta}_m|\mathcal{X}_m)\right) d\boldsymbol{\theta}_m, \tag{8.11}$$

substituting Eq. (8.10) into Eq. (8.11), approximating $f(\boldsymbol{\theta}_m|M_l)$ by its Taylor series expansion, and ignoring the higher-order terms yields

$$U \approx \int_{\mathbb{R}^q} \left(\left[f(\hat{\boldsymbol{\theta}}_m|M_l) + \tilde{\boldsymbol{\theta}}_m^\top \frac{d f(\boldsymbol{\theta}_m|M_l)}{d\boldsymbol{\theta}_m}\bigg|_{\boldsymbol{\theta}_m=\hat{\boldsymbol{\theta}}_m} \right] \right.$$
$$\left. \times\, \mathcal{L}(\hat{\boldsymbol{\theta}}_m|\mathcal{X}_m) \exp\left(-\frac{N_m}{2}\tilde{\boldsymbol{\theta}}_m^\top \hat{\boldsymbol{H}}_m \tilde{\boldsymbol{\theta}}_m \right) d\boldsymbol{\theta}_m \right)$$
$$\approx f(\hat{\boldsymbol{\theta}}_m|M_l)\mathcal{L}(\hat{\boldsymbol{\theta}}_m|\mathcal{X}_m) \int_{\mathbb{R}^q} \exp\left(-\frac{N_m}{2}\tilde{\boldsymbol{\theta}}_m^\top \hat{\boldsymbol{H}}_m \tilde{\boldsymbol{\theta}}_m \right) d\boldsymbol{\theta}_m$$
$$+ \mathcal{L}(\hat{\boldsymbol{\theta}}_m|\mathcal{X}_m) \int_{\mathbb{R}^q} \left(\tilde{\boldsymbol{\theta}}_m^\top \frac{d f(\boldsymbol{\theta}_m|M_l)}{d\boldsymbol{\theta}_m}\bigg|_{\boldsymbol{\theta}_m=\hat{\boldsymbol{\theta}}_m} \right.$$
$$\left. \cdot \exp\left(-\frac{N_m}{2}\tilde{\boldsymbol{\theta}}_m^\top \hat{\boldsymbol{H}}_m \tilde{\boldsymbol{\theta}}_m \right) d\boldsymbol{\theta}_m \right), \tag{8.12}$$

where $\exp\left(-\frac{N_m}{2}\tilde{\boldsymbol{\theta}}_m^\top \hat{\boldsymbol{H}}_m \tilde{\boldsymbol{\theta}}_m\right)$ is a Gaussian kernel with mean $\hat{\boldsymbol{\theta}}_m$ and covariance matrix $\left(N_m \hat{\boldsymbol{H}}_m\right)^{-1}$. Since

$$\mathcal{L}(\hat{\boldsymbol{\theta}}_m|\mathcal{X}_m) \int_{\mathbb{R}^q} \left(\tilde{\boldsymbol{\theta}}_m^\top \frac{d f(\boldsymbol{\theta}_m|M_l)}{d\boldsymbol{\theta}_m}\bigg|_{\boldsymbol{\theta}_m=\hat{\boldsymbol{\theta}}_m} \exp\left(-\frac{N_m}{2}\tilde{\boldsymbol{\theta}}_m^\top \hat{\boldsymbol{H}}_m \tilde{\boldsymbol{\theta}}_m \right) d\boldsymbol{\theta}_m \right)$$

simplifies to $\kappa \mathbb{E}\left[\boldsymbol{\theta}_m - \hat{\boldsymbol{\theta}}_m\right] = 0$, where κ is a constant (see [3, p. 53] for more details), Eq. (8.12) reduces to

$$
\begin{aligned}
U &\approx f(\hat{\boldsymbol{\theta}}_m|M_l)\mathcal{L}(\hat{\boldsymbol{\theta}}_m|\mathcal{X}_m)\int_{\mathbb{R}^q}\exp\left(-\frac{N_m}{2}\tilde{\boldsymbol{\theta}}_m^\top\hat{\boldsymbol{H}}_m\tilde{\boldsymbol{\theta}}_m\right)d\boldsymbol{\theta}_m\\
&= f(\hat{\boldsymbol{\theta}}_m|M_l)\mathcal{L}(\hat{\boldsymbol{\theta}}_m|\mathcal{X}_m)\int_{\mathbb{R}^q}\Bigg((2\pi)^{q/2}\left|N_m^{-1}\hat{\boldsymbol{H}}_m^{-1}\right|^{1/2}\\
&\quad \frac{1}{(2\pi)^{q/2}\left|N_m^{-1}\hat{\boldsymbol{H}}_m^{-1}\right|^{1/2}}\exp\left(-\frac{N_m}{2}\tilde{\boldsymbol{\theta}}_m^\top\hat{\boldsymbol{H}}_m\tilde{\boldsymbol{\theta}}_m\right)d\boldsymbol{\theta}_m\Bigg)\\
&= f(\hat{\boldsymbol{\theta}}_m|M_l)\mathcal{L}(\hat{\boldsymbol{\theta}}_m|\mathcal{X}_m)(2\pi)^{q/2}\left|N_m^{-1}\hat{\boldsymbol{H}}_m^{-1}\right|^{1/2}
\end{aligned}
\tag{8.13}
$$

given that $N_m \to \infty$. Using Eq. (8.13), we can provide an asymptotic approximation to the multidimensional integral in Eq. (8.8).

Now, substituting Eq. (8.13) into Eq. (8.8), we arrive at

$$
\begin{aligned}
\ln p(M_l|\mathcal{X}) &\approx \ln p(M_l) + \sum_{m=1}^{l}\ln\Big(f(\hat{\boldsymbol{\theta}}_m|M_l)\mathcal{L}(\hat{\boldsymbol{\theta}}_m|\mathcal{X}_m)\Big)\\
&\quad + \frac{lq}{2}\ln 2\pi - \frac{1}{2}\sum_{m=1}^{l}\ln\left|\hat{\boldsymbol{J}}_m\right| + \xi,
\end{aligned}
\tag{8.14}
$$

where

$$
\hat{\boldsymbol{J}}_m \triangleq N_m\hat{\boldsymbol{H}}_m = -\frac{d^2\ln\mathcal{L}(\boldsymbol{\theta}_m|\mathcal{X}_m)}{d\boldsymbol{\theta}_m d\boldsymbol{\theta}_m^\top}\bigg|_{\boldsymbol{\theta}_m=\hat{\boldsymbol{\theta}}_m} \in \mathbb{R}^{q\times q}
\tag{8.15}
$$

is the Fisher Information Matrix (FIM) of data from the mth partition.

In the derivation of $\ln p(M_l|\mathcal{X})$, so far, we have made no distributional assumption on the dataset $\mathcal{X}$, except that the log-likelihood function $\ln\mathcal{L}(\boldsymbol{\theta}_m|\mathcal{X}_m)$ and the prior on the parameter vectors $f(\boldsymbol{\theta}_m|M_l)$, for $m = 1, \ldots, l$, should satisfy some mild conditions under each model $M_l \in \mathcal{M}$. Hence, Eq. (8.14) is a general expression of the posterior probability of the model M_l given $\mathcal{X}$ for a general class of data distributions that satisfy assumptions **(A.2)**–**(A.5)**. The BIC is concerned with the computation of the posterior probability of candidate models and, thus, Eq. (8.14) can also be written as

$$
\begin{aligned}
\mathrm{BIC}_{\mathrm{G}}(M_l) &\triangleq \ln p(M_l|\mathcal{X})\\
&\approx \ln p(M_l) + \ln f(\hat{\boldsymbol{\Theta}}_l|M_l) + \ln\mathcal{L}(\hat{\boldsymbol{\Theta}}_l|\mathcal{X})\\
&\quad + \frac{lq}{2}\ln 2\pi - \frac{1}{2}\sum_{m=1}^{l}\ln\left|\hat{\boldsymbol{J}}_m\right| + \xi,
\end{aligned}
\tag{8.16}
$$

where the subscript "G" stands for general because this expression is not derived for a specific data distribution. After calculating $\mathrm{BIC}_{\mathrm{G}}(M_l)$ for each candidate model $M_l \in \mathcal{M}$, the number of clusters in $\mathcal{X}$ is estimated as

$$\hat{K}_{\mathrm{BIC_G}} = \underset{l=L_{\min},\ldots,L_{\max}}{\arg\max} \ \mathrm{BIC_G}(M_l). \tag{8.17}$$

However, calculating $\mathrm{BIC_G}(M_l)$ using Eq. (8.16) is a computationally expensive task as it requires the estimation of the FIM, $\hat{\boldsymbol{J}}_m$, for each cluster $m = 1, \ldots, l$ in the candidate model $M_l \in \mathcal{M}$. In the upcoming subsections, we provide asymptotic and finite sample versions of the above BIC by assuming that $\mathcal{X}_m$ contains iid multivariate Gaussian data points with parameter $\boldsymbol{\theta}$.

8.3.1 Asymptotic Bayesian Information Criterion for Multivariate Gaussian Data

Let $\mathcal{X}$ denote the observed dataset which can be partitioned into K clusters $\{\mathcal{X}_1, \ldots, \mathcal{X}_K\}$, where each cluster $\mathcal{X}_k, k \in \mathcal{K}$, contains N_k data vectors that are realizations of iid Gaussian random variables $\boldsymbol{x}_k \sim \mathcal{N}(\boldsymbol{\mu}_k, \boldsymbol{\Sigma}_k)$. Here, $\boldsymbol{\mu}_k \in \mathbb{R}^{r\times 1}$ and $\boldsymbol{\Sigma}_k \in \mathbb{R}^{r\times r}$ represent the centroid and the covariance matrix of the kth cluster, respectively. Further, let $\mathcal{M} \triangleq \{M_{L_{\min}}, \ldots, M_{L_{\max}}\}$ denote a set of Gaussian candidate models and let there be a clustering algorithm that partitions $\mathcal{X}$ into l independent, mutually exclusive, and non-empty subsets (clusters) $\mathcal{X}_m, m = 1, \ldots, l$, by providing parameter estimates $\hat{\boldsymbol{\theta}}_m = \left[\hat{\boldsymbol{\mu}}_m^\top, \hat{\boldsymbol{u}}_m^\top\right]^\top$ with $\hat{\boldsymbol{u}}_m = \mathrm{vech}(\hat{\boldsymbol{\Sigma}}_m)$ for each candidate model $M_l \in \mathcal{M}$, where $l = L_{\min}, \ldots, L_{\max}$ and $l \in \mathbb{Z}^+$. Here, the symmetry of the covariance matrix $\boldsymbol{\Sigma}_m$ is exploited in order to collect its unique elements into a vector $\boldsymbol{u}_m \in \mathbb{R}^{\frac{1}{2}r(r+1)\times 1}$ by using the vech operator as defined in [22, pp. 56–57]. The duplication matrix $\boldsymbol{D}_r \in \mathbb{R}^{r^2\times\frac{1}{2}r(r+1)}$ can then be used to transform $\boldsymbol{u}_m$ into $\mathrm{vec}(\boldsymbol{\Sigma}_m)$ using the relation $\mathrm{vec}(\boldsymbol{\Sigma}_m) = \boldsymbol{D}_r\boldsymbol{u}_m$ [21].

Theorem 8.1 *The posterior probability of $M_l \in \mathcal{M}$ given $\mathcal{X}$ can be asymptotically approximated as*

$$\boxed{\begin{aligned}\mathrm{BIC_N}(M_l) &\triangleq \ln p(M_l|\mathcal{X}) \\ &\approx \sum_{m=1}^{l} N_m \ln(N_m) - \frac{1}{2}\sum_{m=1}^{l} N_m \ln|\hat{\boldsymbol{\Sigma}}_m| - \frac{q}{2}\sum_{m=1}^{l}\ln(N_m)\end{aligned}} \tag{8.18}$$

assuming that **(A.1)-(A.7)** *are satisfied. N_m is the cardinality of the subset $\mathcal{X}_m$ and it satisfies $N = \sum_{m=1}^{l} N_m$.*

Proving Theorem 8.1 requires finding an asymptotic approximation to $\ln\left|\hat{\boldsymbol{J}}_m\right|$ in Eq. (8.16) and, based on this approximation, deriving an expression for $\mathrm{BIC_N}(M_l)$. A detailed proof is given in [35]. In the sequel, we give a proof sketch, where we begin by expressing the log-likelihood of the data points that belong to the mth cluster as follows:

$$
\begin{aligned}
\ln \mathcal{L}(\boldsymbol{\theta}_m|\mathcal{X}_m) &= \ln \prod_{\boldsymbol{x}_n \in \mathcal{X}_m} p(\boldsymbol{x}_n \in \mathcal{X}_m) f(\boldsymbol{x}_n|\boldsymbol{\theta}_m) \\
&= N_m \ln\left(\frac{N_m}{N}\right) - \frac{rN_m}{2}\ln(2\pi) - \frac{N_m}{2}\ln|\boldsymbol{\Sigma}_m| - \frac{1}{2}\operatorname{Tr}\left(\boldsymbol{\Sigma}_m^{-1}\boldsymbol{\Delta}_m\right),
\end{aligned}
\tag{8.19}
$$

where $\tilde{\boldsymbol{x}}_n \triangleq \boldsymbol{x}_n - \boldsymbol{\mu}_m$ and $\boldsymbol{\Delta}_m \triangleq \sum_{\boldsymbol{x}_n \in \mathcal{X}_m} \tilde{\boldsymbol{x}}_n \tilde{\boldsymbol{x}}_n^\top$. Here, we have assumed that

(A.6) the covariance matrix $\boldsymbol{\Sigma}_m$, $m = 1, \ldots, l$, is positive definite.

Next, we take the first- and second-order derivatives of $\ln \mathcal{L}(\boldsymbol{\theta}_m|\mathcal{X}_m)$ with respect to the elements of $\boldsymbol{\theta}_m$ and simplify the resulting equations to come up with

$$
\begin{aligned}
\frac{d^2 \ln \mathcal{L}(\boldsymbol{\theta}_m|\mathcal{X}_m)}{d\boldsymbol{\theta}_m d\boldsymbol{\theta}_m^\top} &= \frac{N_m}{2}\left(\frac{d\boldsymbol{u}_m}{d\boldsymbol{u}_m}\right)^\top \boldsymbol{D}_r^\top \boldsymbol{V}_m \boldsymbol{D}_r \frac{d\boldsymbol{u}_m}{d\boldsymbol{u}_m^\top} - \left(\frac{d\boldsymbol{u}_m}{d\boldsymbol{u}_m^\top}\right)^\top \boldsymbol{D}_r^\top \boldsymbol{W}_m \boldsymbol{D}_r \frac{d\boldsymbol{u}_m}{d\boldsymbol{u}_m} \\
&\quad - N_m \left(\frac{d\boldsymbol{u}_m}{d\boldsymbol{u}_m}\right)^\top \boldsymbol{D}_r^\top \boldsymbol{Z}_m \operatorname{vec}\left(\frac{d\boldsymbol{\mu}_m^\top}{d\boldsymbol{\mu}_m^\top}\right) - N_m \frac{d\boldsymbol{\mu}_m^\top}{d\boldsymbol{\mu}_m} \boldsymbol{\Sigma}_m^{-1} \frac{d\boldsymbol{\mu}_m}{d\boldsymbol{\mu}_m^\top},
\end{aligned}
\tag{8.20}
$$

with

$$
\boldsymbol{V}_m \triangleq \boldsymbol{\Sigma}_m^{-1} \otimes \boldsymbol{\Sigma}_m^{-1} \in \mathbb{R}^{r^2 \times r^2} \tag{8.21}
$$

$$
\boldsymbol{W}_m \triangleq \boldsymbol{\Sigma}_m^{-1} \otimes \boldsymbol{\Sigma}_m^{-1} \boldsymbol{\Delta}_m \boldsymbol{\Sigma}_m^{-1} \in \mathbb{R}^{r^2 \times r^2} \tag{8.22}
$$

$$
\boldsymbol{Z}_m \triangleq \boldsymbol{\Sigma}_m^{-1} \left(\bar{\boldsymbol{x}}_m - \boldsymbol{\mu}_m\right) \otimes \boldsymbol{\Sigma}_m^{-1} \in \mathbb{R}^{r^2 \times r}. \tag{8.23}
$$

Consistent with Eq. (8.15), the Fisher information matrix is written as

$$
\begin{aligned}
\hat{\boldsymbol{J}}_m &= \begin{bmatrix} -\frac{\partial^2 \ln \mathcal{L}(\hat{\boldsymbol{\theta}}_m|\mathcal{X}_m)}{\partial \boldsymbol{\mu}_m \partial \boldsymbol{\mu}_m^\top} & -\frac{\partial^2 \ln \mathcal{L}(\hat{\boldsymbol{\theta}}_m|\mathcal{X}_m)}{\partial \boldsymbol{\mu}_m \partial \boldsymbol{u}_m^\top} \\ -\frac{\partial^2 \ln \mathcal{L}(\hat{\boldsymbol{\theta}}_m|\mathcal{X}_m)}{\partial \boldsymbol{u}_m \partial \boldsymbol{\mu}_m^\top} & -\frac{\partial^2 \ln \mathcal{L}(\hat{\boldsymbol{\theta}}_m|\mathcal{X}_m)}{\partial \boldsymbol{u}_m \partial \boldsymbol{u}_m^\top} \end{bmatrix} \\
&= \begin{bmatrix} N_m \hat{\boldsymbol{\Sigma}}_m^{-1} & \boldsymbol{0}_{r \times \frac{1}{2}r(r+1)} \\ \boldsymbol{0}_{\frac{1}{2}r(r+1) \times r} & -\frac{N_m}{2} \boldsymbol{D}_r^\top \hat{\boldsymbol{F}}_m \boldsymbol{D}_r \end{bmatrix},
\end{aligned}
\tag{8.24}
$$

where $\boldsymbol{F}_m \triangleq \boldsymbol{\Sigma}_m^{-1} \otimes \left(\boldsymbol{\Sigma}_m^{-1} - \frac{2}{N_m}\boldsymbol{\Sigma}_m^{-1}\boldsymbol{\Delta}_m\boldsymbol{\Sigma}_m^{-1}\right) \in \mathbb{R}^{r^2 \times r^2}$. The determinant of the FIM, $\hat{\boldsymbol{J}}_m$, is computed by

$$
\left|\hat{\boldsymbol{J}}_m\right| = \left|N_m \hat{\boldsymbol{\Sigma}}_m^{-1}\right| \times \left|-\frac{N_m}{2}\boldsymbol{D}_r^\top \hat{\boldsymbol{J}}_m \boldsymbol{D}_r\right|. \tag{8.25}
$$

As $N \to \infty$, $N_m \to \infty$ given that $l << N$, it follows that

$$\left| \frac{1}{N_m} \hat{\boldsymbol{J}}_m \right| \approx \mathcal{O}(1), \tag{8.26}$$

where $\mathcal{O}(1)$ denotes Landau's term, which tends to a constant as $N \to \infty$. Using this result, we provide an asymptotic approximation to Eq. (8.16), for the case in which $\mathcal{X}$ is composed of Gaussian distributed data vectors, as follows:

$$\ln p(M_l|\mathcal{X}) = \ln p(M_l) + \ln f(\hat{\boldsymbol{\Theta}}_l|M_l) + \ln \mathcal{L}(\hat{\boldsymbol{\Theta}}_l|\mathcal{X}) + \frac{lq}{2}\ln(2\pi) - \frac{q}{2}\sum_{m=1}^{l}\ln(N_m) - \frac{1}{2}\sum_{m=1}^{l}\ln\left|\frac{1}{N_m}\hat{\boldsymbol{J}}_m\right| + \xi, \tag{8.27}$$

where $q = \frac{1}{2}r(r+3)$. Assume that

(A.7) $p(M_l)$ and $f(\hat{\boldsymbol{\Theta}}_l|M_l)$ are independent of the data length N.

Finally, ignoring the terms in Eq. (8.27) that do not grow as $N \to \infty$ and ignoring model independent terms results in Eq. (8.18), which concludes the proof.

8.3.2 Bayesian Cluster Enumeration Criterion with Finite Sample Penalty Term

$\mathrm{BIC_N}$, like many model selection criterion in the literature, is derived under asymptotic assumptions on the size of the observed data. However, in most real-world applications the amount of data is rather limited and we operate in the finite sample regime. In the finite sample regime, the asymptotic assumptions made by $\mathrm{BIC_N}$ are violated, which results in a weak penalty term. In consequence, having a weak penalty term makes the criterion prone to overestimation. To alleviate this problem, in this section, we extend the derivation of $\mathrm{BIC_N}$ by providing an exact expression for its penalty term [37].

$$\mathrm{BIC}_{NF}(M_l) = \mathrm{BIC}_N(M_l) + \frac{1}{4}r(r+1)l\ln(2) + \frac{1}{2}\sum_{m=1}^{l}\ln|\hat{\boldsymbol{\Sigma}}_m| - \frac{1}{2}\sum_{m=1}^{l}\ln\left|\boldsymbol{D}_r^\top\left(\hat{\boldsymbol{\Sigma}}_m^{-1}\otimes\hat{\boldsymbol{\Sigma}}_m^{-1}\right)\boldsymbol{D}_r\right|. \tag{8.28}$$

Unlike BIC_N, the penalty term of $\mathrm{BIC_{NF}}$ depends on the covariance matrix of the individual clusters in $M_l \in \mathcal{M}$. This allows $\mathrm{BIC_{NF}}$ to lower the penalty term when the determinant of the covariance matrices are high and penalize more severely when they are low. However, the covariance matrices of individual clusters can be large and their inverses close to zero if the observations span a large range of values. As a result, the penalty term of $\mathrm{BIC_{NF}}$ might go to infinity. In such cases, we recommend normalizing the data prior to the estimation of cluster parameters. In terms of computational complexity, $\mathrm{BIC_{NF}}$ has an excess computational cost of $\mathcal{O}(lr^6)$ compared to $\mathrm{BIC_N}$. It is therefore only recommended for small datasets.

8.4 Bayesian Cluster Enumeration for RES Distributions and M-Estimation

8.4.1 Finite Sample Criterion for RES Distributions and M-Estimation

While $\mathrm{BIC_{NF}}$ that has been defined in Eq. (8.28) assumes that the data may be represented by a multivariate Gaussian distribution, from Eq. (8.16), we can derive a BIC that can be used for any RES distribution and even for non-ML loss functions such as Tukey's M-estimator. M-estimators generalize the concept of ML-estimators to more general loss functions that trade a small loss in optimality at a nominal distribution against near-optimal performance for a broad range of distributions. The generalization to RES distributions and robust M-estimation is important, since a small number of outliers is sufficient to negatively impact $\mathrm{BIC_{NF}}$ [30, 38].

In the following, we will describe how heavy-tailed mixture models from the RES family of distributions can be used to derive robust cluster enumeration criteria. We will show that the derivation leads to weights that depend on the robustly estimated squared Mahalanobis distance from the cluster centroids. In this way, points that can be assigned with more confidence to a cluster are given more weight in the decision compared to far away points that may be outliers. M-estimators build upon this idea and allow for decoupling the weight function from a specific mixture model. Instead, the weight function is designed so that the estimate of the number of clusters is not heavily influenced by a (small) fraction of deviating data points. The robustness can be quantified, for example, by appropriate measures [30, 42], such as the influence function and the breakdown point. Loosely speaking, in the context of cluster enumeration, the influence function shows the bias of the probability of determining the correct number of clusters that is induced by a tiny fraction of outliers as a function of the outlier position in feature space. By contrast, the breakdown point is the fraction of outliers that can be handled by the method before the method completely breaks down, i.e., is dominated by the outliers and no longer provides any information about the correct number of clusters.

The question is, how to arrive at a robust Bayesian cluster enumeration criterion? To answer this question, firstly, let the estimate of the parameter vector be given by

$$\hat{\boldsymbol{\theta}}_m = \left[\hat{\boldsymbol{\mu}}_m^\top, \mathrm{vech}(\hat{\boldsymbol{S}}_m)^\top\right]^\top \in \mathbb{R}^{q\times 1},$$

where $\hat{\boldsymbol{S}}_m$ is the estimate of the scatter matrix (see Chap. 1). Similarly to the Gaussian case, cf. Eq. (8.20), we exploit the fact that $\hat{\boldsymbol{S}}_m$ is symmetric, i.e., it has only $\frac{r}{2}(r+1)$ unique elements. Thus, we apply the vech (vector half) operator that takes a symmetric $r \times r$ matrix and stacks the lower triangular half into a single vector of length $\frac{r}{2}(r+1)$. We also remember that the ML loss function for an RES distribution is denoted by

$$\rho(t) = -\ln\left(g(t)\right), \tag{8.29}$$

where $t = (\boldsymbol{x} - \boldsymbol{\mu})^\top \boldsymbol{S}^{-1} (\boldsymbol{x} - \boldsymbol{\mu})$ denotes the squared Mahalanobis distance and $g(t)$ is the density generator. With these definitions in place, we are ready to state the following theorem.

Theorem 8.2 *The posterior probability of M_l given $\mathcal{X}$, based on any ML or Non-ML loss function $\rho(t)$, can be calculated by*

$$\begin{aligned} BIC_{RobF}(M_l) \approx &- \sum_{m=1}^{l} \left(\sum_{\boldsymbol{x}_n \in \mathcal{X}_m} \rho(\hat{t}_{nm}) \right) + \sum_{m=1}^{l} N_m \ln\left(N_m\right) - l \ln\left(l\right) \\ &- \sum_{m=1}^{l} \frac{N_m}{2} \ln\left(\left|\hat{\boldsymbol{S}}_m\right|\right) + \frac{ql}{2} \ln(2\pi) - \frac{1}{2} \sum_{m=1}^{l} \ln\left(\left|\hat{\boldsymbol{J}}_m\right|\right), \end{aligned} \tag{8.30}$$

where, $\hat{t}_{nm}$ denotes the estimate of the squared Mahalanobis distance for the nth sample and mth cluster, and $\left|\hat{\boldsymbol{J}}_m\right|$ is given in Eq. (8.36), *using Eqs.* (8.32)–(8.35).

Theorem 8.2 is derived from Eq. (8.16) by ignoring model independent terms in the log-likelihood function for an arbitrary RES distribution with associated density generator $g(t)$

$$\begin{aligned} \ln(\mathcal{L}(\hat{\boldsymbol{\theta}}_m|\mathcal{X}_m)) &= \ln\left(\prod_{\boldsymbol{x}_n \in \mathcal{X}_m} p(\boldsymbol{x}_n \in \mathcal{X}_m) f(\boldsymbol{x}_n|\hat{\boldsymbol{\theta}}_m) \right) \\ &= \sum_{\boldsymbol{x}_n \in \mathcal{X}_m} \ln\left(\frac{N_m}{N} \left|\hat{\boldsymbol{S}}_m^{-1}\right|^{\frac{1}{2}} g\left(\hat{t}_{nm}\right) \right) \\ &= - \sum_{\boldsymbol{x}_n \in \mathcal{X}_m} \rho(\hat{t}_{nm}) + N_m \ln\left(N_m\right) - N_m \ln\left(N\right) \\ &\quad - \frac{N_m}{2} \ln\left(\left|\hat{\boldsymbol{S}}_m\right|\right), \end{aligned} \tag{8.31}$$

and computing the FIM

$$\hat{\boldsymbol{J}}_m = \begin{bmatrix} -\hat{\boldsymbol{J}}_{\mu\mu} & -\hat{\boldsymbol{J}}_{\mu S} \\ -\hat{\boldsymbol{J}}_{S\mu} & -\hat{\boldsymbol{J}}_{SS} \end{bmatrix} \in \mathbb{R}^{q \times q}. \tag{8.32}$$

All derivatives are evaluated with the ML estimates of $\boldsymbol{S}_m$ and $\boldsymbol{\mu}_m$, respectively, $\hat{\boldsymbol{S}}_m$ and $\hat{\boldsymbol{\mu}}_m$. The proof of Theorem 8.2 is provided in the appendix of [30]. The associated online supplementary material contains an additional comprehensive step-by-step derivation for all elements of the FIM in Eq. (8.32). The final expressions of the matrices in Eq. (8.32) are

$$\begin{aligned}\hat{\boldsymbol{J}}_{\mu\mu} = -4\hat{\boldsymbol{S}}_m^{-1}\left(\sum_{\boldsymbol{x}_n\in\mathcal{X}_m}\eta(\hat{t}_{nm})\tilde{\boldsymbol{x}}_n\tilde{\boldsymbol{x}}_n^\top\right)\\ \hat{\boldsymbol{S}}_m^{-1} - 2\hat{\boldsymbol{S}}_m^{-1}\sum_{\boldsymbol{x}_n\in\mathcal{X}_m}\psi(\hat{t}_{nm}) \in \mathbb{R}^{r\times r},\end{aligned} \tag{8.33}$$

$$\begin{aligned}\hat{\boldsymbol{J}}_{\mu S} = \hat{\boldsymbol{J}}_{S\mu}^\top = -2\sum_{\boldsymbol{x}_n\in\mathcal{X}_m}\eta(\hat{t}_{nm})\\ \left(\hat{\boldsymbol{S}}_m^{-1}\tilde{\boldsymbol{x}}_n\tilde{\boldsymbol{x}}_n^\top\hat{\boldsymbol{S}}_m^{-1}\otimes\tilde{\boldsymbol{x}}_n^\top\hat{\boldsymbol{S}}_m^{-1}\right)\boldsymbol{D}_r \in \mathbb{R}^{r\times\frac{r}{2}(r+1)},\end{aligned} \tag{8.34}$$

$$\begin{aligned}\hat{\boldsymbol{J}}_{SS} = -\boldsymbol{D}_r^\top\left(\hat{\boldsymbol{S}}_m^{-1}\otimes\hat{\boldsymbol{S}}_m^{-1}\right)\\ \left(\sum_{\boldsymbol{x}_n\in\mathcal{X}_m}\eta(\hat{t}_{nm})\left(\tilde{\boldsymbol{x}}_n\tilde{\boldsymbol{x}}_n^\top\otimes\tilde{\boldsymbol{x}}_n\tilde{\boldsymbol{x}}_n^\top\right)\right)\\ \left(\hat{\boldsymbol{S}}_m^{-1}\otimes\hat{\boldsymbol{S}}_m^{-1}\right)\boldsymbol{D}_r\\ -\frac{N_m}{2}\boldsymbol{D}_r^\top\left(\hat{\boldsymbol{S}}_m^{-1}\otimes\hat{\boldsymbol{S}}_m^{-1}\right)\boldsymbol{D}_r \in \mathbb{R}^{\frac{r}{2}(r+1)\times\frac{r}{2}(r+1)},\end{aligned} \tag{8.35}$$

where $\tilde{\boldsymbol{x}}_n \triangleq \boldsymbol{x}_n - \hat{\boldsymbol{\mu}}_m$. Because the FIM is a partitioned matrix [1, p. 114], the determinant can be easily computed as

$$\left|\hat{\boldsymbol{J}}_m\right| = \left|-\hat{\boldsymbol{J}}_{\mu\mu}\right| \cdot \left| -\hat{\boldsymbol{J}}_{SS} + \hat{\boldsymbol{J}}_{S\mu}\hat{\boldsymbol{J}}_{\mu\mu}^{-1}\hat{\boldsymbol{J}}_{\mu S}\right|. \tag{8.36}$$

Based on (8.30), the number of clusters can be estimated by evaluating

$$\hat{K} = \underset{l=L_{\min},\ldots,L_{\max}}{\arg\max}\ \mathrm{BIC}_{\mathrm{RobF}}(M_l). \tag{8.37}$$

8.4.2 Asymptotic Sample Penalty Term for RES Distributions and M-Estimation

Because the calculation of the FIM that is required to evaluate $\mathrm{BIC}_{\mathrm{RobF}}$ quickly becomes challenging, it may be advantageous to asymptotically approximate the FIM, especially in settings where a large number of samples is available. This approximation is provided in the following theorem.

Theorem 8.3 *Ignoring terms in Eq.* (8.36) *that do not grow as* $N \to \infty$, *the posterior probability of* M_l *given* $\mathcal{X}$ *becomes*

$$\boxed{\begin{aligned} BIC_{RobA}(M_l) \approx &-\sum_{m=1}^{l}\left(\sum_{x_n \in \mathcal{X}_m} \rho(\hat{t}_{nm})\right) + \sum_{m=1}^{l} N_m \ln(N_m) \\ &-\sum_{m=1}^{l} \frac{N_m}{2} \ln\left(\left|\hat{\boldsymbol{S}}_m\right|\right) - \frac{q}{2}\sum_{m=1}^{l} \ln(\varepsilon_m) \end{aligned}} \tag{8.38}$$

with ε_m *given in Eq.* (8.40).

Herein, the scalar variable ε_m must be computed, such that

$$\left|\frac{1}{\varepsilon_m}\hat{\boldsymbol{J}}_m\right| = \text{const}, \tag{8.39}$$

leads to a term that does not grow as the sample size $N \to \infty$. From Eqs. (8.33), (8.34), and (8.35), we can extract three normalization factors to fulfill Eq. (8.39). The maximum of these must be taken, which yields

$$\varepsilon_m = \max\left(\left|\sum_{x_n \in \mathcal{X}_m} \psi(\hat{t}_{nm})\right|, \left|\sum_{x_n \in \mathcal{X}_m} \eta(\hat{t}_{nm})\right|, N_m\right), \tag{8.40}$$

where $\psi(t)$ and $\eta(t)$, respectively, are the first- and second-order derivatives of $\rho(t)$. Based on (8.38), the number of clusters is estimated via

$$\hat{K} = \underset{l=L_{\min},\ldots,L_{\max}}{\arg\max}\ \mathrm{BIC}_{\mathrm{RobA}}(M_l). \tag{8.41}$$

8.4.3 Examples of Robust Bayesian Cluster Enumeration Criteria

8.4.3.1 t-Distribution

We now specify the results of the robust cluster enumeration criteria that have been derived in Eqs. (8.30) and (8.38) for an important member of the RES family, namely the Student's t-distribution. The t-distribution is defined by its density generator

$$g(t) = \frac{\Gamma\left((\nu+r)/2\right)}{\Gamma\left(\nu/2\right)(\pi\nu)^{r/2}}\left(1+\frac{t}{\nu}\right)^{-(\nu+r)/2}, \tag{8.42}$$

where the degree-of-freedom parameter ν controls how heavy-tailed the distribution is.[2] From Eqs. (8.42) and (8.29), the ρ-function, in the case of the t-distribution, as follows:

$$\rho(t) = -\ln\left(\frac{\Gamma\left((\nu+r)/2\right)}{\Gamma\left(\nu/2\right)(\pi\nu)^{r/2}}\right) + \frac{\nu+r}{2}\ln\left(1+\frac{t}{\nu}\right) \tag{8.43}$$

and, subsequently, the associated first- and second-order derivatives are specified as

$$\psi(t) = \frac{1}{2}\cdot\frac{\nu+r}{\nu+t} \tag{8.44}$$

and

$$\eta(t) = -\frac{1}{2}\cdot\frac{\nu+r}{(\nu+t)^2}. \tag{8.45}$$

With these specifications in place, to obtain the versions of BIC_{RobF} and BIC_{RobA} for Student's t-distribution, one simply uses (8.43), (8.44), and (8.45) in (8.30) and (8.38), respectively. A detailed discussion on the derivation of the BIC for the t-distribution can be found in [38].

8.4.3.2 Huber's M-Estimator

As noted in [42, p. 115], Huber's M-estimator is also the ML estimator for a RES distribution, which we will call Huber distribution. In the sequel, using the results from [30], we will show this by deriving the density generator for the Huber distribution. To begin, let us define the following function:

[2] In the context of M-estimation, the aim is usually not to estimate ν. It is simply interpreted as a tuning parameter in the robust loss function that should be chosen small enough that the effect of outliers is suppressed without losing too much efficiency, e.g., at a nominal Gaussian distribution. In practice, the parameter is often chosen as $\nu = 3$.

$$h(t; r, c) = \exp\left(-\frac{1}{2}\rho_{\text{H}}(t; c)\right) \tag{8.46}$$

with

$$\rho_{\text{H}}(t; c) = \begin{cases} \frac{t}{b} & , t \leq c^2 \\ \frac{c^2}{b}\left(\ln\left(\frac{t}{c^2}\right) + 1\right) & , t > c^2. \end{cases} \tag{8.47}$$

To obtain Fisher consistency, we choose

$$b = F_{\chi^2_{r+2}}\left(c^2\right) + \frac{c^2}{r}\left(1 - F_{\chi^2_r}\left(c^2\right)\right), \tag{8.48}$$

where $F_{\chi^2_r}(\cdot)$ is the chi-square cumulative distribution function with degree of freedom r. Finally, to obtain a valid pdf, we must compute a normalization factor, which is given by

$$\begin{aligned} \int_0^\infty u^{r/2-1} h(u; r, c) du &= \int_0^{c^2} u^{r/2-1} e^{-\frac{u}{2b}} du + \int_{c^2}^\infty u^{r/2-1} \left(\frac{eu}{c^2}\right)^{-\frac{c^2}{2b}} du \\ &= (2b)^{r/2}\left(\Gamma\left(\frac{r}{2}\right) - \Gamma\left(\frac{r}{2}, \frac{c^2}{2b}\right)\right) + \frac{2bc^r \exp\left(-\frac{c^2}{2b}\right)}{c^2 - br}, \end{aligned} \tag{8.49}$$

with the gamma function $\Gamma(\cdot)$ and the upper incomplete gamma function $\Gamma(\cdot, \cdot)$. This allows us to write the density generator of a Huber distribution as

$$g(t) = \begin{cases} A_{\text{H}} \exp\left(-\frac{t}{2b}\right) & , t \leq c^2 \\ A_{\text{H}} \left(\frac{t}{c^2}\right)^{-\frac{c^2}{2b}} \exp\left(-\frac{c^2}{2b}\right) & , t > c^2 \end{cases} \tag{8.50}$$

with

$$A_{\text{H}} = \frac{\Gamma\left(\frac{r}{2}\right)}{\pi^{r/2}}\left((2b)^{\frac{r}{2}}\left(\Gamma\left(\frac{r}{2}\right) - \Gamma\left(\frac{r}{2}, \frac{c^2}{2b}\right)\right) + \frac{2bc^r e^{-\frac{c^2}{2b}}}{c^2 - br}\right)^{-1}, \tag{8.51}$$

which shows that Huber's distribution is indeed an RES distribution. Now, the robust cluster enumeration criteria are specified for Huber's distribution by using Huber's $\rho(t)$, and the associated first- and second-order derivatives

$$\psi(t) = \begin{cases} \frac{1}{2b} & , t \leq c^2 \\ \frac{c^2}{2bt} & , t > c^2 \end{cases} \tag{8.52}$$

and

$$\eta(t) = \begin{cases} 0 & , t \leq c^2 \\ -\frac{c^2}{2bt^2} & , t > c^2. \end{cases} \tag{8.53}$$

Specifically, for Huber's versions of $\mathrm{BIC}_{\mathrm{RobF}}$ and $\mathrm{BIC}_{\mathrm{RobA}}$, one simply uses (8.47), (8.52), and (8.53) in (8.30) and (8.38), respectively.

8.4.3.3 Tukey's M-Estimator

While, in many cases, it is sufficient to use a heavy-tailed ML estimator of an RES distribution to obtain robustness against outliers, especially for extreme outliers, it may be advantageous to design a robust M-estimator that is not an ML estimator. This is possible because the key idea of M-estimators is, to decouple the objective function from a specific distribution. For M-estimators, the function $\rho(t)$ can be designed by the user, as long as some general assumptions are fulfilled [42]. One of the most commonly used non-ML loss functions is Tukey's. For the univariate case [10, 42], it is given by

$$\rho(x) = \begin{cases} \frac{x^6}{6c^4} - \frac{x^4}{2c^2} + \frac{x^2}{2} & , |x| \leq c \\ \frac{c^2}{6} & , |x| > c, \end{cases} \tag{8.54}$$

which means that it gives zero weight to values larger than c. Tukey's loss function can be generalized to the multivariate case by using $x^2 = t$ and $|x| = \sqrt{t}$. By adding the constant $\frac{r}{2}\ln(2\pi)$, it can be ensured that for $c \to \infty$, Tukey's loss function is equal to the Gaussian loss function. The resulting expression for Tukey's $\rho(t)$ is given by

$$\rho(t) = \begin{cases} \frac{t^3}{6c^4} - \frac{t^2}{2c^2} + \frac{t}{2} + \frac{r}{2}\ln(2\pi) & , t \leq c^2 \\ \frac{c^2}{6} + \frac{r}{2}\ln(2\pi) & , t > c^2. \end{cases} \tag{8.55}$$

The associated first- and second-order derivatives are given, respectively, by

$$\psi(t) = \begin{cases} \frac{t^2}{2c^4} - \frac{t}{c^2} + \frac{1}{2} & , t \leq c^2 \\ 0 & , t > c^2 \end{cases} \tag{8.56}$$

and

$$\eta(t) = \begin{cases} \frac{t}{c^4} - \frac{1}{c^2} & , t \leq c^2 \\ 0 & , t > c^2. \end{cases} \tag{8.57}$$

To specify Tukey's versions of $\mathrm{BIC}_{\mathrm{RobF}}$ and $\mathrm{BIC}_{\mathrm{RobA}}$, it is sufficient to insert (8.55), (8.56), and (8.57) in (8.30) and (8.38).

8.5 Algorithms

The parameters for the cluster enumeration criteria can be computed using an expectation maximization (EM) algorithm for RES distributions (see Algorithm 8.2). The resulting two-step approach is summarized in Algorithm 8.1, where we provide a unified framework for the robust estimation of the number of clusters and cluster memberships that can be used for all RES distributions. Note that, even for cases when the number of clusters is known, Algorithm 8.2 can be used as a stand-alone algorithm for robust cluster analysis based on mixtures of arbitrary RES distributions.

Algorithm 8.1 Bayesian cluster enumeration algorithm

Input: $\mathcal{X}, L_{\min}, L_{\max}$
Output: $\hat{K}$
for $l = L_{\min}, \ldots, L_{\max}$ **do**
  Compute parameter estimates using Algorithm 2.;
  Hard clustering:;
  for $m = 1, \ldots, l$ **do**
    for $n = 1, \ldots, N$ **do**
      $$\gamma_{nm} = \begin{cases} 1 & , m = \arg\max_{j=1,\ldots,l} \hat{\upsilon}_{nj}^{(i)} \\ 0 & , \text{else} \end{cases}$$
  for $m = 1, \ldots, l$ **do**
    $N_m = \sum_{n=1}^{N} \gamma_{nm}$
  Calculate BIC(M_l) according to (8.30) or (8.38).;
Estimate $\hat{K}$ with Eq. (8.37) or (8.41). ;

Implementations of the algorithms that are presented in this book chapter are available in Python, R, and MATLAB at https://robustdatascience.org/software/.

8.6 Real-Data Experiments and Numerical Robustness Analysis

8.6.1 Example: Gait Data

Using a 24 GHz radar system [2], the experimental data for this example, has been collected in an office environment at Technische Universität Darmstadt [33]. The antenna feed point was positioned at approximately 1.15 m above the floor. Five test subjects were asked to walk toward the radar system starting at approximately 4.5 m in front of the radar, where only one person was present in front of the radar at a time. Data were collected at a 0° angle relative to the radar line-of-sight and with a non-oblique view on the targets. The volunteers were asked to walk slowly and

Algorithm 8.2 EM algorithm for RES distributions

Input: $\mathcal{X}$, $g(t)$, $\psi(t)$

Output: $\hat{\boldsymbol{\mu}}_m$, $\hat{\mathbf{S}}_m$, $\hat{\gamma}_m$

for $m = 1, \ldots, l$ **do**

Initialize $\hat{\boldsymbol{\mu}}_m^{(0)}$ with K-medoids;

$$\hat{\mathbf{S}}_m^{(0)} = \frac{1}{N_m} \sum_{\boldsymbol{x}_n \in \mathcal{X}_m} \left(\boldsymbol{x}_n - \hat{\boldsymbol{\mu}}_m^{(0)}\right)\left(\boldsymbol{x}_n - \hat{\boldsymbol{\mu}}_m^{(0)}\right)^\top$$

$$\hat{\gamma}_m^{(0)} = N_m / N$$

for $i = 1, \ldots, i_{\max}$ **do**

E-step:;

for $m = 1, \ldots, l$ **do**

for $n = 1, \ldots, N$ **do**

$$\hat{v}_{nm}^{(i)} = \frac{\hat{\gamma}_m^{(i-1)} \left|\hat{\mathbf{S}}_m^{(i-1)}\right|^{-\frac{1}{2}} g\left(\hat{t}_{nm}^{(i-1)}\right)}{\sum_{j=1}^{l} \hat{\gamma}_j^{(i-1)} \left|\hat{\mathbf{S}}_j^{(i-1)}\right|^{-\frac{1}{2}} g\left(\hat{t}_{nj}^{(i-1)}\right)}$$

$$\hat{v}_{nm}^{\prime(i)} = \hat{v}_{nm}^{(i)} \psi\left(\hat{t}_{nm}^{(i-1)}\right)$$

M-Step:;

for $m = 1, \ldots, l$ **do**

$$\hat{\boldsymbol{\mu}}_m^{(i)} = \sum_{n=1}^{N} \hat{v}_{nm}^{\prime(i)} \boldsymbol{x}_n \Big/ \sum_{n=1}^{N} \hat{v}_{nm}^{\prime(i)}$$

$$\hat{\mathbf{S}}_m^{(i)} = \frac{2 \sum_{n=1}^{N} \hat{v}_{nm}^{\prime(i)} \left(\boldsymbol{x}_n - \hat{\boldsymbol{\mu}}_m^{(i)}\right)\left(\boldsymbol{x}_n - \hat{\boldsymbol{\mu}}_m^{(i)}\right)^\top}{\sum_{n=1}^{N} \hat{v}_{nm}^{(i)}}$$

$$\hat{\gamma}_m^{(i)} = \frac{1}{N} \sum_{n=1}^{N} \hat{v}_{nm}^{(i)}$$

Calculate log-likelihood:

$$\ln\left(\mathcal{L}(\hat{\boldsymbol{\Phi}}_l^{(i)}|\mathcal{X})\right) = \sum_{n=1}^{N} \ln\left(\sum_{m=1}^{l} \hat{\gamma}_m^{(i)} \left|\hat{\mathbf{S}}_m^{(i)}\right|^{-\frac{1}{2}} g\left(\hat{t}_{nm}^{(i)}\right)\right)$$

if $\left|\ln\left(\mathcal{L}\left(\hat{\boldsymbol{\Phi}}_l^{(i)}|\mathcal{X}\right)\right) - \ln\left(\mathcal{L}\left(\hat{\boldsymbol{\Phi}}_l^{(i-1)}|\mathcal{X}\right)\right)\right| < \delta$ **then**

break loop;

without swinging their arms. In total, 65 radar measurements of 6 seconds duration are considered. The number of measurements is equal among the test subjects, i.e., the dataset contains 13 gait samples per person.

8.6.1.1 Feature Extraction

The recorded radar return signals are processed to obtain the spectrogram, see [33] for more details. In order to detect single strides, the maxima of the envelope signal of the micro-Doppler signatures were utilized. The part of the spectrogram that shows a pair of strides, i.e., a full gait cycle, was extracted and converted to a gray scale image. Examples of extracted stride pairs for two individuals are given in Fig. 8.3. All images were resized to have the same dimension of 100×128. The dimension of the spectrogram images was then reduced using PCA. Thus, a small number of descriptive features is automatically extracted for each radar signature. As an example, Fig. 8.4 shows a scatter plot of principal component scores using three principal components.

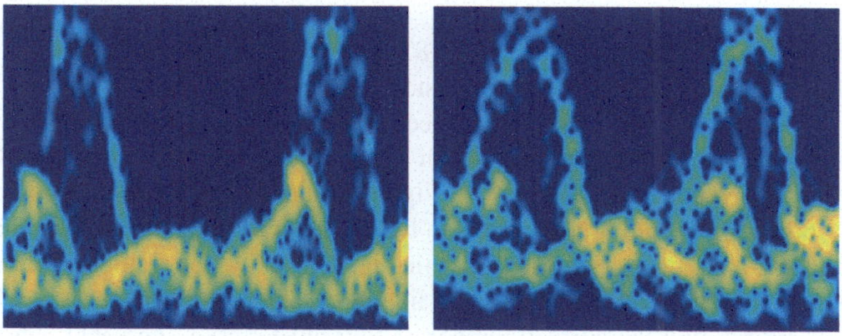

Fig. 8.3 Examples of micro-Doppler stride signatures of two individuals

Fig. 8.4 Scatter plot of principal component scores using three principal components of the radar dataset

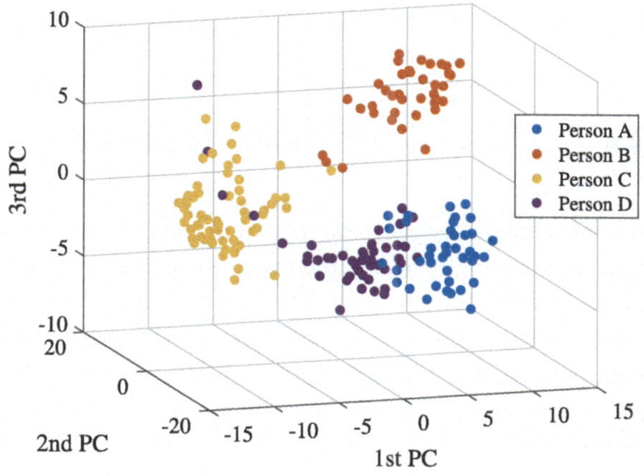

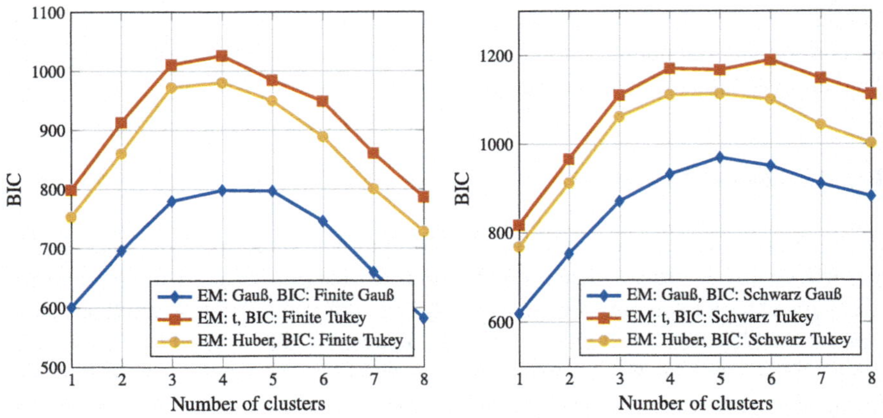

(a) Proposed criteria (b) Schwarz criterion

Fig. 8.5 Cluster enumeration results of the radar-based human gait data

8.6.1.2 Person Enumeration and Labeling

Considering the first four persons, $N = 187$ stride pairs are obtained from 52 radar measurements, where person A, B, C, and D are represented by $N_1 = 40$, $N_2 = 38$, $N_3 = 62$, and $N_4 = 47$ samples, respectively. The first 5 principal components are selected, such that the original set of vectorized spectrogram images of size 12800×187 is reduced to $\mathcal{X} \in \mathbb{R}^{5\times 187}$. Figure 8.5a shows the cluster enumeration results of the radar-based human gait data for the finite sample BIC criteria that are presented in this chapter ("proposed"), while Fig. 8.5b shows the results for Schwarz' BIC. Note that both criteria use the same likelihood function estimates. Clearly, all proposed criteria find the correct number of four clusters (i.e., persons), while the Schwarz criteria overestimate the number of clusters. When comparing the three different versions of the proposed criteria, it becomes clear that the peak in the BIC score is more pronounced for the two robust criteria.

8.6.2 Example: Robustness Measures

Robustness of cluster enumeration criteria can be measured by the empirical influence function, which measures the probability of detecting the correct number of clusters as a function of a single replacement outlier in the data [30]. For a robust criterion, a similar performance should be achieved irrespective of the position of the additional data point that is replaced by an outlier. For non-robust criteria, it is expected that the performance will only be maintained if the replacement outlier is close to one of the cluster centroids.

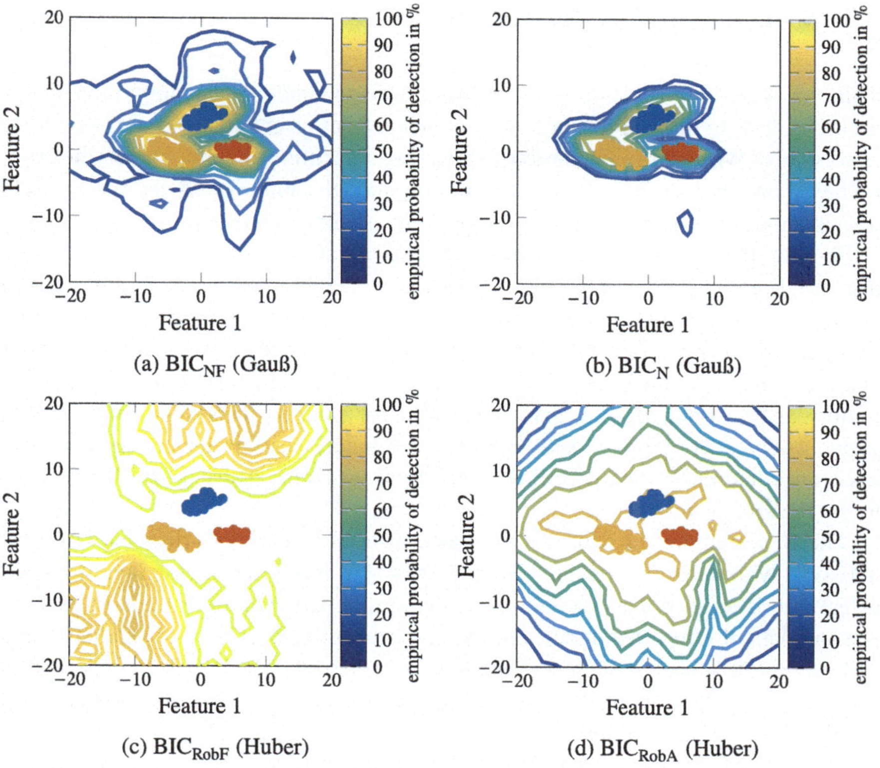

Fig. 8.6 Sensitivity curves for $N_k = 50$ that show six exemplary results for the empirical probability of correctly deciding for $K = 3$ clusters as a function of the single replacement outlier position

Figure 8.6 shows an example with $K = 3$ clusters in a two-dimensional feature space. The $N_k = 50$ feature vectors are colored according to their cluster memberships, i.e., in yellow, blue, and red colors. The empirical influence function is computed by averaging the performance over 500 Monte Carlo trials. In each Monte Carlo experiment, a randomly selected data point is replaced by an outlier that takes values over the range $[-20; 20]$ in the feature space. The first row shows the Gaussian BIC, while the second row displays the robust BIC using Huber's function. The left column shows the finite sample versions, while the right column shows the asymptotic versions. As expected, the Gaussian loss function is not at all robust against outliers, a high probability of detection is only achieved if the data points are replaced by a point that lies within one of the clusters. By contrast, Huber's BIC is not sensitive to a single outlier and shows a high probability of detecting the right number of clusters for all outlier values. Further, overall the finite sample criterion shows better performance. Further examples and an analysis of the breakdown point of robust cluster enumeration criteria can be found in [30].

8.7 Conclusion and Related Work

This chapter discussed, starting from Bayes' theorem and using asymptotic approximations, how to derive robust cluster enumeration criteria that possess closed-form expressions. We also provided robust clustering algorithms that implement the criteria based on arbitrary RES distributed mixture models, and even M-estimators. Real data examples and numerical robustness analysis were also presented.

To date, there exist multiple extensions and variations of the methods that were discussed in this chapter. One such extension is related to data that is both skewed and heavy-tailed. Real Elliptically Skewed (RESK) distributions and associated clustering algorithms that integrate robustness and skewness into a single unified cluster analysis framework have been developed in [29]. The parameters of the RESK distribution can be estimated using an EM algorithm for arbitrary RESK distributions. The paper also presents a robust skew-Huber M-estimator, which is also the approximate maximum-likelihood estimator (MLE) for the skew-Huber distribution, that belongs to the RESK class.

The Bayesian cluster enumeration has also been extended to distributed sensor networks with streaming data in [36, 41]. In [41], robust criteria based on Students' t-distribution have shown to be beneficial in the application of multi-temporal remote sensing data of a forested area in Japan. More recently, in [31], a robust distributed EM algorithm based on RES distributions is proposed, which is combined with a robust data aggregation step to provide robustness against malicious nodes in addition to robustness against outliers in the data.

Finally, in [11], robust detection has been adapted for cluster enumeration to address the challenging cases of severe cluster overlap and cluster imbalance. To take falsely labeled data points into account, in this work, the clusters are modeled as being contaminated with outliers. This leads to an uncertainty model for the cluster densities of the ε-contamination type, whose corresponding minimax optimal robust detector is implemented using the least favorable densities.

References

1. Abadir, K.M., Magnus, J.R.: Matrix Algebra. Econometric Exercises, vol. 1. Cambridge University Press (2005)
2. Ancortek Inc. SDR-KIT 2400AD, http://ancortek.com/sdr-kit-2400ad,retrieved:02/20/2018
3. Ando, T.: Bayesian Model Selection and Statistical Modeling. Chapman & Hall/CRC (2010)
4. Andrews, J.L., McNicholas, P.D.: Model-based clustering, classification, and discriminant analysis via mixtures of multivariate t-distributions. Stat. Comput. **22**(5), 1021–1029 (2012)
5. Arthur, D., Vassilvitskii, S.: K-Means++: the advantages of careful seeding. In: Gabow, H. (ed.) Proceedings of the Eighteenth Annual ACM-SIAM Symposium on Discrete Algorithms. Society for Industrial and Applied Mathematics (2007)
6. Batool, F., Hennig, C.: Clustering by optimizing the average silhouette width
7. Cavanaugh, J., Neath, A.: Generalizing the derivation of the schwarz information criterion. Commun. Stat. - Theory Methods **28**(1), 49–66 (1999)

8. Dasgupta, A., Raftery, A.E.: Detecting features in spatial point processes with clutter via model-based clustering. J. Am. Stat. Assoc. **93**(441), 294 (1998)
9. Dempster, A.P., Laird, N.M., Rubin, D.B.: Maximum likelihood from incomplete data via the EM algorithm. J. Roy. Stat. Soc.: Ser. B (Methodol.) **39**(1), 1–38 (1977)
10. Djurić, P.M.: Asymptotic MAP criteria for model selection. IEEE Trans. Signal Process. **46**(10), 2726–2735 (1998)
11. Fauß, M., Muma, M., Teklehaymanot, F.K., Zoubir, A.M.: Robust detection for cluster analysis. In: ICASSP 2019-2019 IEEE International Conference on Acoustics, Speech and Signal Processing (ICASSP), pp. 5331–5335. IEEE (2019)
12. Fortunati, S., Gini, F., Greco, M.S.: The misspecified Cramér-Rao bound and its application to scatter matrix estimation in complex elliptically symmetric distributions. IEEE Trans. Signal Process. **64**(9), 2387–2399 (2016)
13. Fortunati, S., Gini, F., Greco, M.S., Zoubir, A.M., Rangaswamy, M.: Semiparametric CRB and Slepian-Bangs formulas for complex elliptically symmetric distributions. IEEE Trans. Signal Process. **67**(20), 5352–5364 (2019)
14. Fortunati, S., Renaux, A., Pascal, F.: Robust semiparametric efficient estimators in elliptical distributions
15. Fraley, C.: How many clusters? Which clustering method? Answers via model-based cluster analysis. Comput. J. **41**(8), 578–588 (1998)
16. Gallegos, M.T., Ritter, G.: Trimming algorithms for clustering contaminated grouped data and their robustness. Adv. Data Anal. Classif. **3**(2), 135–167 (2009)
17. Gallegos, M.T., Ritter, G.: Using combinatorial optimization in model-based trimmed clustering with cardinality constraints. Comput. Stat. Data Anal. **54**(3), 637–654 (2010)
18. Huber, P.J., Ronchetti, E.M.: Robust Statistics. Wiley Series in Probability and Statistics. Wiley (2011)
19. Jain, A.K., Dubes, R.C.: Algorithms for Clustering Data. Prentice-Hall, Prentice Hall Advanced Reference Series (1988)
20. Lloyd, S.: Least squares quantization in PCM. IEEE Trans. Inf. Theory **28**(2), 129–137 (1982)
21. Magnus, J.R., Neudecker, H.: The elimination matrix: some lemmas and applications. SIAM J. Algebraic Discret. Methods **1**(4), 422–449 (1980)
22. Magnus, J.R., Neudecker, H.: Matrix Differential Calculus with Applications in Statistics and Econometrics. Wiley Ltd (2007)
23. Maronna, R.A., Martin, D.R., Yohai, V.J., Salibián-Barrera, M.: Robust Statistics: Theory and Methods (with R). Wiley Series in Probability and Statistics. Wiley (2018)
24. McNicholas, P.D., Subedi, S.: Clustering gene expression time course data using mixtures of multivariate t-distributions. J. Stat. Plan. Inference **142**(5), 1114–1127 (2012)
25. Neykov, N., Filzmoser, P., Dimova, R., Neytchev, P.: Robust fitting of mixtures using the trimmed likelihood estimator. Comput. Stat. Data Anal. **52**(1), 299–308 (2007)
26. Ollila, E., Tyler, D.E., Koivunen, V., Poor, H.V.: Complex elliptically symmetric distributions: survey, new results and applications. IEEE Trans. Signal Process. **60**(11), 5597–5625 (2012)
27. Pascal, F., Bombrun, L., Tourneret, J.-Y., Berthoumieu, Y.: Parameter estimation for multivariate generalized Gaussian distributions. IEEE Trans. Signal Process. **61**(23), 5960–5971 (2013)
28. Pascal, F., Chitour, Y., Ovarlez, J.-P., Forster, P., Larzabal, P.: Covariance structure maximum-likelihood estimates in compound Gaussian noise: existence and algorithm analysis. IEEE Trans. Signal Process. **56**(1), 34–48 (2008)
29. Schroth, C.A., Muma, M.: Real elliptically skewed distributions and their application to robust cluster analysis. IEEE Trans. Signal Process. **69**, 3947–3962 (2021)
30. Schroth, C.A., Muma, M.: Robust M-estimation based Bayesian cluster enumeration for real elliptically symmetric distributions. IEEE Trans. Signal Process. **69**, 3525–3540 (2021)
31. Schroth, C.A., Vlaski, S., Zoubir, A.M.: Robust M-estimation based distributed expectation maximization algorithm with robust aggregation. In: ICASSP 2023-2023 IEEE International Conference on Acoustics, Speech and Signal Processing (ICASSP), pp. 1–5. IEEE (2023)
32. Schwarz, G.: Estimating the dimension of a model. Ann. Stat. **6**(2), 461–464 (1978)

33. Seifert, A.-K., Zoubir, A.M., Amin, M.G.: Radar-based human gait recognition in cane-assisted walks. In: Proceedings of the IEEE Radar Conference, pp. 1428–1433. Seattle (2017)
34. Stoica, P., Selen, Y.: Model-order selection: a review of information criterion rules. IEEE Signal Process. Mag. **21**(4), 36–47 (2004)
35. Teklehaymanot, F.K., Muma, M., Zoubir, A.M.: Bayesian cluster enumeration criterion for unsupervised learning. IEEE Trans. Signal Process. **66**(20), 5392–5406 (2018)
36. Teklehaymanot, F.K., Muma, M., Zoubir, A.M.: Diffusion-based Bayesian cluster enumeration in distributed sensor networks. In: 2018 IEEE Statistical Signal Processing Workshop (SSP), pp. 1–5. IEEE (2018)
37. Teklehaymanot, F.K., Muma, M., Zoubir, A.M.: Novel Bayesian cluster enumeration criterion for cluster analysis with finite sample penalty term. In: 2018 IEEE International Conference on Acoustics, Speech and Signal Processing (ICASSP), pp. 4274–4278. IEEE (2018)
38. Teklehaymanot, F.K., Muma, M., Zoubir, A.M.: Robust Bayesian cluster enumeration based on the t distribution. Signal Process. **182**, 107870 (2020)
39. Wang, M., Abrams, Z.B., Kornblau, S.M., Coombes, K.R.: Thresher: determining the number of clusters while removing outliers. BMC Bioinf. 19 (2018)
40. Dongkuan, X., Tian, Y.: A comprehensive survey of clustering algorithms. Ann. Data Sci. **2**(2), 165–193 (2015)
41. Zhang, Y., Teklehaymanot, F.K., Muma, M., Zoubir, A.M.: Distributed robust Bayesian cluster enumeration criterion for unsupervised learning. In: 2019 IEEE 8th International Workshop on Computational Advances in Multi-Sensor Adaptive Processing (CAMSAP), pp. 176–180. IEEE (2019)
42. Zoubir, A.M., Koivunen, V., Ollila, E., Muma, M.: Robust Statistics for Signal Processing. Cambridge University Press (2018)

Chapter 9
FEMDA: A Unified Framework for Discriminant Analysis

Pierre Houdouin, Matthieu Jonckheere, and Frédéric Pascal

Abstract Although linear and quadratic discriminant analysis are widely recognized classical methods, they can encounter significant challenges when dealing with non-Gaussian distributions or contaminated datasets. This is primarily due to their reliance on the Gaussian assumption, which lacks robustness. We first explain and review the classical methods to address this limitation and then present a novel approach that overcomes these issues. In this new approach, the model considered is an arbitrary Elliptically Symmetrical (ES) distribution per cluster with its own arbitrary scale parameter. This flexible model allows for potentially diverse and independent samples that may not follow identical distributions. By deriving a new decision rule, we demonstrate that maximum-likelihood parameter estimation and classification are simple, efficient, and robust compared to state-of-the-art methods.

9.1 Introduction

In machine learning, the classification task involves assigning an observation to a group of similar elements that share homogeneous properties. During the training phase, labeled observations are used to determine all possible groups or clusters, along with their associated features. These identified features are then used to build a decision rule for classifying new observations. Classification is a research field of great interest as such problems can arise across various scientific domains. Intensive

---

P. Houdouin (✉)
Université Paris-Saclay, CentraleSupélec, Laboratoire des signaux et systèmes, 91190 Gif-sur-Yvette & RTE, La Défense, Gif-sur-Yvette, France
e-mail: pierre.houdouin@centralesupelec.fr

M. Jonckheere
LAAS, CNRS, 31400 Toulouse, France
e-mail: matthieu.jonchkeere@gmail.com

F. Pascal
Université Paris-Saclay, CentraleSupélec, Laboratoire des signaux et systèmes, 91190 Gif-sur-yvette, France
e-mail: frederic.pascal@centralesupelec.fr

J.-P. Delmas et al. (eds.), *Elliptically Symmetric Distributions in Signal Processing and Machine Learning*, https://doi.org/10.1007/978-3-031-52116-4_9

efforts have been devoted to developing diverse algorithms capable of addressing the large variety of classification problems [16, 25, 28, 33]. In this chapter, we will focus on discriminant analysis, a popular statistical-based family of classification methods.

Discriminant Analysis

Fisher presented the first historical work on discriminant analysis in [14]. Fisher discriminant analysis (FDA) aims to find a linear transformation of the input data that maximizes the separation between different classes. The key idea behind the FDA is to project the original high-dimensional data onto a lower-dimensional subspace while maximizing the ratio of between-class scatter to within-class scatter. In other words, FDA seeks a projection that maximizes the separation between classes while minimizing the variation within each class.

For the two-class problem, the FDA method consists of finding the best line onto which the data should be projected to achieve linear separability. Such an optimal line is estimated using the labeled data and involves estimating the mean vector and covariance matrix for each cluster. The classification rule is then derived using the projection of the observation onto the optimal line [14].

Another popular approach in discriminant analysis involves employing a discrimination rule based on estimating the likelihood that an observation belongs to a specific cluster. In this framework, we assume that each observation is generated according to a probability density function p_k that only depends on the cluster C_k it belongs to. We typically suppose that all p_k belong to a parametric family of distributions: $\forall k,\ p_k \in \{p_{\boldsymbol{\theta}}, \boldsymbol{\theta} \in \boldsymbol{\Theta}\}$. Discriminant analysis methods then boil down to two steps:

- The first one is a supervised learning problem. It consists of finding the optimal parameter $\hat{\boldsymbol{\theta}}_k \in \boldsymbol{\Theta}$ that describes best each cluster using the labeled data of each cluster.
- The second step is the derivation of the decision rule to classify the unlabeled observations. The goal is to choose the cluster C_k that maximizes the likelihood $p_k(\mathbf{x})$ conditioned by the cluster.

In the first works that employed a discrimination rule based on likelihood estimation, it was assumed that all the p_k followed a multivariate Gaussian distribution. As discussed by Huberty in [22] in its review of the different discriminant analysis techniques, the supervised learning part involves using the labeled data to estimate each cluster's mean and covariance matrix. The decision rule for classifying unlabeled observations then boils down to maximizing the likelihood, computed using the estimated parameters and the multivariate Gaussian probability density function. When we introduce the assumption of homoscedasticity (i.e., same covariance matrix for all clusters), the clusters become linearly separable, and the resulting decision rule is close to the one obtained by Fisher [14]. It is referred to as Linear Discriminant Analysis (LDA), where a linear combination of the features is used to predict the label of an observation. If the homoscedasticity assumption, however, does not hold, the clusters are no longer linearly separable, and the resulting decision rule is referred to as Quadratic Discriminant Analysis (QDA), where classification is performed by considering quadratic combinations of the features [40].

Robust Discriminant Analysis

One of the main drawbacks of traditional discriminant analysis methods is their sensitivity to certain assumptions. When one underlying assumption fails to hold, it can significantly impact the results. This issue was already studied in 1966 by Lachenbruch [26], who investigated how misclassified samples in the training set affect the performances. In 1979, together with Goldstein [27], they examined the influence of data contamination on the maximum likelihood approach and provided formulas for the error rates. In the same year, Lachenbruch [8] also published an article on how the non-normality assumption affects QDA, concluding that misclassification rates increase with high skewness and, to a smaller extent, high kurtosis. In answer to this sensitivity, robust methods were developed to improve robustness. To alleviate the sensitivity of mean vectors and covariance matrix estimation in FDA, Kim [24] proposed to explicitly incorporate a model of data uncertainty and optimize the worst-case scenario. This model, however, is only suitable for binary classification. In 1978, Randles [34] suggested improving the robustness of the FDA by assigning weights to observations based on their proximity to overlapping regions, again in the case of binary classification. The goal was to reduce the influence of observations that are far from the average of mean vectors, thereby mitigating the impact of outliers. The author also provides LDA and QDA approaches using robust M-estimators, developed by Huber in [20], to reduce the sensitivity of mean vectors and covariance matrix estimation to outliers. The main limitation of M-estimators is their low breakpoint (proportion of outliers estimator can handle) in high dimension, as highlighted by Maronna [29]. This led to recent research focusing on higher breakpoint options, such as Chork and Rousseeuw in [7] with the minimum volume ellipsoid estimator, Hawkins and McLachlan in [17] with the minimum within-group covariance determinant and more recently Croux and Dehon in 2001 [10] that examined the performances of discriminant analysis with high breakpoint S-estimators. In 2004, Hubert and Van Driessen [21] studied the performances of the minimum covariance discriminant estimator, which offers robustness and computational efficiency, making it suitable for large datasets. Finally, in 2019, Wen and Fang [43] proposed an extension of FDA, called robust sparse linear discriminant analysis, which incorporates feature selection along with the projection.

Generalized Discriminant Analysis

Many studies have been conducted to address the limitations associated with the Gaussian underlying model assumption and reduce its sensitivity. In 1992, Wakaki [41] investigated the expression of Fisher's linear discriminant function for the very large family of elliptical symmetric distributions with a common covariance matrix, aiming to extend the applicability of discriminant analysis beyond the Gaussian distributions. Different approaches such as marginal Fisher analysis [19], discriminative locality alignment [44], and manifold partition discriminant analysis [45] were also developed. These three methods use both nearby observations and label information to come up with a more general projection that can deal more efficiently with non-linear structures. Regarding likelihood-based approaches, namely LDA and QDA, Andrews et al. [1] examined in 2010 the classification with a mixture of multivariate t-distributions. They developed an algorithm for parameter estimation across four distinct scenarios, which involved considering equality or inequality of covariance matrices and variations in degrees of freedom across different clusters. By considering more flexible distributions, these approaches aimed to enhance the robustness and performance of discriminant analysis methods.

Bose et al. [4] in 2015 attempted to generalize LDA and QDA to elliptically symmetric distributions. The proposed method is called Generalized Quadratic Discriminant Analysis (GQDA) and is semi-parametric. It involves estimating a threshold parameter c whose optimal value is fixed for all sub-families of distributions. For example, when $c = 1$, GQDA corresponds to the classic QDA for a mixture of multivariate Gaussian distributions, while $c = 0$ boils down to LDA. In the case of multivariate t-distributions, the optimal value of c is determined by the degrees of freedom of each cluster's distribution. Of course, in practical applications, the sub-family is unknown, necessitating the estimation of the threshold parameter. Finally, Ghosh et al. [15] in 2020 used Bose's work to handle situations where the sensitive Gaussianity hypothesis was not fulfilled. They also explored using different robust estimators for the mean and covariance matrix to handle outliers. Their work resulted in Robust Generalized Quadratic Discriminant Analysis (RGQDA), a new robust discriminant analysis method.

With the rise of neural networks, novel approaches to discriminant analysis have been developed. In 2016, Dorfer et al. [12] introduced deep LDA, a new deep neural network method. Deep LDA can be considered as a nonlinear extension of LDA, aiming to minimize intra-cluster variance while maximizing extra-cluster variance, able to capture complex relationships and patterns in the data. A last approach that deserves to be highlighted is the study of distribution-free discriminant analysis by Burr and Doak in [6]. This approach uses the kernel method developed by Silverman in [38] to estimate the probability density function of each cluster. Although the approach is radically different from the one proposed in this article, the goal is the same: build a decision rule for discriminant analysis without relying on prior assumptions about the distributions and handle clusters with different distributions.

Our Contribution

This work introduces a novel, robust, and model-free discriminant analysis algorithm. Our method does not assume that points within the same cluster are drawn from the same distribution. Instead, we allow each observation to be drawn from its own elliptically symmetric distribution; see Chap. 1, [3, 30] for more information on this family of distributions. This provides greater flexibility in the modeling as we only assume that observations are independent.

Since each point can have its own distribution, observations inside the same cluster are not necessarily identically distributed. We are no longer in the framework where each $\mathbf{x} \in C_k$ is drawn from p_k; in this framework, each $\mathbf{x}_i$ is drawn from $p_{i,k}$ [35]. The conditions to belong to the same cluster are consequently redefined. The precise definition is provided in the next section.

The counterpart of such flexibility in the modeling does reverberate in the more permissive clusters' membership conditions. Observations within the same cluster only fulfill proportional covariance matrices: their covariance matrix share the same direction, but the scale may differ. While this may appear insufficient for describing a cluster in low-dimensional cases, it remains very satisfying in higher dimensions. This increased flexibility also poses challenges for the likelihood estimation, as there are many more unknown parameters to consider. To deal with this issue, we will look to maximize the marginal likelihood, which is an average of the likelihood computed over the scale parameter of the covariance matrix. The structure of the paper is as follows: in Sect. 9.2, we present the model along with the theoretical framework, providing the main derivations of this work. Section 9.3 encompasses the experiments conducted on synthetic data. Section 9.4 contains the real data experiments, and conclusions and perspectives are drawn in Sect. 9.5.

9.2 Model and Theoretical Contribution

9.2.1 Presentation of the Statistical Modeling

Let $\mathbf{x}_1, \ldots, \mathbf{x}_n \in \mathbb{R}^m$ be n **independent** observations split into K different classes. Each cluster C_k comprises n_k observations. Previous works in discriminant analysis usually assume that points in the same cluster are drawn from the same distribution. In our contribution, we discharge this hypothesis. We allow each observation $\mathbf{x}_i$ to:

- be drawn from **its own elliptically symmetric distribution** with density generator function $g_{i,k}$,
- have its own covariance matrix, written as $\mathbf{\Sigma}_{i,k} = \tau_{i,k}\mathbf{\Sigma}_k$, $\tau_{i,k}$ being an unknown scale factor of the observation and $\mathbf{\Sigma}_k$ the common scatter matrix of points of cluster C_k,
- have a prior probability density function $h_{i,k}$ on the scale factor.

In this new modeling, the joint probability density function is

$$f(\mathbf{x}_i \in C_k, \tau_{i,k}) = \frac{\Gamma(\frac{m}{2}) h_{i,k}(\tau_{i,k})}{\pi^{\frac{m}{2}} \int_0^{+\infty} t^{\frac{m}{2}-1} g_{i,k}(t)dt} |\mathbf{\Sigma}_k|^{-\frac{1}{2}} \tau_{i,k}^{-\frac{m}{2}} g_{i,k}\left(\frac{(\mathbf{x}-\boldsymbol{\mu}_k)^T \mathbf{\Sigma}_k^{-1}(\mathbf{x}-\boldsymbol{\mu}_k)}{\tau_{i,k}}\right)$$

$$\mathbf{x}_i \in C_k \sim \boldsymbol{\mu}_k + \sqrt{\tau_{i,k}}\sqrt{Q_{i,k}}\mathbf{\Sigma}_k^{\frac{1}{2}}\mathbf{u_i}.$$

The second equation is the stochastic representation of $\mathbf{x}_i$. The vector $\mathbf{u}_i$ is drawn using the uniform distribution over the m-dimensional unit sphere, while $Q_{i,k}$ is a positive random variable directly linked to $g_{i,k}$ by the relation: $p_{Q_{i,k}}(t) = \frac{t^{\frac{m}{2}-1} g_{i,k}(t)}{\int_0^{\infty} t^{\frac{m}{2}-1} g_{i,k}(t)dt}$. The double dependence regarding i and k for a parameter $\boldsymbol{\theta} \in \mathbf{\Theta}$ implies that each point of the dataset has its own value $\boldsymbol{\theta}_{i,k}$, belonging to a subset of $\mathbf{\Theta}$ depending on k: $\boldsymbol{\theta}_{i,k} \in \mathbf{\Theta}_k \subseteq \mathbf{\Theta}$. Since both the cluster and the observation influence the joint probability density function $p_{i,k}$, belonging to the same cluster can no longer be defined as sharing the same probability density function. Belonging to cluster C_k is now characterized as follows:

- respecting the constraints on the density generator function $g_{i,k}$ and the prior $h_{i,k}$;
- have the mean vector equal to $\boldsymbol{\mu}_k$;
- have the scatter matrix equal to $\mathbf{\Sigma}_k$.

This new modeling is very flexible and provides a high level of generality. Observations are not i.i.d anymore, but only independent. The family of distributions used is significantly larger than the traditional multivariate Gaussian or student distribution. The density generator function $g_{i,k}$ can be any function $\mathbb{R}^+ \longrightarrow \mathbb{R}^+$ that fulfills the condition $\int_{\mathbb{R}^+} t^{\frac{m}{2}} g_{i,k}(t)dt < +\infty$ and $h_{i,k}$ is any prior defined on $\mathbb{R}^{+*}$. In this framework, the constraints on the density generator and the prior are fixed by the user, whereas $\boldsymbol{\mu}_k$ and $\mathbf{\Sigma}_k$ are unknown parameters that must be estimated. Each choice of $(g_{i,k}, h_{i,k})$ leads to a different discriminant analysis method. Besides, all the usual discriminant analysis methods (QDA, t-QDA...) can be retrieved with the appropriate choice on the density generator and the prior. Details for many famous methods are provided later on. To obtain our classification algorithm, we aim to stick as long as possible to the general case of unknown $(g_{i,k}, h_{i,k})$. In the next subsection, we are first going to derive the estimators of $\boldsymbol{\mu}_k$ and $\mathbf{\Sigma}_k$ and the decision rule in this general case, and then add an assumption on $h_{i,k}$ in order to obtain a tractable algorithm.

9.2.2 General Case Estimators and Decision Rule

To obtain estimators for the unknown scatter and location parameters, one has to efficiently deal with the last unknown parameter, $\tau_{i,k}$. Two options are possible:

- Use the completed likelihood, the likelihood where $\tau_{i,k}$ is replaced by its maximum-likelihood estimator.
- Use a Bayesian approach with the maximum marginal likelihood estimators.

The authors have already applied the first approach [18] in a less general case. In this previous work, there are more stringent hypotheses (the density generator function only depended on i, and no prior was assumed on $\tau_{i,k}$) and used the completed likelihood to come up with the Flexible EM-inspired Discriminant Analysis (FEMDA) algorithm. In this paper, we will focus on the second Bayesian approach. The marginal likelihood of $\mathbf{x}_1, \ldots, \mathbf{x}_{n_k} \in C_k$ relatively to $\boldsymbol{\mu}_k$ and $\boldsymbol{\Sigma}_k$ is

$$\mathcal{L}(\mathbf{x}; \boldsymbol{\mu}_k, \boldsymbol{\Sigma}_k) = \int_{]0,+\infty[^{n_k}} \prod_{i=1}^{n_k} p_{i,k}\left(\mathbf{x}_i, \tau_{i,k}; \boldsymbol{\mu}_k, \boldsymbol{\Sigma}_k\right) d\tau_{i,k}.$$

Proposition 9.1 *Assuming that $h_{i,k}(t)$ is an integrable function with respect to the Lebesgue measure, the marginal likelihood of the sample $\mathbf{x}_k = \left(\mathbf{x}_1, \ldots \mathbf{x}_{n_k}\right)$ belonging to cluster C_k relatively to $\boldsymbol{\mu}_k$ and $\boldsymbol{\Sigma}_k$ in this theoretical framework is*

$$\mathcal{L} = \prod_{i=1}^{n_k} \frac{\Gamma(\frac{m}{2})|\boldsymbol{\Sigma}_k|^{-\frac{1}{2}} \left((\mathbf{x}_i - \boldsymbol{\mu}_k)^T \boldsymbol{\Sigma}_k^{-1} (\mathbf{x}_i - \boldsymbol{\mu}_k)\right)^{1-\frac{m}{2}}}{\pi^{\frac{m}{2}} \int_0^{+\infty} t^{\frac{m}{2}-1} g_{i,k}(t)dt}$$
$$\int_0^{+\infty} h_{i,k}\left(\frac{(\mathbf{x}_i - \boldsymbol{\mu}_k)^T \boldsymbol{\Sigma}_k^{-1} (\mathbf{x}_i - \boldsymbol{\mu}_k)}{t}\right) t^{\frac{m}{2}-2} g_{i,k}(t)dt.$$

Proof See Sect. 9.6.

Remark 9.1 Choosing the Dirac measure $h_{i,k}(\tau_{i,k})d\tau = \delta_1(\tau_{i,k})$ allows us to come back to the traditional discriminant analysis modeling. In this particular case, the framework, however, changes since $h_{i,k}(\tau_{i,k})$ is no longer a probability density function defined over the Lebesgue measure. In this particular case, the use of the Maximal Marginal Likelihood Estimator (MMLE) boils down to the classic Maximum-Likelihood Estimator (MLE). We will now derive the analytical expression of the MMLE by maximizing the marginal likelihood of the sample with respect to $\boldsymbol{\mu}_k$ and $\boldsymbol{\Sigma}_k$. Once the estimators are obtained, we can obtain the expression of the decision rule for a new observation $\mathbf{x}$. It consists in choosing the cluster that will maximize its marginal likelihood $\mathcal{L}(\mathbf{x}|\mathbf{x} \in C_k)$. According to Neyman–Pearson's lemma, as distributions here are not degenerate, the unique Uniformly Most Powerful (UMP) statistical test is used to classify the observation [42].

Theorem 9.1 *Assuming that $h_{i,k}$ is differentiable, the expressions of the MMLE estimators of the location and scale parameters, as well as the decision rule, are*

$$\hat{\boldsymbol{\mu}}_k = \frac{\sum_{i=1}^{n_k} w_{i,k}\mathbf{x}_i}{\sum_{i=1}^{n_k} w_{i,k}}$$

$$\hat{\boldsymbol{\Sigma}}_k = \frac{m-2}{n_k}\sum_{i=1}^{n_k} w_{i,k}(\mathbf{x}_i - \hat{\boldsymbol{\mu}}_k)(\mathbf{x}_i - \hat{\boldsymbol{\mu}}_k)^T$$

$$w_{i,k} = \frac{1}{(\mathbf{x}_i - \hat{\boldsymbol{\mu}}_k)^T \hat{\boldsymbol{\Sigma}}_k^{-1} (\mathbf{x}_i - \hat{\boldsymbol{\mu}}_k)} - \frac{2}{m-2}\frac{\int_0^{+\infty} h'_{i,k}\left(\frac{(\mathbf{x}_i - \hat{\boldsymbol{\mu}}_k)^T \hat{\boldsymbol{\Sigma}}_k^{-1} (\mathbf{x}_i - \hat{\boldsymbol{\mu}}_k)}{t}\right) t^{\frac{m}{2}-3} g_{i,k}(t)dt}{\int_0^{+\infty} h_{i,k}\left(\frac{(\mathbf{x}_i - \hat{\boldsymbol{\mu}}_k)^T \hat{\boldsymbol{\Sigma}}_k^{-1} (\mathbf{x}_i - \hat{\boldsymbol{\mu}}_k)}{t}\right) t^{\frac{m}{2}-2} g_{i,k}(t)dt}$$

$$k = \arg\max_{k\in[1,K]} \frac{|\hat{\boldsymbol{\Sigma}}_k|^{-\frac{1}{2}}\left((\mathbf{x}_i - \hat{\boldsymbol{\mu}}_k)^T \hat{\boldsymbol{\Sigma}}_k^{-1} (\mathbf{x}_i - \hat{\boldsymbol{\mu}}_k)\right)^{1-\frac{m}{2}}}{\int_0^{+\infty} t^{\frac{m}{2}-1} g_{i,k}(t)dt} \int_0^{+\infty} h_{i,k}\left(\frac{(\mathbf{x}_i - \hat{\boldsymbol{\mu}}_k)^T \hat{\boldsymbol{\Sigma}}_k^{-1} (\mathbf{x}_i - \hat{\boldsymbol{\mu}}_k)}{t}\right) t^{\frac{m}{2}-2} g_{i,k}(t)dt.$$

Proof See Sect. 9.7.

Remark 9.2 At this stage, let us underline the specific parametric case of $h_{i,k} = h_k = h_{\alpha_k}(.)$ and $g_{i,k} = g$ where α_k is a class parameter, which will be used later on to deal with the cases of compound Gaussian distributions. In this situation, the expression of the estimators of $\boldsymbol{\mu}_k$ and $\boldsymbol{\Sigma}_k$ becomes tractable if we can estimate α_k. This can be done by solving the following optimization problem.

$$\hat{\alpha}_k = \arg\max_{\alpha_k\in\boldsymbol{\Theta}} \sum_{i=1}^{n_k} \log\left(\int_0^{\infty} h_{\alpha_k}\left(\frac{(\mathbf{x}_i - \hat{\boldsymbol{\mu}}_k)^T \hat{\boldsymbol{\Sigma}}_k^{-1} (\mathbf{x}_i - \hat{\boldsymbol{\mu}}_k)}{t}\right) t^{\frac{m}{2}-2} g(t)dt\right).$$

At this point, both the estimators and the decision rule are intractable in the general case. We will add assumptions on the prior distribution of $\tau_{i,k}$ to obtain our algorithm.

9.2.3 Derivation of the Algorithm

9.2.3.1 Non-informative Prior $h_{i,k}$

If we do not have specific information on the nuisance parameters $\tau_{i,k}$, then using the non-informative Jeffrey's prior [23]. In the particular case of a uni-dimensional parameter, Jeffrey's prior, which may be an improper prior, is proportional to the square root of the Fisher information. Based on [2], we obtain the following result.

Proposition 9.2 *The expression of the non-informative Jeffrey's prior on $\tau_{i,k}$ is given by*

$$h_{i,k}(t) \propto \frac{1}{t}.$$

Proof By definition, the expression of Jeffrey's prior for a one-dimensional parameter τ is

$$h_{i,k}(\tau) \propto \sqrt{\mathcal{I}_{i,k}(\tau)}, \text{ with } \mathcal{I}_{i,k}(\tau) = -\mathbb{E}_{\mathbf{x_i}}\left[\frac{\partial^2 \left(\log \circ p_{i,k}(\mathbf{x_i}, \tau)\right)}{\partial \tau^2}\right]$$

being the Fisher information.

We define $d_{i,k} = (\mathbf{x_i} - \boldsymbol{\mu}_k)^T \boldsymbol{\Sigma}_k^{-1}(\mathbf{x_i} - \boldsymbol{\mu}_k)$ and $q_{i,k} = \frac{(\mathbf{x_i} - \boldsymbol{\mu}_k)^T \boldsymbol{\Sigma}_k^{-1}(\mathbf{x_i} - \boldsymbol{\mu}_k)}{\tau}$. Using the stochastic form, one can write $\forall \mathbf{x_i} \in \mathbb{R}^m$

$$\begin{aligned}
&\mathbf{x_i} \sim \boldsymbol{\mu}_k + \sqrt{\tau}\sqrt{\mathcal{Q}_{i,k}}\boldsymbol{\Sigma}_k^{\frac{1}{2}}\boldsymbol{u}_i \\
\iff &\tau^{-\frac{1}{2}}\boldsymbol{\Sigma}_k^{-\frac{1}{2}}(\mathbf{x_i} - \boldsymbol{\mu}_k) \sim \sqrt{\mathcal{Q}_{i,k}}\boldsymbol{u}_i \\
\iff &q_{i,k} \sim \mathcal{Q}_{i,k} \text{ since } \boldsymbol{u}_i{}^T\boldsymbol{u}_i = 1.
\end{aligned}$$

We can now compute the Fisher information:

$$\begin{aligned}
\mathcal{I}_{i,k}(\tau) =&\mathbb{E}_{\mathbf{x_i}}\left[\left(\frac{\partial \left(\log \circ p_{i,k}(\mathbf{x_i}, \tau)\right)}{\partial \tau}\right)^2\right] = -\mathbb{E}_{\mathbf{x_i}}\left[\frac{\partial^2 \left(\log \circ p_{i,k}(\mathbf{x_i}, \tau)\right)}{\partial \tau^2}\right] \\
=& -\mathbb{E}_{\mathcal{Q}_{i,k}}\left[\frac{m}{2\tau^2} + \frac{2}{\tau^2}\frac{q_{i,k}g'_{i,k}(q_{i,k})}{g_{i,k}(q_k)} + \frac{1}{\tau^2}q_{i,k}^2\frac{g''_{i,k}(q_{i,k})g_{i,k}(q_{i,k}) - g'_{i,k}(q_{i,k})^2}{g_{i,k}(q_{i,k})^2}\right] \\
=&\frac{-m}{2\tau^2} - \frac{2}{\tau^2}\int_0^\infty \frac{qg'_{i,k}(q)}{g_{i,k}(q)}\frac{q^{\frac{m}{2}-1}g_{i,k}(q)}{\int_0^\infty t^{\frac{m}{2}-1}g_{i,k}(t)dt}dq \\
&-\frac{1}{\tau^2}\int_0^\infty q^2\frac{g''_{i,k}(q)g_{i,k}(q) - g'_{i,k}(q)^2}{g_{i,k}(q)^2}\frac{q^{\frac{m}{2}-1}g_{i,k}(q)}{\int_0^\infty t^{\frac{m}{2}-1}g_{i,k}(t)dt}dq
\end{aligned}$$

$$
\begin{aligned}
&= \frac{-m}{2\tau^2} - \frac{2}{\tau^2}\int_0^\infty \frac{q^{\frac{m}{2}} g'_{i,k}(q)}{\int_0^\infty t^{\frac{m}{2}-1} g_{i,k}(t)dt} dq \\
&\quad - \frac{1}{\tau^2}\int_0^\infty q^{\frac{m}{2}+1} \frac{g''_{i,k}(q) g_{i,k}(q) - g'_{i,k}(q)^2}{g_{i,k}(q)\int_0^\infty t^{\frac{m}{2}-1} g_{i,k}(t)dt} dq \\
&= \frac{-m}{2\tau^2} + \frac{m}{\tau^2}\int_0^\infty \frac{q^{\frac{m}{2}-1} g_{i,k}(q)}{\int_0^\infty t^{\frac{m}{2}-1} g_{i,k}(t)dt} dq \\
&\quad - \frac{1}{\tau^2}\int_0^\infty q^{\frac{m}{2}+1} \frac{g''_{i,k}(q) g_{i,k}(q) - g'_{i,k}(q)^2}{g_{i,k}(q)\int_0^\infty t^{\frac{m}{2}-1} g_{i,k}(t)dt} dq \\
&= \frac{1}{\tau^2}\left(\frac{m}{2} - \int_0^\infty q^{\frac{m}{2}+1} \frac{g''_{i,k}(q) g_{i,k}(q) - g'_{i,k}(q)^2}{g_{i,k}(q)\int_0^\infty t^{\frac{m}{2}-1} g_{i,k}(t)dt} dq\right).
\end{aligned}
$$

Thus, we do indeed have

$$
h_{i,k}(t) \propto \frac{\sqrt{\frac{m}{2} - \int_0^\infty q^{\frac{m}{2}+1} \frac{g''_{i,k}(q) g_{i,k}(q) - g'_{i,k}(q)^2}{g_{i,k}(q)\int_0^\infty t^{\frac{m}{2}-1} g_{i,k}(t)dt} dq}}{t}.
$$

Let us make a few remarks:

- Since $t \longrightarrow t^{\frac{m}{2}-1} g_{i,k}(t)$ is integrable on $\mathbb{R}^+$, then $lim_{\infty} t^{\frac{m}{2}} g_{i,k}(t) = 0$, which justifies the integration by part to reach the last line.
- The Fisher information is always positive since it can also be written as the expected value of a positive random variable, so the square root is always well-defined.

9.2.3.2 Decision Rule and Estimators Under Jeffrey's Prior

We can now derive the expression of the estimators as well as the decision rule using Jeffrey's prior. Since $h(t) \propto \frac{C}{t}$ is differentiable, Theorem 9.1 applies.

Theorem 9.2 *If we suppose that we have the non-informative Jeffrey's prior on the* $\tau_{i,k}$, *the decision rule and the estimators of location and scale parameters are*

$$\hat{\boldsymbol{\mu}}_k = \frac{\sum_{i=1}^{n_k} w_{i,k}\mathbf{x}_i}{\sum_{i=1}^{n_k} w_{i,k}}$$

$$\hat{\boldsymbol{\Sigma}}_k = \frac{m}{n_k}\sum_{i=1}^{n_k} w_{i,k}(\mathbf{x}_i - \hat{\boldsymbol{\mu}}_k)(\mathbf{x}_i - \hat{\boldsymbol{\mu}}_k)^T$$

$$w_{i,k} = \frac{1}{(\mathbf{x}_i - \hat{\boldsymbol{\mu}}_k)^T \hat{\boldsymbol{\Sigma}}_k^{-1}(\mathbf{x}_i - \hat{\boldsymbol{\mu}}_k)}$$

$$k = \arg\min_{k\in[1,K]} \log\left((\mathbf{x}_i - \hat{\boldsymbol{\mu}}_k)^T \hat{\boldsymbol{\Sigma}}_k^{-1}(\mathbf{x}_i - \hat{\boldsymbol{\mu}}_k)\right) + \frac{1}{m}\log(|\hat{\boldsymbol{\Sigma}}_k|).$$

Proof See Sect. 9.8.

At this stage, one can notice that the resulting estimators are M-estimators that are intrinsically robust despite being derived from an MMLE approach. They can be obtained using $\rho(\mathbf{x}, \boldsymbol{\mu}, \boldsymbol{\Sigma}) = \log(|\boldsymbol{\Sigma}|) + \log\left((\mathbf{x} - \boldsymbol{\mu})^T \boldsymbol{\Sigma}^{-1}(\mathbf{x} - \boldsymbol{\mu})\right)$. Both estimators are also very close to Tyler's M-estimators. The only difference arises in the estimator of the location parameter, as Tyler obtains a square root at the weight's denominator. Another interesting remark is that first, $\hat{\boldsymbol{\mu}}_k$ is insensitive to the scale of $\hat{\boldsymbol{\Sigma}}_k$ and if $(\boldsymbol{\mu}, \boldsymbol{\Sigma})$ is a solution to this coupled fixed-point equation, $(\boldsymbol{\mu}, \lambda\boldsymbol{\Sigma})$ is also a solution. The same remark can be drawn for the decision rule: replacing $\boldsymbol{\Sigma}$ by $\lambda\boldsymbol{\Sigma}$ does not change the classification. To conclude, both our estimators and decision rule are scale-insensitive. Although the approach is different, using the MMLE instead of the completed likelihood, we achieve the same estimators as in [18].

9.2.4 Retrieval of Other Usual Discriminant Analysis Methods

Jeffrey's prior choice leads to the FEMDA algorithm's derivation. However, different choices on $g_{i,k}$ and $h_{i,k}$ allow the retrieval of many other well-known discriminant analysis methods.

9.2.4.1 Quadratic Discriminant Analysis

QDA assumes that the points of each cluster are drawn from a multivariate Gaussian distribution with parameters $\boldsymbol{\mu}_k$ and $\boldsymbol{\Sigma}_k$. The MLE of the parameters and the decision rule are

$$\hat{\boldsymbol{\mu}}_k = \frac{1}{n_k}\sum_{i=1}^{n_k}\mathbf{x}_i$$
$$\hat{\boldsymbol{\Sigma}}_k = \frac{1}{n_k}\sum_{i=1}^{n_k}(\mathbf{x}_i - \hat{\boldsymbol{\mu}}_k)(\mathbf{x}_i - \hat{\boldsymbol{\mu}}_k)^T$$
$$k = \arg\min_{k\in[1,K]} (\mathbf{x}_i - \hat{\boldsymbol{\mu}}_k)^T \hat{\boldsymbol{\Sigma}}_k^{-1}(\mathbf{x}_i - \hat{\boldsymbol{\mu}}_k).$$

One can write

$$\mathbf{x}_i \in C_k \sim \boldsymbol{\mu}_k + \boldsymbol{\Sigma}_k^{\frac{1}{2}}\mathcal{N}(0, I)$$
$$p(\mathbf{x}_i \in C_k) = (2\pi)^{-\frac{m}{2}}|\boldsymbol{\Sigma}_k|^{-\frac{1}{2}}e^{-\frac{1}{2}(\mathbf{x}_i-\boldsymbol{\mu}_k)^T\boldsymbol{\Sigma}_k^{-1}(\mathbf{x}_i-\boldsymbol{\mu}_k)}.$$

This corresponds to the choices:

- $h_{i,k}(\tau)d\tau = \delta_1(\tau)$,
- $g_{i,k}(t) = g(t) = e^{-\frac{t}{2}}$.

Here, the prior h is no longer defined over the Lebesgue measure but over the Dirac measure. Thus, neither Proposition 9.1 nor Theorem 9.1 hold.

9.2.4.2 *t*-QDA

t-QDA assumes that the points of each cluster are drawn from a multivariate student distribution with parameters ν_k, $\boldsymbol{\mu}_k$ and $\boldsymbol{\Sigma}_k$. The MLE of the parameters are

$$\hat{\boldsymbol{\mu}}_k = \frac{\sum_{i=1}^{n_k} w_{i,k}\mathbf{x}_i}{\sum_{i=1}^{n_k} w_{i,k}}$$
$$\hat{\boldsymbol{\Sigma}}_k = \frac{\hat{\nu}_k + m}{n_k}\sum_{i=1}^{n_k} w_{i,k}(\mathbf{x}_i - \hat{\boldsymbol{\mu}}_k)(\mathbf{x}_i - \hat{\boldsymbol{\mu}}_k)^T$$
$$\hat{\nu}_k = \arg\min_{\nu\in\mathbb{R}^+} \frac{m}{2}\log(\nu) + \log\Gamma\left(\frac{\nu}{2}\right) - \log\Gamma\left(\frac{\nu+m}{2}\right) + \frac{\nu+m}{2n_k}\sum_{i=1}^{n_k}\log\left(1 + \frac{(\mathbf{x}_i - \hat{\boldsymbol{\mu}}_k)^T\hat{\boldsymbol{\Sigma}}_k^{-1}(\mathbf{x}_i - \hat{\boldsymbol{\mu}}_k)}{\nu}\right)$$
$$w_{i,k} = \frac{1}{\hat{\nu}_k + (\mathbf{x}_i - \hat{\boldsymbol{\mu}}_k)^T\hat{\boldsymbol{\Sigma}}_k^{-1}(\mathbf{x}_i - \hat{\boldsymbol{\mu}}_k)}.$$

Multivariate student distribution belongs to the family of compound Gaussian distributions. Thus, one can write

$$\mathbf{x}_i \in C_k \sim \boldsymbol{\mu}_k + \sqrt{\frac{1}{\Gamma(\frac{\nu_k}{2}, \frac{2}{\nu_k})}} \boldsymbol{\Sigma}_k^{\frac{1}{2}} \mathcal{N}(0, I)$$

$$p(\mathbf{x}_i \in C_k) = \frac{\Gamma(\frac{\nu_k+m}{2})}{\Gamma(\frac{\nu_k}{2}) \nu_k^{\frac{m}{2}} \pi^{\frac{m}{2}}} |\boldsymbol{\Sigma}_k|^{-\frac{1}{2}} \left(1 + \frac{(\mathbf{x}_i - \boldsymbol{\mu}_k)^T \boldsymbol{\Sigma}_k^{-1} (\mathbf{x}_i - \boldsymbol{\mu}_k)}{\nu_k}\right)^{-\frac{\nu_k+m}{2}}.$$

This corresponds to the choices:

- $h_{i,k}(\tau) = \frac{(\frac{\nu_k}{2})^{\frac{\nu_k}{2}} e^{-\frac{\nu_k}{2\tau}}}{\Gamma(\frac{\nu_k}{2}) \tau^{1+\frac{\nu_k}{2}}}$,
- $g_{i,k}(t) = g(t) = e^{-\frac{x}{2}}$.

Here, $h_{i,k}$ is differentiable and has the form of $h_{\alpha_k}(.)$ so both Theorem 9.1 and Remark 9.2 apply. It does allow us to retrieve the estimators of the degree of freedom, scale, and location parameters, as well as the decision rule:

$$\begin{aligned}
\hat{\boldsymbol{\mu}}_k &= \frac{\sum_{i=1}^{n_k} w_{i,k} \mathbf{x}_i}{\sum_{i=1}^{n_k} w_{i,k}} \\
\hat{\boldsymbol{\Sigma}}_k &= \frac{m-2}{n_k} \sum_{i=1}^{n_k} w_{i,k} (\mathbf{x}_i - \hat{\boldsymbol{\mu}}_k)(\mathbf{x}_i - \hat{\boldsymbol{\mu}}_k)^T \\
w_{i,k} &= \frac{\nu_k + m}{(m-2)\left(\nu_k + (\mathbf{x}_i - \hat{\boldsymbol{\mu}}_k)^T \hat{\boldsymbol{\Sigma}}_k^{-1} (\mathbf{x}_i - \hat{\boldsymbol{\mu}}_k)\right)} \\
\hat{\nu}_k &= \arg\min_{\nu \in \mathbb{R}^+} \frac{m}{2} \log(\nu) + \log\Gamma\left(\frac{\nu}{2}\right) - \log\Gamma\left(\frac{\nu+m}{2}\right) \\
&\quad + \frac{\nu+m}{2n_k} \sum_{i=1}^{n_k} \log\left(1 + \frac{(\mathbf{x}_i - \hat{\boldsymbol{\mu}}_k)^T \hat{\boldsymbol{\Sigma}}_k^{-1} (\mathbf{x}_i - \hat{\boldsymbol{\mu}}_k)}{\nu}\right) \\
k &= \arg\min_{k \in [1,K]} \left(1 + \frac{\hat{\nu}_k}{m}\right) \log\left(1 + \frac{(\mathbf{x}_i - \hat{\boldsymbol{\mu}}_k)^T \hat{\boldsymbol{\Sigma}}_k^{-1} (\mathbf{x}_i - \hat{\boldsymbol{\mu}}_k)}{\hat{\nu}_k}\right) \\
&\quad + \log(\hat{\nu}_k) + \frac{1}{m} \log\left(|\hat{\boldsymbol{\Sigma}}_k|\right) + \frac{2}{m} \log\left(\frac{\Gamma\left(\frac{\hat{\nu}_k}{2}\right)}{\Gamma\left(\frac{\hat{\nu}_k+m}{2}\right)}\right).
\end{aligned}$$

Proof See Sect. 9.9.

9.2.5 Implementation Details and Numerical Considerations

The structure of the proposed algorithm is similar to the t-QDA or RQDA algorithm. It involves recursively estimating the parameters of each cluster and then classifying test observations based on the derived decision rule.

Algorithm 9.1 FEMDA algorithm

Inputs:

- $\{\mathbf{x}_i\}_{i=1_{tr}^n} \in \mathbb{R}^m$ observations in train set
- $\{y_i\}_{i=1_{tr}^n} \in \{1, \ldots, K\}$ the train labels
- $\{\mathbf{x}_i\}_{i=1_{te}^n} \in \mathbb{R}^m$ observations in test set

Output: $\{y_i\}_{i=1_{te}^n}$ the test labels
Hyperparameters:

- N_{iter} maximum number of iterations for the estimation
- ϵ related to the condition of convergence
- λ regularization parameter

for $k \in [1, K]$ **do**
  $it, convergence \longleftarrow 0, False$
  $\boldsymbol{\mu}_k^0, \boldsymbol{\Sigma}_k^0 = \boldsymbol{\mu}_k^{MLE}, \boldsymbol{\Sigma}_k^{MLE} + \lambda I_m$, initialization with the regularized MLE estimator
  while not $convergence$ **and** $it < N_{iter}$ **do**
    for $i \in [1, n_k]$ **do**
      Compute $w_{i,k} = min\left(0.5, \frac{1}{(\mathbf{x}_i - \boldsymbol{\mu}_k^{it})^T \boldsymbol{\Sigma}_k^{-1it} (\mathbf{x}_i - \boldsymbol{\mu}_k^{it})}\right)$
    $\boldsymbol{\mu}_k^{it+1} = \frac{\sum_{i=1}^{n_k} w_{i,k}\mathbf{x}_i}{\sum_{i=1}^{n_k} w_{i,k}}$
    $\boldsymbol{\Sigma}_k^{it+1} = \frac{m}{n_k} \sum_{i=1}^{n_k} w_{i,k}(\mathbf{x}_i - \boldsymbol{\mu}_k^{it})(\mathbf{x}_i - \boldsymbol{\mu}_k^{it})^T$
    Regularization: $\boldsymbol{\Sigma}_k^{it+1} = \boldsymbol{\Sigma}_k^{it+1} + \lambda I_m$
    $it, convergence \longleftarrow it + 1, ||\boldsymbol{\Sigma}_k^{it+1} - \boldsymbol{\Sigma}_k^{it}||_1 + ||\boldsymbol{\mu}_k^{it+1} - \boldsymbol{\mu}_k^{it}||_1 < \epsilon$
for $i \in [1, n_{te}]$ **do**
  for $k \in [1, K]$ **do**
    Compute the score $s_{i,k} = \log\left((\mathbf{x}_i - \hat{\boldsymbol{\mu}}_k)^T \hat{\boldsymbol{\Sigma}}_k^{-1} (\mathbf{x}_i - \hat{\boldsymbol{\mu}}_k)\right) + \frac{1}{m}\log\left(|\hat{\boldsymbol{\Sigma}}_k|\right)$
  Label the observation $\mathbf{x}_i$ with $y_i = \arg\min_{k \in [1,K]} s_{i,k}$

In our algorithm, the estimated parameters are initialized using a regularized MLE. This helps to obtain more stable and reliable initial parameter values. Additionally, to avoid singularity problems, we regularize the estimated scatter matrix at each iteration. Inspired by Huber [20], we apply a trimming to avoid diverging weights that occur when the estimated mean gets too close to a training observation. The last point that needs to be discussed is the convergence of this fixed-point equation. In 1976, Maronna [29] studied the convergence of fixed-point equations defining M-estimators and proved it under restrictive assumptions that are not fulfilled here. We sometimes encounter convergence issues during our experiments, which suggests that convergence is not guaranteed here. To address this problem, we limit the number of iterations when solving the equation. This algorithm ends up being close to the robust version of classic QDA, except that we compare the logarithm of the squared Mahalanobis distances rather than the squared Mahalanobis distances.

9.3 Experiments on Synthetical Data

9.3.1 *Classification Methods Used for Comparison*

To evaluate the performances of the FEMDA algorithm, we perform classification tasks on various simulated and real datasets. We compare the performances with other discriminant analysis methods and more machine learning-based techniques. All the results presented are averaged over N simulations. We shuffle the data in each simulation and randomly select a new train and test set. We allocate a comparable training time for each method to ensure a fair comparison. Since some methods are intrinsically faster than others, we gradually reduce the size of the train set until the training time reaches at most T times the fastest method's training time. The goal of such a process is to compare the performances of the different methods for real-time applications, where the tradeoff between computation time and performance is crucial. We choose $N = 5$ and $T = 30$.

9.3.1.1 Discriminant Analysis Methods

During the experiments, FEMDA will be compared to the methods recalled above and the robust version of QDA implemented with Tyler's M-estimators [29]. To avoid convergence issues, singular matrix estimations, and diverging weights, we apply the respective following modifications:

- For fixed-point equations solving, we limit the number of iterations to $N_{iter} = 10$ and choose for the convergence condition $\epsilon = 10^{-5}$.
- We regularize the estimated matrix with $\lambda = 10^{-5}$.
- We use the trimmed version $\tilde{w}_{ik} = min\,(0.5, w_{ik})$.

9.3.1.2 Machine Learning Methods

We also compare our performances with Support Vector Classifier (SVC) [5], k-Nearest Neighbors (KNN) [9], and Random Forest (RF) [11] algorithms. The train set used by discriminant analysis methods for parameter estimation is split into a smaller train set and a validation set for the machine learning methods. We then perform a hyperparameter optimization using the Sklearn [32] RandomizedGridSearchCV function on the validation set. We use the following sets of values to perform the optimization:

Hyperparameter optimization					
SVC		**KNN**		**RF**	
Regularization	$[0.01, 0.1, 1]$	Nb neighbors	$[5, 11, 15, 21]$	Nb trees	$[50, 100, 200]$
Kernel function	[rbf, poly, sigm]	Tree leaf size	$[20, 30, 40]$	Max depth	$[5, 10, 15, 20]$
Kernel parameter	$[0.01, 0.1, 1]$	Minkowski power	$[1, 2, 3]$	Min sample split	$[2, 4, 5]$
				Max features	$[1, 2, 3]$
				Max leaf nodes	$[None, 4, 5]$
Nb samples	10	Nb samples	20	Nb samples	10

9.3.2 *Data Generation*

Our main goal is to compare the performances of different discriminant analysis methods. Since all these methods assume that clusters have an underlying specific statistical model, we will support such a hypothesis by generating the data with one elliptically symmetric distribution per cluster. Thus, for a given number of points n split into clusters K with priors $\pi_1, \ldots, \pi_K$, we need to construct randomly K covariance matrix $\mathbf{\Sigma}_1, \ldots, \mathbf{\Sigma}_K \in \mathbb{R}^{m \times m}$, mean vectors $\boldsymbol{\mu}_1, \ldots, \boldsymbol{\mu}_K \in \mathbb{R}^m$ and define the density generator functions $g_1, \ldots, g_K$ that will then be used to generate the points.

9.3.2.1 Mean Vector Random Generation

For each $\boldsymbol{\mu}_i$, we generate a random point on the m-sphere of radius r. Regarding the norm of the covariance matrix, we choose a radius that, on the one hand, does not make the classification too easy (clusters too far away from one another) and, on the other hand, ensures that it remains possible (clusters do not overlap too much).

9.3.2.2 Covariance Matrix Random Generation

The generation of the covariance matrix is trickier as poor choices for the eigenvalues will significantly impact the methods' performances. While a too-large eigenvalue is completely non-informative and is useless for the classification, a too-small eigenvalue will be too discriminant. To generate $\mathbf{\Sigma}_i$, we first generate $\lambda_{i1}, \ldots, \lambda_{im}$, then a random rotation matrix P_i and finally compute $\mathbf{\Sigma}_i = P_i Diag(\lambda_{i1}, \ldots, \lambda_{im}) P_i^{-1}$.

The random rotation matrix P_i is drawn using the Haar distribution [39] implemented in Scipy.

We generate the eigenvalues using a χ^2 distribution trimmed by a minimum and maximum value. The higher the degree of freedom of the χ_2 distribution is, the less informative the eigenvalues become. To get some heterogeneity in the clusters, we choose a random degree of freedom using a Poisson distribution of parameter ξ:

$$\lambda_{ij} \sim \min\left(\max\left(\chi^2\left(P(\xi)\right), \lambda_{\min}\right), \lambda_{\max}\right).$$

In practice, we choose:

- $\xi = 1$,
- $\lambda_{\min} = 1$,
- $\lambda_{\max} = 20$.

9.3.2.3 Density Generator Function Definition

To generate the data of each cluster, we choose the density generator function among two parametric families:

- Multivariate generalized Gaussian distributions [31] of shape parameter β.
- Multivariate student distributions [36] with degree of freedom ν.

9.3.3 Performances of the Estimators

We start by comparing the training time of each method. We display both the training time and the percentage of data effectively exploited during the training phase. Since the percentage of data used is gradually decreased to ensure a similar training time for all methods, a lower percentage of processed data necessarily implies a slower method (Fig. 9.1).

Among the discriminant analysis methods, QDA is the fastest. Indeed, FEMDA, RQDA, and t-QDA all involve solving iteratively a fixed-point equation for the estimation part, which takes more time. FEMDA and RQDA have similar computation times since their fixed-point equation have close expressions. However, t-QDA is significantly slower because it requires solving an optimization problem at each iteration of the fixed-point equation. Despite using only a fraction of the available training data, t-QDA is the slowest method to train. The prediction time is roughly the same for all discriminant analysis methods, as it always consists in evaluating the likelihood using the estimated parameters. For machine learning methods, it is

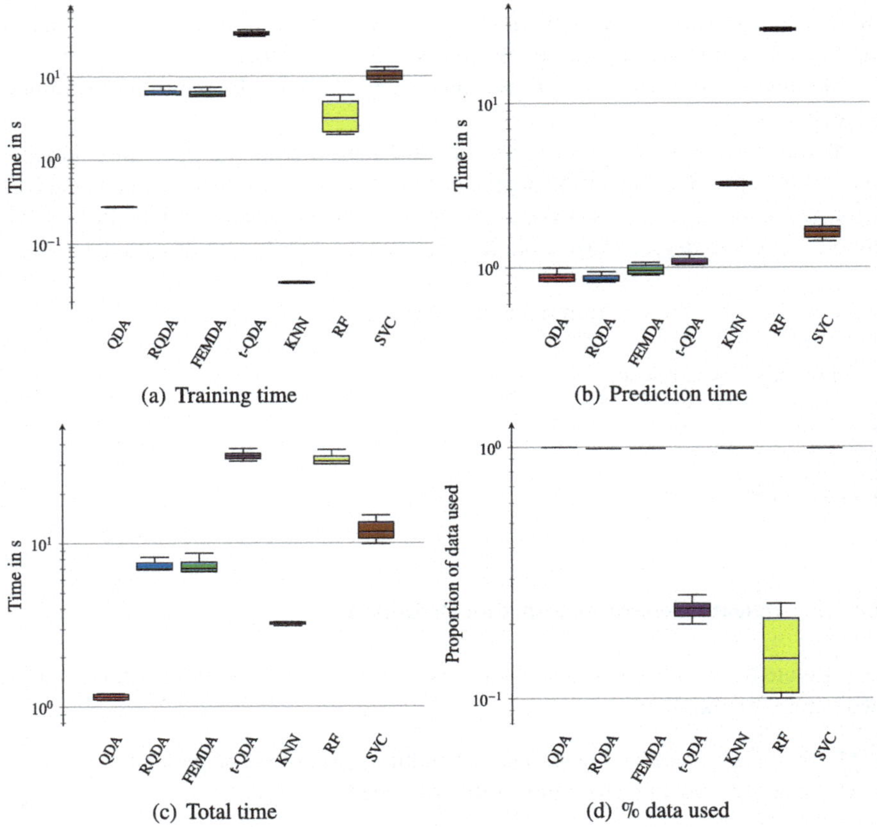

Fig. 9.1 Training and testing computation times comparison between the different algorithms, and study of the percentage of data used for training on simulated dataset

relevant to display the prediction time as some methods may be fast to train but slow to predict (such as KNN). When comparing the total time, which includes both training and prediction time, machine learning methods, on average, tend to be slower than discriminant analysis methods. Among all the methods, only QDA and KNN are faster than FEMDA. In particular, FEMDA is ten times faster than t-QDA while processing three times more data.

Since we know the mean and covariance matrix of the elliptically symmetric distributions used to generate the data, we can compare the accuracy of the different estimators. For the mean vector, we compute the Relative Mean Square Error (RMSE) between the real and the estimated parameter: $error(\boldsymbol{\mu}) = \frac{||\boldsymbol{\mu}-\hat{\boldsymbol{\mu}}||}{||\boldsymbol{\mu}||}$. This error is then averaged across the different clusters. However, such a metric is less relevant for the covariance matrix as FEMDA's fixed point equation does not even impose a specific scale on the solution. A more suitable metric for comparing the covariance matrix estimates is the Riemannian distance over the positive-definite matrix [37]. Once again, we average the Riemannian distance across the different clusters (Fig. 9.2).

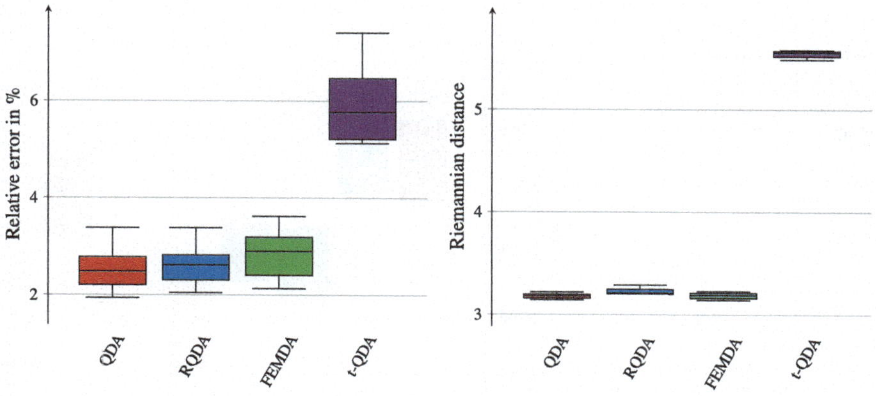

(a) RMSE on μ (b) Riemannian distance with Σ

Fig. 9.2 Estimation error comparison between the estimators of the different discriminant analysis methods

The relative error on the estimation of the mean vector is approximately the same for QDA, RQDA, and FEMDA, at around 3%. t-QDA, however, happens to be the least accurate, with a relative error exceeding 6%, which is twice as high as the other methods. Similarly, when considering the estimation of the covariance matrix, QDA, RQDA, and FEMDA obtain very similar average Riemannian distances to Σ, around 3.2. Again, t-QDA's estimator is the least accurate, with an average distance of around 5.6. Given that t-QDA already had a lower accuracy on the mean vector, its lower performance in estimating the covariance matrix is expected.

9.3.4 Performances on Clean Data

We now compare the classification performances of the different methods on test data simulated in the same way. Data generation parameters are the following:

Number of clusters	3
Dimension	10
Number of points	3000
Radius for μ generation	2
λ parameter for eigenvalues	1
Minimum eigenvalue value	1
Maximum eigenvalue value	20
Cluster 1	Prior of 0.33 and Generalized gaussian with $\beta = 0.8$
Cluster 2	Prior of 0.33 and Generalized gaussian with $\beta = 1.5$
Cluster 3	Prior of 0.34 and multivariate student with $\nu = 10$

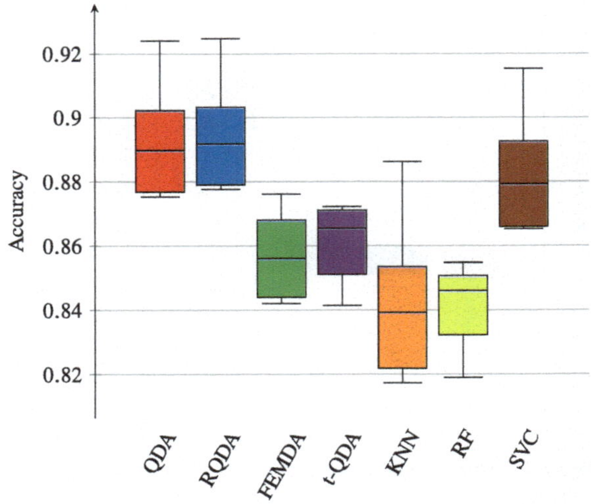

Fig. 9.3 Classification accuracy comparison between the different machine learning methods on simulated dataset

We obtain the following results (Fig. 9.3):

Overall, discriminant analysis methods outperform machine learning methods, probably due to the elliptical structure of the simulated data. As there is no noise and the distributions used for each cluster are close to Gaussian distributions, QDA and RQDA are expected to perform above other methods. FEMDA performs better than KNN and RF but is below SVC and t-QDA, although the latter two methods require a longer total time for training and prediction.

9.3.5 Performances on Noisy Data

To conclude the study on simulated data, we will now evaluate the robustness of each method to outliers. We replace a fraction of the train set by a uniform random noise. These points are randomly generated within the hypercube containing the original train set. We focus on the evolution in accuracy as the percentage of contaminated data increases (Fig. 9.4).

Intrinsically non-robust, QDA is the most affected method by the noise. Its performance decreases slightly above a random classifier at the maximum contamination rate. However, since the noise introduced is random and has no specific pattern, most methods still have sufficient data to capture each cluster's inherent structure accurately. Machine learning methods handle the contamination well, as they lose an average of 8% accuracy at the maximum contamination rate of 35%. Among the robust methods, RQDA is the most impacted, with a 15% decrease in accuracy. On the other hand, FEMDA and t-QDA show more robustness and only lose around 5% in accuracy. FEMDA is the method with the smallest performance decrease and becomes the most accurate discriminant analysis method at a 10%.

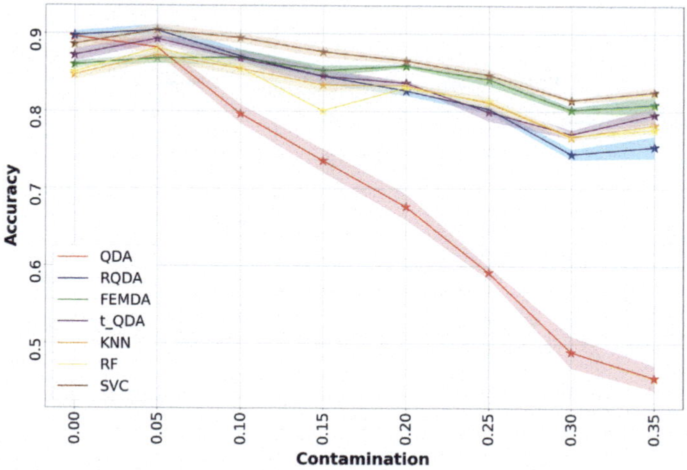

Fig. 9.4 Accuracy evolution of the different algorithms when a percentage of the simulated train set is replaced by mild Gaussian noise

9.4 Experiments on Real Data

In this section, our objective is to compare the performances of our classification methods on real datasets. Along with simulated data, we first study the accuracy of clean data before adding noise to evaluate the robustness of the methods. To optimize the performances of all methods, we perform a Principal Component Analysis (PCA) before training to reduce the input dimension. Our experiments focus on four real datasets obtained from the UCI machine learning repository [13]:

Name	Avila	Frogs species	Glass	Sonar
Number of data	20867	7195	214	208
% used in train set	70%	70%	70%	70%
Dimension	10	22	9	60
Dimension after PCA	8	14	5	16
Number of clusters	12	10	6	2
Prior	[0.41, 0.00, 0.00, 0.03, 0.1, 0.18, 0.04, 0.04, 0.08, 0.04, 0.05, 0.03]	[0.09, 0.48, 0.08, 0.04, 0.07, 0.15, 0.04, 0.02, 0.01, 0.02]	[0.32, 0.35, 0.08, 0.06, 0.05, 0.14]	[0.54, 0.46]

The computation time analysis (Fig. 9.5) is roughly the same as for simulated data. QDA is the fastest method, while t-QDA is the slowest. Again, FEMDA and RQDA exhibit similar computation times, with a slight advantage for FEMDA. When dealing with large datasets, KNN tends to be the fastest method, whereas SVC becomes slower. However, SVC proves to be the fastest method for smaller datasets. Concerning the percentage of data used, all methods, except for t-QDA and RF, use almost

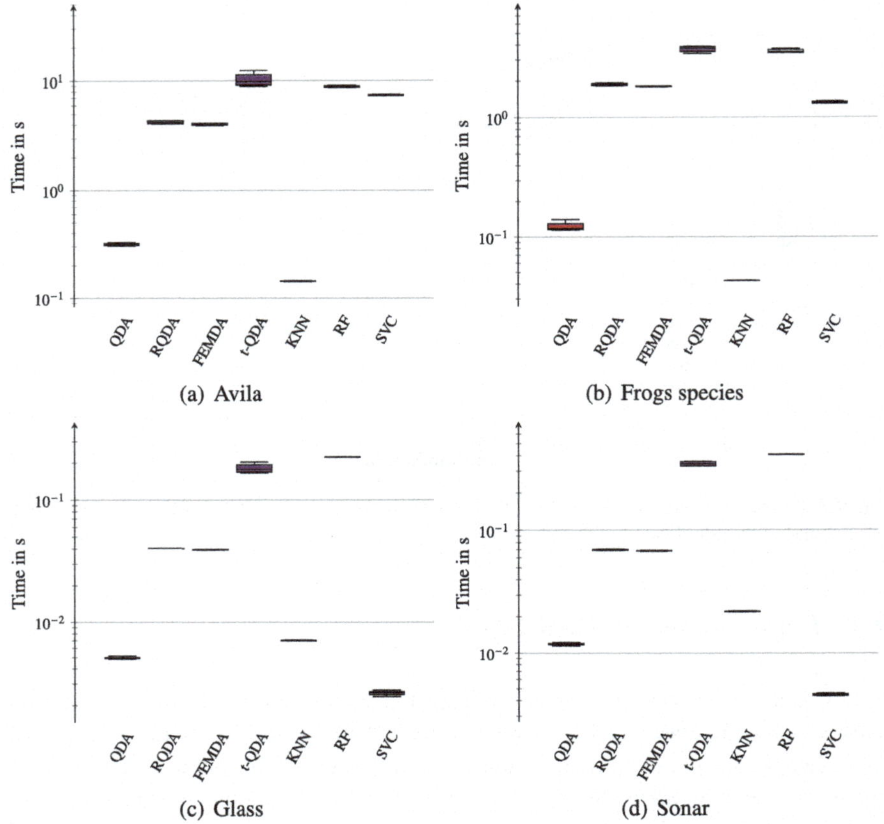

Fig. 9.5 Comparison of the time required by each method for training and testing on four real datasets

the entire train set. On large datasets, t-QDA leaves out most of the training data, respectively, more than 90% and 99% being left out for the Frog species and Avila datasets. On the contrary, RF is close to exploiting all the available data. However, the situation is reversed for small datasets. t-QDA uses a larger portion of the train set while RF's prediction time becomes too high to even consider using the train set, which is thus unused by RF (Fig. 9.6).

Despite training involving all data on Avila, QDA (Fig. 9.7) is slightly above t-QDA, indicating that this dataset's structure is far from Gaussianity. FEMDA and RQDA, however, perform well on Avila and Frogs species, their robustness allowing them to handle the non-Gaussianity better. As expected for big datasets, machine learning methods outperform discriminant analysis methods. We observe the opposite on smaller datasets: discriminant analysis methods perform better than machine learning methods, probably because the underlying statistical model compensates for the lack of data. The terrible performances of RF were expected since no training data could be used. On Glass and Sonar datasets, QDA retrieves good performances,

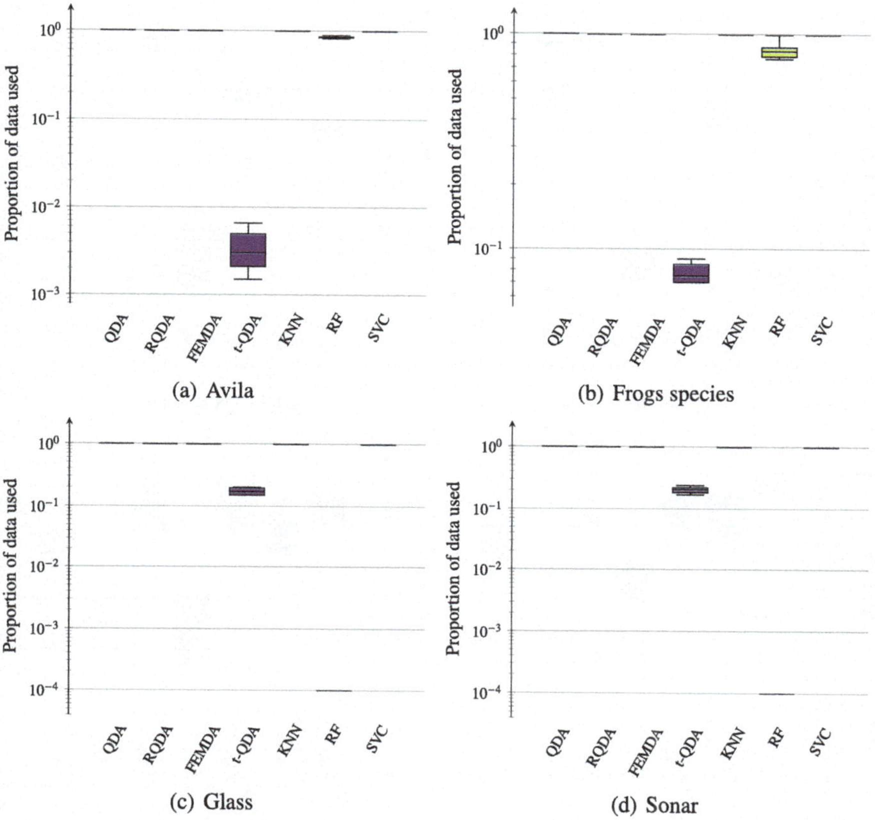

Fig. 9.6 Comparison of the percentage of data used by each method for training and testing on four real datasets

while t-QDA is still penalized by the low amount of processed data during the training part. Again, RQDA and FEMDA obtain very similar performances, FEMDA being the best method on the Sonar dataset.

On both Avila and Frogs species, t-QDA performs very poorly because of the very low amount of data processed during training. Despite training involving all data, QDA is only slightly above t-QDA, indicating that these datasets' structures are far from Gaussianity. In both cases, FEMDA is the best discriminant analysis method but gets overwhelmed by machine learning methods. On the smaller datasets, Glass and Sonar, t-QDA has access to more data for the training part, which leads to much better performances. It is consistent with the fact that it is considered the best discriminant analysis method when the training time is not restricted. However, it still remains the least efficient discriminant analysis method in this framework. As expected with the inexistent training part, RF has a terrible performance on Glass. The two other machine learning methods perform well but still get outperformed by QDA, RQDA, and FEMDA on low sample sizes.

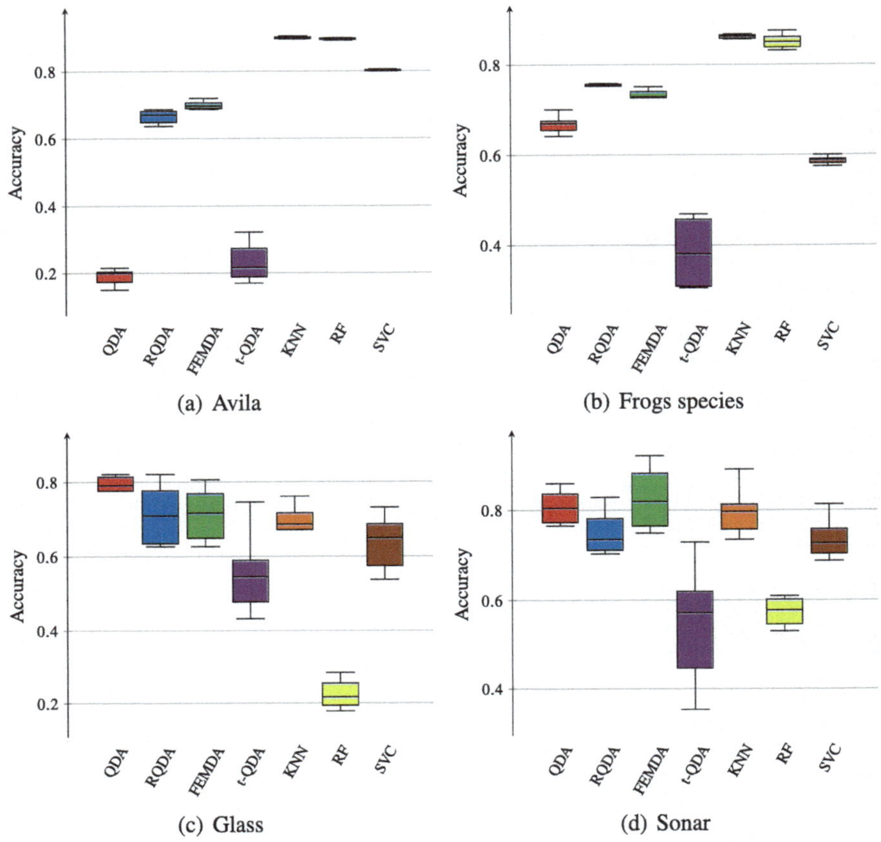

Fig. 9.7 Classification accuracy comparison between the different machine learning methods on four real datasets

Finally, to conclude the experiments, we contaminate the training data with random noise like for simulated data (Fig. 9.8).

Similarly to simulated data on Avila and Frogs species, QDA is highly impacted by noise and quickly reaches a performance comparable to a random classifier. While having similar performances to FEMDA on clean datasets, RQDA appears to be significantly more impacted by contamination. On the Avila dataset, it almost reaches the performances of the non-robust QDA method. Although this unstructured noise does not significantly impact machine learning methods, FEMDA is again the method that handles the noise best, with the smallest decrease in performance. On smaller datasets, Glass and Sonar, QDA shows a better resilience to noise. Most methods aren't very affected by noise, and FEMDA emerges as the top-performing method when data is contaminated. While the authors [18] showed that t-QDA and FEMDA achieved similar performances and robustness on both simulated and real datasets, t-QDA's efficiency drops quickly when the computational time is limited. To conclude, FEMDA proves to be a fast and robust method. FEMDA offers the best tradeoff

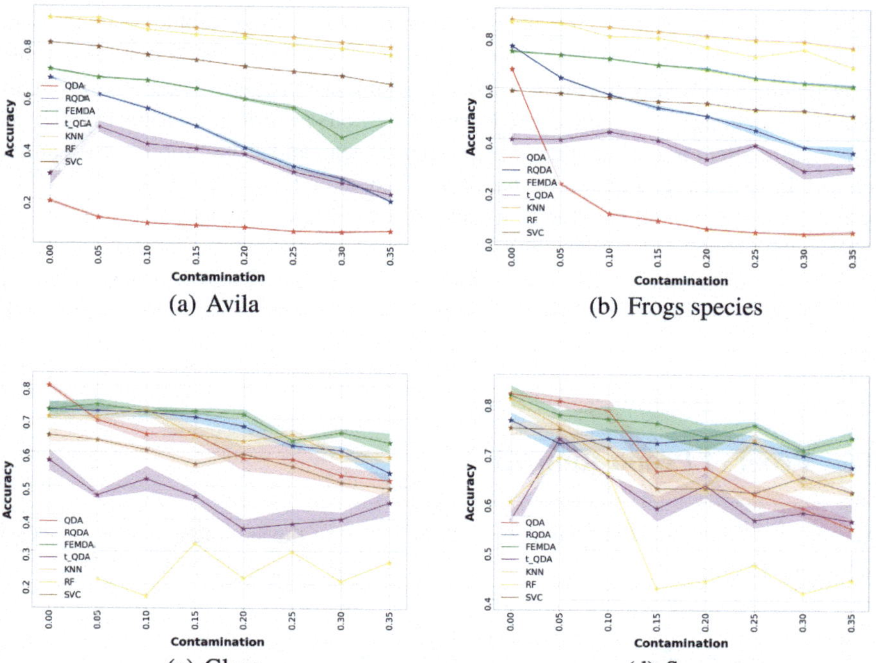

(a) Avila (b) Frogs species

(c) Glass (d) Sonar

Fig. 9.8 Accuracy evolution of the different algorithms when a percentage of the real train set is replaced by mild Gaussian noise

between computation time, robustness, and accuracy on real datasets and a real-time application framework. It can even compete with machine learning models when the dataset is small compared to the dimension.

9.5 Conclusion and Perspectives

This chapter aims to provide readers with a tutorial presentation of discriminant analysis. In the second part, we introduce a general Bayesian framework that models the data of each cluster with an elliptically symmetric distribution. In this framework, each data point is allowed to have its own elliptically symmetric distribution with its own scale on the covariance matrix and a prior on the scale. The cluster's membership only imposes the mean vector and the scatter matrix. We demonstrate that we can retrieve the usual discriminant analysis methods by choosing the appropriate prior and density generator function. This chapter also provides a tractable way to deal with the general case using Jeffrey's prior, leading to the FEMDA algorithm. The third part is devoted to experiments. First, we conduct experiments on synthetic data, allowing us to evaluate various estimators' performances. Then, we focus on the classification performance for both simulated and real datasets. We gradually contaminate the data with uniform random noise to evaluate the robustness of each method. Experiments show that FEMDA outperforms the other discriminant analysis methods in most cases and can even compete with machine learning methods in some scenarios. FEMDA is also very resilient to outliers thanks to its robust mean and scatter matrix estimators, making it particularly suitable for handling noisy data. Faster than t-QDA, FEMDA is the statistical method with the best tradeoff between performance, robustness, and computation time.

In high-dimensional regimes, Mahalanobis distance usually grows toward infinity, and t-QDA estimators and decision rules become very close to FEMDA. Future work aims to quantify the difference between these two methods in such a regime.

9.6 Proof of Proposition 9.1

According to the definition, the marginal likelihood of the sample $\mathbf{x} = (\mathbf{x}_1, \ldots, \mathbf{x}_{n_k})$ belonging to cluster C_k is

$$
\begin{aligned}
\mathcal{L}(\mathbf{x}; \boldsymbol{\mu}_k, \boldsymbol{\Sigma}_k) &= \int\limits_{]0,+\infty[^{n_k}} \prod_{i=1}^{n_k} p_{ik}\left(\mathbf{x}_i, t; \boldsymbol{\mu}_k, \boldsymbol{\Sigma}_k\right) dt. \\
&= \int\limits_{]0,+\infty[^{n_k}} \prod_{i=1}^{n_k} \frac{\Gamma(\frac{m}{2}) h_{i,k}(t)}{\pi^{\frac{m}{2}} \int_0^{+\infty} u^{\frac{m}{2}-1} g_{i,k}(u) du} |\boldsymbol{\Sigma}_k|^{-\frac{1}{2}} t^{-\frac{m}{2}} \\
&\quad g_{i,k}\left(\frac{(\mathbf{x}_i - \boldsymbol{\mu}_k)^T \boldsymbol{\Sigma}_k^{-1} (\mathbf{x}_i - \boldsymbol{\mu}_k)}{t}\right) dt.
\end{aligned}
$$

Since we suppose that the observations are independent of each other, we have

$$
\begin{aligned}
\mathcal{L}(\mathbf{x}; \boldsymbol{\mu}_k, \boldsymbol{\Sigma}_k) &= \prod_{i=1}^{n_k} \frac{\Gamma(\frac{m}{2}) |\boldsymbol{\Sigma}_k|^{-\frac{1}{2}}}{\pi^{\frac{m}{2}} \int_0^{+\infty} u^{\frac{m}{2}-1} g_{i,k}(u) du} \int\limits_0^{+\infty} h_{i,k}(t) t^{-\frac{m}{2}} \\
&\quad g_{i,k}\left(\frac{(\mathbf{x}_i - \boldsymbol{\mu}_k)^T \boldsymbol{\Sigma}_k^{-1} (\mathbf{x}_i - \boldsymbol{\mu}_k)}{t}\right) dt \\
&= \prod_{i=1}^{n_k} \frac{\Gamma(\frac{m}{2}) |\boldsymbol{\Sigma}_k|^{-\frac{1}{2}}}{\pi^{\frac{m}{2}} \int_0^{+\infty} u^{\frac{m}{2}-1} g_{i,k}(u) du} \\
&\quad \int\limits_0^{\infty} h_{i,k}(t) t^{-\frac{m}{2}} g_{i,k}\left(\frac{(\mathbf{x}_i - \boldsymbol{\mu}_k)^T \boldsymbol{\Sigma}_k^{-1} (\mathbf{x}_i - \boldsymbol{\mu}_k)}{t}\right) dt.
\end{aligned}
$$

We then define $d_{i,k} = (\mathbf{x}_i - \boldsymbol{\mu}_k)^T \boldsymbol{\Sigma}_k^{-1}(\mathbf{x}_i - \boldsymbol{\mu}_k)$:

$$\mathcal{L}(\mathbf{x}; \boldsymbol{\mu}_k, \boldsymbol{\Sigma}_k) = \prod_{i=1}^{n_k} \frac{\Gamma(\frac{m}{2})|\boldsymbol{\Sigma}_k|^{-\frac{1}{2}} d_{i,k}^{-\frac{m}{2}}}{\pi^{\frac{m}{2}} \int_0^{+\infty} u^{\frac{m}{2}-1} g_{i,k}(u)du} \int_0^{\infty} h_{i,k}\left(d_{i,k}\frac{t}{d_{i,k}}\right)\left(\frac{d_{i,k}}{t}\right)^{\frac{m}{2}} g_{i,k}\left(\frac{d_{i,k}}{t}\right)dt$$

$$= \prod_{i=1}^{n_k} \frac{\Gamma(\frac{m}{2})|\boldsymbol{\Sigma}_k|^{-\frac{1}{2}} d_{i,k}^{1-\frac{m}{2}}}{\pi^{\frac{m}{2}} \int_0^{+\infty} u^{\frac{m}{2}-1} g_{i,k}(u)du} \int_0^{\infty} \frac{d_{i,k}}{t^2} h_{i,k}\left(d_{i,k}\frac{t}{d_{i,k}}\right)\left(\frac{d_{i,k}}{t}\right)^{\frac{m}{2}-2} g_{i,k}\left(\frac{d_{i,k}}{t}\right)dt$$

$$= \prod_{i=1}^{n_k} \frac{\Gamma(\frac{m}{2})|\boldsymbol{\Sigma}_k|^{-\frac{1}{2}} d_{i,k}^{1-\frac{m}{2}}}{\pi^{\frac{m}{2}} \int_0^{+\infty} u^{\frac{m}{2}-1} g_{i,k}(u)du} \int_0^{\infty} h_{i,k}\left(\frac{d_{i,k}}{t}\right) t^{\frac{m}{2}-2} g_{i,k}(t)dt.$$

9.7 Proof of Theorem 9.1

We derive the estimators of location and scale by voiding the differential of the log-marginal likelihood l by the parameter of interest. Assuming that $h_{i,k}$ is differentiable and that the integrals converge, one can write

$$\frac{\partial l}{\partial \boldsymbol{\mu}_k} = \sum_{i=1}^{n_k} \left(1 - \frac{m}{2}\right) \frac{2\boldsymbol{\Sigma}_k^{-1}(\mathbf{x}_i - \boldsymbol{\mu}_k)}{(\mathbf{x}_i - \boldsymbol{\mu}_k)^T \boldsymbol{\Sigma}_k^{-1}(\mathbf{x}_i - \boldsymbol{\mu}_k)} + \frac{\int_0^{\infty} \frac{2\boldsymbol{\Sigma}_k^{-1}(\mathbf{x}_i - \boldsymbol{\mu}_k)}{t} h'_{i,k}\left(\frac{(\mathbf{x}_i - \boldsymbol{\mu}_k)^T \boldsymbol{\Sigma}_k^{-1}(\mathbf{x}_i - \boldsymbol{\mu}_k)}{t}\right) t^{\frac{m}{2}-2} g_{i,k}(t)dt}{\int_0^{\infty} h_{i,k}\left(\frac{(\mathbf{x}_i - \boldsymbol{\mu}_k)^T \boldsymbol{\Sigma}_k^{-1}(\mathbf{x}_i - \boldsymbol{\mu}_k)}{t}\right) t^{\frac{m}{2}-2} g_{i,k}(t)dt}$$

$$= \boldsymbol{\Sigma}_k^{-1} \sum_{i=1}^{n_k} (\mathbf{x}_i - \boldsymbol{\mu}_k) \left[\frac{2 - m}{(\mathbf{x}_i - \boldsymbol{\mu}_k)^T \boldsymbol{\Sigma}_k^{-1}(\mathbf{x}_i - \boldsymbol{\mu}_k)} + \frac{2\int_0^{\infty} h'_{i,k}\left(\frac{(\mathbf{x}_i - \boldsymbol{\mu}_k)^T \boldsymbol{\Sigma}_k^{-1}(\mathbf{x}_i - \boldsymbol{\mu}_k)}{t}\right) t^{\frac{m}{2}-3} g_{i,k}(t)dt}{\int_0^{\infty} h_{i,k}\left(\frac{(\mathbf{x}_i - \boldsymbol{\mu}_k)^T \boldsymbol{\Sigma}_k^{-1}(\mathbf{x}_i - \boldsymbol{\mu}_k)}{t}\right) t^{\frac{m}{2}-2} g_{i,k}(t)dt} \right].$$

Thus,

$$\frac{\partial l}{\partial \boldsymbol{\mu}_k}(\hat{\boldsymbol{\mu}}_k) = 0 \iff \hat{\boldsymbol{\mu}}_k = \frac{\sum_{i=1}^{n_k} w_{i,k} \mathbf{x}_i}{\sum_{i=1}^{n_k} w_{i,k}}, \text{ with}$$

$$w_{i,k} = \frac{1}{(\mathbf{x}_i - \hat{\boldsymbol{\mu}}_k)^T \hat{\boldsymbol{\Sigma}}_k^{-1} (\mathbf{x}_i - \hat{\boldsymbol{\mu}}_k)} - \frac{2}{m-2} \frac{\int_0^\infty h'_{i,k}\left(\frac{(\mathbf{x}_i-\hat{\boldsymbol{\mu}}_k)^T \hat{\boldsymbol{\Sigma}}_k^{-1}(\mathbf{x}_i-\hat{\boldsymbol{\mu}}_k)}{t}\right) t^{\frac{m}{2}-3} g_{i,k}(t)dt}{\int_0^\infty h_{i,k}\left(\frac{(\mathbf{x}_i-\hat{\boldsymbol{\mu}}_k)^T \hat{\boldsymbol{\Sigma}}_k^{-1}(\mathbf{x}_i-\hat{\boldsymbol{\mu}}_k)}{t}\right) t^{\frac{m}{2}-2} g_{i,k}(t)dt}$$

$$\begin{aligned}\frac{\partial l}{\partial \boldsymbol{\Sigma}_k^{-1}} =& \frac{n_k}{2}\hat{\boldsymbol{\Sigma}}_k + \sum_{i=1}^{n_k} \frac{1-\frac{m}{2}}{(\mathbf{x}_i - \hat{\boldsymbol{\mu}}_k)^T \hat{\boldsymbol{\Sigma}}_k^{-1} (\mathbf{x}_i - \hat{\boldsymbol{\mu}}_k)} (\mathbf{x}_i - \hat{\boldsymbol{\mu}}_k)^T (\mathbf{x}_i - \hat{\boldsymbol{\mu}}_k)^T \\ &+ (\mathbf{x}_i - \hat{\boldsymbol{\mu}}_k)(\mathbf{x}_i - \hat{\boldsymbol{\mu}}_k)^T \frac{\int_0^\infty h'_{i,k}\left(\frac{(\mathbf{x}_i-\hat{\boldsymbol{\mu}}_k)^T \hat{\boldsymbol{\Sigma}}_k^{-1}(\mathbf{x}_i-\hat{\boldsymbol{\mu}}_k)}{t}\right) t^{\frac{m}{2}-3} g_{i,k}(t)dt}{\int_0^\infty h_{i,k}\left(\frac{(\mathbf{x}_i-\hat{\boldsymbol{\mu}}_k)^T \hat{\boldsymbol{\Sigma}}_k^{-1}(\mathbf{x}_i-\hat{\boldsymbol{\mu}}_k)}{t}\right) t^{\frac{m}{2}-2} g_{i,k}(t)dt} \\ =& \frac{n_k}{2}\left(\hat{\boldsymbol{\Sigma}}_k - \frac{m-2}{n_k}\sum_{i=1}^{n_k} w_{ik}(\mathbf{x}_i - \hat{\boldsymbol{\mu}}_k)(\mathbf{x}_i - \hat{\boldsymbol{\mu}}_k)^T\right).\end{aligned}$$

Similarly,

$$\frac{\partial l}{\partial \boldsymbol{\Sigma}_k}(\hat{\boldsymbol{\Sigma}}_k^{-1}) = 0 \iff \hat{\boldsymbol{\Sigma}}_k = \frac{m-2}{n_k}\sum_{i=1}^{n_k} w_{i,k}(\mathbf{x}_i - \hat{\boldsymbol{\mu}}_k)(\mathbf{x}_i - \hat{\boldsymbol{\mu}}_k)^T.$$

MMLE provides two coupled fixed-point equations that we solve iteratively, using the previous estimations $\hat{\boldsymbol{\mu}}_k^{(t)}$ and $\hat{\boldsymbol{\Sigma}}_k^{(t)}$ to estimate the weights at iteration $t+1$. Following Neyman–Pearson's lemma, the decision rule consists of choosing the cluster that maximizes the marginal likelihood computed using previously estimated parameters. The result is retrieved immediately after removing the multiplicative constants independent of k in the likelihood obtained by Proposition 9.1.

9.8 Proof of Theorem 9.2

This theorem is simply the particular case of Theorem 9.1 with $h_{i,k}(t) \propto \frac{1}{t}$ and $h'_{i,k}(t) = -\frac{C}{t^2}$. Let us compute the new expression of $w_{i,k}$:

$$w_{i,k} = \frac{1}{(\mathbf{x}_i - \hat{\boldsymbol{\mu}}_k)^T \hat{\boldsymbol{\Sigma}}_k^{-1} (\mathbf{x}_i - \hat{\boldsymbol{\mu}}_k)} - \frac{2}{m-2} \frac{\int_0^{+\infty} h'_{i,k}\left(\frac{(\mathbf{x}_i-\hat{\boldsymbol{\mu}}_k)^T \hat{\boldsymbol{\Sigma}}_k^{-1}(\mathbf{x}_i-\hat{\boldsymbol{\mu}}_k)}{t}\right) t^{\frac{m}{2}-3} g_{i,k}(t)dt}{\int_0^{+\infty} h_{i,k}\left(\frac{(\mathbf{x}_i-\hat{\boldsymbol{\mu}}_k)^T \hat{\boldsymbol{\Sigma}}_k^{-1}(\mathbf{x}_i-\hat{\boldsymbol{\mu}}_k)}{t}\right) t^{\frac{m}{2}-2} g_{i,k}(t)dt}$$

$$
\begin{aligned}
&= \frac{1}{(\mathbf{x}_i - \hat{\boldsymbol{\mu}}_k)^T \hat{\boldsymbol{\Sigma}}_k^{-1} (\mathbf{x}_i - \hat{\boldsymbol{\mu}}_k)} \\
&\quad + \frac{2}{m-2} \frac{\int_0^{+\infty} C \left(\frac{t}{(\mathbf{x}_i - \hat{\boldsymbol{\mu}}_k)^T \hat{\boldsymbol{\Sigma}}_k^{-1} (\mathbf{x}_i - \hat{\boldsymbol{\mu}}_k)} \right)^2 t^{\frac{m}{2}-3} g_{i,k}(t) dt}{\int_0^{+\infty} C \frac{t}{(\mathbf{x}_i - \hat{\boldsymbol{\mu}}_k)^T \hat{\boldsymbol{\Sigma}}_k^{-1} (\mathbf{x}_i - \hat{\boldsymbol{\mu}}_k)} t^{\frac{m}{2}-2} g_{i,k}(t) dt} \\
&= \frac{1}{(\mathbf{x}_i - \hat{\boldsymbol{\mu}}_k)^T \hat{\boldsymbol{\Sigma}}_k^{-1} (\mathbf{x}_i - \hat{\boldsymbol{\mu}}_k)} + \frac{2}{(m-2)(\mathbf{x}_i - \hat{\boldsymbol{\mu}}_k)^T \hat{\boldsymbol{\Sigma}}_k^{-1} (\mathbf{x}_i - \hat{\boldsymbol{\mu}}_k)} \\
&= \frac{m}{m-2} \frac{1}{(\mathbf{x}_i - \hat{\boldsymbol{\mu}}_k)^T \hat{\boldsymbol{\Sigma}}_k^{-1} (\mathbf{x}_i - \hat{\boldsymbol{\mu}}_k)}.
\end{aligned}
$$

This allows us to retrieve the new expression of $\hat{\boldsymbol{\mu}}_k$ and $\hat{\boldsymbol{\Sigma}}_k$. Concerning the decision rule, we have

$$
\begin{aligned}
&\frac{|\hat{\boldsymbol{\Sigma}}_k|^{-\frac{1}{2}} \left((\mathbf{x}_i - \hat{\boldsymbol{\mu}}_k)^T \hat{\boldsymbol{\Sigma}}_k^{-1} (\mathbf{x}_i - \hat{\boldsymbol{\mu}}_k) \right)^{1-\frac{m}{2}}}{\int_0^{+\infty} t^{\frac{m}{2}-1} g_{i,k}(t) dt} \\
&\int_0^{+\infty} h_{i,k} \left(\frac{(\mathbf{x}_i - \hat{\boldsymbol{\mu}}_k)^T \hat{\boldsymbol{\Sigma}}_k^{-1} (\mathbf{x}_i - \hat{\boldsymbol{\mu}}_k)}{t} \right) t^{\frac{m}{2}-2} g_{i,k}(t) dt \\
&= \frac{C |\hat{\boldsymbol{\Sigma}}_k|^{-\frac{1}{2}} \left((\mathbf{x}_i - \hat{\boldsymbol{\mu}}_k)^T \hat{\boldsymbol{\Sigma}}_k^{-1} (\mathbf{x}_i - \hat{\boldsymbol{\mu}}_k) \right)^{1-\frac{m}{2}}}{\int_0^{+\infty} t^{\frac{m}{2}-1} g_{i,k}(t) dt} \\
&\int_0^{+\infty} \frac{t}{(\mathbf{x}_i - \hat{\boldsymbol{\mu}}_k)^T \hat{\boldsymbol{\Sigma}}_k^{-1} (\mathbf{x}_i - \hat{\boldsymbol{\mu}}_k)} t^{\frac{m}{2}-2} g_{i,k}(t) dt \\
&= C |\hat{\boldsymbol{\Sigma}}_k|^{-\frac{1}{2}} \left((\mathbf{x}_i - \hat{\boldsymbol{\mu}}_k)^T \hat{\boldsymbol{\Sigma}}_k^{-1} (\mathbf{x}_i - \hat{\boldsymbol{\mu}}_k) \right)^{\frac{m}{2}}.
\end{aligned}
$$

Removing the constant C that does not depend on k and taking the logarithm allows us to retrieve the new expression of the decision rule.

9.9 Retrieval of t-QDA

We will apply Theorem 9.1 and Remark 9.2 to retrieve the t-QDA method. First, we will derive the expression of the weight $w_{i,k}$. One has

- $h_k(t) = \frac{(\frac{\nu_k}{2})^{\frac{\nu_k}{2}} e^{-\frac{\nu_k}{2t}}}{\Gamma(\frac{\nu_k}{2}) t^{1+\frac{\nu_k}{2}}}$,
- $h'_k(t) = \frac{(\frac{\nu_k}{2})^{\frac{\nu_k}{2}} e^{-\frac{\nu_k}{2t}}}{2\Gamma(\frac{\nu_k}{2}) t^{3+\frac{\nu_k}{2}}} (\nu_k - (\nu_k + 2)\, t)$.

We define $d_{i,k} = (\mathbf{x}_i - \hat{\boldsymbol{\mu}}_k)^T \hat{\boldsymbol{\Sigma}}_k^{-1} (\mathbf{x}_i - \hat{\boldsymbol{\mu}}_k)$.

- $h_k\left(\frac{d_{i,k}}{t}\right) = \frac{(\frac{\nu_k}{2})^{\frac{\nu_k}{2}} e^{-\frac{\nu_k t}{2d_{i,k}}}}{\Gamma(\frac{\nu_k}{2}) d_{i,k}^{1+\frac{\nu_k}{2}}} t^{1+\frac{\nu_k}{2}}$,
- $h_k'\left(\frac{d_{i,k}}{t}\right) = \frac{(\frac{\nu_k}{2})^{\frac{\nu_k}{2}} e^{-\frac{\nu_k t}{2d_{i,k}}}}{2\Gamma(\frac{\nu_k}{2}) d_{i,k}^{2+\frac{\nu_k}{2}}} t^{2+\frac{\nu_k}{2}} \left(\frac{\nu_k}{d_{i,k}} t - (\nu_k + 2)\right)$.

We now focus on computing the ratio I of the two integrals in the expression of $w_{i,k}$:

$$
\begin{aligned}
I &= \frac{\int_0^\infty \frac{(\frac{\nu_k}{2})^{\frac{\nu_k}{2}} e^{-\frac{\nu_k t}{2d_{i,k}}}}{2\Gamma(\frac{\nu_k}{2}) d_{i,k}^{2+\frac{\nu_k}{2}}} t^{2+\frac{\nu_k}{2}} \left(\frac{\nu_k}{d_{i,k}} t - (\nu_k + 2)\right) t^{\frac{m}{2}-3} e^{-\frac{1}{2}t} dt}{\int_0^\infty \frac{(\frac{\nu_k}{2})^{\frac{\nu_k}{2}} e^{-\frac{\nu_k t}{2d_{i,k}}}}{\Gamma(\frac{\nu_k}{2}) d_{i,k}^{1+\frac{\nu_k}{2}}} t^{1+\frac{\nu_k}{2}} t^{\frac{m}{2}-2} e^{-\frac{1}{2}t} dt} \\
&= \frac{1}{2d_{i,k}} \frac{\int_0^\infty e^{-\frac{\nu_k + d_{i,k}}{2d_{i,k}} t} t^{\frac{\nu_k+m}{2}-1} \left(\frac{\nu_k}{d_{i,k}} t - (\nu_k + 2)\right) dt}{\int_0^\infty e^{-\frac{\nu_k + d_{i,k}}{2d_{i,k}} t} t^{\frac{\nu_k+m}{2}-1} dt} \\
&= \frac{1}{2d_{i,k}} \left(-(\nu_k + 2) + \frac{\nu_k}{d_{i,k}} \frac{\int_0^\infty e^{-\frac{\nu_k + d_{i,k}}{2d_{i,k}} t} t^{\frac{\nu_k+m}{2}} dt}{\int_0^\infty e^{-\frac{\nu_k + d_{i,k}}{2d_{i,k}} t} t^{\frac{\nu_k+m}{2}-1} dt} \right) \\
&= \frac{1}{2d_{i,k}} \left(-(\nu_k + 2) + \nu_k \frac{\nu_k + m}{\nu_k + d_{i,k}} \right).
\end{aligned}
$$

Now we can obtain the expression of $w_{i,k}$ as follows:

$$
\begin{aligned}
w_{i,k} &= \frac{1}{d_{i,k}} - \frac{2}{m-2} I \\
&= \frac{1}{d_{i,k}(m-2)} \left(m - 2 + \nu_k + 2 - \nu_k \frac{\nu_k + m}{\nu_k + d_{i,k}} \right) \\
&= \frac{\nu_k + m}{d_{i,k}(m-2)} \left(1 - \frac{\nu_k}{\nu_k + d_{i,k}} \right) \\
&= \frac{\nu_k + m}{(\nu_k + d_{i,k})(m-2)}.
\end{aligned}
$$

Now, we are going to use Remark 9.2 to obtain an estimator for ν_k:

$$\begin{aligned}\hat{\nu}_k &= \arg\max_{\nu\in\mathbb{R}^{+*}} \sum_{i=1}^{n_k} \log\left(\int_0^\infty \frac{\left(\frac{\nu}{2}\right)^{\frac{\nu}{2}} e^{-\frac{\nu t}{2d_{i,k}}}}{\Gamma\left(\frac{\nu}{2}\right) d_{i,k}^{1+\frac{\nu}{2}}} t^{1+\frac{\nu}{2}} t^{\frac{m}{2}-2} e^{-\frac{t}{2}} dt\right)\\
&= \arg\max_{\nu\in\mathbb{R}^{+*}} \sum_{i=1}^{n_k} \log\left(\frac{1}{\Gamma\left(\frac{\nu}{2}\right)}\left(\frac{\nu}{2d_{i,k}}\right)^{\frac{\nu}{2}} \int_0^\infty e^{-\frac{\nu+d_{i,k}}{2d_{i,k}}t} t^{\frac{\nu+m}{2}-1} dt\right)\\
&= \arg\max_{\nu\in\mathbb{R}^{+*}} \sum_{i=1}^{n_k} \log\left(\frac{(2d_{i,k})^{\frac{m}{2}} \nu^{\frac{\nu}{2}}}{\Gamma\left(\frac{\nu}{2}\right)(\nu+d_{i,k})^{\frac{\nu+m}{2}}}\right.\\
&\qquad \left.\int_0^\infty \frac{\nu+d_{i,k}}{2d_{i,k}} e^{-\frac{\nu+d_{i,k}}{2d_{i,k}}t}\left(\frac{\nu+d_{i,k}}{2d_{i,k}}t\right)^{\frac{\nu+m}{2}-1} dt\right)\\
&= \arg\min_{\nu\in\mathbb{R}^{+*}} \sum_{i=1}^{n_k} \log\left(\frac{\Gamma\left(\frac{\nu}{2}\right)\nu^{\frac{m}{2}}\left(1+\frac{d_{i,k}}{\nu}\right)^{\frac{\nu+m}{2}}}{\Gamma\left(\frac{\nu+m}{2}\right)}\right)\\
&= \arg\min_{\nu\in\mathbb{R}^{+}} \frac{m}{2}\log(\nu) + \log\Gamma\left(\frac{\nu}{2}\right) - \log\Gamma\left(\frac{\nu+m}{2}\right) + \frac{\nu+m}{2n_k}\sum_{i=1}^{n_k}\log\left(1+\frac{d_{i,k}}{\nu}\right).\end{aligned}$$

Finally, we retrieve the t-QDA decision rule using Theorem 3:

$$\begin{aligned}k &= \arg\max_{k\in[1,K]} \frac{|\hat{\boldsymbol{\Sigma}}_k|^{-\frac{1}{2}} d_{i,k}^{1-\frac{m}{2}}}{\int_0^{+\infty} t^{\frac{m}{2}-1} e^{-\frac{t}{2}} dt} \int_0^\infty \frac{\left(\frac{\hat{\nu}_k}{2}\right)^{\frac{\hat{\nu}_k}{2}} e^{-\frac{\hat{\nu}_k t}{2d_{i,k}}}}{\Gamma\left(\frac{\hat{\nu}_k}{2}\right) d_{i,k}^{1+\frac{\hat{\nu}_k}{2}}} t^{1+\frac{\hat{\nu}_k}{2}} t^{\frac{m}{2}-2} e^{-\frac{t}{2}} dt\\
&= \arg\max_{k\in[1,K]} \frac{|\hat{\boldsymbol{\Sigma}}_k|^{-\frac{1}{2}}\left(\frac{\hat{\nu}_k}{2}\right)^{\frac{\hat{\nu}_k}{2}} d_{i,k}^{-\frac{\hat{\nu}_k+m}{2}}}{\Gamma\left(\frac{\hat{\nu}_k}{2}\right)} \int_0^\infty e^{-\frac{\hat{\nu}_k+m}{2d_{i,k}}t} t^{\frac{\hat{\nu}_k+m}{2}-1} dt\\
&= \arg\max_{k\in[1,K]} \frac{|\hat{\boldsymbol{\Sigma}}_k|^{-\frac{1}{2}}\left(\frac{\hat{\nu}_k}{2}\right)^{\frac{\hat{\nu}_k}{2}} d_{i,k}^{-\frac{\hat{\nu}_k+m}{2}}}{\Gamma\left(\frac{\hat{\nu}_k}{2}\right)} \left(\frac{2d_{i,k}}{\hat{\nu}_k+d_{i,k}}\right)^{\frac{\hat{\nu}_k+m}{2}} \Gamma\left(\frac{\hat{\nu}_k+m}{2}\right)\\
&= \arg\max_{k\in[1,K]} \frac{\Gamma\left(\frac{\hat{\nu}_k+m}{2}\right)\hat{\nu}_k^{-\frac{m}{2}}}{\Gamma\left(\frac{\hat{\nu}_k}{2}\right)} |\boldsymbol{\Sigma}_k|^{-\frac{1}{2}}\left(1+\frac{d_{i,k}}{\hat{\nu}_k}\right)^{-\frac{\hat{\nu}_k+m}{2}}\\
&= \arg\min_{k\in[1,K]} \left(1+\frac{\hat{\nu}_k}{m}\right)\log\left(1+\frac{(\mathbf{x}_i-\hat{\boldsymbol{\mu}}_k)^T\hat{\boldsymbol{\Sigma}}_k^{-1}(\mathbf{x}_i-\hat{\boldsymbol{\mu}}_k)}{\hat{\nu}_k}\right)\\
&\qquad + \log(\hat{\nu}_k) + \frac{1}{m}\log\left(|\hat{\boldsymbol{\Sigma}}_k|\right) + \frac{2}{m}\log\left(\frac{\Gamma\left(\frac{\hat{\nu}_k}{2}\right)}{\Gamma\left(\frac{\hat{\nu}_k+m}{2}\right)}\right).\end{aligned}$$

References

1. Andrews, J.L., McNicholas, P.D., Subedi, S.: Model-based classification via mixtures of multivariate t-distributions. Comput. Stat. Data Anal. **55**(1), 520–529 (2011). https://doi.org/10.1016/j.csda.2010.05.019. www.sciencedirect.com/science/article/pii/S0167947310002203
2. Besson, O., Abramovich, Y.I.: On the fisher information matrix for multivariate elliptically contoured distributions. IEEE Signal Process. Lett. **20**(11), 1130–1133 (2013)
3. Boente, G., Salibián Barrera, M., Tyler, D.E.: A characterization of elliptical distributions and some optimality properties of principal components for functional data. J. Multivar. Anal. **131**, 254–264 (2014). https://doi.org/10.1016/j.jmva.2014.07.006. https://www.sciencedirect.com/science/article/pii/S0047259X14001638
4. Bose, S., Pal, A., SahaRay, R., Nayak, J.: Generalized quadratic discriminant analysis. Pattern Recogn. **48**(8), 2676–2684 (2015). https://doi.org/10.1016/j.patcog.2015.02.016. www.sciencedirect.com/science/article/pii/S003132031500076X
5. Brereton, R.G., Lloyd, G.R.: Support vector machines for classification and regression. Analyst **135**, 230–267 (2010). https://doi.org/10.1039/B918972F. https://dx.doi.org/10.1039/B918972F
6. Burr, T., Doak, J.: Distribution-free discriminant analysis. Intell. Data Anal. **11** (1999). https://doi.org/10.3233/IDA-2007-11605
7. Chork, C., Rousseeuw, P.: Integrating a high-breakdown option into discriminant analysis in exploration geochemistry. J. Geochem. Explor. **43**(3), 191–203 (1992). https://doi.org/10.1016/0375-6742(92)90105-H. www.sciencedirect.com/science/article/pii/037567429290105H
8. Clarke, W.R., Lachenbruch, P.A., Broffitt, B.: How non-normality affects the quadratic discriminant function. Commun. Stat.-Theory Methods **8**(13), 1285–1301 (1979). https://doi.org/10.1080/03610927908827830
9. Cover, T., Hart, P.: Nearest neighbor pattern classification. IEEE Trans. Inf. Theory **13**(1), 21–27 (1967). https://doi.org/10.1109/TIT.1967.1053964
10. Croux, C., Dehon, C.: Robust linear discriminant analysis using s-estimators. Can. J. Stat./La Revue Canadienne de Statistique **29**(3), 473–493 (2001). http://www.jstor.org/stable/3316042
11. Dessì, N., Milia, G., Pes, B.: Enhancing random forests performance in microarray data classification. In: N. Peek, R. Marín Morales, M. Peleg (eds.) Artificial Intelligence in Medicine, pp. 99–103. Springer Berlin Heidelberg, Berlin, Heidelberg (2013)
12. Dorfer, M., Kelz, R., Widmer, G.: Deep linear discriminant analysis (2016)
13. Dua, D., Graff, C.: UCI machine learning repository (2017). http://archive.ics.uci.edu/ml
14. FISHER, R.A.: The use of multiple measurements in taxonomic problems. Ann. Eugen. **7**(2), 179–188 (1936). https://doi.org/10.1111/j.1469-1809.1936.tb02137.x. https://onlinelibrary.wiley.com/doi/abs/10.1111/j.1469-1809.1936.tb02137.x
15. Ghosh, A., SahaRay, R., Chakrabarty, S., Bhadra, S.: Robust generalised quadratic discriminant analysis (2020)
16. Harper, P.R.: A review and comparison of classification algorithms for medical decision making. Health Policy **71**(3), 315–331 (2005). https://doi.org/10.1016/j.healthpol.2004.05.002. www.sciencedirect.com/science/article/pii/S016885100400096X
17. Hawkins, D.M., McLachlan, G.J.: High-breakdown linear discriminant analysis. J. Am. Stat. Assoc. **92**(437), 136–143 (1997). http://www.jstor.org/stable/2291457
18. Houdouin, P., Pascal, F., Jonckheere, M., Wang, A.: Robust classification with flexible discriminant analysis in heterogeneous data (2022). https://doi.org/10.48550/ARXIV.2201.02967. arXiv:2201.02967
19. Huang, Z., Zhu, H., Zhou, J.T., Peng, X.: Multiple marginal fisher analysis. IEEE Trans. Industr. Electron. **66**(12), 9798–9807 (2019). https://doi.org/10.1109/TIE.2018.2870413
20. Huber, P.J.: Robust covariances. In: Statistical Decision Theory and Related Topics, pp. 165–191. Elsevier (1977)
21. Hubert, M., Driessen, K.: Fast and robust discriminant analysis. Comput. Stat. Data Anal. **45**, 301–320 (2004). https://doi.org/10.1016/S0167-9473(02)00299-2

22. Huberty, C.J.: Discriminant analysis. Rev. Educ. Res. **45**(4), 543–598 (1975). https://doi.org/10.3102/00346543045004543
23. Jeffreys, H.: An invariant form for the prior probability in estimation problems. Proc. R. Soc. London. Ser. A. Math. Phys. Sci. **186**(1007), 453–461 (1946)
24. Kim, S.J., Magnani, A., Boyd, S.: Robust fisher discriminant analysis. In: Y. Weiss, B. Schölkopf, J. Platt (eds.) Advances in Neural Information Processing Systems, vol. 18. MIT Press (2006). https://proceedings.neurips.cc/paper/2005/file/1264a061d82a2edae1574b07249800d6-Paper.pdf
25. Kowsari, K., Jafari Meimandi, K., Heidarysafa, M., Mendu, S., Barnes, L., Brown, D.: Text classification algorithms: a survey. Information **10**(4) (2019). https://doi.org/10.3390/info10040150. https://www.mdpi.com/2078-2489/10/4/150
26. Lachenbruch, P.A.: Discriminant analysis when the initial samples are misclassified. Technometrics **8**(4), 657–662 (1966). http://www.jstor.org/stable/1266637
27. Lachenbruch, P.A., Goldstein, M.: Discriminant analysis. Biometrics **35**(1), 69–85 (1979). http://www.jstor.org/stable/2529937
28. Lotte, F., Congedo, M., Lécuyer, A., Lamarche, F., Arnaldi, B.: A review of classification algorithms for EEG-based brain-computer interfaces. J. Neural Eng. **4**(2), R1 (2007). https://doi.org/10.1088/1741-2560/4/2/R01
29. Maronna, R.A.: Robust M-estimators of multivariate location and scatter. Ann. Stat. **4**(1), 51–67 (1976). https://doi.org/10.1214/aos/1176343347
30. Ollila, E., Tyler, D., Koivunen, V., Poor, H.V.: Complex elliptically symmetric distributions: survey, new results and applications. IEEE Trans. Signal Process. **60**, 5597–5625 (2012). https://doi.org/10.1109/TSP.2012.2212433
31. Pascal, F., Bombrun, L., Tourneret, J.Y., Berthoumieu, Y.: Parameter estimation for multivariate generalized gaussian distributions. IEEE Trans. Signal Process. **61**(23), 5960–5971 (2013). https://doi.org/10.1109/TSP.2013.2282909. https://hal.science/hal-00879851
32. Pedregosa, F., Varoquaux, G., Gramfort, A., Michel, V., Thirion, B., Grisel, O., Blondel, M., Prettenhofer, P., Weiss, R., Dubourg, V., Vanderplas, J., Passos, A., Cournapeau, D., Brucher, M., Perrot, M., Duchesnay, E.: Scikit-learn: machine learning in Python. J. Mach. Learn. Res. **12**, 2825–2830 (2011)
33. Pérez-Ortiz, M., Jiménez-Fernández, S., Gutiérrez, P.A., Alexandre, E., Hervás-Martínez, C., Salcedo-Sanz, S.: A review of classification problems and algorithms in renewable energy applications. Energies **9**(8) (2016). https://doi.org/10.3390/en9080607. https://www.mdpi.com/1996-1073/9/8/607
34. Randles, R.H., Broffitt, J.D., Ramberg, J.S., Hogg, R.V.: Generalized linear and quadratic discriminant functions using robust estimates. J. Am. Stat. Assoc. **73**(363), 564–568 (1978). https://doi.org/10.1080/01621459.1978.10480055. www.tandfonline.com/doi/abs/10.1080/01621459.1978.10480055
35. Roizman, V., Jonckheere, M., Pascal, F.: A flexible EM-like clustering algorithm for noisy data (2020)
36. Roth, M.: On the multivariate t distribution. Technical Report 3059, Linköping University, Automatic Control (2012)
37. Shahbazi, M., Shirali, A., Aghajan, H., Nili, H.: Using distance on the Riemannian manifold to compare representations in brain and in models. Neuroimage **239**, 118271 (2021). https://doi.org/10.1016/j.neuroimage.2021.118271. www.sciencedirect.com/science/article/pii/S1053811921005474
38. Silverman, B.W.: Density Estimation for Statistics and Data Analysis (1986). https://doi.org/10.1201/9781315140919
39. Stewart, G.W.: The efficient generation of random orthogonal matrices with an application to condition estimators. SIAM J. Numer. Anal. **17**(3), 403–409 (1980)
40. Tharwat, A.: Linear versus quadratic discriminant analysis classifier: a tutorial. Int. J. Appl. Pattern Recognit. **3**(2), 145–180 (2016). https://doi.org/10.1504/IJAPR.2016.079050. https://www.inderscienceonline.com/doi/abs/10.1504/IJAPR.2016.079050. PMID: 79050

41. Wakaki, H.: Discriminant analysis under elliptical populations. Hiroshima Math. J. **24**(2), 257–298 (1994). https://doi.org/10.32917/hmj/1206128025
42. Wald, A.: Chapter ii: The Neyman-Pearson theory of testing a statistical hypothesis. In: On the Principles of Statistical Inference, vol. 1, pp. 10–21. University of Notre Dame (1942)
43. Wen, J., Fang, X., Cui, J., Fei, L., Yan, K., Chen, Y., Xu, Y.: Robust sparse linear discriminant analysis. IEEE Trans. Circuits Syst. Video Technol. **29**(2), 390–403 (2019). https://doi.org/10.1109/TCSVT.2018.2799214
44. Zhang, T., Tao, D., Yang, J.: Discriminative locality alignment. In: Forsyth, D., Torr, P., Zisserman, A. (eds.) Computer Vision—ECCV 2008, pp. 725–738. Springer, Berlin Heidelberg, Berlin, Heidelberg (2008)
45. Zhou, Y., Sun, S.: Manifold partition discriminant analysis. IEEE Trans. Cybern. **47**(4), 830–840 (2017). https://doi.org/10.1109/tcyb.2016.2529299. https://dx.doi.org/10.1109/TCYB.2016.2529299

Chapter 10
Learning Graphs from Heavy-Tailed Data

José Vinícius de Miranda Cardoso, Jiaxi Ying, and Daniel P. Palomar

Abstract In this chapter, we present advancements in graph learning for elliptical distributed data. More precisely, we model the inverse scatter matrix of a multivariate Student's t-distribution as a Laplacian matrix associated to a graph whose node features (or signals) are observable. We design numerical algorithms, via the alternating direction method of multipliers, to learn connected, k-component, bipartite, and k-component bipartite graphs suitable to be applied in datasets that contain outliers or whose assumption on the data-generating process is that it follows a multivariate Student's t-distribution. We measure the performance of graph learning algorithms in terms of graph modularity and node accuracy.

10.1 Introduction

Graph learning from data has been a problem of critical importance for the statistical graph learning and graph signal processing fields [21, 27, 36, 40, 54, 70, 79], with direct impact on applied areas such as unsupervised learning, clustering [31, 34, 38, 39, 49, 63, 66], applied finance [17, 42, 43], network topology inference [15, 44, 57, 58], and community detection [10, 26, 41]. In addition, graph matrices play a fundamental role in graph neural networks [52, 73].

The basic idea behind graph learning is to answer the following question: given a data matrix $\mathbf{X}$ whose columns represent signals (observations) measured at the graph nodes, how can one estimate a graph structure that best fits such data matrix without possibly any (or with at most partial) knowledge of the underlying graph structure? This question is often answered on the basis of the assumption that the observed graph signals are Gaussian distributed [20, 21, 27, 36, 39, 40, 59, 76, 79].

J. V. de Miranda Cardoso (✉) · J. Ying · D. P. Palomar
Hong Kong University of Science and Technology, Clear Water Bay, Hong Kong SAR, China
e-mail: jvdmc@connect.ust.hk

J. Ying
e-mail: jx.ying@connect.ust.hk

D. P. Palomar
e-mail: palomar@ust.hk

J.-P. Delmas et al. (eds.), *Elliptically Symmetric Distributions in Signal Processing and Machine Learning*, https://doi.org/10.1007/978-3-031-52116-4_10

While such an assumption for graphical models has found great success in many practical areas, which includes brain network analysis [60], psychological networks [22], and single-cell sequencing [62], it inherently neglects scenarios where there may exist outliers or the underlying data is naturally heavy-tailed distributed, such as financial time series [14, 18, 24, 30, 32, 67]. As a consequence, those methods often lack robustness and may not succeed in capturing a meaningful representation of the underlying graph [25].

Motivated by practical challenging applications such as time-series clustering and network estimation, this chapter investigates the problem of learning graph matrices whose structure follows that of a Laplacian matrix of an undirected weighted graph for which the data generating process is assumed to be Student's t-distributed. In particular, the main contributions of this chapter, which are largely based on novel material published at [8, 9], are as follows:

- We discuss a novel formulation for learning undirected weighted graphs under the assumption that the data-generating process is Student's t-distributed. We solve the underlying learning problem via a carefully designed numerical algorithm based on the alternating direction method of multipliers (ADMM). We note that this algorithm can be easily extended to account for additional linear constraints on graph weights.
- We extend the previously discussed framework to account for heavy tails, k-component graphs, as well as bipartite graphs, which enables a novel method for clustering financial time series.
- We present extensive practical results, with real-world data from the US stock market, foreign exchanges, and cryptocurrencies, that showcase clear advantages of including heavy-tail assumptions into graph learning frameworks when compared to state-of-the-art, Gaussian-based methods.
- Along with the methods discussed in this chapter, we release R packages named fingraph[1] and bipartite,[2] containing fast, unit-tested code that implements the algorithms discussed thereafter.

10.2 Background

An undirected, weighted graph is denoted as a triple $\mathcal{G} = (\mathcal{V}, \mathcal{E}, \mathbf{W})$, where $\mathcal{V} = \{1, 2, \ldots, p\}$ is the node set, $\mathcal{E} \subseteq \{\{u, v\} : u, v \in \mathcal{V}, u \neq v\}$ is the edge set, that is, a subset of the set of all possible unordered pairs of nodes such that $\{u, v\} \in \mathcal{E}$ iff there exists a link between nodes u and v. $\mathbf{W} \in \mathbb{R}_{+}^{p \times p}$ is the symmetric weighted adjacency matrix that satisfies $W_{ii} = 0$, $W_{ij} > 0$ iff $\{i, j\} \in \mathcal{E}$ and $W_{ij} = 0$, otherwise. The combinatorial, unnormalized graph Laplacian matrix $\mathbf{L}$ is defined, as $\mathbf{L} \triangleq \mathbf{D} - \mathbf{W}$, where $\mathbf{D} \triangleq \mathsf{Diag}(\mathbf{W1})$ is the degree matrix.

[1] https://github.com/convexfi/fingraph.

[2] https://github.com/convexfi/bipartite.

A p-dimensional, real-valued, Gaussian random vector $\mathbf{x}$, with mean vector $\mathbb{E}[\mathbf{x}] \triangleq \boldsymbol{\mu}$ and rank-deficient precision matrix $\mathbf{L}$, is said to form a Laplacian constrained Gaussian Markov random field (LGMRF) [38, 75–77] of rank $p-k, k \geq 1$, with respect to a graph $\mathcal{G}$, when its probability density function is given as

$$p(\mathbf{x}) \propto \sqrt{\det^*(\mathbf{L})} \exp\left\{-\frac{1}{2}(\mathbf{x}-\boldsymbol{\mu})^\top \mathbf{L}(\mathbf{x}-\boldsymbol{\mu})\right\}, \tag{10.1}$$

where $\det^*(\mathbf{L})$ is the pseudo-determinant of $\mathbf{L}$, i.e., the product of its positive eigenvalues [37].

Assume we are given n observations of $\mathbf{x}$, i.e., $\mathbf{X} = [\mathbf{x}_1, \mathbf{x}_2, \ldots, \mathbf{x}_n]^\top, \mathbf{X} \in \mathbb{R}^{n \times p}$, $\mathbf{x}_i \in \mathbb{R}^{p \times 1}$. The goal of graph learning algorithms is to learn a Laplacian matrix, or equivalently an adjacency matrix, given only the data matrix $\mathbf{X}$, i.e., often without any knowledge of $\mathcal{E}$.

To that end, the penalized maximum likelihood estimator (MLE) of the Laplacian constrained precision matrix of $\mathbf{x}$, on the basis of the observed data $\mathbf{X}$, is

$$\begin{array}{ll}\underset{\mathbf{L} \succeq \mathbf{0}}{\mathsf{minimize}} & \operatorname{tr}(\mathbf{LS}) - \log\det^*(\mathbf{L}) + h(\mathbf{L}), \\ \mathsf{subject\ to} & \mathbf{L1} = \mathbf{0},\ L_{ij} = L_{ji} \leq 0,\end{array} \tag{10.2}$$

where $\mathbf{S}$ is a similarity matrix, e.g., the sample covariance (or correlation) matrix $\mathbf{S} \propto \mathbf{X}^\top\mathbf{X}$, and h is a regularization function to promote certain properties on $\mathbf{L}$ such as sparsity or low-rankness.

Even though Problem (10.2) is convex, provided we assume a convex choice for h, it is not adequate to be solved by disciplined convex programming languages, such as cvxpy [19], particularly due to scalability issues related to the computation of the term $\log\det^*(\mathbf{L})$ [21, 79]. Indeed, recently, considerable efforts have been directed toward the design of scalable, iterative algorithms based on block coordinate descent [71], majorization–minimization (MM) [35, 65], and ADMM [5] to solve Problem (10.2) in an efficient fashion, e.g., [21, 79].

Estimators based on Gaussian assumptions have been proposed for connected graphs [20, 21, 36, 40, 79]. Some of their properties, such as sparsity, are yet being investigated [76, 77]. The authors in [38, 49] proposed optimization programs for learning the class of k-component graphs, as such class is an appealing model for clustering tasks due to the spectral properties of the Laplacian matrix. However, a major shortcoming in their formulations is the lack of constraints on the degrees of the nodes, which allows for trivial solutions, i.e., graphs with isolated nodes.

Elliptical losses along with linear structural constraints that retain the positive-definiteness of the estimated covariance matrix have been proposed in the literature [64, 68]. In this work, however, we address the case of Laplacian constraints, which lead to positive-semidefinite precision matrices, and nonconvex k-component structural constraints.

10.3 Brief Detour: Algorithmic Frameworks for Optimization

The graph learning algorithms that will be discussed in the preceding section rely heavily on two well-established iterative numerical optimization frameworks, namely: Alternating Direction Method of Multipliers (ADMM) [5] and Majorization–Minimization (MM) [51, 65]. In the following subsections, we provide a brief overview about ADMM and MM.

10.3.1 Alternating Direction Method of Multipliers (ADMM)

ADMM is a primal–dual framework designed to solve the following class of optimization problems:

$$\begin{array}{ll}\underset{\mathbf{x},\mathbf{z}}{\mathsf{minimize}} & f(\mathbf{x}) + g(\mathbf{z}) \\ \mathsf{subject\ to} & \mathbf{A}\mathbf{x} + \mathbf{B}\mathbf{z} = \mathbf{c},\end{array} \tag{10.3}$$

where $\mathbf{x} \in \mathbb{R}^n$ and $\mathbf{z} \in \mathbb{R}^m$ are the optimization variables; $\mathbf{A} \in \mathbb{R}^{p\times n}$, $\mathbf{B} \in \mathbb{R}^{p\times m}$, and, $\mathbf{c} \in \mathbb{R}^p$ are parameters; and f and g are convex, proper, closed, and possibly non-differentiable functions.

The central object in the ADMM framework is the augmented Lagrangian function, which is given as

$$L_\rho(\mathbf{x}, \mathbf{z}, \mathbf{y}) = f(\mathbf{x}) + g(\mathbf{z}) + \mathbf{y}^\top(\mathbf{A}\mathbf{x} + \mathbf{B}\mathbf{z} - \mathbf{c}) + \frac{\rho}{2}\left\|\mathbf{A}\mathbf{x} + \mathbf{B}\mathbf{z} - \mathbf{c}\right\|_2^2, \tag{10.4}$$

where ρ is a penalty parameter.

The basic workflow of the ADMM algorithm is summarized in Algorithm 10.1.

Algorithm 10.1 ADMM framework

Data: $\mathbf{z}^0, \mathbf{y}^0, \mathbf{A}, \mathbf{B}, \mathbf{c}, \rho > 0$
Result: $\mathbf{x}^\star, \mathbf{z}^\star, \mathbf{y}^\star$

$l \leftarrow 0$
while *not converged* **do**
    $\mathbf{x}^{l+1} \leftarrow \underset{\mathbf{x}\in\mathcal{X}}{\mathsf{argmin}}\, L_\rho\left(\mathbf{x}, \mathbf{z}^l, \mathbf{y}^l\right)$
    $\mathbf{z}^{l+1} \leftarrow \underset{\mathbf{z}\in\mathcal{Z}}{\mathsf{argmin}}\, L_\rho\left(\mathbf{x}^{l+1}, \mathbf{z}, \mathbf{y}^l\right)$
    $\mathbf{y}^{l+1} \leftarrow \mathbf{y}^l + \rho\left(\mathbf{A}\mathbf{x}^{l+1} + \mathbf{B}\mathbf{z}^{l+1} - \mathbf{c}\right)$
    $l \leftarrow l + 1$
end

The convergence of ADMM algorithms is attained provided that the following conditions are met:

1. $\text{epi}(f) = \{(\mathbf{x}, t) \in \mathbb{R}^n \times \mathbb{R} : f(\mathbf{x}) \leq t\}$ and $\text{epi}(g) = \{(\mathbf{z}, s) \in \mathbb{R}^m \times \mathbb{R} : g(\mathbf{z}) \leq s\}$ are both closed nonempty convex sets.
2. The unaugmented Lagrangian function L_0 has a saddle point.

We refer readers to [5] where elaborate convergence results are discussed.

10.3.2 Majorization–Minimization (MM)

The MM framework seeks to solve the following general optimization problem:

$$\begin{array}{ll} \underset{\mathbf{x}}{\text{minimize}} & f(\mathbf{x}) \\ \text{subject to} & \mathbf{x} \in \mathcal{X}, \end{array} \tag{10.5}$$

where we consider f a smooth, possibly non-convex function.

The general idea behind MM is to find a sequence of feasible points $\{\mathbf{x}^i\}_{i \in \mathbb{N}}$ by minimizing a sequence of carefully constructed global upper-bounds of f. The popular expectation–maximization (EM) algorithm is a special case of MM [72].

At point $\mathbf{x}^i$, we design a continuous global upper-bound function $g\left(\cdot, \mathbf{x}^i\right) : \mathcal{X} \rightarrow \mathbb{R}$ such that

$$g\left(\mathbf{x}, \mathbf{x}^i\right) \geq f(\mathbf{x}), \ \forall\, \mathbf{x} \in \mathcal{X}. \tag{10.6}$$

Then, in the minimization step, we update $\mathbf{x}$ as

$$\mathbf{x}^{i+1} \in \underset{\mathbf{x} \in \mathcal{X}}{\arg\min}\, g(\mathbf{x}, \mathbf{x}^i). \tag{10.7}$$

The global upper-bound function $g(\cdot, \mathbf{x}^i)$ must satisfy the following conditions in order to guarantee convergence:

1. $g\left(\mathbf{x}, \mathbf{x}^i\right) \geq f(\mathbf{x})\ \forall\, \mathbf{x} \in \mathcal{X}$,
2. $g\left(\mathbf{x}^i, \mathbf{x}^i\right) = f\left(\mathbf{x}^i\right)$,
3. $\nabla g\left(\mathbf{x}^i, \mathbf{x}^i\right) = \nabla f\left(\mathbf{x}^i\right)$,
4. $g(\mathbf{x}, \mathbf{x}^i)$ is continuous on both $\mathbf{x}$ and $\mathbf{x}^i$.

A thorough discussion about MM, along with a significant number of its extensions, with practical examples, can be found in [65].

10.4 Algorithms

In this section, we discuss optimization formulations and an iterative algorithm to learn a Laplacian matrix from heavy-tailed assumptions. With that goal, we express the Laplacian matrix via its linear operator, *i.e.*, $\mathbf{L} = \mathcal{L}\mathbf{w}$ [38], where $\mathbf{w} \in \mathbb{R}^{p(p-1)/2}$ is the vectorized form of the upper triangular part of the adjacency matrix, also known as the vector of graph weights. The formal definition of the Laplacian operator $\mathcal{L}$ is as follows:

Definition 10.1 (*Laplacian operator*) The Laplacian operator [39] $\mathcal{L} : \mathbb{R}^{p(p-1)/2} \to \mathbb{R}^{p\times p}$, which takes a nonnegative vector $\mathbf{w}$ and outputs a Laplacian matrix $\mathcal{L}\mathbf{w}$, is defined as

$$[\mathcal{L}\mathbf{w}]_{ij} = \begin{cases} -w_{i+s(j)}, & \text{if } \ i > j, \\ [\mathcal{L}\mathbf{w}]_{ji}, & \text{if } \ i < j, \\ -\sum_{i\neq j} [\mathcal{L}\mathbf{w}]_{ij}, & \text{if } \ i = j, \end{cases} \tag{10.8}$$

where $s(j) = \frac{j-1}{2}(2p - j) - j$.

In addition, we use the fact that, for connected graphs, it follows that $\det^*(\mathcal{L}\mathbf{w}) = \det(\mathcal{L}\mathbf{w} + \mathbf{J})$, $\mathbf{J} \triangleq \frac{1}{p}\mathbf{1}\mathbf{1}^\top$ [21].

In order to address the inherent heavy-tailed nature of financial market data [55], we consider the Student's t-distribution under the improper Markov random field assumption [56] with Laplacian structural constraints, that is, we assume the data generating process to be modeled as a multivariate zero-mean Student's t-distribution, whose probability density function can be written as

$$p(\mathbf{x}) \propto \sqrt{\det^*(\boldsymbol{\Theta})}\left(1 + \frac{\mathbf{x}^\top\boldsymbol{\Theta}\mathbf{x}}{\nu}\right)^{-\frac{\nu+p}{2}}, \ \nu > 2, \tag{10.9}$$

where $\boldsymbol{\Theta}$ is a positive-semidefinite inverse scatter matrix modeled as a combinatorial graph Laplacian matrix and ν is the number of degrees of freedom, which measures the rate of decay of the tails.

This results in a robustified version of the MLE for connected graph learning, i.e.,

$$\begin{aligned} &\underset{\mathbf{w}\geq\mathbf{0},\boldsymbol{\Theta}\succeq\mathbf{0}}{\text{minimize}} \ \frac{p+\nu}{n}\sum_{i=1}^{n}\log\left(1 + \frac{\mathbf{x}_i^\top\mathcal{L}\mathbf{w}\mathbf{x}_i}{\nu}\right) - \log\det\left(\boldsymbol{\Theta} + \mathbf{J}\right), \\ &\text{subject to } \boldsymbol{\Theta} = \mathcal{L}\mathbf{w}, \ \mathfrak{d}\mathbf{w} = \mathbf{d}, \end{aligned} \tag{10.10}$$

where $\mathfrak{d} : \mathbb{R}^{p(p-1)/2} \to \mathbb{R}^{p}$ is the degree operator defined as follows:

Definition 10.2 (*Degree operator*) The degree operator $\mathfrak{d} : \mathbb{R}^{p(p-1)/2} \to \mathbb{R}^{p}$, which takes a nonnegative vector $\mathbf{w}$ and outputs the diagonal of a Laplacian matrix, is defined as

$$\mathfrak{d}\mathbf{w} = \text{diag}(\mathcal{L}\mathbf{w}). \tag{10.11}$$

The constraint $\mathfrak{d}\mathbf{w} = \mathbf{d}$ in Problem (10.10) enables the learning of additional graph structures such as weighted regular graphs and it is crucial for k-component graphs.

From a theoretical perspective, the Student's t model naturally yields sparse graphs. Comparing the objective function in Problem (10.10) to that of Problem (10.2), we note that the Student's t contains a $\log(\cdot)$ term in place of a linear term of the graph weights. The usage of a log function to promote sparsity is closely related to the iteratively reweighted ℓ_1-norm as an approximation for the ℓ_0-norm problem [6]. Problem (10.10) is, in general, nonconvex due to the summation of log terms and hence it is challenging to be considered directly. Hence, we design an iterative algorithm based on the ADMM framework.

10.4.1 ADMM Solution

The partial augmented Lagrangian function of Problem (10.10) is given as

$$L_\rho(\boldsymbol{\Theta}, \mathbf{w}, \mathbf{Y}, \mathbf{y}) = \frac{p+\nu}{n} \sum_{i=1}^{n} \log\left(1 + \frac{\mathbf{x}_i^\top \mathcal{L}\mathbf{w}\mathbf{x}_i}{\nu}\right) - \log\det(\boldsymbol{\Theta} + \mathbf{J})$$
$$+ \langle \mathbf{y}, \mathfrak{d}\mathbf{w} - \mathbf{d} \rangle + \frac{\rho}{2} \|\mathfrak{d}\mathbf{w} - \mathbf{d}\|_2^2 + \langle \mathbf{Y}, \boldsymbol{\Theta} - \mathcal{L}\mathbf{w} \rangle + \frac{\rho}{2} \|\boldsymbol{\Theta} - \mathcal{L}\mathbf{w}\|_F^2, \tag{10.12}$$

where $\mathbf{Y}$ and $\mathbf{y}$ are the dual variables associated with the equality constraints $\boldsymbol{\Theta} = \mathcal{L}\mathbf{w}$ and $\mathfrak{d}\mathbf{w} = \mathbf{d}$, respectively. Note that we deal with the constraints $\mathbf{w} \geq \mathbf{0}$ and $\boldsymbol{\Theta} \succeq \mathbf{0}$ directly, hence there are no dual variables associated with them.

Given $\mathbf{w}^l$ and $\mathbf{Y}^l$, the subproblem for $\boldsymbol{\Theta}$ can be written as

$$\boldsymbol{\Theta}^{l+1} = \underset{\boldsymbol{\Theta} \succeq \mathbf{0}}{\arg\min} \; -\log\det(\boldsymbol{\Theta} + \mathbf{J}) + \langle \boldsymbol{\Theta}, \mathbf{Y}^l \rangle + \frac{\rho}{2} \left\|\boldsymbol{\Theta} - \mathcal{L}\mathbf{w}^l\right\|_F^2, \tag{10.13}$$

whose closed-form solution is given by Lemma 10.1.

Lemma 10.1 *The global minimizer of Problem* (10.13) *is [16, 69]*

$$\boldsymbol{\Theta}^{l+1} = \frac{1}{2\rho} \mathbf{U} \left(\boldsymbol{\Gamma} + \sqrt{\boldsymbol{\Gamma}^2 + 4\rho \mathbf{I}}\right) \mathbf{U}^\top - \mathbf{J}, \tag{10.14}$$

where $\mathbf{U}\boldsymbol{\Gamma}\mathbf{U}^\top$ *is the eigenvalue decomposition of* $\rho\left(\mathcal{L}\mathbf{w}^l + \mathbf{J}\right) - \mathbf{Y}^l$.

Given $\boldsymbol{\Theta}^{l+1}$, $\mathbf{Y}^{l}$, and $\mathbf{y}^{l}$, the subproblem for $\mathbf{w}$ can be formulated as

$$\begin{aligned}\underset{\mathbf{w}\geq\mathbf{0}}{\mathsf{minimize}}\ & \frac{\rho}{2}\mathbf{w}^{\top}\left(\mathfrak{d}^{*}\mathfrak{d}+\mathcal{L}^{*}\mathcal{L}\right)\mathbf{w}-\left\langle \mathbf{w},\mathcal{L}^{*}\left(\mathbf{Y}^{l}+\rho\boldsymbol{\Theta}^{l+1}\right)-\mathfrak{d}^{*}\left(\mathbf{y}^{l}-\rho\mathbf{d}\right)\right\rangle \\ & +\frac{p+\nu}{n}\sum_{i=1}^{n}\log\left(1+\frac{\mathbf{x}_{i}^{\top}\mathcal{L}\mathbf{w}\mathbf{x}_{i}}{\nu}\right),\end{aligned} \tag{10.15}$$

where $\mathfrak{d}^{*}$ and $\mathcal{L}^{*}$ are the adjoint operators of the degree and Laplacian operators, respectively, which are defined as follows:

Definition 10.3 (*Adjoint of Laplacian operator*) The adjoint of Laplacian operator [39] $\mathcal{L}^{*}:\mathbb{R}^{p\times p}\rightarrow\mathbb{R}^{p(p-1)/2}$ is defined as

$$\left(\mathcal{L}^{*}\mathbf{P}\right)_{s(i,j)}=\mathbf{P}_{i,i}-\mathbf{P}_{i,j}-\mathbf{P}_{j,i}+\mathbf{P}_{j,j}, \tag{10.16}$$

where $s(i,j)=i-j+\frac{j-1}{2}(2p-j), i>j$.

Definition 10.4 (*Adjoint of degree operator*) The adjoint of degree operator $\mathfrak{d}^{*}:\mathbb{R}^{p}\rightarrow\mathbb{R}^{p(p-1)/2}$ is given as

$$\left(\mathfrak{d}^{*}\mathbf{y}\right)_{s(i,j)}=\mathbf{y}_{i}+\mathbf{y}_{j}, \tag{10.17}$$

where $s(i,j)=i-j+\frac{j-1}{2}(2p-j), i>j$.

In general, subproblem (10.15) is nonconvex due to the concave nature of the logarithm function. Hence, we resort to the MM method [65] to find a stationary point of subproblem (10.15). We proceed by constructing a global upper-bound of the objective function of (10.15) at point $\mathbf{w}^{j}\in\mathbb{R}_{+}^{p(p-1)/2}$ as

$$g(\mathbf{w},\mathbf{w}^{j})=g(\mathbf{w}^{j},\mathbf{w}^{j})+\langle\mathbf{w}-\mathbf{w}^{j},\nabla_{\mathbf{w}}f(\mathbf{w}^{j})\rangle+\frac{\mu}{2}\left\|\mathbf{w}-\mathbf{w}^{j}\right\|_{2}^{2}, \tag{10.18}$$

where f is the objective function of subproblem (10.15), its gradient is given as $\nabla_{\mathbf{w}}f(\mathbf{w}^{j})=\mathbf{a}^{j}+\mathbf{b}^{j}$, where

$$\mathbf{a}^{j}=\mathcal{L}^{*}\left(\tilde{\mathbf{S}}^{j}-\mathbf{Y}^{l}-\rho\left(\boldsymbol{\Theta}^{l+1}-\mathcal{L}\mathbf{w}^{j}\right)\right), \tag{10.19}$$

$$\mathbf{b}^{j}=\mathfrak{d}^{*}\left(\mathbf{y}^{l}-\rho\left(\mathbf{d}-\mathfrak{d}\mathbf{w}^{j}\right)\right), \tag{10.20}$$

where $\tilde{\mathbf{S}}^{j}\triangleq\frac{1}{n}\sum_{i=1}^{n}\dfrac{(p+\nu)\mathbf{x}_{i}\mathbf{x}_{i}^{\top}}{\langle\mathbf{w}^{j},\mathcal{L}^{*}(\mathbf{x}_{i}\mathbf{x}_{i}^{\top})\rangle+\nu}$ is a weighted sample covariance matrix, and $\mu=\rho\lambda_{\mathsf{max}}\left(\mathfrak{d}^{*}\mathfrak{d}+\mathcal{L}^{*}\mathcal{L}\right)$, and the maximum eigenvalue of $\mathfrak{d}^{*}\mathfrak{d}+\mathcal{L}^{*}\mathcal{L}$ is given by Lemma 10.2 as follows:

Lemma 10.2 *The maximum eigenvalue of the matrix* $\mathfrak{d}^{*}\mathfrak{d}+\mathcal{L}^{*}\mathcal{L}$ *is given as*

$$\lambda_{\mathsf{max}}\left(\mathfrak{d}^*\mathfrak{d} + \mathcal{L}^*\mathcal{L}\right) = 2(2p - 1). \tag{10.21}$$

The vector of graph weights $\mathbf{w}$ can then be updated by minimizing the function g constructed in (10.18), which is tantamount to solving the following nonnegative, quadratic-constrained, strictly convex problem:

$$\mathbf{w}^{j+1} = \underset{\mathbf{w}\geq \mathbf{0}}{\text{arg min}}\ \rho(2p-1)\left\|\mathbf{w} - \mathbf{w}^j\right\|_2^2 + \langle \mathbf{w}, \mathbf{a}^j + \mathbf{b}^j\rangle, \tag{10.22}$$

whose unique solution can be readily obtained via its KKT optimality conditions and is given as

$$\mathbf{w}^{j+1} = \left(\mathbf{w}^j - \frac{\mathbf{a}^j + \mathbf{b}^j}{2\rho(2p-1)}\right)^+, \tag{10.23}$$

that is, a projected gradient descent step with learning rate $(2\rho(2p-1))^{-1}$. Thus, we iterate (10.23) in order to obtain a stationary point, $\mathbf{w}^{l+1}$, of Problem (10.15). In practice, we observe that a few iterations are sufficient to retrieve $\mathbf{w}^{l+1}$.

The dual variables $\mathbf{Y}$ and $\mathbf{y}$ are updated via gradient ascent steps. Algorithm 10.2 summarizes the implementation to find a stationary point of Problem (10.10).

Algorithm 10.2 Student-*t* Graph Learning

Data: Data matrix $\mathbf{X} \in \mathbb{R}^{n\times p}$, initial estimate of the graph weights $\mathbf{w}^0$, desired degree vector $\mathbf{d}$, penalty parameter $\rho > 0$, degrees of freedom $\nu > 2$, convergence tolerance $\epsilon > 0$

Result: Graph Laplacian estimation: $\mathcal{L}\mathbf{w}^\star$

initialize $\mathbf{Y} = \mathbf{0}$, $\mathbf{y} = \mathbf{0}$

$l \leftarrow 0$

while $\left\|\mathbf{r}^l\right\|_\infty > \epsilon$ *or* $\left\|\mathbf{s}^l\right\|_\infty > \epsilon$ **do**

- ▷ update $\boldsymbol{\Theta}^{l+1}$ via (10.14)
- ▷ update $\mathbf{w}^{l+1}$ by iterating (10.23)
- ▷ update $\mathbf{Y}^{l+1} = \mathbf{Y}^l + \rho\left(\boldsymbol{\Theta}^{l+1} - \mathcal{L}\mathbf{w}^{l+1}\right)$
- ▷ update $\mathbf{y}^{l+1} = \mathbf{y}^l + \rho\left(\mathfrak{d}\mathbf{w}^{l+1} - \mathbf{d}\right)$
- ▷ compute residual $\mathbf{r}^{l+1} = \boldsymbol{\Theta}^{l+1} - \mathcal{L}\mathbf{w}^{l+1}$
- ▷ compute residual $\mathbf{s}^{l+1} = \mathfrak{d}\mathbf{w}^{l+1} - \mathbf{d}$
- $l \leftarrow l + 1$

end

10.4.2 Extension to k-Component Graphs

The graph learning formulation presented in (10.10) is applicable to learn connected graphs. Learning graphs with k components, k assumed to be known, poses a considerably higher challenge, as the dimension of the nullspace of the Laplacian matrix $\mathcal{L}\mathbf{w}$ is equal to the number of components of the graph [11]. One way to achieve this requirement is by imposing the constraint $\operatorname{rank}(\mathcal{L}\mathbf{w}) = p - k$, which is nonconvex and nondifferentiable, in the maximum likelihood problem generated by (10.44). We instead relax this rank constraint by noting that via Fan's theorem [23], we have

$$\sum_{i=1}^{k} \lambda_i (\mathcal{L}\mathbf{w}) = \underset{\mathbf{V}\in\mathbb{R}^{p\times k}, \mathbf{V}^\top\mathbf{V}=\mathbf{I}}{\text{minimize}} \operatorname{tr}\left(\mathbf{V}^\top \mathcal{L}\mathbf{w}\mathbf{V}\right). \tag{10.24}$$

Thus, by using the right-hand side of (10.24) as a regularization term, we are able to formulate the following optimization problem to learn a Student's t k-component graph:

$$\begin{aligned} &\underset{\mathbf{w}\geq\mathbf{0}, \boldsymbol{\Theta}\succeq 0, \mathbf{V}}{\text{minimize}} \frac{p+\nu}{n} \sum_{i=1}^{n} \log\left(1 + \frac{\mathbf{x}_i^\top \mathcal{L}\mathbf{w}\mathbf{x}_i}{\nu}\right) - \log\det{}^*(\boldsymbol{\Theta}) + \eta\operatorname{tr}(\mathcal{L}\mathbf{w}\mathbf{V}\mathbf{V}^\top), \\ &\text{subject to } \boldsymbol{\Theta} = \mathcal{L}\mathbf{w},\ \operatorname{rank}(\boldsymbol{\Theta}) = p-k, \mathfrak{d}\mathbf{w} = \mathbf{d},\ \mathbf{V}^\top\mathbf{V} = \mathbf{I},\ \mathbf{V}\in\mathbb{R}^{p\times k}. \end{aligned} \tag{10.25}$$

The partial augmented Lagrangian function of Problem (10.25) can be expressed as

$$\begin{aligned} L_\rho(\boldsymbol{\Theta}, \mathbf{w}, \mathbf{V}, \mathbf{Y}, \mathbf{y}) = &\frac{p+\nu}{n} \sum_{i=1}^{n} \log\left(1 + \frac{\mathbf{x}_i^\top \mathcal{L}\mathbf{w}\mathbf{x}_i}{\nu}\right) - \log\det{}^*(\boldsymbol{\Theta}) + \eta\operatorname{tr}\left(\mathcal{L}\mathbf{w}\mathbf{V}\mathbf{V}^\top\right) \\ &+ \langle \mathbf{y}, \mathfrak{d}\mathbf{w} - \mathbf{d}\rangle + \frac{\rho}{2}\|\mathfrak{d}\mathbf{w} - \mathbf{d}\|_2^2 + \langle \mathbf{Y}, \boldsymbol{\Theta} - \mathcal{L}\mathbf{w}\rangle + \frac{\rho}{2}\|\boldsymbol{\Theta} - \mathcal{L}\mathbf{w}\|_F^2. \end{aligned} \tag{10.26}$$

Given $\mathbf{w}^l$ and $\mathbf{Y}^l$, the subproblem for $\boldsymbol{\Theta}$ can be written as

$$\boldsymbol{\Theta}^{l+1} = \underset{\operatorname{rank}(\boldsymbol{\Theta})=p-k}{\arg\min} \; -\log\det{}^*(\boldsymbol{\Theta}) + \langle \boldsymbol{\Theta}, \mathbf{Y}^l\rangle + \frac{\rho}{2}\left\|\boldsymbol{\Theta} - \mathcal{L}\mathbf{w}^l\right\|_F^2, \tag{10.27}$$

which is nearly the same as (10.13). Its solution is obtained as

$$\boldsymbol{\Theta}^{l+1} = \frac{1}{2\rho}\mathbf{U}\left(\boldsymbol{\Gamma} + \sqrt{\boldsymbol{\Gamma}^2 + 4\rho\mathbf{I}}\right)\mathbf{U}^\top, \tag{10.28}$$

except that now $\mathbf{U}\boldsymbol{\Gamma}\mathbf{U}^\top$ is the eigenvalue decomposition of $\rho\mathcal{L}\mathbf{w}^l - \mathbf{Y}^l$, with $\boldsymbol{\Gamma}$ having the largest $p-k$ eigenvalues along its diagonal and $\mathbf{U}\in\mathbb{R}^{p\times(p-k)}$ contains the corresponding eigenvectors.

The subproblem to obtain $\mathbf{w}^{l+1}$ is virtually the same as in (10.15) except for the additional linear term $\eta\operatorname{tr}(\mathcal{L}\mathbf{w}\mathbf{V}^l\mathbf{V}^{l\top})$. Hence, its update is also a projected gradient descent step, alike (10.23) where

$$\mathbf{a}^j \triangleq \mathcal{L}^*\left(\tilde{\mathbf{S}}^j + \eta\mathbf{V}^l\mathbf{V}^{l\top} - \mathbf{Y}^l - \rho\left(\boldsymbol{\Theta}^{l+1} - \mathcal{L}\mathbf{w}^j\right)\right). \tag{10.29}$$

Given $\mathbf{w}^{l+1}$, we have the following subproblem for $\mathbf{V}$:

$$\underset{\mathbf{V}\in\mathbb{R}^{p\times k},\mathbf{V}^\top\mathbf{V}=\mathbf{I}}{\text{minimize}} \quad \operatorname{tr}\left(\mathbf{V}^\top \mathcal{L}\mathbf{w}^{l+1}\mathbf{V}\right), \tag{10.30}$$

whose closed-form solution is given by the k eigenvectors associated with the k smallest eigenvalues of $\mathcal{L}\mathbf{w}^{l+1}$ [1, 33]. Algorithm 10.3 summarizes the implementation to find a stationary point of Problem (10.25).

Algorithm 10.3 k-component Student-t graph learning

Data: Data matrix $\mathbf{X} \in \mathbb{R}^{n\times p}$, initial estimate of the graph weights $\mathbf{w}^0$, number of graph components k, desired degree vector $\mathbf{d}$, degrees of freedom ν, rank hyperparameter $\eta > 0$, penalty parameter $\rho > 0$, tolerance $\epsilon > 0$

Result: Laplacian estimation: $\mathcal{L}\mathbf{w}^\star$

initialize $\mathbf{Y} = \mathbf{0}$, $\mathbf{y} = \mathbf{0}$
$l \leftarrow 0$
while $\|\mathbf{r}^l\|_\infty > \epsilon$ *or* $\|\mathbf{s}^l\|_\infty > \epsilon$ **do**
  ▷ update $\mathbf{\Theta}^{l+1}$ via (10.28)
  ▷ update $\mathbf{w}^{l+1}$ as in (10.23) with $\mathbf{a}^j$ given in (10.29)
  ▷ update $\mathbf{V}^{l+1}$ as in (10.30)
  ▷ update $\mathbf{Y}^{l+1} = \mathbf{Y}^l + \rho\left(\mathbf{\Theta}^{l+1} - \mathcal{L}\mathbf{w}^{l+1}\right)$
  ▷ update $\mathbf{y}^{l+1} = \mathbf{y}^l + \rho\left(\mathfrak{d}\mathbf{w}^{l+1} - \mathbf{d}\right)$
  ▷ compute residual $\mathbf{r}^{l+1} = \mathbf{\Theta}^{l+1} - \mathcal{L}\mathbf{w}^{l+1}$
  ▷ compute residual $\mathbf{s}^{l+1} = \mathfrak{d}\mathbf{w}^{l+1} - \mathbf{d}$
  $l \leftarrow l + 1$
end

10.5 Numerical Experiments

In order to evaluate the performance of the graph learning algorithms, we perform experiments using historical daily price time series data, available in Yahoo! Finance™, from financial instruments in three scenarios: (i) stocks belonging to the S&P500 index, (ii) foreign exchange markets, and (iii) cryptocurrencies. We start by constructing the log-returns data matrix, i.e., a matrix $\mathbf{X} \in \mathbb{R}^{n\times p}$, where n is the number of log-returns observations and p is the number of instruments, as

$$X_{i,j} = \log P_{i,j} - \log P_{i-1,j}, \tag{10.31}$$

where $P_{i,j}$ is the closing price of the j-th instrument at the i-th day.

We note that, in all the experiments, we have $\frac{n}{p} > 1$, meaning that we are not in a high-dimensional setting.

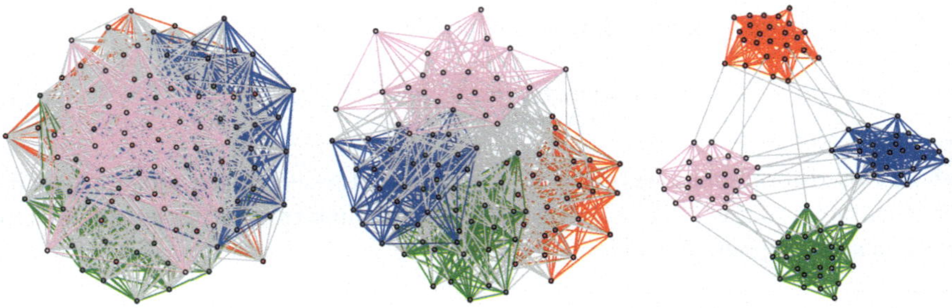

(a) modularity = 0.1. (b) modularity = 0.37. (c) modularity = 0.7.

Fig. 10.1 Visualizations of graphs with different values of modularity

Benchmarks: We compare algorithms developed in this chapter with state-of-the-art, Gaussian distribution-based methods for connected graphs, namely GLE [21, 79] and NGL [76], which use ℓ_1-norm and minimax concave penalty regularizations for sparsity inducing solutions, respectively; and CLR [49] and SGL [38] that consider k-component graphs. For a fair comparison among algorithms, we set the degree vector $\mathbf{d}$ equal to $\mathbf{1}$, i.e., we do not consider any prior information on the degree of nodes.

The figure of merit used to quantify the performance of the graph learning methods is the modularity of the estimated graph.

Definition 10.5 (*Modularity*) The modularity of a graph $\mathcal{G}$ [47] is defined as

$$\mathsf{modularity}(\mathcal{G}) \triangleq \frac{1}{2|\mathcal{E}|} \sum_{i,j \in \mathcal{V}} \left(\mathbf{W}_{ij} - \frac{d_i d_j}{2|\mathcal{E}|} \right) \mathbb{I}(t_i = t_j), \tag{10.32}$$

where d_i is the weighted degree of the i-th node, i.e., $d_i \triangleq [\mathfrak{d}(\mathbf{w})]_i$, t_i is the type (or label) of the i-th node, and $\mathbb{I}(\cdot)$ is the indicator function.

The graph modularity measures the strength of separability of a graph into groups [47]. For illustrative purposes, Fig. 10.1 shows the structure of graphs with different modularity values. It is observed that highly modular graphs present a community-like structure, which is often the expected behavior for networks of financial instruments such as stocks.

In our ADMM algorithms, we set the penalty parameter to $\rho = 1$ and the hyperparameter η in (10.25) is adaptively increased until the rank constraint is satisfied. For GLE and NGL, we use grid search on the sparsity hyperparameter such that the resulting graph yields the highest modularity value. The graph weights in Algorithms 10.2 and 10.3 are initialized using the same procedure as in [39]. Finally, the choice of k is guided by the number of sectors that we consider in the S&P500. For other experiments, the choice of k is somewhat arbitrary, but cross-validation approaches could be applied.

Our goal with the experiments that follow is to verify whether the heavy-tail assumption provides an improved version of the learned graph, which is evaluated based on the modularity of the estimated graph and the graph visualization. In addition, for the task of clustering stocks, we analyze whether the learned graphs agree with industry standards of sector classification set by the Global Industry Classification Standard (GICS)[3] [46, 61].

10.5.1 Communities in S&P500 Stocks

In this experiment, we consider S&P500 stocks belonging to three sectors, namely, Communication Services (red), Utilities (blue), and Real Estate (green), totaling $p = 82$ stocks, during the time horizon from January 3, 2014, to December 29, 2017, resulting in $n = 1006$ observations. In order to obtain descriptive insights on this dataset, we measure its degree of heavy-tailedness and annualized volatility.[4] The former is obtained by fitting the degrees of freedom of a Student's t-distribution to the matrix of log-returns, whereby we obtain $\nu \approx 5.5$ and $\sigma \approx 21\%$. This scenario can be considered as having a moderate amount of heavy-tailedness.

Figure 10.2 depicts the learned connected graphs on the aforementioned time periods. It can be readily noticed that the graph learned with the Student's t-distribution (Fig. 10.2c) is sparser than those learned with the Gaussian assumption (Fig. 10.2a and b), which results from the fact that the Gaussian distribution is more sensitive to outliers. Moreover, the Student's t graph presents a higher degree of interpretability

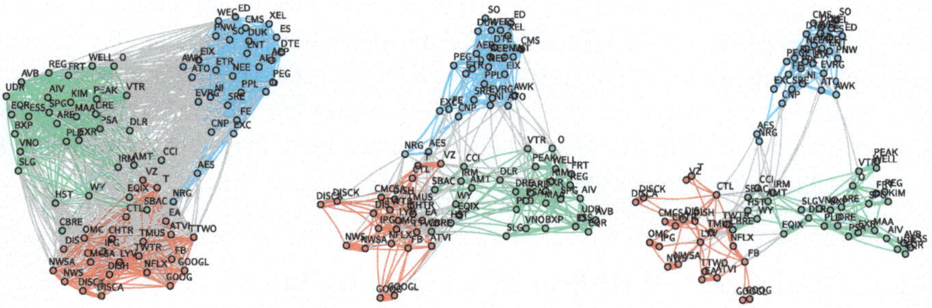

(a) GLE, modularity = 0.31. (b) NGL, modularity = 0.49. (c) Algorithm 10.2, modularity = 0.54.

Fig. 10.2 Learned graphs of S&P500 stocks

---

[3] More often than not, stocks have impacts on multiple sectors, e.g., the evident case of technology companies, whose influence not only affects the prices of stocks not only in their own sector but also spans across multiple sectors. Therefore, GICS or other sector classification systems might not always be representative systems for ground-truth labels.

[4] The annualized volatility is computed as $\sigma = \frac{\sqrt{252}}{p} \sum_{i=1}^{p} \sigma_i$, where σ_i is the daily sample standard deviation of the i-th stock.

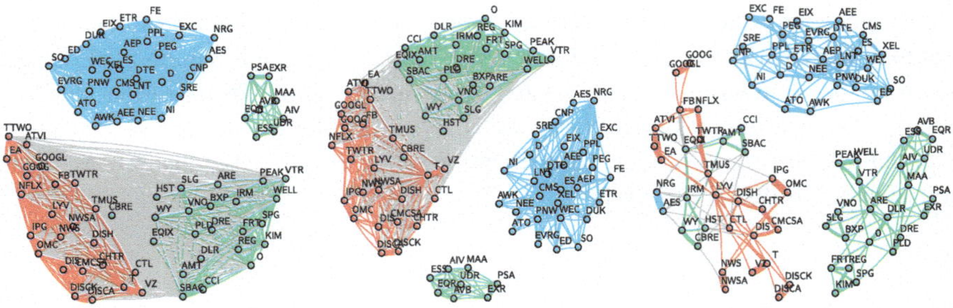

(a) SGL, modularity = 0.29. (b) CLR, modularity = 0.33. (c) Algorithm 10.3, modularity = 0.56.

Fig. 10.3 Learned 3-component graphs of S&P500 stocks

as measured by its modularity value. In addition, a larger number of inter-sector connections, as indicated by gray-colored edges, which are often spurious from a practical perspective, are present in the graphs learned by NGL and GLE. Sparsity regularization provides a means to remove edges between nodes in the presence of data with outliers and possibly increasing the modularity of the resulting graph. However, they bring the additional task of tunning hyperparameters, which is often repetitive and impractical for real-time applications. A cleaner graph, without the need for postprocessing or additional regularization, is obtained directly by using the Student's t assumption.

Figure 10.3 illustrates the learned 3-component graphs during the time from January 3, 2014, to December 29, 2017. We can notice that SGL (Fig. 10.3a) and CLR (Fig. 10.3b) are unable to separate the stocks in a way that agrees with their sector information as given by GICS. In addition, the high number of spurious connections (gray-colored edges) is uncharacteristic of the actual expected behavior in stock markets. Figure 10.3c displays the graph learned by Algorithm 10.3, where it presents not only a higher modularity value but also a sparser, more plausible representation of an actual network of stocks with three sectors.

10.5.1.1 Impact of COVID-19 in the US Stock Market

We collect price data of $p = 85$ stocks belonging to three sectors, namely, Communication Services (red), Utilities (blue), and Real Estate (green) from April 22, 2019, to December 30, 2020, resulting in $n = 429$ observations. The degrees of freedom and annualized volatility during this period, which includes the financial crisis caused by the COVID-19 pandemic, were $\nu \approx 2.89$ and $\sigma \approx 41\%$, indicating a strongly heavy-tailed scenario.

In Fig. 10.4, we observe that the modularity values of the Gaussian-based graphs are severely reduced as compared to the results presented in Fig. 10.2, while the opposite occurs with the Student's t graph. That may be explained by the increase in

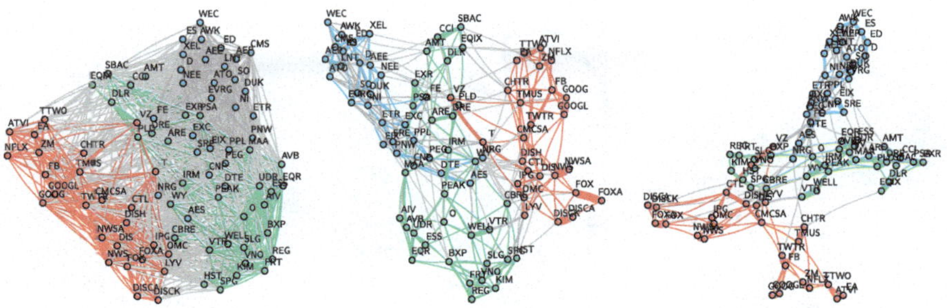

(a) GLE, modularity = 0.23.

(b) NGL, modularity = 0.46.

(c) Algorithm 10.2, modularity = 0.56.

Fig. 10.4 Learned graphs of S&P500 stocks during the COVID-19 pandemic

variability and outliers in the data during the financial crisis caused by the COVID-19 pandemic, which is better modeled by a Student's t-distribution. In addition, the learned graph in Fig. 10.4c clearly displays expected behaviors such as the strong correlation between pairs of companies including {Zoom Video Communications Inc., Netflix Inc.}, and {Twitter Inc., Facebook Inc.}, which are not as evident in Fig. 10.4a and b.

10.5.2 Communities in Foreign Exchange Markets

We query foreign exchange data from the 34 most traded currencies between the period from January 2, 2019, to December 31, 2020, totaling $n = 522$ observations. The data matrix is composed by the log-returns of the currency prices with respect to the United States Dollar. Similar to the previous experiment, we compute the degrees of freedom of a Student's t-distribution fitted to the log-returns data matrix, whereby we obtain $\nu \approx 4.6$, which represents a scenario with considerable amount of heavy-tailedness. Unlike in the experiment involving S&P500 stocks, there are no classification standard for currencies, hence we rely on a community detection algorithm [12] in order to create classes within the learned graph. In particular, the algorithm in [12] takes as input the learned Laplacian matrix of the graph and outputs a membership assignment that maximizes the modularity of the graph.

Figure 10.5 displays the learned graphs. As it can be observed, the Student's t graph (Fig. 10.5c) is sparser, more interpretable, and has a higher modularity value than that of the Gaussian-based graphs (Fig. 10.5a and b). In addition, the expected correlation between currencies of locations geographically close to each other, e.g., {Hong Kong SAR, China}, {Taiwan, South Korea}, and {Poland, Czech Republic} are significantly more evident for the Student's t graph.

Figure 10.6 depicts the learned 9-component graphs of currencies during the time window from January 2, 2019, to December 31, 2020. It can be observed that the

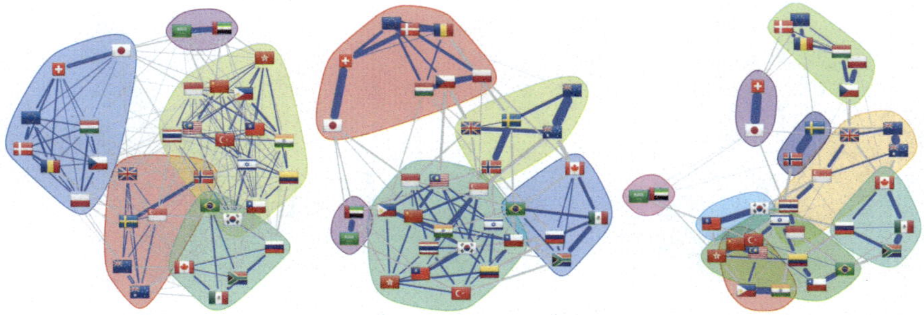

(a) GLE, modularity = 0.34. (b) NGL, modularity = 0.46. (c) Algorithm 10.2, modularity = 0.58.

Fig. 10.5 Learned connected graphs of currencies

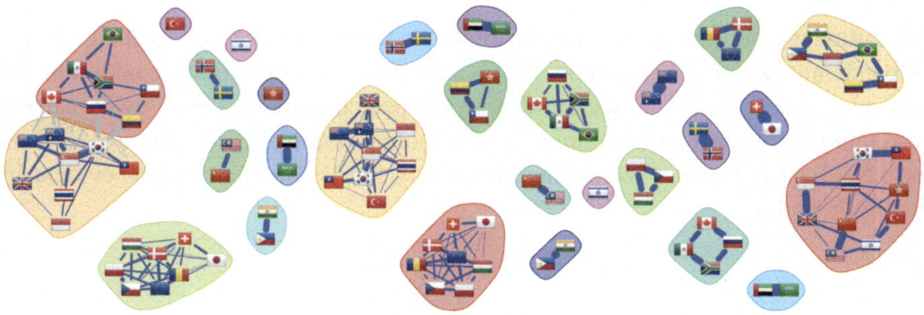

(a) SGL, modularity = 0.62. (b) CLR, modularity = 0.79. (c) Algorithm 10.3, modularity = 0.84.

Fig. 10.6 Learned 9-component graphs of currencies

graph learned by Algorithm 10.3 (Fig. 10.6c) presents a finer structure and a higher modularity value than those learned by SGL (Fig. 10.6a) and CLR (Fig. 10.6b). In addition, the learned graph in Fig. 10.6c presents more reasonable clusters such as {New Zealand, Australia} and {Poland, the Czech Republic, Hungary}, which are not separated in the Gaussian-based graphs. More critically, we can observe that SGL and CLR allow the existence of isolated nodes in the learned graphs. Algorithm 10.3 avoids such solutions by imposing linear constraints on the degree of the graph.

10.5.3 Communities in Cryptocurrencies

We query daily prices of the $p = 41$ most traded cryptocurrencies during the period starting from August 1, 2017, to December 1, 2020, which amounts to $n = 1218$ observations. The degrees of freedom during this time frame was measured as $\nu \approx 3$, which is tantamount to a strong heavy-tail scenario.

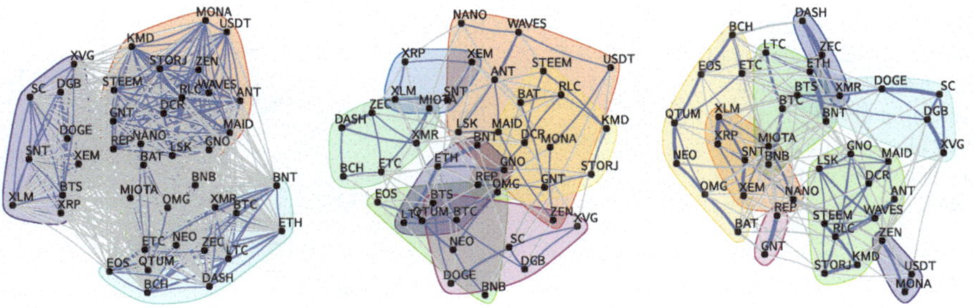

(a) GLE, modularity = 0.19. (b) NGL, modularity = 0.40. (c) Algorithm 10.2, modularity = 0.52.

Fig. 10.7 Learned connected graphs of cryptocurrencies

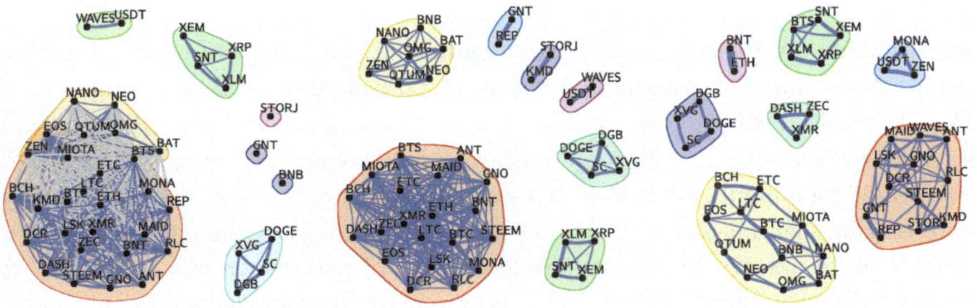

(a) SGL, modularity = 0.36 (b) CLR, modularity = 0.66. (c) Algorithm 10.3, modularity = 0.79.

Fig. 10.8 Learned 7-component graphs of cryptocurrencies

Figure 10.7 shows the learned graphs by GLE, NGL, and Algorithm 10.2. While the graphs in Fig. 10.7a and b present a small modularity value and contain a large number of edges, which impairs interpretability, the resulting graph in Fig. 10.7c reveals a refined representation of the interactions between pairs of cryptocurrencies, which is possibly more aligned with the actual market scenario. As an example, the link between Bitcoin (BTC) and Litecoin (LTC), a Bitcoin spinoff established in 2011, is substantially more evident in Fig. 10.7c.

Figure 10.8 shows the learned 7-component graphs of cryptocurrencies during the aforementioned time window. As in the previous experiments with foreign exchange data, SGL shows isolated nodes in the learned graph (Fig. 10.8a) and CLR contains a large number of spurious connections in the main cluster (Fig. 10.8b), whereas the graph learned via Algorithm 10.3 (Fig. 10.8c) has the largest modularity value. Interestingly, while all three methods agree to cluster {Dogecoin (DOGE), Verge (XVG), Siacoin (SC), and DigiByteCoin (DGB)}, which may be related to their similar initial release dates, only Algorithm 10.3 clusters together the coins that mainly focus on privacy and anonymity features, i.e., {Monero (XMR), Zcash (ZEC), DASH}.

Fig. 10.9 Bipartite graph illustrating the modeling of dependencies between a collection of objects and their classes. The edges correspond to nonzero elements of **W**

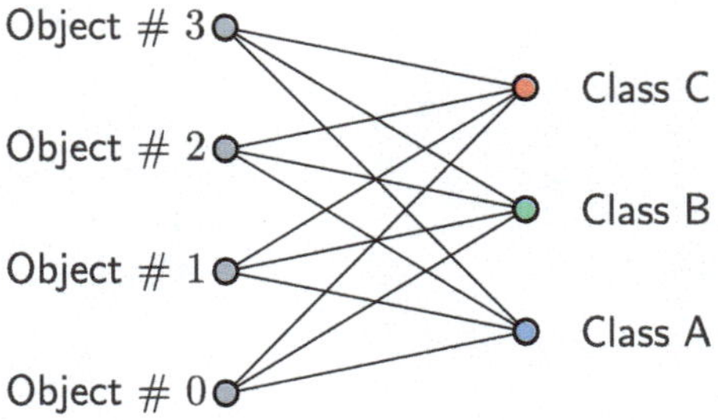

(a) Connected (single-component) bipartite graph.

10.6 Learning Bipartite Graphs

In this section, we focus our attention on learning a special class of graphs: bipartite graphs. An undirected, weighted, bipartite graph can be defined as a 4-tuple $\mathcal{G} \triangleq \left\{\mathcal{V}_r, \mathcal{V}_q, \mathcal{E}, \mathbf{W}\right\}$, where $\mathcal{V}_r \triangleq \{1, 2, \ldots, r\}$ and $\mathcal{V}_q \triangleq \{r+1, r+2, \ldots, r+q\}$ are the node sets associated to a group of objects and classes, respectively, where $p \triangleq r+q$ is the total number of nodes, and we assume that $r \gg q$, i.e., there exist many more objects than classes; $\mathcal{E} \subseteq \left\{\{u, v\} : u \in \mathcal{V}_r, v \in \mathcal{V}_q\right\}$ is the edge set, that is, a subset of the set of all possible unordered pairs of nodes such that $\{u, v\} \in \mathcal{E}$ iff nodes u and v are connected; $\mathbf{W} \in \mathbb{R}_+^{p\times p}$ is the symmetric weighted adjacency matrix that satisfies $W_{ii} = 0$, $W_{ij} > 0$ iff $\{i, j\} \in \mathcal{E}$ and $W_{ij} = 0$, otherwise. The weighted Laplacian matrix of a graph is defined as $\mathbf{L} \triangleq \mathsf{Diag}(\mathbf{W}\mathbf{1}) - \mathbf{W}$. Figure 10.9 displays an instance of this model.

The Laplacian matrix $\mathbf{L}$ of a bipartite graph can be written as

$$\mathbf{L} = \begin{bmatrix} \mathsf{Diag}\left(\mathbf{B}\mathbf{1}_q\right) & -\mathbf{B} \\ -\mathbf{B}^\top & \mathsf{Diag}\left(\mathbf{B}^\top\mathbf{1}_r\right) \end{bmatrix}, \tag{10.33}$$

where $\mathbf{B} \in \mathbb{R}_+^{r\times q}$ contains the edge weights between the nodes of objects and the nodes of classes. Note that $\mathbf{L}$ satisfies $\mathbf{L}\mathbf{1} = \mathbf{0}$ and $L_{ij} = L_{ji} \leq 0 \; \forall\, i \neq j$ by definition.

In what follows, we briefly discuss the two main approaches proposed in the literature to tackle the estimation of bipartite graphs. They basically follow two avenues: (*i*) exploiting the bipartite structure via (10.33) as in [48] and (*ii*) exploiting the structure of the spectral properties of graph matrices as in [39].

Bipartite Structure. The authors in [50] proposed the following optimization problem to learn a k-component bipartite graph from a given bipartite graph weights $\mathbf{A} \in \mathbb{R}^{r\times q}$. More precisely, their estimator is obtained as the solution of the following optimization problem:

$$\begin{array}{ll} \underset{\mathbf{B},\mathbf{V}\in\mathbb{R}^{p\times k}}{\mathsf{minimize}} & \|\mathbf{B}-\mathbf{A}\|_{\mathrm{F}}^2 + \eta\,\mathrm{tr}\left(\mathbf{V}^\top\mathbf{L}\mathbf{V}\right), \\ \mathsf{subject\ to} & \mathbf{B} \geq \mathbf{0},\ \mathbf{B}\mathbf{1}_q = \mathbf{1}_r,\ \mathbf{V}^\top\mathbf{V} = \mathbf{I}_k, \end{array} \tag{10.34}$$

where $\mathbf{L}$ depends on $\mathbf{B}$ through (10.33), $\eta > 0$ is a hyperparameter that promotes the rank of $\mathbf{L}$ to be $p-k$, and $\mathbf{A}$ can be constructed from the correlation between nodes of objects and classes.

An alternating minimization approach was proposed in [50] to solve Problem (10.34). For a fixed value of $\mathbf{V}$, it reduces to a constrained quadratic program, which can be solved efficiently via standard quadratic programming solvers. On the other hand, for a fixed $\mathbf{B}$, the optimal solution of $\mathbf{V}$ are the k eigenvectors of $\mathbf{L}$ associated to its smallest eigenvalues. A shortcoming of this formulation is it lacks statistical support, which makes it difficult to draw statistical conclusions about the learned graph as well as to consider this model under different statistical assumptions.

Spectral Regularization. Properties associated with the spectral decomposition of graph matrices have demonstrated prolific advantages that enable learning graphs with specific structures, such as bipartite and k-component graphs [38, 39, 48–50]. Let $\mathbf{A} = \mathsf{UDiag}(\boldsymbol{\psi})\mathbf{U}^\top$ and $\mathbf{L} = \mathbf{V}\mathsf{Diag}(\boldsymbol{\lambda})\mathbf{V}^\top$ be the spectral decomposition of the adjacency and Laplacian matrices, respectively. For a k-component graph, the eigenvalues of $\mathbf{L}$ belong to the set

$$C_\lambda = \left\{\boldsymbol{\lambda} \in \mathbb{R}^p : \{\lambda_i = 0\}_{i=1}^k, c_1 \leq \lambda_{k+1} \leq \lambda_{k+2} \leq \cdots \leq \lambda_p \leq c_2\right\},$$

where c_1 and c_2 are positive constants. On the other hand, the adjacency matrix of a bipartite graph satisfies the following spectral properties [28]: (*i*) $\boldsymbol{\psi}$ contains exactly $r-q$ zero elements and (*ii*) $\boldsymbol{\psi}$ is anti-symmetric, i.e., $\boldsymbol{\psi} \in C_{\boldsymbol{\psi}}$, where

$$C_{\boldsymbol{\psi}} = \left\{\boldsymbol{\psi} \in \mathbb{R}^p : \psi_i = -\psi_{p+1-i},\ i = 1, 2, \ldots, p,\ \psi_1 \geq \psi_2 \geq \cdots \geq \psi_p\right\}.$$

By combining these facts, the authors in [39] proposed the following optimization program to learn a k-component bipartite graph based on the GMRF framework:

$$\begin{aligned}
\underset{\mathbf{w}\geq 0, \mathbf{V}, \mathbf{U}, \boldsymbol{\psi}, \boldsymbol{\lambda}}{\mathsf{minimize}}\ & \operatorname{tr}(\mathcal{L}\mathbf{w}\mathbf{S}) - \log\det{}^*(\mathcal{L}\mathbf{w}) + \tfrac{\gamma}{2}\left\|\mathcal{A}\mathbf{w} - \mathsf{UDiag}(\boldsymbol{\psi})\mathbf{U}^\top\right\|_F^2 \\
& + \tfrac{\beta}{2}\left\|\mathcal{L}\mathbf{w} - \mathbf{V}\mathsf{Diag}(\boldsymbol{\lambda})\mathbf{V}^\top\right\|_F^2, \\
\mathsf{subject\ to}\ & \mathbf{U}^\top\mathbf{U} = \mathbf{I},\ \mathbf{U} \in \mathbb{R}^{p\times p},\ \psi \in C_\psi,\ \mathbf{V}^\top\mathbf{V} = \mathbf{I},\ \mathbf{V} \in \mathbb{R}^{p\times p},\ \boldsymbol{\lambda} \in C_\lambda,
\end{aligned} \tag{10.35}$$

where $\mathcal{L}$ and $\mathcal{A}$ are the Laplacian and adjacency operators and $\mathbf{w}$ is the vector of graph weights [39]. Although this model is statistically sound, we note that this approach does not always guarantee that $\mathcal{A}\mathbf{w}^\star$ is a bipartite graph because the equality $\mathcal{A}\mathbf{w}^\star = \mathbf{U}^\star\mathsf{Diag}(\boldsymbol{\psi}^\star)\mathbf{U}^{\star\top}$ is often not met, further requiring postprocessing of $\mathcal{A}\mathbf{w}^\star$, which is often problematic in practical settings. In addition, the lack of constraints on the node degrees allows trivial solutions with isolated nodes, which is also undesired in practice. The method in [39] does not require knowledge of the partition of the node set, however, it does assume that the number of nodes in each set is known.

10.6.1 Gaussian Bipartite Graphs

In what follows, we discuss methods and algorithms for learning bipartite graphs, starting off with the simple case of a connected bipartite graph under Gaussian settings. It is important to look at the Gaussian scenario first, as this case will serve as a building block for more elaborate formulations that include additional statistical assumptions, such as heavy tails, and its extension to k-component bipartite graphs.

Noting that the adjacency matrix of a bipartite graph can be partitioned as $\mathbf{W} = \begin{bmatrix} \mathbf{0} & \mathbf{B} \\ \mathbf{B}^\top & \mathbf{0} \end{bmatrix}$ [28], where $\mathbf{B} \in \mathbb{R}_+^{r \times q}$, we can formulate its MLE as the following convex optimization program:

$$\begin{aligned} \underset{\mathbf{L},\mathbf{B}\geq\mathbf{0}}{\mathsf{minimize}} \quad & \mathrm{tr}\,(\mathbf{LS}) - \log\det\,(\mathbf{L}+\mathbf{J}), \\ \mathsf{subject\ to} \quad & \mathbf{L} = \begin{bmatrix} \mathsf{Diag}\,(\mathbf{B}\mathbf{1}_q) & -\mathbf{B} \\ -\mathbf{B}^\top & \mathsf{Diag}\,(\mathbf{B}^\top\mathbf{1}_r) \end{bmatrix}, \end{aligned} \tag{10.36}$$

where we have used the fact that, for a connected graph, $\det^*(\mathbf{L}) = \det\,(\mathbf{L}+\mathbf{J})$, $\mathbf{J} \triangleq \frac{1}{p}\mathbf{1}\mathbf{1}^\top$[21].

We can reformulate Problem (10.36) by taking advantage of the symmetry of the similarity matrix $\mathbf{S}$, i.e., $\begin{bmatrix} \mathbf{S}_{rr} & \mathbf{S}_{rq} \\ \mathbf{S}_{rq}^\top & \mathbf{S}_{qq} \end{bmatrix}$, where $\mathbf{S}_{rq} \in \mathbb{R}^{r\times q}$ contains the pairwise similarities between the nodes in the set of objects $\mathcal{V}_r$ and the nodes in the set of classes $\mathcal{V}_q$. In particular, we have

$$\mathrm{tr}\,(\mathbf{LS}) = \mathsf{diag}(\mathbf{S})^\top \begin{bmatrix} \mathbf{B}\mathbf{1}_q \\ \mathbf{B}^\top\mathbf{1}_r \end{bmatrix} - 2\mathrm{tr}(\mathbf{B}\mathbf{S}_{rq}^\top) = \mathrm{tr}\left(\mathbf{B}\left(\mathbf{1}_q\mathbf{s}_{1:r}^\top + \mathbf{s}_{r+1:p}\mathbf{1}_r^\top - 2\mathbf{S}_{rq}^\top\right)\right), \tag{10.37}$$

where $\mathbf{s} \triangleq \mathsf{diag}(\mathbf{S})$.

Plugging (10.33) and (10.37) in (10.36), we arrive at the following formulation to learn a connected bipartite graph:

$$\begin{aligned} \underset{\mathbf{B}\geq\mathbf{0}}{\mathsf{minimize}}\ & \mathrm{tr}\left(\mathbf{B}\left(\mathbf{1}_q\mathbf{s}_{1:r}^\top + \mathbf{s}_{r+1:p}\mathbf{1}_r^\top - 2\mathbf{S}_{rq}^\top\right)\right) \\ & - \log\det\left(\begin{bmatrix} \mathsf{Diag}(\mathbf{B}\mathbf{1}_q) + \mathbf{J}_{rr} & -\mathbf{B} + \mathbf{J}_{rq} \\ -\mathbf{B}^\top + \mathbf{J}_{qr} & \mathsf{Diag}(\mathbf{B}^\top\mathbf{1}_r) + \mathbf{J}_{qq} \end{bmatrix}\right). \end{aligned} \tag{10.38}$$

We discuss an optimization algorithm based on the PGD framework [2, 3] that amounts to two steps: (1) computation of a descent update and (2) projection onto the feasible set.

A common way to compute a descent direction is using the gradient of the objective function. The gradient of the objective function in Problem (10.38) would involve the computation of the inverse of the matrix inside the log-determinant term, which in general costs $O(p^3)$. By leveraging the block structure of the Laplacian matrix of bipartite graphs, and using the classical matrix determinant lemma for block matrices that relates the determinant of a block matrix with its Schur complement [78, p. 4], we have the following equality:

$$\log\det\left(\begin{bmatrix}\mathsf{Diag}(\mathbf{B}\mathbf{1}_q)+\mathbf{J}_{rr} & -\mathbf{B}+\mathbf{J}_{rq}\\ -\mathbf{B}^\top+\mathbf{J}_{qr} & \mathsf{Diag}(\mathbf{B}^\top\mathbf{1}_r)+\mathbf{J}_{qq}\end{bmatrix}\right)=\log\det\left(\mathsf{Diag}(\mathbf{B}\mathbf{1}_q)+\mathbf{J}_{rr}\right)+$$
$$\log\det\left(\mathsf{Diag}(\mathbf{B}^\top\mathbf{1}_r)+\mathbf{J}_{qq}-(-\mathbf{B}^\top+\mathbf{J}_{qr})(\mathsf{Diag}(\mathbf{B}\mathbf{1}_q)+\mathbf{J}_{rr})^{-1}(-\mathbf{B}+\mathbf{J}_{rq})\right). \quad (10.39)$$

In practice, we also would like to normalize the degrees of the nodes in the objects group, so that the edge weights can be directly interpreted as the degree of membership of an object to a particular class. Mathematically, this can be achieved by a simple linear constraint $\mathbf{B}\mathbf{1}_q=\mathbf{1}_r$.

Plugging the above constraint and equality (10.39) into (10.38), we have the following formulation:

$$\underset{\mathbf{B}\geq\mathbf{0},\mathbf{B}\mathbf{1}_q=\mathbf{1}_r}{\mathsf{minimize}}\ \mathrm{tr}\,(\mathbf{B}\mathbf{C})-\log\det\left(\mathsf{Diag}(\mathbf{B}^\top\mathbf{1}_r)+\mathbf{J}_{qq}-(\mathbf{B}-\mathbf{J}_{rq})^\top(\mathbf{I}_r+\mathbf{J}_{rr})^{-1}(\mathbf{B}-\mathbf{J}_{rq})\right), \quad (10.40)$$

where $\mathbf{C}\triangleq\left(\mathbf{s}_{r+1:p}\mathbf{1}_r^\top-2\mathbf{S}_{rq}^\top\right)$.

Let $f(\mathbf{B})\triangleq-\log\det(g(\mathbf{B}))+\mathrm{tr}\,(\mathbf{B}\mathbf{C})$ be the objective function of Problem (10.40), where $g(\mathbf{B})\triangleq\mathsf{Diag}(\mathbf{B}^\top\mathbf{1}_r)+\mathbf{J}_{qq}-(\mathbf{B}-\mathbf{J}_{rq})^\top(\mathbf{I}_r+\mathbf{J}_{rr})^{-1}(\mathbf{B}-\mathbf{J}_{rq})$. Then the gradient of $f(\mathbf{B})$ can be computed as

$$\nabla f(\mathbf{B})=\mathbf{1}_r\left[\mathsf{diag}\left(-g(\mathbf{B})^{-1}\right)\right]^\top-2(\mathbf{I}_r+\mathbf{J}_{rr})^{-1}(\mathbf{B}-\mathbf{J}_{rq})\left(-g(\mathbf{B})^{-1}\right)^\top+\mathbf{C}^\top. \quad (10.41)$$

The PGD update is formulated as

$$\mathbf{B}^{l+1}=\underset{\mathbf{B}\geq\mathbf{0},\mathbf{B}\mathbf{1}_q=\mathbf{1}_r}{\arg\min}\ \left\|\mathbf{B}-\left(\mathbf{B}^l-\alpha_l\nabla f(\mathbf{B}^l)\right)\right\|_{\mathrm{F}}^2\triangleq P_\Delta\left(\mathbf{B}^l-\alpha_l\nabla f(\mathbf{B}^l)\right). \quad (10.42)$$

Problem (10.42) is an Euclidean projection of the rows of $\mathbf{B}^l-\alpha_l\nabla f(\mathbf{B}^l)$ onto the probability simplex. The unique solution to Problem (10.42) can be found efficiently via several algorithms [4, 45, 53] whose theoretical worst-case complexity is $O(rq^2)$ but their observed practical complexity is $O(rq)$ [13].

Finally, to validate the point $\mathbf{B}^{l+1}$ and to adaptively update the learning rate α_l, we check the following backtracking condition [2]:

$$f\left(\mathbf{B}^{l+1}\right)\leq f\left(\mathbf{B}^l\right)+\left\langle\nabla f\left(\mathbf{B}^l\right),\mathbf{B}^{l+1}-\mathbf{B}^l\right\rangle+\frac{1}{2\alpha_l}\|\mathbf{B}^{l+1}-\mathbf{B}^l\|_{\mathrm{F}}^2. \quad (10.43)$$

In practice, we decrease the learning rate $\alpha_l\in(0,1)$ until condition (10.43) is satisfied or convergence has been achieved. Algorithm 10.4 summarizes the scheme for learning a bipartite graph from a Gaussian MRF assumption. Since the objective function of Problem (10.40) is convex and its feasible set is compact, Algorithm 10.4 converges to a global minimum [3].

Algorithm 10.4 Gaussian bipartite graph learning (GBG)

Data: Similarity matrix $\mathbf{S}$, initial feasible estimate of the graph weights $\mathbf{B}^0$, initial learning rate $\alpha_0 > 0$, tolerance $\epsilon > 0$.

Result: Optimal bipartite graph: $\mathbf{B}^\star$

```
while l ≤ maxiter do
    ▷ B^{l+1} ← P_△(B^l − α_l ∇f(B^l)), where α_l is chosen such that (10.43) is satisfied
    if ‖B^{l+1} − B^l‖_F / ‖B^l‖_F ≤ ϵ then
        return B^{l+1}
    end
end
```

10.6.2 Student's t Bipartite Graphs

Outliers and heavy-tailed events are pervasive in modern datasets [55]. To account for this phenomenon, instead of assuming a Gaussian generative process, as done in [39] and in Problem (10.36), we assume the data-generating process to be modeled by a multivariate zero-mean Student's t-distribution, whose probability density function can be written as

$$p(\mathbf{x}) \propto \sqrt{\det^*(\boldsymbol{\Theta})}\left(1 + \frac{\mathbf{x}^\top \boldsymbol{\Theta}\mathbf{x}}{\nu}\right)^{-(\nu+p)/2}, \tag{10.44}$$

where $\boldsymbol{\Theta}$ is a positive-semidefinite inverse scatter matrix modeled as a combinatorial graph Laplacian matrix, and $\nu > 2$ is the number of degrees of freedom that controls the rate of decay of the tails. More precisely, as $\nu \to \infty$ the tails of (10.44) approach those of a Gaussian distribution.

This results in a robustified version of the MLE for connected graph learning, i.e.,

$$\begin{aligned} &\underset{\mathbf{L},\mathbf{B}}{\text{minimize}} \quad -\log\det(\mathbf{L}+\mathbf{J}) + \frac{p+\nu}{n}\sum_{i=1}^{n}\log\left(1+\tfrac{1}{\nu}\mathbf{x}_i^\top\mathbf{L}\mathbf{x}_i\right), \\ &\text{subject to } \mathbf{L} = \begin{bmatrix} \mathbf{I}_r & -\mathbf{B} \\ -\mathbf{B}^\top & \mathsf{Diag}\left(\mathbf{B}^\top\mathbf{1}_r\right)\end{bmatrix}, \ \mathbf{B} \geq \mathbf{0}, \ \mathbf{B}\mathbf{1}_q = \mathbf{1}_r. \end{aligned} \tag{10.45}$$

We can further simplify the objective function of Problem (10.45) by noting that

$$\mathbf{x}_i^\top\mathbf{L}\mathbf{x}_i = \mathbf{x}_i^\top\begin{bmatrix} \mathbf{I}_r & -\mathbf{B} \\ -\mathbf{B}^\top & \mathsf{Diag}\left(\mathbf{B}^\top\mathbf{1}_r\right)\end{bmatrix}\mathbf{x}_i = h_i + \operatorname{tr}\left(\mathbf{B}\mathbf{G}_i\right), \tag{10.46}$$

where $\mathbf{G}_i \triangleq \mathsf{diag}(\mathbf{x}_i\mathbf{x}_i^\top)_{r+1:p}\mathbf{1}_r^\top - 2\left(\mathbf{x}_i\mathbf{x}_i^\top\right)_{rq}^\top$ and $h_i \triangleq \mathbf{1}_r^\top\mathsf{diag}(\mathbf{x}_i\mathbf{x}_i^\top)_{1:r}$.

Plugging (10.39) and (10.46) into (10.45), we have the following optimization program to learn a heavy-tailed bipartite graph:

$$\underset{\mathbf{B}\geq\mathbf{0},\ \mathbf{B}\mathbf{1}_q=\mathbf{1}_r}{\mathsf{minimize}} \ -\log\det\left(\mathsf{Diag}(\mathbf{B}^\top\mathbf{1}_r)+\mathbf{J}_{qq}-(\mathbf{B}-\mathbf{J}_{rq})^\top(\mathbf{I}_r+\mathbf{J}_{rr})^{-1}(\mathbf{B}-\mathbf{J}_{rq})\right) \\ +\frac{p+\nu}{n}\sum_{i=1}^{n}\log\left(1+\frac{h_i+\mathrm{tr}\,(\mathbf{B}\mathbf{G}_i)}{\nu}\right). \tag{10.47}$$

Problem (10.47) is, in general, nonconvex due to the summation over concave terms and hence it is difficult to be dealt with directly. To tackle this issue, we leverage the MM framework that generates a sequence of updates, where in each iteration an upper-bounded subproblem is solved [64, 65].

To find such upper-bound, we use the fact that the logarithm term in Problem (10.47) can be globally upper-bounded by its first-order Taylor expansion, i.e., for all $a\geq 0, t\geq 0, b>0$, we have $\log\left(1+\frac{t}{b}\right)\leq\log\left(1+\frac{a}{b}\right)+\frac{t-a}{a+b}$, hence, at a point $\mathbf{B}^j$, we have that

$$\log\left(1+\frac{h_i+\mathrm{tr}\,(\mathbf{B}\mathbf{G}_i)}{\nu}\right)\leq\frac{\mathrm{tr}\,(\mathbf{B}\mathbf{G}_i)}{\nu+h_i+\mathrm{tr}\left(\mathbf{B}^j\mathbf{G}_i\right)}+c,$$

where $c\triangleq\log\left(1+\frac{h_i+\mathrm{tr}\left(\mathbf{B}^j\mathbf{G}_i\right)}{\nu}\right)-\frac{\mathrm{tr}\left(\mathbf{B}^j\mathbf{G}_i\right)}{\nu+h_i+\mathrm{tr}\left(\mathbf{B}^j\mathbf{G}_i\right)}$ is a constant that does not depend on $\mathbf{B}$.

Hence, to find an update $\mathbf{B}^{l+1}$, we iterate the solution of the following upper-bounded convex subproblem:

$$\mathbf{B}^{j+1}=\begin{array}{ll}\underset{\mathbf{B}}{\mathsf{arg\,min}} & -\log\det\left(\mathsf{Diag}(\mathbf{B}^\top\mathbf{1}_r)+\mathbf{J}_{qq}-(\mathbf{B}-\mathbf{J}_{rq})^\top(\mathbf{I}_r+\mathbf{J}_{rr})^{-1}(\mathbf{B}-\mathbf{J}_{rq})\right) \\ & +\,\mathrm{tr}\left(\mathbf{B}\mathbf{M}^j\right), \\ \mathsf{subject\ to} & \mathbf{B}\geq\mathbf{0},\ \mathbf{B}\mathbf{1}_q=\mathbf{1}_r,\end{array} \tag{10.48}$$

where $\mathbf{M}^j\triangleq\frac{p+\nu}{n}\sum_{i=1}^{n}\frac{\mathbf{G}_i}{\nu+h_i+\mathrm{tr}\left(\mathbf{B}^j\mathbf{G}_i\right)}$. We observe in practice that a few (≈ 5) iterations are sufficient for convergence.

Problem (10.48) has the same format as Problem (10.40) and it can be readily solved by Algorithm 10.4. The learning scheme for Student's t bipartite graphs is summarized in Algorithm 10.5. The sequence $\left\{\mathbf{B}^l\right\}$ generated by Algorithm 10.5 converges to the set of stationary points of Problem (10.47). This result directly follows from the convergence results of the MM framework in [65].

Algorithm 10.5 Student-t bipartite graph learning (SBG)

Data: Data matrix $\mathbf{X}$, initial feasible estimate of the graph weights $\mathbf{B}^0$, initial learning rate $\alpha_0 > 0$, tolerance $\epsilon > 0$, degree of freedoms $\nu > 2$.
Result: Learned bipartite graph: $\mathbf{B}^{l+1}$

```
while l ≤ maxiter do
    ▷ update B^{l+1} by iterating (10.48)
    if ‖B^{l+1} − B^l‖_F / ‖B^l‖_F ≤ ε then
        return B^{l+1}
    end
end
```

10.6.3 k-Component Student's t Bipartite Graphs

Graphs with multiple components can be learned by introducing rank constraints on the Laplacian matrix [39, 49, 50]. More precisely, the number of zero eigenvalues of the Laplacian matrix amounts to the number of graph components, i.e., $\mathsf{rank}(\mathbf{L}) = p - k$, where k is the number of components.

The approach taken by [50] approximates the rank constraint by applying Fan's theorem [23] as follows:

$$\sum_{i=1}^{k} \lambda_i(\mathbf{L}) = \underset{\mathbf{V}\in\mathbb{R}^{p\times k}, \mathbf{V}^\top\mathbf{V}=\mathbf{I}_k}{\mathsf{minimize}} \ \mathsf{tr}\left(\mathbf{V}^\top\mathbf{L}\mathbf{V}\right), \tag{10.49}$$

where the right-hand side of (10.49) can be used as a regularization term (*cf.* Problem (10.34)). On the other hand, the authors in [39] introduced the quadratic relaxation $\left\|\mathbf{L} - \mathbf{U}\boldsymbol{\Lambda}\mathbf{U}^\top\right\|_\mathrm{F}^2$. Those formulations [39, 50] can be conveniently solved by coordinate descent methods. However, they introduce additional variables to be optimized, extra hyperparameters to be tuned, and do not guarantee that the rank constraint on $\mathbf{L}$ is satisfied.

In the method presented here, we directly apply the rank constraint into the optimization formulation. More precisely, using the Student's t-case, we formulate the following optimization problem to learn a k-component bipartite graph:

$$\begin{array}{ll} \underset{\mathbf{L}\succeq\mathbf{0},\mathbf{B}}{\mathsf{minimize}} & \dfrac{p+\nu}{n}\displaystyle\sum_{i=1}^{n}\log\left(1+\dfrac{h_i+\mathrm{tr}\left(\mathbf{B}\mathbf{G}_i\right)}{\nu}\right) - \log\det{}^*\left(\mathbf{L}\right), \\ \mathsf{subject\ to} \ \mathbf{L} = \begin{bmatrix} \mathbf{I}_r & -\mathbf{B} \\ -\mathbf{B}^\top & \mathsf{Diag}\left(\mathbf{B}^\top\mathbf{1}_r\right)\end{bmatrix}, & \mathsf{rank}(\mathbf{L}) = p-k, \ \mathbf{B}\geq\mathbf{0}, \mathbf{B}\mathbf{1}_q = \mathbf{1}_r. \end{array} \tag{10.50}$$

Unlike the case for connected bipartite graphs in Problem (10.38), the pseudo-determinant term in (10.50) cannot be simplified through the block matrix determinant lemma [78]. Therefore, we design an iterative algorithm based on the ADMM framework [5] for Problem (10.50).

10.6.3.1 ADMM Solution

The augmented Lagrangian of Problem (10.50) can be written as

$$L_\rho(\mathbf{L}, \mathbf{B}, \mathbf{Y}) = \frac{p+\nu}{n} \sum_{i=1}^{n} \log\left(1 + \frac{h_i + \operatorname{tr}(\mathbf{B}\mathbf{G}_i)}{\nu}\right) - \log\det{}^*(\mathbf{L})$$
$$+ \left\langle \mathbf{L} - \begin{bmatrix} \mathbf{I}_r & -\mathbf{B} \\ -\mathbf{B}^\top & \mathsf{Diag}(\mathbf{B}^\top \mathbf{1}_r) \end{bmatrix}, \mathbf{Y} \right\rangle + \frac{\rho}{2} \left\| \mathbf{L} - \begin{bmatrix} \mathbf{I}_r & -\mathbf{B} \\ -\mathbf{B}^\top & \mathsf{Diag}(\mathbf{B}^\top \mathbf{1}_r) \end{bmatrix} \right\|_{\mathrm{F}}^2, \quad (10.51)$$

where $\rho > 0$ is a penalty hyperparameter.

For fixed $\mathbf{B}$ and $\mathbf{Y}$, the subproblem for $\mathbf{L}$ can be written as

$$\mathbf{L}^\star = \underset{\operatorname{rank}(\mathbf{L})=p-k}{\mathsf{arg\,min}} \; -\log\det{}^*(\mathbf{L}) + \langle \mathbf{L}, \mathbf{Y} \rangle + \frac{\rho}{2} \left\| \mathbf{L} - \begin{bmatrix} \mathbf{I}_r & -\mathbf{B} \\ -\mathbf{B}^\top & \mathsf{Diag}(\mathbf{B}^\top \mathbf{1}_r) \end{bmatrix} \right\|_{\mathrm{F}}^2, \quad (10.52)$$

whose closed-form solution is given by Lemma 10.3.

Lemma 10.3 *The global minimizer of Problem* (10.52) *is [16, 69]*

$$\mathbf{L}^\star = \frac{1}{2\rho} \mathbf{R} \left(\mathbf{\Gamma} + \sqrt{\mathbf{\Gamma}^2 + 4\rho \mathbf{I}} \right) \mathbf{R}^\top, \quad (10.53)$$

where $\mathbf{R}\mathbf{\Gamma}\mathbf{R}^\top$ *is the eigenvalue decomposition of* $\rho \begin{bmatrix} \mathbf{I}_r & -\mathbf{B} \\ -\mathbf{B}^\top & \mathsf{Diag}(\mathbf{B}^\top \mathbf{1}_r) \end{bmatrix} - \mathbf{Y}$, *with* $\mathbf{\Gamma}$ *having the largest* $p-k$ *eigenvalues along its diagonal and* $\mathbf{R} \in \mathbb{R}^{p \times (p-k)}$ *containing the corresponding eigenvectors.*

Fixing $\mathbf{L}$ and $\mathbf{Y}$, the subproblem for $\mathbf{B}$ is

$$\mathbf{B}^\star = \underset{\mathbf{B} \geq \mathbf{0}, \mathbf{B}\mathbf{1}=\mathbf{1}}{\mathsf{arg\,min}} \; \frac{p+\nu}{n} \sum_{i=1}^{n} \log\left(1 + \frac{h_i + \operatorname{tr}(\mathbf{B}\mathbf{G}_i)}{\nu}\right) + \frac{\rho}{2} \left\| \mathbf{L} - \begin{bmatrix} \mathbf{I}_r & -\mathbf{B} \\ -\mathbf{B}^\top & \mathsf{Diag}\left(\mathbf{B}^\top \mathbf{1}_r\right) \end{bmatrix} \right\|_{\mathrm{F}}^2$$
$$- \left\langle \begin{bmatrix} \mathbf{I}_r & -\mathbf{B} \\ -\mathbf{B}^\top & \mathsf{Diag}\left(\mathbf{B}^\top \mathbf{1}_r\right) \end{bmatrix}, \mathbf{Y} \right\rangle, \quad (10.54)$$

which can be simplified as

$$\mathbf{B}^\star = \underset{\mathbf{B} \geq \mathbf{0}, \mathbf{B}\mathbf{1}=\mathbf{1}}{\mathsf{arg\,min}} \; \frac{p+\nu}{n} \sum_{i=1}^{n} \log\left(1 + \frac{h_i + \operatorname{tr}(\mathbf{B}\mathbf{G}_i)}{\nu}\right) + \operatorname{tr}(\mathbf{B}\mathbf{H}) + \rho \|\mathbf{B}\|_{\mathrm{F}}^2 + \frac{\rho}{2} \mathbf{1}_r^\top \mathbf{B}\mathbf{B}^\top \mathbf{1}_r, \quad (10.55)$$

where $\mathbf{H} \triangleq 2\mathbf{Y}_{rq}^\top - \mathbf{y}_{r+1:p}\mathbf{1}_r^\top - \rho\left(\mathbf{l}_{r+1:p}\mathbf{1}_r^\top - 2\mathbf{L}_{rq}^\top\right)$, $\mathbf{y}_{r+1:p} \triangleq \mathsf{diag}(\mathbf{Y})_{r+1:p}$, and $\mathbf{l}_{r+1:p} \triangleq \mathsf{diag}(\mathbf{L})_{r+1:p}$.

Leveraging the MM framework to Problem (10.55), like it was applied to Problem (10.47), we can find an upper-bounded subproblem at point $\mathbf{B}^j$ as follows:

$$\mathbf{B}^{j+1} = \underset{\mathbf{B}\geq\mathbf{0},\mathbf{B}\mathbf{1}=\mathbf{1}}{\mathsf{arg\ min}}\ \mathrm{tr}\left(\mathbf{B}\left(\mathbf{H}+\mathbf{M}^j\right)\right) + \rho\,\|\mathbf{B}\|_{\mathrm{F}}^2 + \frac{\rho}{2}\mathbf{1}_r^\top\mathbf{B}\mathbf{B}^\top\mathbf{1}_r. \tag{10.56}$$

Subproblem (10.56) is strongly convex, thus it can be solved efficiently via convex solvers. Finally, to find an update $\mathbf{B}^\star$, we iterate the solution of Problem (10.56). The dual variable $\mathbf{Y}$ is updated approximately via gradient ascent.

Algorithm 10.6 summarizes the discussed scheme to learn k-component bipartite graphs.

Algorithm 10.6 Student-t k-component bipartite graph learning (kSBG)

Data: Data matrix $\mathbf{X}$, number of components k, initial feasible estimate of the graph weights $\mathbf{B}^0$, penalty hyperparameter $\rho > 0$, tolerance $\epsilon > 0$, degrees of freedom $\nu > 2$.

Result: Learned bipartite graph: $\mathbf{B}^{l+1}$

while $l \leq$ maxiter **do**
- ▷ update $\mathbf{L}^{l+1}$ via (10.53)
- ▷ update $\mathbf{B}^{l+1}$ by iterating (10.56)
- ▷ update $\mathbf{Y}^{l+1} = \mathbf{Y}^l - \rho\left(\mathbf{L}^{l+1} - \begin{bmatrix} \mathbf{I}_r & -\mathbf{B}^{l+1} \\ -\mathbf{B}^{l+1\top} & \mathsf{Diag}\left(\mathbf{B}^{l+1\top}\mathbf{1}_r\right)\end{bmatrix}\right)$
- **if** $\|\mathbf{B}^{l+1} - \mathbf{B}^l\|_{\mathrm{F}}/\|\mathbf{B}^l\|_{\mathrm{F}} \leq \epsilon$ **then**
  - **return** $\mathbf{B}^{l+1}$
- **end**

end

10.6.4 *Experimental Results*

We conduct experiments to illustrate the performance of the discussed algorithms in terms of quantitative measures such as node label accuracy and graph modularity. Accuracy is computed as the ratio between the number of correctly predicted node labels and the number of nodes in the objects set, whereas modularity measures the strength of division of a graph into groups [47]. A high modularity value means that the nodes from the same group are more likely to be connected.
Benchmarks: We compare the presented algorithms with state-of-the-art methods in two settings: (*i*) connected bipartite graphs, where SOBG ($k = 1$) [50] and SGA [39] are the benchmarks, and (*ii*) k-component bipartite graphs, where SOBG ($k > 1$) and SGLA [39] act as benchmarks. In our Algorithm 10.6, we set the hyperparameter $\rho = 1$ and the relative tolerance $\epsilon = 10^{-5}$. The hyperparameter of SOBG was tunned according to the heuristic procedure described in [50]. The presented algorithms and benchmarks were implemented using the R programming language.

For SGA and SGLA, we relied on their official implementation via the package spectralGraphTopology [7].

10.6.4.1 Returns of S&P500 Stocks

We perform experiments using publicly available, historical daily prices, queried from Yahoo! Finance™, of $r = 333$ stocks listed in the S&P500 Index and $q = 8$ sectors indexes, namely: Consumer Staples, Energy, Financials, Health Care, Industrials, Materials, Real Estate, and Utilities. Following our notation, the node set of classes $\mathcal{V}_q$ represents the sectors indexes, whereas the stocks are represented by the node set of objects $\mathcal{V}_r$.

Stock sector labels are available through the Global Industry Classification Standard (GICS) [46, 61]. Arguably the most well-recognized, industry-standard stock classification system, GICS relies on a fixed classification structure that may not fully capture the actual impact of a particular stock at particular time. With the intent of solving this shortcoming, we employ the learned bipartite graphs to extend the current industry stock classification system from a static classification to a dynamic soft clustering, in which a particular node (stock) may belong to different clusters (sectors) with different degrees of membership that can change over time. Our goal is to perform soft clustering of stocks according to the market sectors. To that end, we recall that each row of $\mathbf{B}$ is a point belonging to the probability simplex in $\mathbb{R}^q$, which represents the level of membership of a stock to a particular set of sectors. The final sector assigned to the i-th stock corresponds to $\arg\max_{j \in 1,\ldots,q} B_{ij}$.

We start by constructing the log-returns data matrix, i.e., a matrix $\mathbf{X} \in \mathbb{R}^{n \times p}$, where n is the number of log-returns observations and p is the number of instruments, as $X_{i,j} = \log P_{i,j} - \log P_{i-1,j}$, where $P_{i,j}$ is the closing price of the j-th instrument at the i-th day. We collect data from January 5, 2016, to January 5, 2021, totaling $n = 1291$ daily observations. Then, we proceed to learn connected bipartite graphs via the benchmarks and discussed methods using the observations $\mathbf{X}$ in a rolling-window fashion.[5] More precisely, we use a window of length 504 d (2 years in terms of stock market days) and shift this window by 63 d (3 months in terms of stock market days).

Figure 10.10 shows quantitative measurements that reveal the higher performance of SBG algorithm both in terms of accuracy and modularity. That may be explained by the fact that SBG adopts a multivariate Student's t-distribution, which is well known to better model financial datasets in comparison to Gaussian assumptions.

Furthermore, we evaluate the performance of the k-component bipartite estimators under the same settings as previously described, where we set $k = 8$, i.e., k is the number of sectors. Figure 10.11 reveals that kSBG presents a better performance across the whole time period both in terms of accuracy and modularity when com-

[5] An estimate for the degree of freedoms ν, required by Algorithms 10.5 and 10.6, is obtained by fitting a univariate Student's t-distribution to the log-returns of the S&P500 index during the corresponding time period.

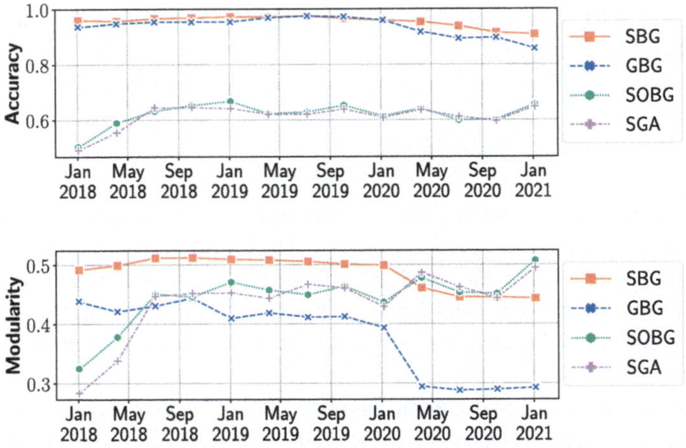

Fig. 10.10 Performance of the estimators for connected bipartite graphs of S&P500 stocks

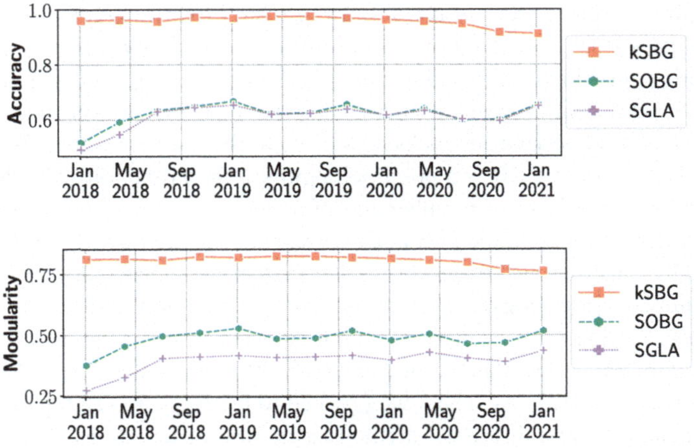

Fig. 10.11 Performance of the estimators for 8-component bipartite graphs of S&P500 stocks

pared to SGLA and SOBG. Interestingly, in both Figs. 10.10 and 10.11, we observe a decline in accuracy and modularity of the methods GBG, SBG, and kSBG, starting in January 2020, which can be explained by the impact of the COVID-19 pandemic on the US stock market. This reveals that our methods may be able to identify periods of economic turmoil and suggests that, during such times, standards such as GICS may not be the best reference for stock market sector classification. This insight may be critical for investors who rely on diversification of their portfolios through information on the sectors [74].

Figure 10.12 shows the k-component bipartite models learned using the whole dataset. We observe that SGLA allows the existence of isolated nodes, which degrades performance. The method kSBG avoids that issue by imposing a linear

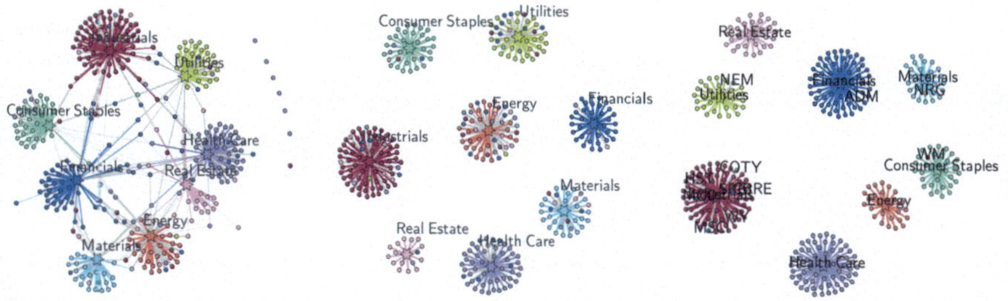

(a) SGLA (Kumar et al., 2020). accuracy = 0.77, modularity = 0.56.

(b) SOBG (Nie et al., 2017). accuracy = 0.75, modularity = 0.61.

(c) kSBG. accuracy = 0.97, modularity = 0.82.

Fig. 10.12 Graphs learned by different algorithms in the S&P500 stock returns dataset

constraint on the node degrees. In addition, kSBG yields a more intuitive representation of the financial network than SOBG as verified by the accuracy and modularity values. Finally, we note that kSGD algorithm "misclassifies" a few nodes such as the company ADM, which is originally classified by GICS as a Consumer Staples company but our algorithm places it in the Financials sector. A closer look at the fundamentals of ADM reveals that it is an active commodities trading participant, which then justifies it being classified as a Financials company. This application showcases the benefits of using data-driven classification algorithms, such as the one discussed in this chapter, which can be used to inform investors that prioritize diversification of their portfolios [29].

10.6.4.2 Robustness to Initialization

The performance of iterative algorithms may be affected by their initial point, especially when dealing with nonconvex problems as in Algorithms 10.5 and 10.6. Hence, we perform an experiment to compare two initialization schemes: (*i*) uniform, the graph weights are randomly sampled, i.e., $B^0_{ij} \sim \mathsf{Uniform}(0, 1)$, and (*ii*) default, $\mathbf{B}^0 \propto \max\left(\mathbf{0}, -\Xi_{rq}\right)$, where $\mathbf{S}^{-1} = \left[\Xi_{rr}\ \Xi_{rq};\ \Xi_{rq}^\top\ \Xi_{qq}\right]$ is the inverse of the sample correlation matrix, as discussed in more details in [39]. In both schemes, we normalize the rows of $\mathbf{B}^0$ such that $\mathbf{B}^0\mathbf{1} = \mathbf{1}$ holds. For this experiment, we consider log-returns of $r = 362$ stocks and $q = 9$ stock sectors from October 5, 2005, to December 30, 2015, totaling $n = 2577$ observations. Figure 10.13 shows the result of 100 realizations of Algorithms 10.5 and 10.6. It can be observed that Algorithms 10.5 and 10.6 are not sensitive to the choice of an initial point, which is an important feature for practical applications. Finally, Fig. 10.14a and b show the learned graphs from kSBG with different initial points, where we observe that the performance of kSBG is largely unaffected.

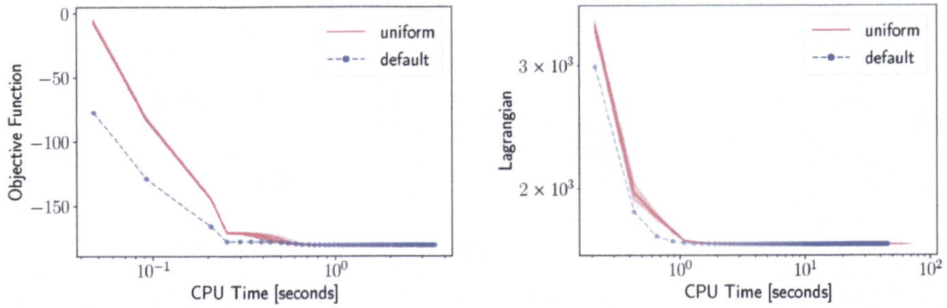

(a) Algorithm 10.5 (SBG). (b) Algorithm 10.6 (kSBG).

Fig. 10.13 Convergence trend of the presented algorithms for different initial points

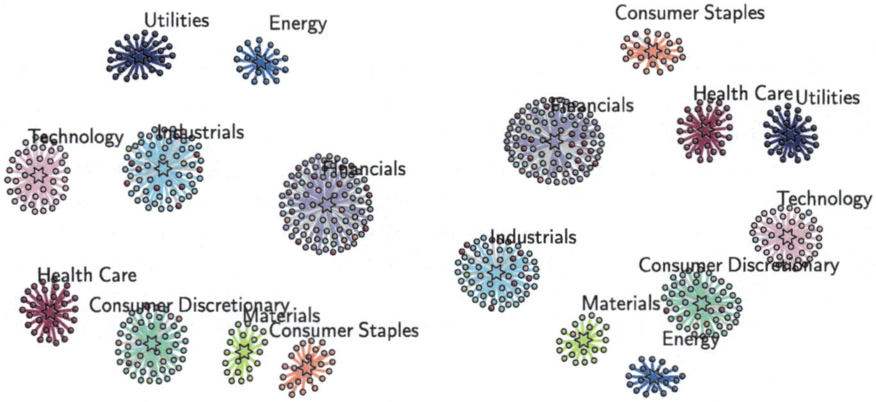

(a) default initial point. accuracy = 0.825, modularity = 0.691. (b) uniform initial point. accuracy = 0.820, modularity = 0.687.

Fig. 10.14 Visualizations of the graphs learned by kSBG with distinct initial points

10.7 Conclusions

Heavy tails are prevalent in time series of financial markets. Yet, they have been little explored in the context of Laplacian constrained graphical models. In this chapter, we have discussed optimization programs to learn graphical models with Laplacian constraints assuming that the data generating process is Student's t-distributed. The formulations follow a maximum-likelihood approach of a Markov random field, for which we designed ADMM algorithms. The algorithms presented in this chapter showed significant gains, measured via the modularity values of the estimated graphs as well as node accuracy, when compared to state-of-the-art counterparts in real-world scenarios that involved data from the US stock market, foreign exchange markets, and cryptocurrencies.

Acknowledgements This work was supported by the Hong Kong GRF 16207820 research grant.

References

1. Absil, P.-A., Mahony, R., Sepulchre, R.: Optimization Algorithms on Matrix Manifolds. Princeton University Press, Princeton, NJ (2007)
2. Beck, A.: First-Order Methods in Optimization. Society for Industrial and Applied Mathematics (SIAM), MOS-SIAM Series on Optimization (2017)
3. Bertsekas, D.P.: Nonlinear Programming. Athena Scientific (1999)
4. Boyd, S., Vandenberghe, L.: Convex Optimization. Cambridge University Press (2004)
5. Boyd, S., Parikh, N., Chu, E., Peleato, B., Eckstein, J.: Distributed optimization and statistical learning via the alternating direction method of multipliers. Found. Trends Mach. Learn. **3**(1), 1–122 (2011)
6. Candès, E.J., Wakin, M.B., Boyd, S.P.: Enhancing sparsity by reweighted ℓ_1 minimization. J. Fourier Anal. Appl. **14**(5), 877–905 (2008)
7. Cardoso, J.V.M., Palomar, D.: spectralGraphTopology: learning graphs from data via spectral constraints. In: The Comprehensive R Archive Network (2019). https://CRAN.R-project.org/package=spectralGraphTopology
8. Cardoso, J.V.M., Ying, J., Palomar, D.P.: Graphical models in heavy-tailed markets. In: Advances in Neural Information Processing Systems (NeurIPS'21) (2021)
9. Cardoso, J.V.M., Ying, J., Palomar, D.P.: Learning bipartite graphs: heavy tails and multiple components. In: Advances in Neural Information Processing Systems (NeurIPS'22) (2022)
10. Chen, Z., Li, L., Bruna, J.: Supervised community detection with graph neural networks. In: International Conference on Learning Representations (ICLR'19) (2019)
11. Chung, F.R.K.: Spectral Graph Theory. In: CBMS Regional Conference Series in Mathematics, vol. 92 (1997)
12. Clauset, A., Newman, M.E.J., Moore, C.: Finding community structure in very large networks. Phys. Rev. E **70** (2004)
13. Condat, L.: Fast projection onto the simplex and the ℓ_1 ball. Math. Program. **158** (2016)
14. Cont, R.: Empirical properties of asset returns: stylized facts and statistical issues. Quant. Financ. **1**, 223–236 (2001)
15. Coutino, M., Isufi, E., Maehara, T., Leus, G.: State-space network topology identification from partial observations (2019). arXiv:1906.10471
16. Danaher, P., Wang, P., Witten, D.M.: The joint graphical lasso for inverse covariance estimation across multiple classes. J. Roy. Stat. Soc. B **76**(2), 373–397 (2014)
17. de Prado, M.L.: Building diversified portfolios that outperform out of sample. J. Portf. Manag. **42**(4), 59–69 (2016)
18. Dehouche, N.: Scale matters: the daily, weekly and monthly volatility and predictability of bitcoin, gold, and the S&P 500 (2021). arXiv:2103.00395
19. Diamond, S., Boyd, S.: CVXPY: a Python-embedded modeling language for convex optimization. J. Mach. Learn. Res. **17**(83), 1–5 (2016)
20. Dong, X., Thanou, D., Frossard, P., Vandergheynst, P.: Learning Laplacian matrix in smooth graph signal representations. IEEE Trans. Signal Process. **64**(23), 6160–6173 (2016)
21. Egilmez, H.E., Pavez, E., Ortega, A.: Graph learning from data under Laplacian and structural constraints. IEEE J. Sel. Top. Signal Process. **11**(6), 825–841 (2017)
22. Epskamp, S., Borsboom, D., Fried, E.I.: Estimating psychological networks and their accuracy: a tutorial paper. Behav. Res. Methods **50**, 195–212 (2018)
23. Fan, K.: On a theorem of Weyl concerning eigenvalues of linear transformations I. Proc. Natl. Acad. Sci. **35**(11), 652–655 (1949)
24. Feng, Y., Palomar, D.: A signal processing perspective on financial engineering. Found. Trends Signal Process. **9**, 1–231 (2015)
25. Finegold, M., Drton, M.: Robust graphical modeling with t-distributions. In: Conference on Uncertainty in Artificial Intelligence (2009)
26. Fortunato, S.: Community detection in graphs. Phys. Rep. **486**(3), 75–174 (2010)
27. Friedman, J., Hastie, T., Tibshirani, R.: Sparse inverse covariance estimation with the graphical lasso. Biostatistics **9**, 432–41 (2008)

28. Godsil, C., Royle, G.: Algebraic Graph Theory. In: Graduate Texts in Mathematics, Springer Science (2001)
29. Goetzmann, N.W., Kumar, A.: Equity Portfolio Diversification. Rev. Financ. **12**(3), 433–463 (2008)
30. Gourieroux, C., Monfort, A.: Time Series and Dynamic Models. In: Themes in Modern Econometrics. Cambridge University Press (1997)
31. Hao, B., Sun, W.W., Liu, Y., Cheng, G.: Simultaneous clustering and estimation of heterogeneous graphical models. J. Mach. Learn. Res. **18**(217), 1–58 (2018)
32. Harvey, A.C.: Dynamic Models for Volatility and Heavy Tails: With Applications to Financial and Economic Time Series. Cambridge University Press (2013)
33. Horn, R.A., Johnson, C.R.: Matrix Analysis. Cambridge University Press (1985)
34. Hsieh, C., Banerjee, A., Dhillon, I.S., Ravikumar, P.K.: A divide-and-conquer method for sparse inverse covariance estimation. In: Advances in Neural Information Processing Systems (NeurIPS'12), pp. 2330–2338 (2012)
35. Hunter, D., Lange, K.: A tutorial on MM algorithms. Am. Stat. **58**(1), 30–37 (2004)
36. Kalofolias, V.: How to learn a graph from smooth signals. In: Proceedings of the 19th International Conference on Artificial Intelligence and Statistics, vol. 51, pp. 920–929. PMLR (2016)
37. Knill, O.: Cauchy-Binet for pseudo-determinants. Linear Algebra Appl. **459**, 522–547 (2014)
38. Kumar, S., Ying, J., Cardoso, J.V.M., Palomar, D.P.: Structured graph learning via Laplacian spectral constraints. In: Advances in Neural Information Processing Systems (NeurIPS) (2019)
39. Kumar, S., Ying, J., Cardoso, J.V.M., Palomar, D.P.: A unified framework for structured graph learning via spectral constraints. J. Mach. Learn. Res. **21**, 1–60 (2020)
40. Lake, B.M., Tenenbaum, J.B.: Discovering structure by learning sparse graph. In: Proceedings of the 33rd Annual Cognitive Science Conference (2010)
41. Li, Y., Sha, C., Huang, X., Zhang, Y.: Community detection in attributed graphs: an embedding approach. In: AAAI Conference on Artificial Intelligence (AAAI'18) (2018)
42. Mantegna, R.N.: Hierarchical structure in financial markets. Eur. Phys. J. B **11**(1), 193–197 (1999). ISSN 1434-6028
43. Marti, G., Nielsen, F., Bińkowski, M., Donnat, P.: A review of two decades of correlations, hierarchies, networks and clustering in financial markets (2017). arXiv: 1703.00485
44. Mateos, G., Segarra, S., Marques, A.G., Ribeiro, A.: Connecting the dots: identifying network structure via graph signal processing. IEEE Signal Process. Mag. **36**(3), 16–43 (2019)
45. Michelot, C.: A finite algorithm for finding the projection of a point onto the canonical simplex of $\mathbb{R}^n$. J. Optim. Theory Appl. **50**(1), 195–200 (1986). July
46. Morgan Stanley Capital International and S&P Dow Jones. Revisions to the global industry classification standard (GICS) structure (2018)
47. Newman, M.E.J.: Modularity and community structure in networks. In: Proceedings of the National Academy of Sciences of the United States of America **103** (2006)
48. Nie, F., Wang, X., Huang, H.: Clustering and projected clustering with adaptive neighbors. In: Proceedings of the 20th ACM SIGKDD International Conference on Knowledge Discovery & Data Mining (KDD'14) (2014)
49. Nie, F., Wang, X., Jordan, M.I., Huang, H.: The constrained Laplacian rank algorithm for graph-based clustering. In: Proceedings of the Thirtieth AAAI Conference on Artificial Intelligence, AAAI'16, pp. 1969–1976. AAAI Press (2016)
50. Nie, F., Wang, X., Deng, C., Huang, H.: Learning a structured optimal bipartite graph for co-clustering. In: Advances on Neural Information Processing Systems (NeurIPS'17), pp. 4132–4141 (2017)
51. Ortega, J.M., Rheinboldt, W.C.: Iterative Solution of Nonlinear Equations in Several Variables. In: Society for Industrial and Applied Mathematics (2000)
52. Pal, S., Regol, F., Coates, M.: Bayesian graph convolutional neural networks using non-parametric graph learning. In: International Conference on Learning Representations (ICLR'19) (2019)
53. Palomar, D.P., Fonollosa, J.R.: Practical algorithms for a family of water filling solutions. IEEE Trans. Signal Process. **53**(2), 686–695 (2005)

54. Pavez, E., Egilmez, H.E., Ortega, A.: Learning graphs with monotone topology properties and multiple connected components. IEEE Trans. Signal Process. **66**(9), 2399–2413 (2018)
55. Resnick, S.I.: Heavy-Tail Phenomena: Probabilistic and Statistical Modeling. Springer, New York (2007)
56. Rue, H., Held, L.: Gaussian Markov Random Fields: Theory And Applications (Monographs on Statistics and Applied Probability). Chapman & Hall/CRC (2005)
57. Segarra, S., Marques, A.G., Mateos, G., Ribeiro, A.: Network topology inference from spectral templates. IEEE Trans. Signal Inf. Process. Netw. **3**(3), 467–483 (2017)
58. Shafipour, R., Mateos, G.: Online topology inference from streaming stationary graph signals with partial connectivity information. Algorithms **13**(9) (2020)
59. Slawski, M., Hein, M.: Estimation of positive definite M-matrices and structure learning for attractive Gaussian Markov random fields. In: Linear Algebra and its Applications, pp. 145–179 (2015)
60. Smith, S.M., Miller, K.L., Salimi-Khorshidi, G., Webster, M., Beckmann, C.F., Nichols, T.E., Ramsey, J.D., Woolrich, M.W.: Network modelling methods for FMRI. Neuroimage **54**(2), 875–891 (2011)
61. Standard & Poor's. Global Industry Classification Standard (GICS). Technical Report (2006)
62. Stegle, O., Teichmann, S.A., Marioni, J.C.: Computational and analytical challenges in single-cell transcriptomics. Nat. Rev. Genet. **16**, 133–145 (2015)
63. Sun, S., Zhu, Y., Xu, J.: Adaptive variable clustering in Gaussian graphical models. In: Proceedings of the Seventeenth International Conference on Artificial Intelligence and Statistics, vol. 33, pp. 931–939. PMLR (2014)
64. Sun, Y., Babu, P., Palomar, D.P.: Robust estimation of structured covariance matrix for heavy-tailed elliptical distributions. IEEE Trans. Signal Process. **64**(14), 3576–3590 (2016)
65. Sun, Y., Babu, P., Palomar, D.P.: Majorization-minimization algorithms in signal processing, communications, and machine learning. IEEE Trans. Signal Process. **65**(3), 794–816 (2017). ISSN 1941-0476
66. Tan, K.M., Witten, D., Shojaie, A.: The cluster graphical Lasso for improved estimation of Gaussian graphical models. Comput. Stat. Data Anal. **85**, 23–36 (2015). ISSN 0167-9473
67. Tsay, R.S.: Analysis of Financial Time Series. Wiley, 3rd edn (2010)
68. Wald, Y., Noy, N., Elidan, G., Wiesel, A.: Globally optimal learning for structured elliptical losses. In: Advances in Neural Information Processing Systems (NeurIPS'19) (2019)
69. Witten, D.M., Tibshirani, R.: Covariance-regularized regression and classification for high dimensional problems. J. R. Stat. Society. Ser. B (Stat. Methodol.) **71**(3), 615–636 (2009)
70. Witten, D.M., Friedman, J.H., Simon, N.: New insights and faster computations for the graphical lasso. J. Comput. Graph. Stat. **20**(4), 892–900 (2011)
71. Wright, S.J.: Coordinate descent algorithms. Math. Program. **151**, 3–34 (2015)
72. Wu, T.T., Lange, K.: The MM alternative to EM. Stat. Sci. **25**(4), 492–505 (2010)
73. Wu, Z., Pan, S., Chen, F., Long, G., Zhang, C., Yu., P.S.: A comprehensive survey on graph neural networks (2019). arXiv: 1901.00596
74. Yang, Y., Zhao, L., Chen, L., Wang, C., Han, J.: Portfolio optimization with idiosyncratic and systemic risks for financial networks (2021). arXiv: 2111.11286
75. Ying, J., Cardoso, J.V.M., Palomar, D.P.: Does the ℓ_1-norm learn a sparse graph under Laplacian constrained graphical models? (2020a). arXiv: 2006.14925
76. Ying, J., Cardoso, J.V.M., Palomar, D.P.: Nonconvex sparse graph learning under Laplacian-structured graphical model. In: Advances in Neural Information Processing Systems (NeurIPS), vol. 33, pp. 7101–7113 (2020)
77. Ying, J., Cardoso, J.V.M., Palomar, D.: Minimax estimation of Laplacian constrained precision matrices. In: 24th International Conference on Artificial Intelligence and Statistics (2021)
78. Zhang, F.: The Schur Complement and Its Applications. Springer (2005)
79. Zhao, L., Wang, Y., Kumar, S., Palomar, D.P.: Optimization algorithms for graph Laplacian estimation via ADMM and MM. IEEE Trans. Signal Process. **67**(16), 4231–4244 (2019)

Index

J.-P. Delmas et al. (eds.), *Elliptically Symmetric Distributions in Signal Processing and Machine Learning*, https://doi.org/10.1007/978-3-031-52116-4

GPSR Compliance

The European Union's (EU) General Product Safety Regulation (GPSR) is a set of rules that requires consumer products to be safe and our obligations to ensure this.

If you have any concerns about our products, you can contact us on ProductSafety@springernature.com

In case Publisher is established outside the EU, the EU authorized representative is:

Springer Nature Customer Service Center GmbH
Europaplatz 3
69115 Heidelberg, Germany

The manufacturer's authorised representative in the EU is Springer Nature Customer Service Centre GmbH, Europaplatz 3, 69115 Heidelberg, Germany. If you have any concerns regarding our products, please contact ProductSafety@springernature.com

Printed and bound by CPI Group (UK) Ltd, Croydon, CR0 4YY
01/10/2026
02225914-0011